National Center for Construction Education and Research

Pipeline Mechanical Level One

Prentice Hall
Upper Saddle River, NJ
Columbus, OH

contren™
Learning Series

American Petroleum Institute
Pipeliner Training and Assessment Program (PTAP)

This information is general in nature and intended for training purposes only. Actual performance of activities described in this manual requires compliance with all applicable operating, service, maintenance, and safety procedures under the direction of qualified personnel. References in this manual to patented or proprietary devices do not constitute a recommendation of their use.

Prentice Hall
nccer
contren™ Learning Series

10 9 8 7
ISBN 0-13-046670-0

Preface

This volume was developed by the National Center for Construction Education and Research (NCCER) in response to the training needs of the construction, maintenance, and pipeline industries. It is one of many in NCCER's *Contren™ Learning Series*. The program, covering training for close to 40 construction and maintenance areas, and including skills assessments, safety training, and management education, was developed over a period of years by industry and education specialists.

NCCER also maintains a National Registry that provides transcripts, certificates, and wallet cards to individuals who have successfully completed modules of NCCER's *Contren™ Learning Series*, when the training program is delivered by an NCCER Accredited Training Sponsor.

The NCCER is a not-for-profit 501(c)(3) education foundation established in 1995 by the world's largest and most progressive construction companies and national construction associations. It was founded to address the severe workforce shortage facing the industry and to develop a standardized training process and curricula. Today, NCCER is supported by hundreds of leading construction and maintenance companies, manufacturers, and national associations, including the following partnering organizations:

PARTNERING ASSOCIATIONS

- American Fire Sprinkler Association
- American Petroleum Institute
- American Society for Training & Development
- American Welding Society
- Associated Builders & Contractors, Inc.
- Associated General Contractors of America
- Association for Career and Technical Education
- Carolinas AGC, Inc.
- Carolinas Electrical Contractors Association
- Citizens Democracy Corps
- Construction Industry Institute
- Construction Users Roundtable
- Design-Build Institute of America
- Merit Contractors Association of Canada
- Metal Building Manufacturers Association
- National Association of Minority Contractors
- National Association of State Supervisors for Trade and Industrial Education
- National Association of Women in Construction
- National Insulation Association
- National Ready Mixed Concrete Association
- National Utility Contractors Association
- National Vocational Technical Honor Society
- North American Crane Bureau
- Painting & Decorating Contractors of America
- Portland Cement Association
- SkillsUSA-VICA
- Steel Erectors Association of America
- Texas Gulf Coast Chapter ABC
- U.S. Army Corps of Engineers
- University of Florida
- Women Construction Owners & Executives, USA

Some features of NCCER's *Contren™ Learning Series* are:

- An industry-proven record of success
- Curricula developed by the industry for the industry
- National standardization providing portability of learned job skills and educational credits
- Credentials for individuals through NCCER's National Registry
- Compliance with Apprenticeship, Training, Employer, and Labor Services (ATELS) requirements for related classroom training (CFR 29:29)
- Well-illustrated, up-to-date, and practical information

Acknowledgments

This curriculum was revised as a result of the farsightedness and leadership of the following sponsors:

Alyeska Pipeline Service Company
Colonial Pipeline Company
Equistar Pipeline Operations
Gulf South Pipeline Company
Lakehead Pipeline
Marathon Ashland Pipeline LLC
Minnesota Pipeline
Plains All American Pipeline
Plantation Pipe Line Company
Sun Pipeline Company
TEPPCO

This curriculum would not exist were it not for the dedication and unselfish energy of those volunteers who served on the Authoring Team. A sincere thanks is extended to:

Billy Copes
Bill Coulter
Greg Crume
Jamie Ferguson
Ron Gunter
Todd Hendricks
Jordan Janak
Grady McWhorter
Scott Neujahr
Rich OBeid
Steve Sanders
Dave Yeager

A final note: This book is the result of a collaborative effort involving the production, editorial, and development staff at Prentice-Hall, Inc., and the National Center for Construction Education and Research. Thanks to all of the dedicated people involved in the many stages of this project.

Contents

Module 66101-02

Introduction to the Pipeline Industry

Introduction to the Pipeline Operator Qualification Rule, API, and NCCER

It could be said that petroleum products keep America running. Petroleum products like oil and natural gas are the principal sources of energy in the United States and are used in many manufacturing and chemical processes. Petroleum products are primarily transported by pipeline. There are more than a million miles of liquid and natural gas pipeline in this country, ranging in diameter from two inches to four feet. These pipelines bring crude oil and natural gas from production fields to processing plants and refineries—and ultimately to the consumer.

Pipeline Operator Qualification

Petroleum products are considered hazardous, so pipeline personnel must take extra precautions and follow specific procedures when working with them. To help ensure the safe handling and transportation of petroleum products, the United States Department of Transportation (DOT) mandates safe practices as federal regulations.

One such regulation, the Operator Qualification (OQ) rule, requires pipeline operators to develop a written qualification plan to evaluate their personnel and their contract personnel (see *49 CFR Part 192 Subpart N—Gas*, and *49 CFR Part 195 Subpart G—Liquid*). The evaluation hinges on the operator's ability to perform certain covered tasks and to be able to recognize and respond to abnormal operating conditions that may be encountered while performing these covered tasks.

Covered Tasks

A pipeline performance task is deemed *covered* if the activity passes the following four-part test:

1. Is it performed on a pipeline facility?
2. Is it an operations or maintenance task?
3. Is it performed as a requirement of *49 CFR Part 195 (Liquid)* or *49 CFR Part 192 (Gas)*?
4. Does it affect the operation or integrity of the pipeline facility?

If the answer to all four questions is *yes*, then the activity is considered *covered* and must figure into the evaluation process of pipeline personnel.

American Petroleum Institute

The OQ rule and its related covered tasks provided the catalyst to develop a unified pipeline training and assessment program—so that pipeline operators and owners have ready access to the assessment and training tools they need. This effort was the direct result of the vision and leadership provided by The American Petroleum Institute (API) and API members.

Efforts to unify the industry are nothing new to API. Indeed, the impetus for forming the Institute in 1919 was the need to standardize engineering specifications for drilling and production equipment. Today, the most pressing issues revolve around public perceptions and government policies toward the industry. Speaking with one voice on these issues has become as essential as having interchangeable parts in the field. API's members determine the petroleum industry's positions on public policy as well as the standards that govern its day-to-day operations worldwide.

Consortium on Operator Qualification

API assembled a focus group whose primary purpose was to identify covered tasks and provide compliance recommendations to pipeline operators.

The Consortium on Operator Qualification (COOQ) became the guiding force for the identification and development of covered tasks for the liquid pipeline industry. Their work product, API publication 1161, *Guidance Document for the Qualification of Liquid Pipeline Personnel,* formed the basis for the resulting Pipeliner Training and Assessment Program.

The covered task list created in API publication 1161 has now evolved to include gas covered tasks and is maintained by NCCER. The most current covered task list can be downloaded at www.nccer.org.

National Center for Construction Education and Research

With the guidance and support of API, its members, and the Consortium on Operator Qualification, the National Center for Construction Education and Research (NCCER) was able to design and implement the Pipeliner Training and Assessment Program.

NCCER is a not-for-profit 501(c)(3) education foundation established in 1995 by the world's largest and most progressive construction and maintenance contractors and national contractor associations. NCCER was founded to address the

severe workforce shortage facing the construction and maintenance industries and to develop a standardized training process and curricula. Today, NCCER is supported by hundreds of leading construction and maintenance companies, manufacturers, and national associations and developed more than 35 construction and maintenance titles.

NCCER also maintains a National Registry, which contains the records of trainees and instructors who have received NCCER training. All completion information for every trainee is sent to the NCCER and filed within the Registry. The Registry can then confirm training and skills for workers as they move from company to company, state to state, or even within their own company.

Standardized Curriculum (NCCER's Contren™ Learning Series)

NCCER curricula are developed using the expertise of industry subject matter experts (SMEs). The use of SMEs as an integral part of every new development or revision ensures that the curriculum is current, industry-driven, and indicative of industry practices. The development and/or revision process is facilitated from start to finish by NCCER project managers, who work closely with the SMEs and a select team of technical writer(s) on the development and presentation of the curriculum.

Pipeline: Five Training Paths

The pipeline curriculum focuses on essential skills and competencies within five skill areas:

- Corrosion Control
- Electrical and Instrumentation
- Maintenance
- Mechanical
- Operations (Field and Control Center)

SMEs from the pipeline industry have determined the skills and competencies needed based on API's covered tasks and the consensus of the group/industry. Specific company information is generally not incorporated into the curriculum unless the SMEs determine that the information is the industry standard.

Training Modules and Covered Tasks

The smallest unit of instruction in the curriculum is a *module*. Each curriculum is composed of modules (the equivalent of stand-alone chapters) that are logically sequenced into complete *levels* of instruction. Every module is task-oriented and requires the successful completion of a written (knowledge-based) examination and a performance (hands-on) test. Modules may be individually selected and/or combined for efficient task training. Modules in the pipeline curricula, to the extent possible, will align with API-developed covered tasks. While not all modules will relate to covered tasks, all covered tasks will relate to at least one module.

Benefits of Training to the Pipeline Marketplace

Applying NCCER's standardized training process allows pipeline owners and contractors in the United States to establish the benchmark for acceptable skill and competency levels within the pipeline industry. This process will assist training organizations in developing and/or improving standardized pipeline training programs that effectively match their resources and operating conditions while responding to regulatory demands and to the needs of the industry. The process will also recognize training organizations that meet the minimum standards of competence, warranting regulatory, public, and professional confidence. This in turn will foster national unity within the pipeline industry for the mutual benefit of individual technicians, owners, contractors, and ultimately pipeline-users and society.

Additional Resources

To review the Pipeliner Training and Assessment Program Implementation and Resource Guide and the latest version of pipeline covered tasks, visit NCCER's Web site at www.nccer.org, or contact:

National Center for Construction Education and Research
PO Box 141104
Gainesville, FL 32614-1104
(888) 622-3720
Web site: www.nccer.org

For more information on API, please contact:

American Petroleum Institute
1220 L Street, Northwest
Washington, D.C. 20005
(202) 682-8000
Web site: www.api.org

To order *API Pub 1161*, contact Global Engineering Documents by calling 1-800-854-7149 or visiting their Web site at www.global.his.com.

Additional sources of information on the pipeline operator qualification regulation include:

DOT – Office of Pipeline Safety
www.ops.dot.gov

Transportation Safety Institute
www.tsi.dot.gov

COURSE MAP

This course map shows each curriculum path in the Pipeliner Training program. The suggested training order begins on the left with the Pipeline Core modules and proceeds to the right. Which curriculum path you follow depends on your areas of specialty and interest. On a given curriculum path, skills levels increase as you advance to the right on the course map, although some are single-level curricula. The local Training Program Sponsor may adjust the training order.

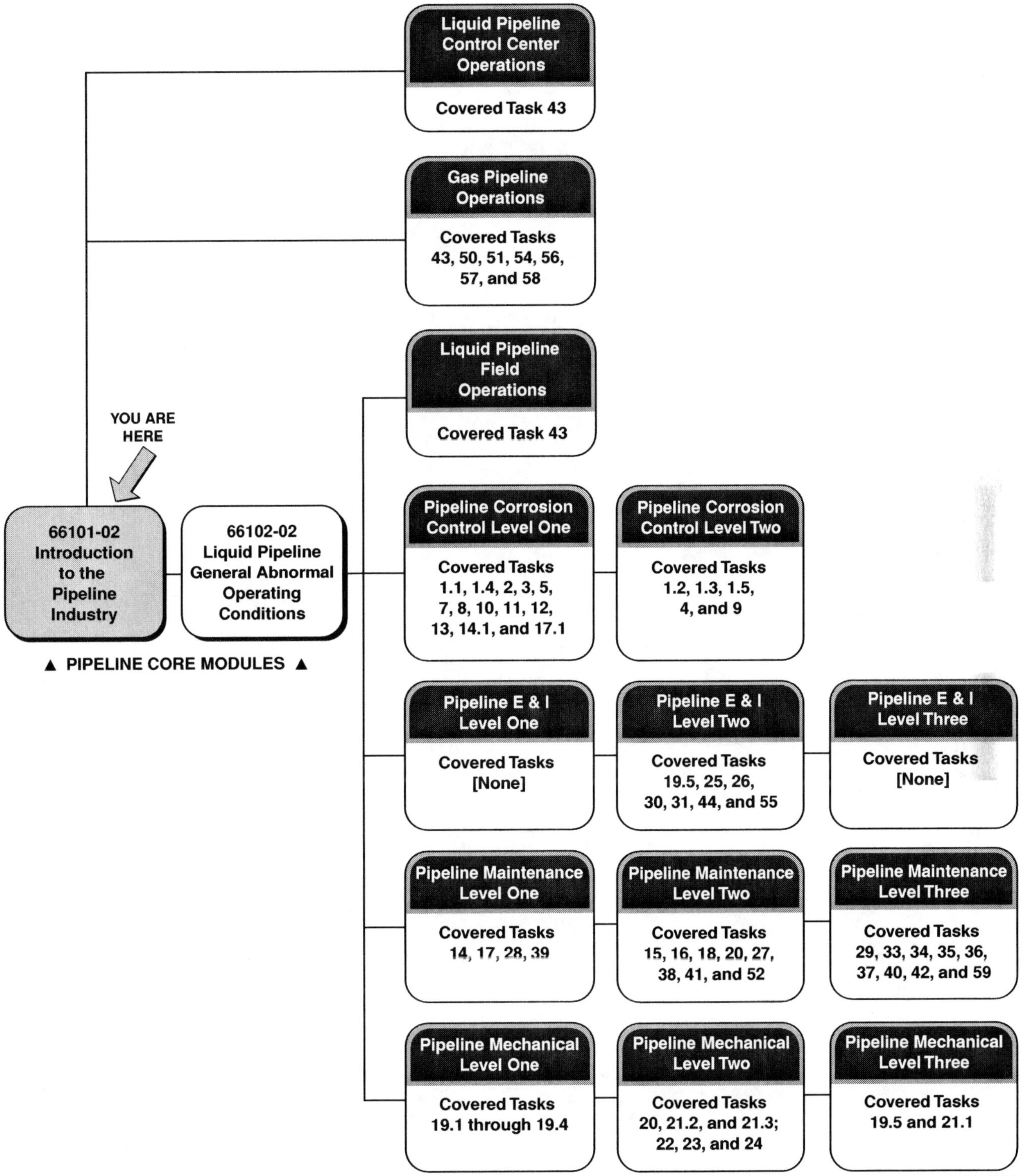

101CMAP.EPS

MODULE 66101-02 CONTENTS

Figures

Tables

Introduction to the Pipeline Industry

Objectives

When you have completed this module, you will be able to do the following:

1. Explain the basic functions and purposes of pipelines and facilities and identify the characteristics and hazards of common pipeline products.
2. Identify maps and drawings used to depict pipelines and facilities.
3. Explain the roles of control personnel and equipment in the overall operation of a pipeline.
4. Explain liquid pipeline hydraulics and gas pipeline pneumatics.
5. Explain the types and purposes of pipeline equipment.
6. Explain pipeline electrical power systems and corrosion control.
7. Review operations, maintenance, and emergency procedures and perform documentation required for pipeline operations.

Prerequisites

Before you begin this module, it is recommended that you successfully complete the following: Core Curriculum.

1.0.0 ◆ INTRODUCTION

Pipelines are used to transport **petroleum**, petroleum **products**, liquid carbon dioxide, and other liquids and gases over great distances. In spite of the costs of constructing pipelines, they are a relatively economical means of transport. A pipeline system consists of pipes, pump stations, compressor stations, and other facilities to support the safe, continuous movement of liquids and gases. Pipelines carry petroleum from the source, the oil and gas fields, to the **refineries**, to and from gas processing plants, and eventually to consumers. Pipes are most often buried in the ground, but they are also laid under water and sometimes erected above the ground.

Today, a network of gas and liquid pipelines crisscrosses the United States. The first pipeline on a continental scale was completed in the 1930s, but the real boom in cross-country lines came during World War II when the Big Inch and the Little Big Inch were constructed to provide fuel for wartime industry and the military. The Big Inch began operating in 1943, was 24 inches in diameter, and stretched 1,341 miles from Longview, Texas, to Phoenixville, Pennsylvania, and Linden, New Jersey. This made it the largest **crude oil** pipeline in the world. The Little Inch, opened in 1944, was 20 inches in diameter and 1,475 miles long. It carried **refined products** from Beaumont, Texas, to Linden, New Jersey.

The most famous crude oil pipeline in the U.S. crosses 800 miles of permanently frozen ground in Alaska. The Trans-Alaska Pipeline, which began operations in 1977, is 48 inches in diameter and carries crude oil from the Prudhoe Bay field to the shipping terminal at Valdez, Alaska. Along the way, it crosses 34 major rivers and streams and three major mountain ranges, going from sea level to as high as 4,800 feet and back to sea level. Protection of the delicate tundra, permafrost, wilderness environment, and wildlife in its path were and still are critical concerns.

All pipeline operators are concerned with safety and environmental impact, particularly those carrying petroleum products that are hazardous to the environment as well as to the people who live near the pipeline. Good construction practices, good maintenance and operations practices, continuous monitoring, and rapid response are necessary to prevent or to minimize the effects of pipeline leaks. Operations of a pipeline must be conducted in a fashion that will ensure the safety of employees and the general public and protect the environment.

The petroleum pipeline system can be divided into the following functions:

- Origination
- Transport
- Storage
- Delivery

1.1.0 Origination

All petroleum products produced or shipped originate from the domestic and foreign crude oil or natural gas from the oil and gas fields. Exploration teams from petroleum companies use geological and geophysical methods to find possible deposits of oil and gas deep under the layers of the earth. Drilling confirms and accesses the find; the well is completed and production begins.

1.1.1 Exploration

Oil and gas are formed in the earth from decaying organic matter that was buried long ago. The combination of time, pressure, and heat causes the matter to undergo a chemical change into hydrocarbons. These fluids and gases then tend to move, or migrate, upward in the earth until they either escape at the surface or are trapped in the earth's layers. It is this trapped fluid and gas that is sought by exploration.

The oil/gas-bearing formation is basically the layer of earth in which the hydrocarbon deposit is trapped until it is drilled. The formation must have the porosity to hold the fluid and gas and have the structure to keep it from moving further upward. Traps can have a variety of structures, and some of the most typical are the anticline trap, stratigraphic trap, fault trap, and salt dome trap, as shown in *Figure 1*.

1.1.2 Drilling and Completion

When a possible deposit is indicated, a well must be drilled to try to find it. Many wells have been drilled without finding the sought-after oil and gas. There may not be enough to warrant production costs or it cannot be drawn out economically. However, when a workable deposit is found, the well is completed for production. Completion includes casing the well with concrete, perforating the casing with

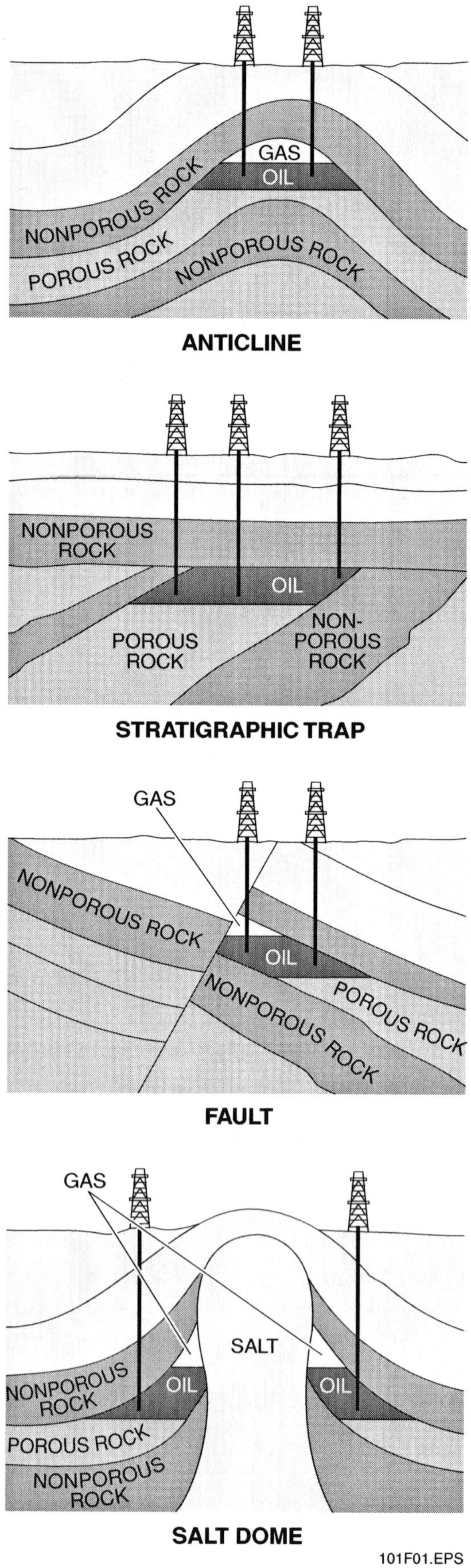

Figure 1 ◆ Types of traps.

holes for the product to flow through tubing within the casing, and installing pumps or valves on the top of the hole to control the flow. Tests of the producing well may show a large enough deposit to warrant other wells being drilled and completed in the area. All of these wells comprise a field. It is from these wells that the oil or gas is gathered to transport through the pipeline.

Many crude oil reservoirs have a gas cap. The oil produced from these reservoirs often contains various amounts of associated gas. Some formations, however, trap only gas. This trapped gas forms a gas or gas-condensate reservoir, which produces only gas or gas with varying amounts of liquid gas condensate. Drilling gas wells is very similar to drilling and completing crude oil wells. The primary difference between crude oil and gas wells is in the type of surface facility equipment needed to process the oil or gas after it is produced.

1.1.3 Production

Different types of pumps or valves, such as a Christmas tree and a beam pump, are used to cap the well depending on the properties of the product being withdrawn and the formation producing it (*Figure 2*). Oil and gas are often found in the same formation along with some amount of water. The amount of each may vary in relation to each other, and this may affect the production of the well. Since gas is lighter than oil, and oil is lighter than water, they are layered in that order: gas on the top, then oil, then water on the bottom.

Because the formation had to be closed off in some way for the product to accumulate, pressure is usually also trapped in the formation. When it is drilled, that pressure can push the oil, gas, and water into the hole with some force. When it is controlled, this pressure is good because it can push the oil and gas up to the surface with little or no pumping. Generally a newer well will have more pressure than an older one, so in time, pumping power may be needed to pull the product out. However, some wells start with only a little pressure and must be pumped from the beginning. Control of the flow is important since producing a well too fast can affect the life of the well. These conditions, along with other properties, will determine how the well is capped and produced. A Christmas tree of valves may be installed to control the flow, and a beam pump may be used to pull the product out. Gas wells are completed with only a Christmas tree.

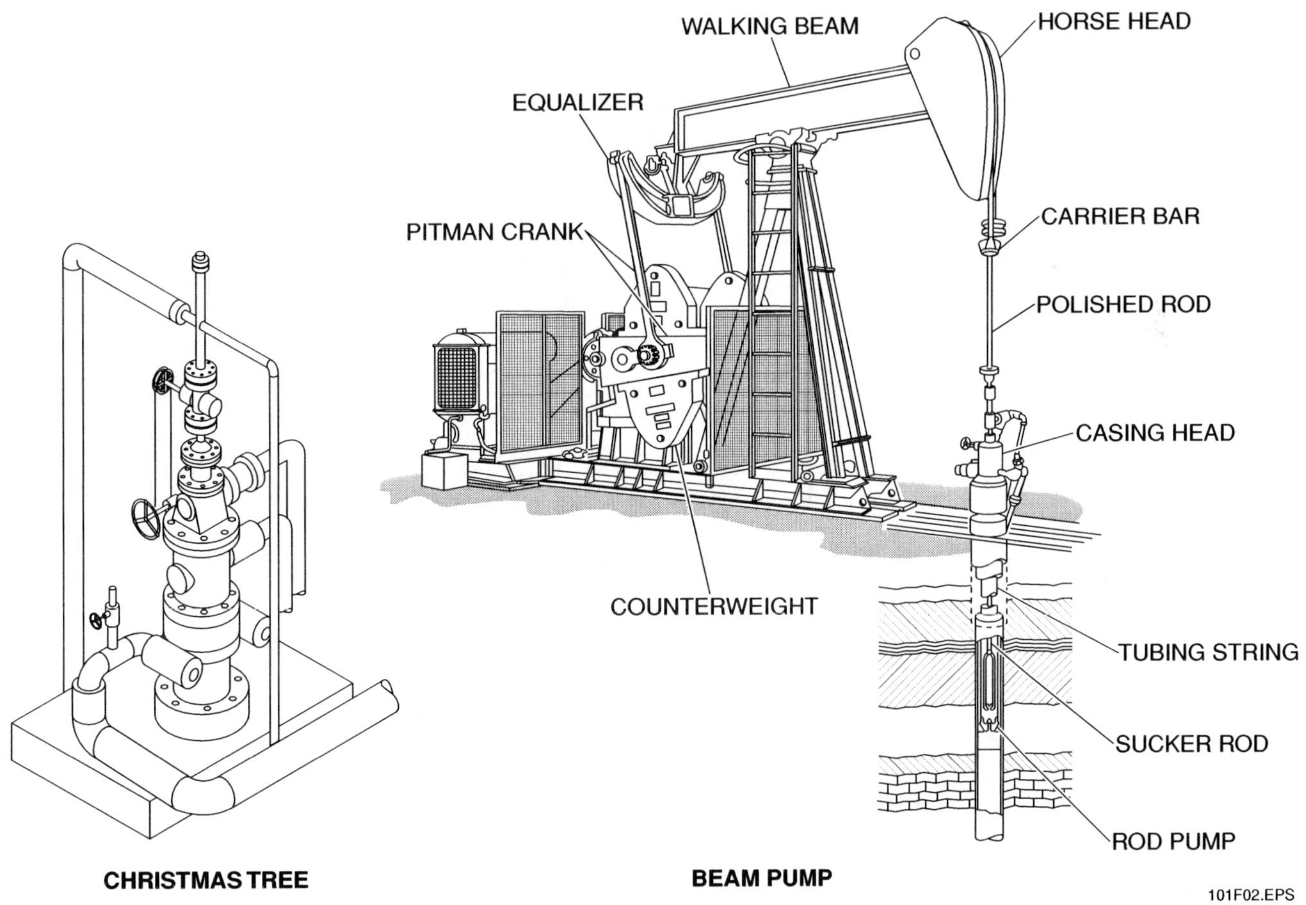

Figure 2 ◆ Christmas tree and beam pump.

1.2.0 Transport

Lines from the oil and gas fields are called *gathering lines* or *flow lines*. These gathering lines bring the oil and gas together into a facility for separation and processing. The vapor gas must travel in a different pipeline from the crude oil, so the gas and oil are separated in huge high-pressure **tanks**, called *gas/oil separators*. *Figure 3* shows the high-pressure separators.

From the separation or gas processing facility, the crude oil and gas is directed into a large **main line**, or trunk line, for the longest part of the journey to a refinery or a shipping terminal. The gas may be stored in tanks or routed through gas pipelines to market. Gas main lines transport gas to a gas distribution center for distribution to consumers, industrial facilities, or storage caverns. From the refinery, a large main line, or trunk line, will carry products across country with distribution lines branching off the main line along the way. A large distribution line may carry the liquid to a central distribution point, such as a regional or city station, from which smaller lines or vehicles carry it to consumers. Pipelines are constructed to meet *ASME Piping Standard, B31.4-1998, Pipeline Transportation Systems for Liquid Hydrocarbons, and other Liquids*. *Figure 4* shows the transportation chain.

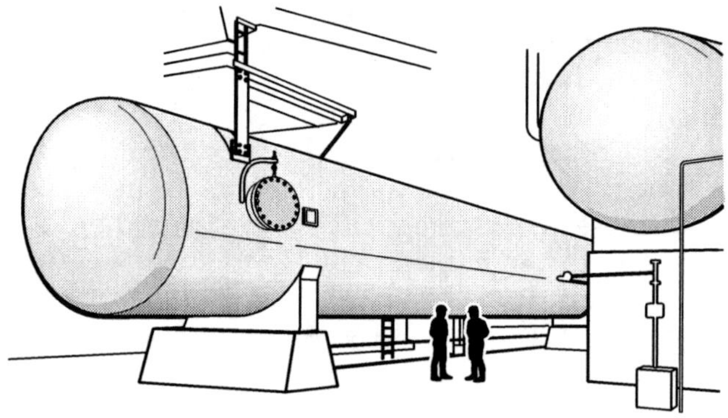

Figure 3 ◆ High-pressure separators.

1.2.1 Pipe

The size of pipe used in pipelines varies from 2-inch diameter pipe for gathering lines or small distribution lines, to 48-inch pipe for main lines or trunk lines. Most modern pipe is made of steel, either seamless or with a welded longitudinal seam, that is produced in lengths that are 40-60 feet long. These are often welded into 80-foot sections before being delivered to the construction site. Steel pipe is vulnerable to **corrosion**, which can cause damage to the pipe and reduce wall thickness, compromising the integrity of the pipe. Pipe coating and wrapping are the primary

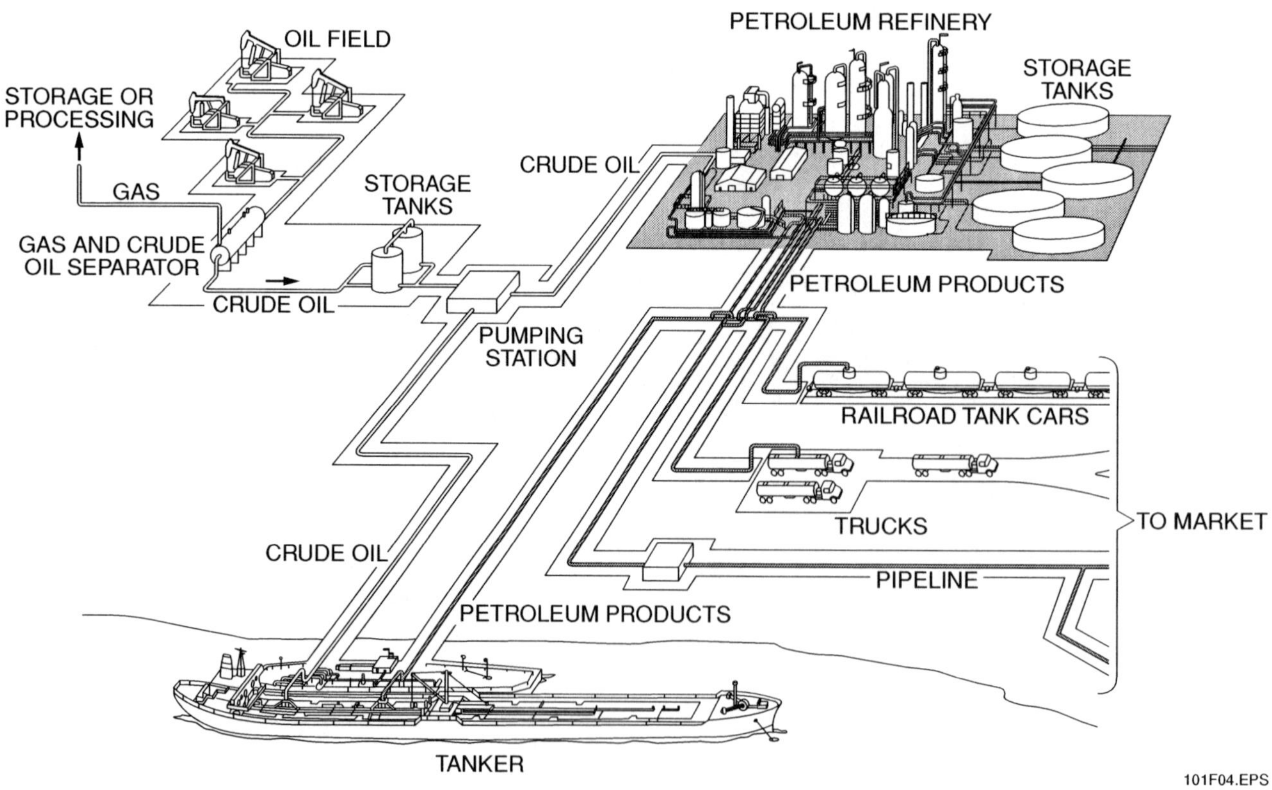

Figure 4 ◆ Transportation chain.

means of protecting the pipe. **Cathodic protection** is used after the pipe is buried to inhibit a corrosive environment.

1.2.2 Construction

At the time of construction, permits are obtained for a specified right-of-way for the pipeline. Earthmoving equipment clears the land, an access road is built along the route, and trenches are excavated for the pipe. Trenches require a minimum depth of coverage as stated in *Pipeline Safety Regulations, Part 195.248* for liquid and *Part 192.327* for gas.

Pipe is hauled to the right-of-way and welded in strings by teams of welders. After welding, radiographic inspection is used to check the welds. Sometimes, pipe must be bent by bending machines to fit the terrain or to go around obstacles. To inhibit corrosion, the pipe must be coated and wrapped before burial. This can be done at the mill or at the site.

The trench is dug and then prepared for the pipe with a layer of protective bedding. The pipe is strung or laid in the trench by sideboom tractors. Long sections are tied-in or welded together to form a continuous conduit. The trench is then filled with backfill to cover the pipe (see *Figure 5*). The pipe must be **hydrostatically tested** or checked for leaks. This involves filling the line with water and pressurizing it to see if the system will maintain the pressure intended.

When piping crosses lakes and rivers, it may be laid and weighted with concrete or steel anchors to keep it on the bottom. *Figure 6* shows an add-on weight installation. In most rivers, the crossings are drilled under the river bed without disturbing the bottom.

Pipelines are usually buried except at pump and compressor stations, some valve stations, and terminals. However, pipe is sometimes erected aboveground by using H-shaped supports, called bents, to hold the pipe. These bents are located at specified intervals along the line. The uprights on the bents are vertical support members (VSMs). In areas like Alaska, the VSMs have to be self-cooling to avoid heating the permafrost. This is not the normal installation, but it is necessary in frozen environments. *Figure 7* shows a bent with self-cooling VSMs.

Supports have sliding plates, or shoes, to allow the pipe to contract or expand with weather or to move with **seismic activity**. The shoes allow the pipe to move as much as 20 feet back and forth and 10 feet up and down. Aboveground lines are also constructed in a zigzag pattern instead of a straight line to allow for movement. *Figure 8* shows a common elevated pipeline section.

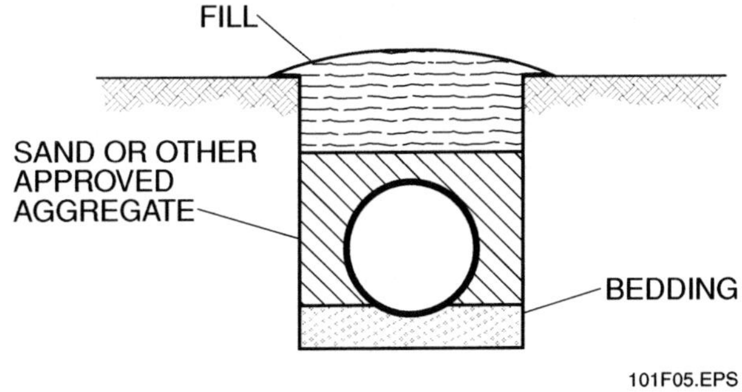

Figure 5 ◆ Pipe burial.

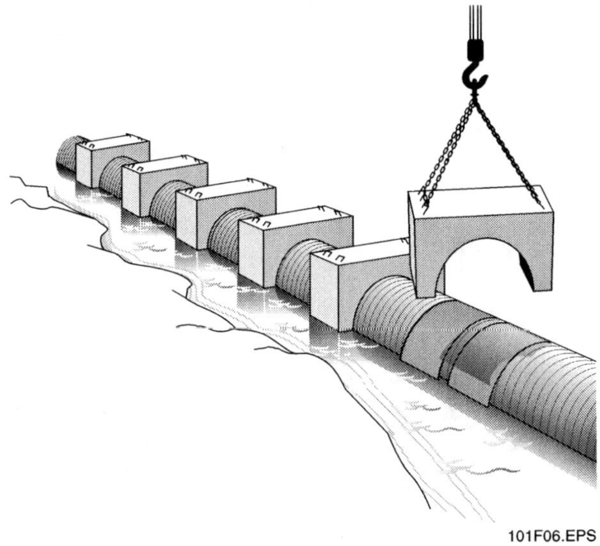

Figure 6 ◆ Add-on weight installation.

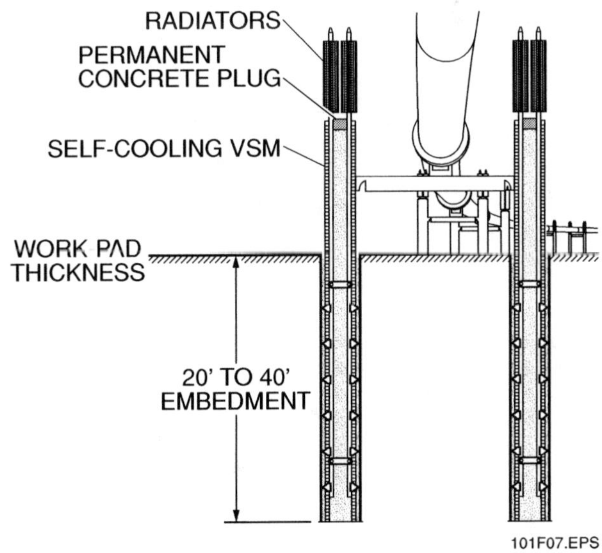

Figure 7 ◆ Bent with self-cooling VSMs.

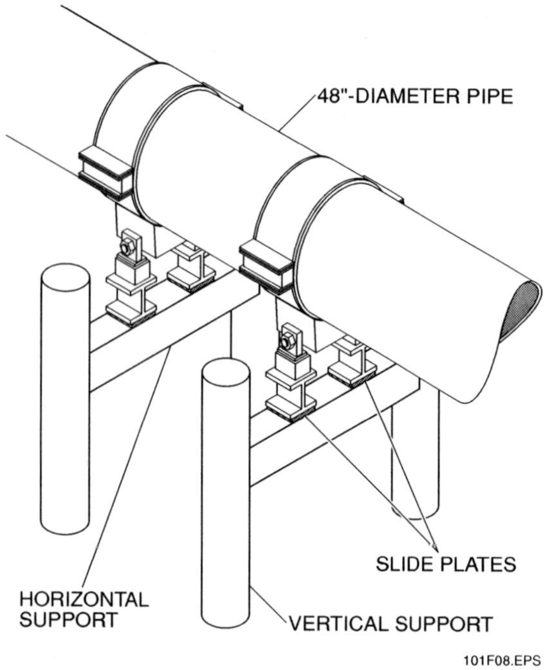

Figure 8 ◆ Common elevated pipeline section.

Pump and compressor stations are spaced strategically along the pipeline at intervals that are dependent on ground elevation, terrain, and other pipeline design factors to maintain the desired flow rate in the line. The pressure that causes the fluid or gas to flow is lost mostly through friction of the product but also by changes in elevation. This pressure must be changed by pumps and compressors to maintain flow. Drag-reducing agents (DRA) can also be added to the product in the line to reduce the friction loss.

A pump or compressor station facility includes pumps or compressors for moving great volumes of liquid or gas, methods of controlling the flow and pressure, means of measuring and accounting for the commodities being moved, storage facilities, and other devices for quality control. *Figure 9* shows an example of a liquid pipeline pump station layout. *Figure 10* shows an example of a gas pipeline compressor station layout.

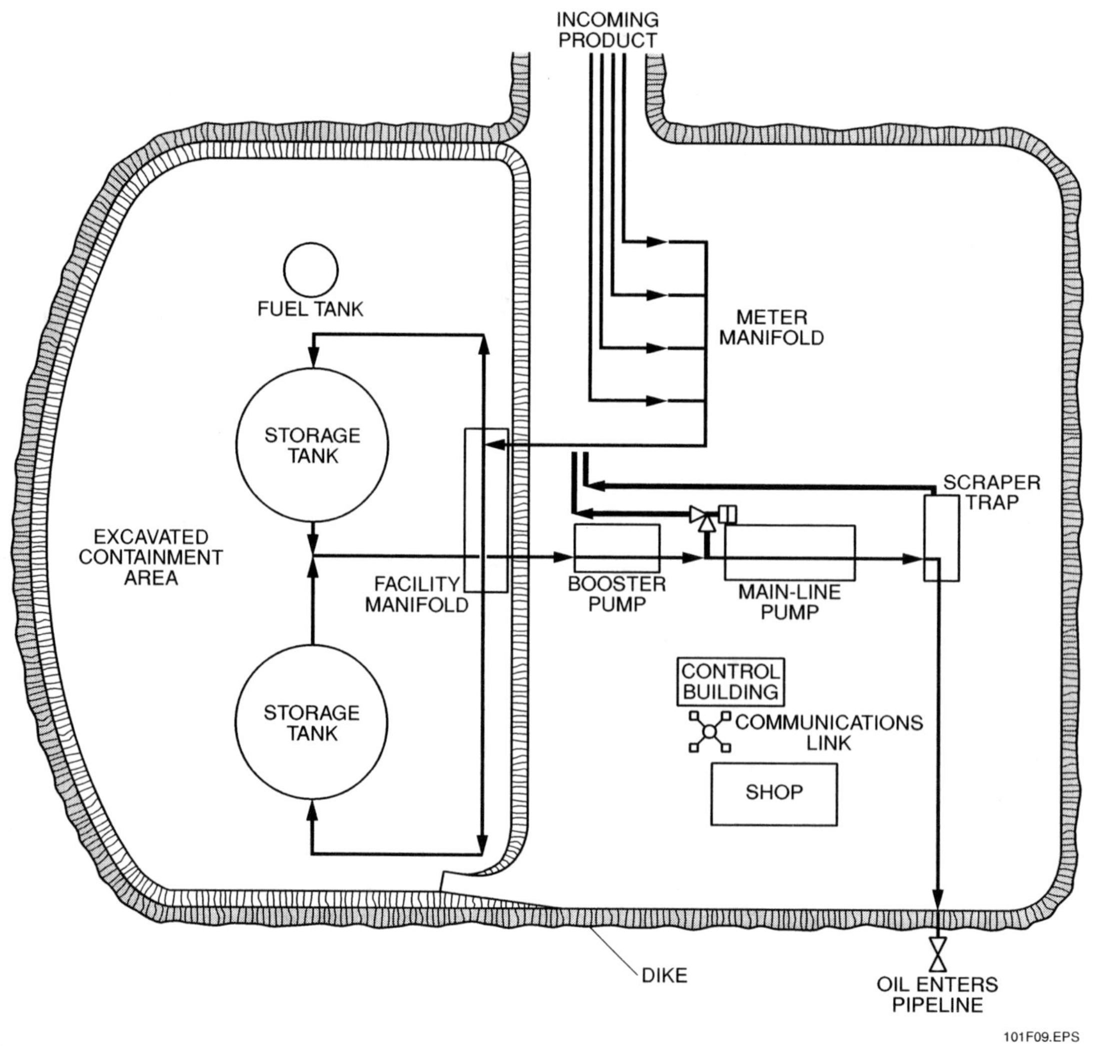

Figure 9 ◆ Example of a pump station layout.

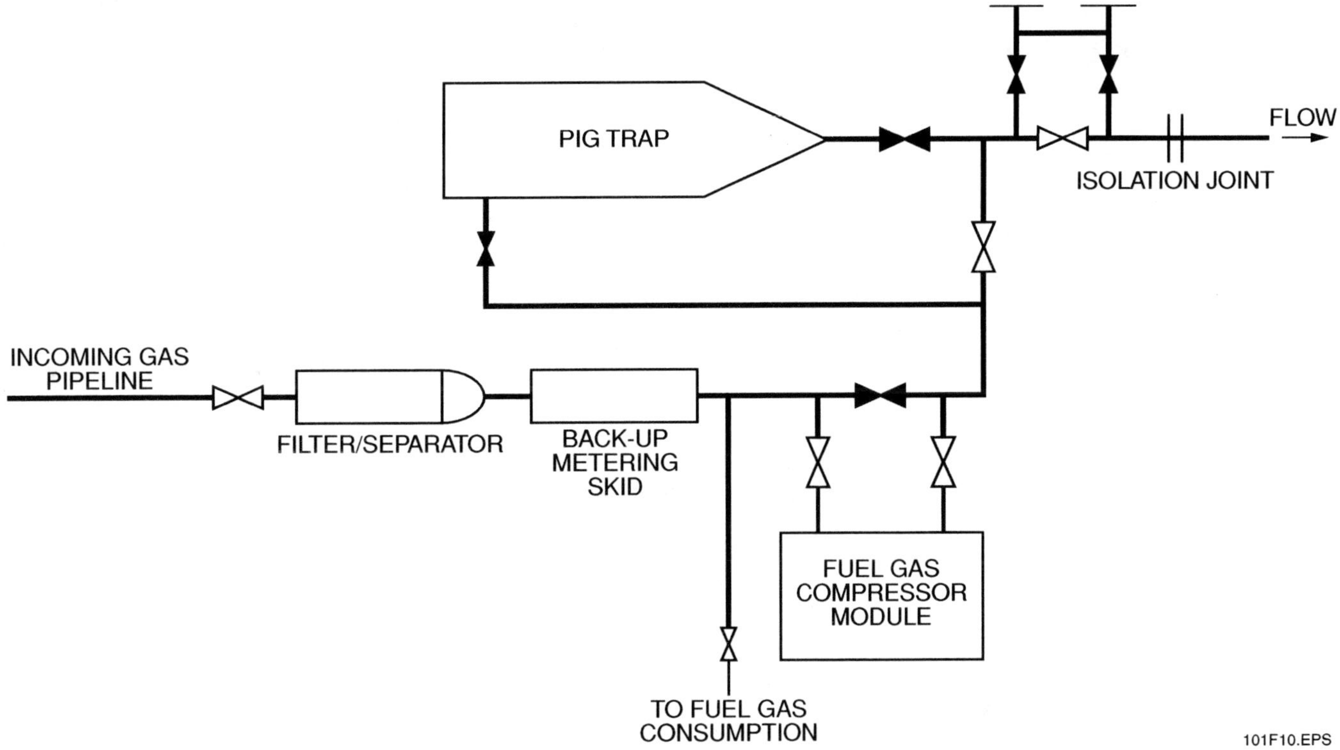

Figure 10 ◆ Example of a gas pipeline compressor station layout.

1.2.3 Operation and Maintenance of the Pipeline

The pipeline is monitored closely and inspected for potential equipment failure and line integrity. Crews identify problems and dig up the pipe to perform repairs. Small **anomalies** may be repaired temporarily while large anomalies may require replacing the pipe. In this case, a bypass line may be constructed so that flow continues while the damaged line is repaired.

The inside of the pipe must also be cleaned periodically to remove wax, scale, and debris. A scraper **pig** is inserted into the line at one point and flows with the liquid or gas to another point, where it is removed. The pig scrapes the inside walls using the hydraulic force of the flow behind it. An instrument pig, also called a smart pig, is inserted into the line periodically to check for anomalies.

1.2.4 Petroleum Products

Different petroleum products can be shipped consecutively from the point of origin in the same pipeline. This is called batching. To maintain quality, batches of each product are arranged in the line and pumped under pressure so that little or no mixing occurs during transit.

1.3.0 Storage

Storage facilities may be located along the pipeline route, at the origination sites, at pump stations, and at terminals. Storage is necessary to meet seasonal needs and other demand fluctuations. Storage is also necessary for the efficient operation of production equipment, pipelines, tankers, refineries, and for emergency use. Large volumes of natural gas are typically stored in underground storage fields throughout the United States.

The trend towards giant tankers created a need for large storage facilities at the loading and unloading ends of tanker runs. A large crude oil terminal typically has several tanks, each with capacities up to half a million **barrels**. A small pump station might have only one tank with a capacity of about 50,000 barrels. Stations along a product line may need separate tanks for storage of each product coming through. These are sometimes called relief tanks or surge tanks. *Figure 11* shows some storage tanks.

1.4.0 Delivery

A wide range of delivery methods is used for gas and petroleum products depending on their destination, intended use, and transport properties.

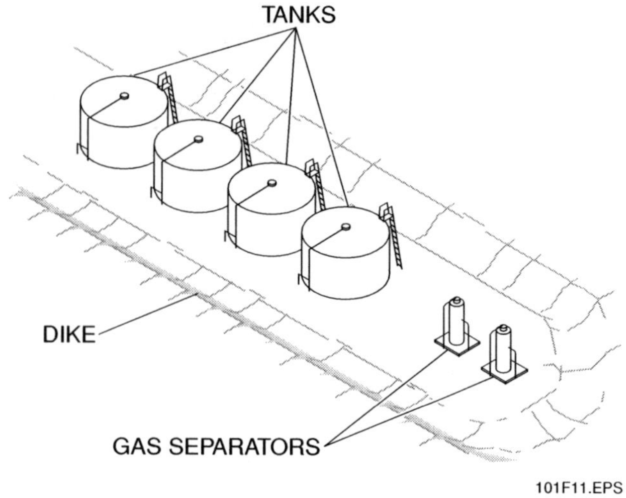

Figure 11 ◆ Storage tanks.

Petroleum is carried chiefly by pipeline, tanker, barge, tank truck, and railroad tank car. Nearly all product moves through pipelines for at least part of the route. Some products are delivered to city or regional stations by pipeline. From there they may be piped or delivered to local consumers or distribution sites. Gasoline, diesel, and kerosene are delivered by tank truck to gas stations for purchase by users. Industrial products may be shipped to companies by pipeline, truck, or railroad. Supertankers are used to move product on ocean routes, while barges are used on rivers and canals. While liquid petroleum products are typically measured in barrels, gas is typically measured using cubic feet. Depending on the volume being measured, gas quantities can be measured using billion cubic feet (BCF), thousand cubic feet (MCF), or hundred cubic feet (CCF). *Table 1* shows typical capacities that can be moved by different means of transport.

Table 1 Means of Transport

Transport	Capacity
Tanker Truck	200 barrels
Railroad Tanker	1,500 barrels
Barge	15,000 barrels
Supertanker	1 million barrels+
Large Liquid Pipeline	2 million barrels+/day
Large Gas Pipeline	1 BCF/day

101T01.EPS

2.0.0 ◆ PIPELINE PRODUCTS

A pipeline can carry a wide range of petroleum and gas products. Each of these products has its own characteristics and rules for handling. Each of these products also presents its own set of hazards for the pipeline workers, for the public, and for the environment. An understanding of the products in the pipeline, the hazards involved, and protective measures to be taken is essential to safe and secure operations.

Petroleum products are produced from crude oil in three basic stages. The first stage is separation and consists of separating the oil into its various fractions, or parts. The main process in this stage is fractional distillation, which separates light, medium, and heavy fractions. The second stage is conversion, in which less useful fractions are converted into more valuable ones. The third stage is treatment, which improves the quality and performance of petroleum products. *Figure 12* shows a diagram of the refinery process.

Products in a typical pipeline are discussed in terms of their characteristics and associated hazards. These products may include:

- Natural hydrocarbons
  - Crude oil
  - Liquefied petroleum gas
  - Wellhead natural gas
- Refined products
  - Gasoline
  - Diesel fuel
  - Fuel oil
  - Jet fuel
  - Consumer-ready natural gas
- Special products
  - Chlorine gas
  - Hydrogen chloride gas
  - Propylene
  - Acetone
  - Benzene
  - Ethane and ethylene
  - Anhydrous ammonia
  - Methane gas
  - Carbon dioxide

2.1.0 Natural Hydrocarbons

Natural hydrocarbons include crude oil, liquefied petroleum gas, and natural gas.

2.1.1 Crude Oil

Crude oil is unrefined petroleum—the source of the myriad of petroleum products. Its specific gravity ranges from 9 to 55 as compared with the specific gravity of water, which is 10. A light crude is one with a higher gravity, and a heavy crude has a lower gravity.

Because the petroleum may have different ingredients or fractions, its color can range from yellow

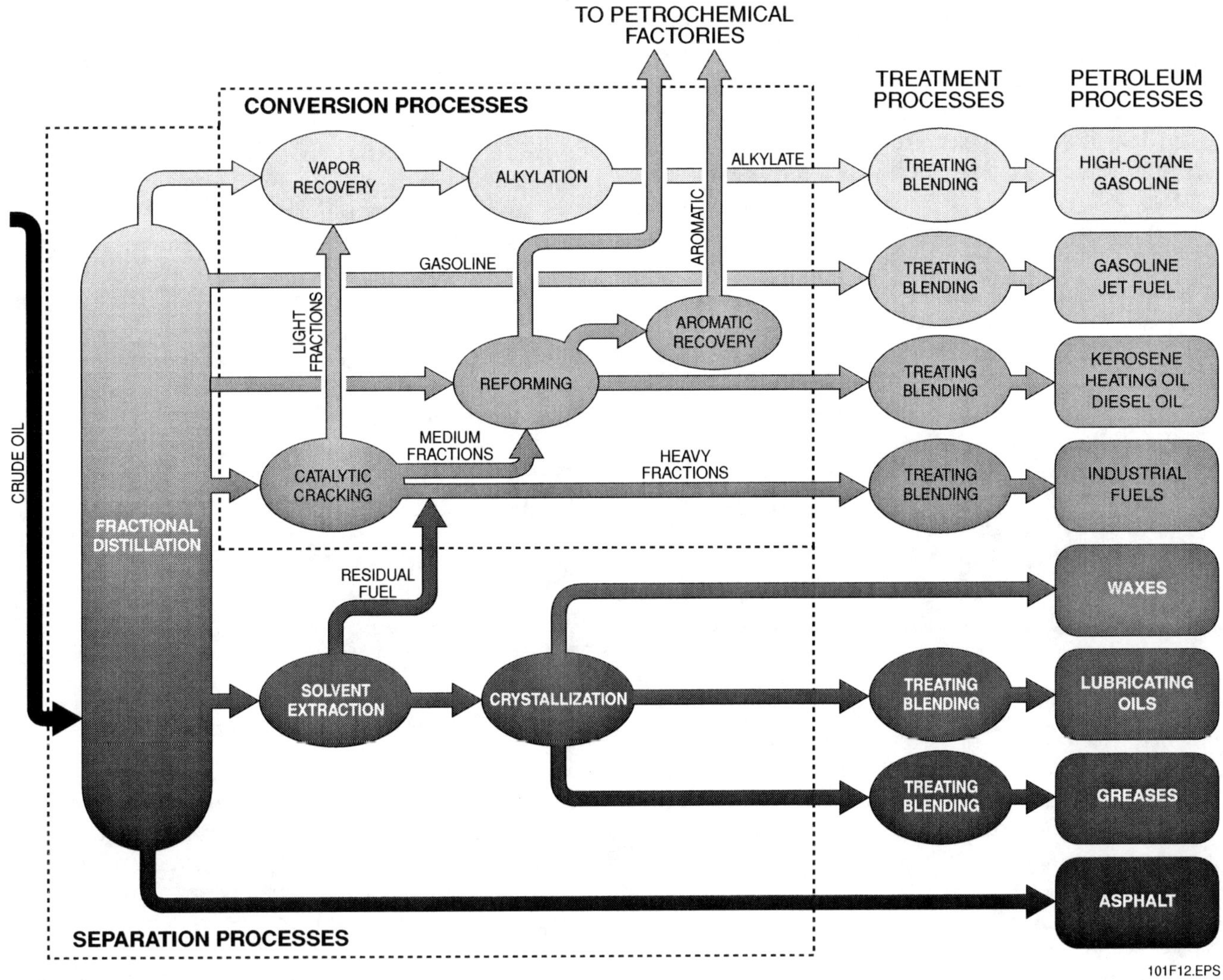

Figure 12 ◆ Diagram of the refinery process.

to black. It may have a paraffin, asphalt, or mixed base. Crude oil also contains substances other than hydrocarbons. These impurities include metallic compounds and sulfur and may make up as much as 10% of the oil. Oil containing a large amount of sulfur compound is called a sour crude. A sweet crude contains little or no sulfur.

2.1.2 Liquefied Petroleum Gas

Liquefied petroleum gas (LPG) is a mixture of heavier, gaseous, paraffinic hydrocarbons, principally butane and propane. These gases, easily liquefied at moderate pressures, may be transported as liquids and converted to gases on release of the pressure. Thus, LPG is a portable source of thermal energy that finds wide application in areas in which it is impractical to distribute natural gas. LPG is used in industry for cutting and welding metals and on farms for operating various kinds of machinery. The principal sources are natural and refinery gas, from which the liquefied petroleum gases are separated by fractionation.

Butane/butylene mix, a form of LPG, is extremely flammable. It releases vapors at well below ambient temperatures and readily forms flammable mixtures with air. If ignited, it will burn in open air or explode in confined spaces. Its vapors are heavier than air and may travel long distances to a point of ignition and then flash back. Alkane/chlorine gas mixtures have produced explosions. Closed systems will form a white frost at the point of the release. Liquid spills will vaporize, forming cold, dense vapor clouds that should be avoided even with respiratory equipment.

2.1.3 Wellhead Natural Gas

Natural gas is the name given to the naturally occurring hydrocarbon mixtures found in many underground wells. Natural gas is an unrefined product that is a naturally occurring gaseous mixture of hydrocarbons that may also contain nitrogen, helium, carbon dioxide, hydrogen sulfide, and water vapor. Because the content or makeup of natural gas can vary from well to well and region to region, it is normally referred to as wellhead natural gas. Natural gas is processed to recover the **natural gas liquids (NGLs)**, the heavier hydrocarbon components that include butane, propane, and natural gasoline.

WARNING!

Exposure to petroleum products can cause eye, skin, and respiratory tract irritation. Always wear proper personal protective equipment (PPE) and refer to company procedures and the product material safety data sheet (MSDS) for specific details.

Natural gas is highly compressible, highly expandable, and has a low specific gravity. Although gaseous at normal temperatures and pressures, the gases making up the mixture are variable in form and may be found as gases or as liquids under varying temperatures and pressures.

2.2.0 Refined Products

Refined petroleum and gas products include gasoline, jet fuel, diesel fuel, heating oil, and kerosene. Other refined products may include propane, butane, consumer-ready natural gas, and blending components.

2.2.1 Gasoline

Gasoline is a volatile, flammable liquid hydrocarbon refined from crude oil and used universally as a fuel for internal-combustion, spark-ignition engines. Natural gasoline, or casinghead gasoline, is the liquid hydrocarbon recovered from wet natural gas. Gasoline is classified as regular, premium, and aviation grades according to how smoothly it burns in an engine.

2.2.2 Diesel Fuel

Diesel fuel is a light hydrocarbon mixture for diesel engines, similar to furnace fuel oil. It has a boiling range just above that of kerosene. Diesel fuel requires less refining than gasoline.

2.2.3 Fuel Oil

Refined fuel oil comes in two grades. The specifications for both are defined by *American Society for Testing and Materials (ASTM) Specification D 396*.

No. 1 distillate, such as kerosene, is a light fuel. No. 2 fuel oil diesel is a distillate fuel oil prepared for use in atomizing-type burners or for smaller industrial burner units. Lesser grades of fuel oil, called *residual fuel oils*, are made from the topped crude of refinery operations. Residual fuel oil is used for the production of electric power, space heating, and various industrial purposes.

2.2.4 Jet Fuel

Jet fuel is a highly refined kerosene petroleum distillate. Jet fuel has good flow properties in low-temperature conditions, and it burns efficiently and cleanly. It must be delivered pure and free of rust, water, and sediment.

Military jet fuel is a highly refined kerosene distillate that meets stricter military specifications.

2.2.5 Consumer-Ready Natural Gas

Wellhead natural gas is normally run through a refining process to remove any undesired contents, remove contents that can be used for other commercial purposes, and achieve a more consistent safe product and heat content. The final product is known as commercial natural gas or consumer-ready natural gas.

Consumer-ready natural gas is still a mixture of different gases but is made up primarily of methane. Other gases, such as ethane, propane, nitrogen, and carbon dioxide can also be found in trace quantities. The composition of the consumer-ready natural gas remains fairly consistent over time. The heat value range of typical consumer-ready natural gas is 1,020 to 1,045 Btus per cubic foot.

2.3.0 Special Products

Special products include other by-products that are produced and carried through pipelines.

2.3.1 Chlorine Gas

Chlorine gas is a product used for water purification, chemical processes, and the manufacture of plastic. Chlorine gas is a strong oxidizing agent; it can react explosively with organic products and it is poisonous. Liquid and gas chlorine under pressure can be hazardous. Chlorine is nonflammable and nonexplosive, but like oxygen, it can support combustion of certain substances.

It can react violently or form explosive compounds with many chemicals, such as acetylene, turpentine, ether, ammonia gas, and hydrogen, and finely divided metals.

2.3.2 Hydrogen Chloride Gas

Hydrogen chloride gas is a fuming, poisonous gas used to make hydrochloric acid, an important industrial chemical used in many chemical compounds, in metallurgy, in food processing, and as a cleaning agent. Hydrogen chloride is nonflammable and noncombustible. However, if dissolved in water, it produces hydrochloric acid, which attacks most metals.

2.3.3 Hydrogen Sulfide

Hydrogen sulfide is a flammable, colorless, gaseous compound of hydrogen and sulfur, which in small amounts has the odor of rotten eggs. It is formed naturally by the decomposition of organic animal and vegetable matter by bacterial action. Hydrogen sulfide is found in crude oil, natural gas, sewers, and swamps and is created by a number of industrial processes. It is extremely corrosive and poisonous. Hydrogen sulfide causes damage to the eyes, skin, breathing passages, and lungs, attacks and paralyzes the central nervous system, especially the parts that control the lungs and heart. In large amounts, it deadens the sense of smell. Hydrogen sulfide is an extremely dangerous compound, and exposure is deadly.

2.3.4 Propylene

Propylene is used in plastics manufacturing. It is extremely flammable and will readily ignite at ambient temperatures. The flammable vapors, being heavier than air, may travel long distances along the ground to a source of ignition and then flash back. Propylene presents the danger of rocketing cylinders, ruptured containers, and explosion. Reignition or explosion can occur if the flames are extinguished without stopping the source flow. An electrostatic charge may build up during handling if the equipment is not grounded.

2.3.5 Acetone

Acetone is a clear, colorless liquid with a sweet, pungent odor that is used as a solvent. It is extremely flammable and may cause flash fire. It is incompatible with a variety of materials, including inorganic acids, caustic, amines, alkanolamines, oxygen, halogens, aldehydes, ammonia, oxidizing agents, and chlorinated compounds. Contact with these should be avoided during transport, handling, and storage.

2.3.6 Benzene

Benzene is an extremely flammable liquid, readily igniting at ambient temperatures. The vapor forms an explosive mixture with air. Flammable vapors, being heavier than air, may travel long distances to a point of ignition and flash back. Incomplete combustion can produce carbon monoxide.

2.3.7 Ethane and Ethylene

Ethane and ethylene are simple asphyxiants. They are dangerous when exposed to heat or flame and can react violently with oxidizing materials. Both are flammable over a wide range in air, with an even broader range in pure oxygen.

2.3.8 Anhydrous Ammonia

Anhydrous ammonia is the liquid form of pure ammonia gas. It is used as a fertilizer because of its high nitrogen content. It is also used as an industrial refrigerant in cold-storage and ice manufacturing plants, since it absorbs a large amount of heat when going from a liquid back to a gas. Anhydrous ammonia is made by compressing dry ammonia gas.

Ammonia is a colorless alkaline gas made up of one part nitrogen and three parts hydrogen. It is lighter than air and has a sharp, stinging odor. Ammonia can be safely inhaled if it is greatly diluted in air, but concentrated ammonia gas can cause suffocation and death. It does not burn in air, but it burns in oxygen with a weak yellow flame. Ammonia is made by combining hydrogen and nitrogen, and as a by-product in the production of coal and coke gas.

2.3.9 Methane Gas

Methane is a member of the alkaline or paraffin series of hydrocarbons. It is called marsh gas because it forms by anaerobic bacterial decomposition of vegetable matter in swampy land. Coal miners know it as *firedamp* because it is combustible when mixed with air. It is a major component of natural gas and of coal gas, and forms in large amounts in sewage disposal processes.

In addition to its use as a fuel, methane is important as a source of organic chemicals and of hydrogen. In the presence of steam and catalysts, it yields carbon monoxide and hydrogen, which can be converted to liquid hydrocarbons or to methanol and other alcohols. The incomplete combustion of methane with air produces carbon black, which is used in rubber and ink. Chlorination of methane yields various chlorides and chloroform.

2.3.10 Carbon Dioxide

Carbon dioxide (CO_2) is a colorless, odorless, tasteless gas which is about 15 times heavier than air. Under normal conditions, it is stable, inert, and nontoxic. The decay of all organic matter and the burning of hydrocarbon fuels produces carbon dioxide. It is obtained commercially from four sources: naturally occurring underground CO_2 wells, fermentation, combustion of carbonaceous fuels, and as a by-product of chemical processing.

Carbon dioxide gas may be liquid or solid based on compression and lowered temperatures. It may be used as a refrigerant, inerting medium, chemical reactant, neutralizing agent for alkalies, and pressurizing agent. Liquid carbon dioxide is an effective fire-extinguishing agent because it rapidly reduces the temperature of burning materials below the ignition point. The gas, being heavy and inert, displaces air, which is the source of the fire's oxygen, and blankets the flames. Oil tankers, barges, storage vessels, and pipelines are quickly and safely made inert with carbon dioxide to allow for welding and other repairs.

2.4.0 Protective Measures

Material safety data sheets (MSDSs) are good sources of information on these products and must be made available to all employees of the company. There is an MSDS for each hazardous substance, listing its properties and hazards, including a hazard rating for health, fire, reactivity, and other hazards. MSDSs also detail safe handling, personal protective equipment requirements, and first aid/emergency procedures.

All products pose hazards to the environment if spilled or disposed of improperly. Areas of possible spills should have containment structures and/or procedures. These include dikes around storage and transfer facilities and containment booms around tankers being filled. The marine environment is particularly sensitive to oil spills because the tides and currents spread the oil and complex marine life is so vulnerable. *Figure 13* shows temporary containment of a spill from a tank.

Oil may be harmful on land, too. It can seep into the groundwater or streams where it contaminates plant and animal life and eventually human

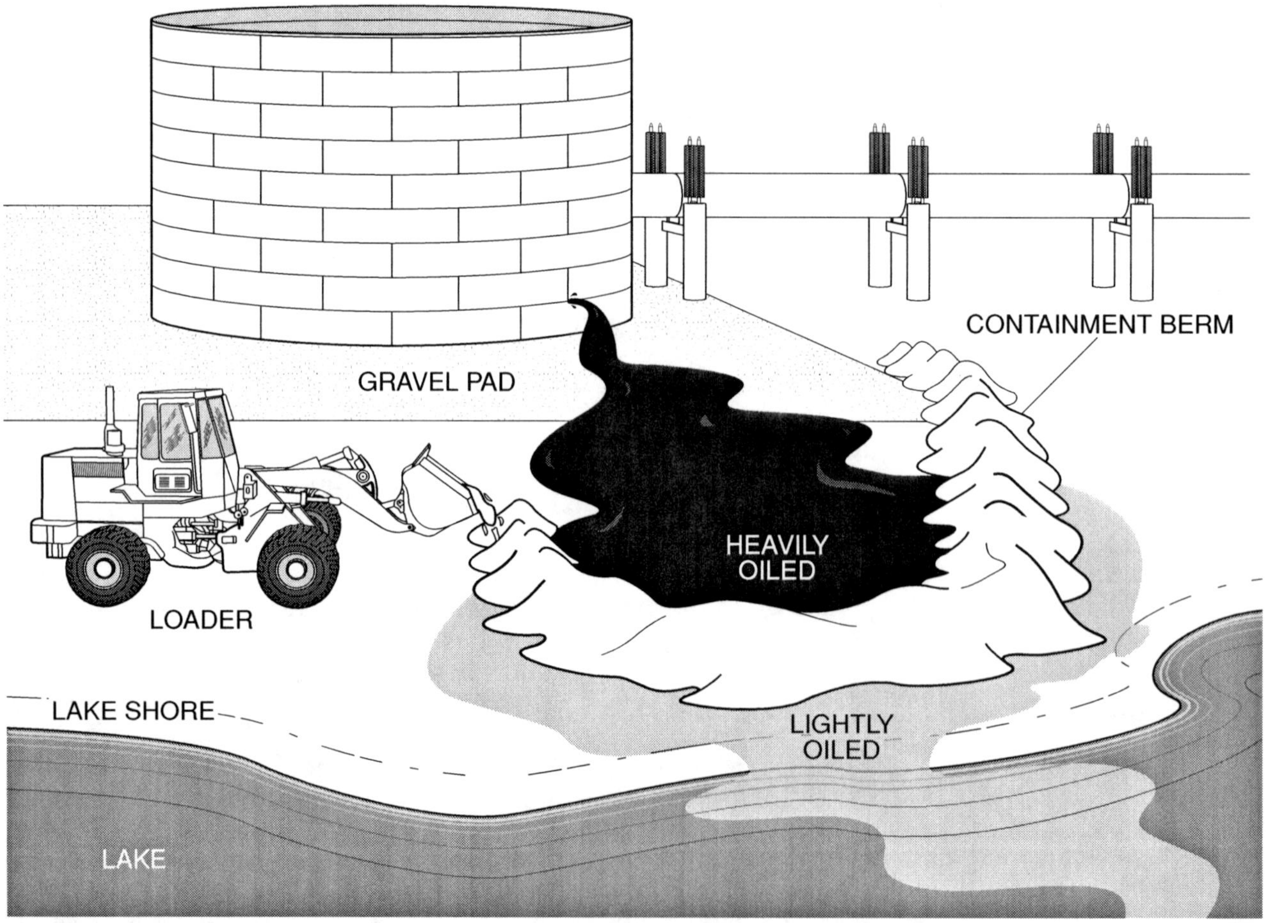

Figure 13 ◆ Temporary containment of a tank spill.

life. Handling and disposal of hazardous chemicals are addressed in company procedures and MSDSs. Detailed procedures exist for handling all petroleum products, and the observance of these procedures is strictly regulated and enforced by federal, state, and local agencies.

3.0.0 ◈ MAPS AND DRAWINGS

Maps and drawings are an important means of communicating information about pipeline equipment, flow, and processes. Some of these are piping and instrumentation diagrams (P&IDs), strip maps, blueprints, and alignment sheets.

3.1.0 Piping and Instrumentation Diagrams (P&IDs)

A P&ID depicts the functional layout of a piping system and its components, valves, and instrumentation as clearly and accurately as possible. It is accurate to the extent that all components are connected to each other as shown in relation to flow-path orientation. A P&ID does not represent the actual physical layout of equipment; for example, a valve that may appear to be right at the discharge of a pump can actually be located some distance away or even on another floor. Many times, a P&ID will use a broken line around groups of equipment to indicate that they are all in the same building (see *Figure 14*).

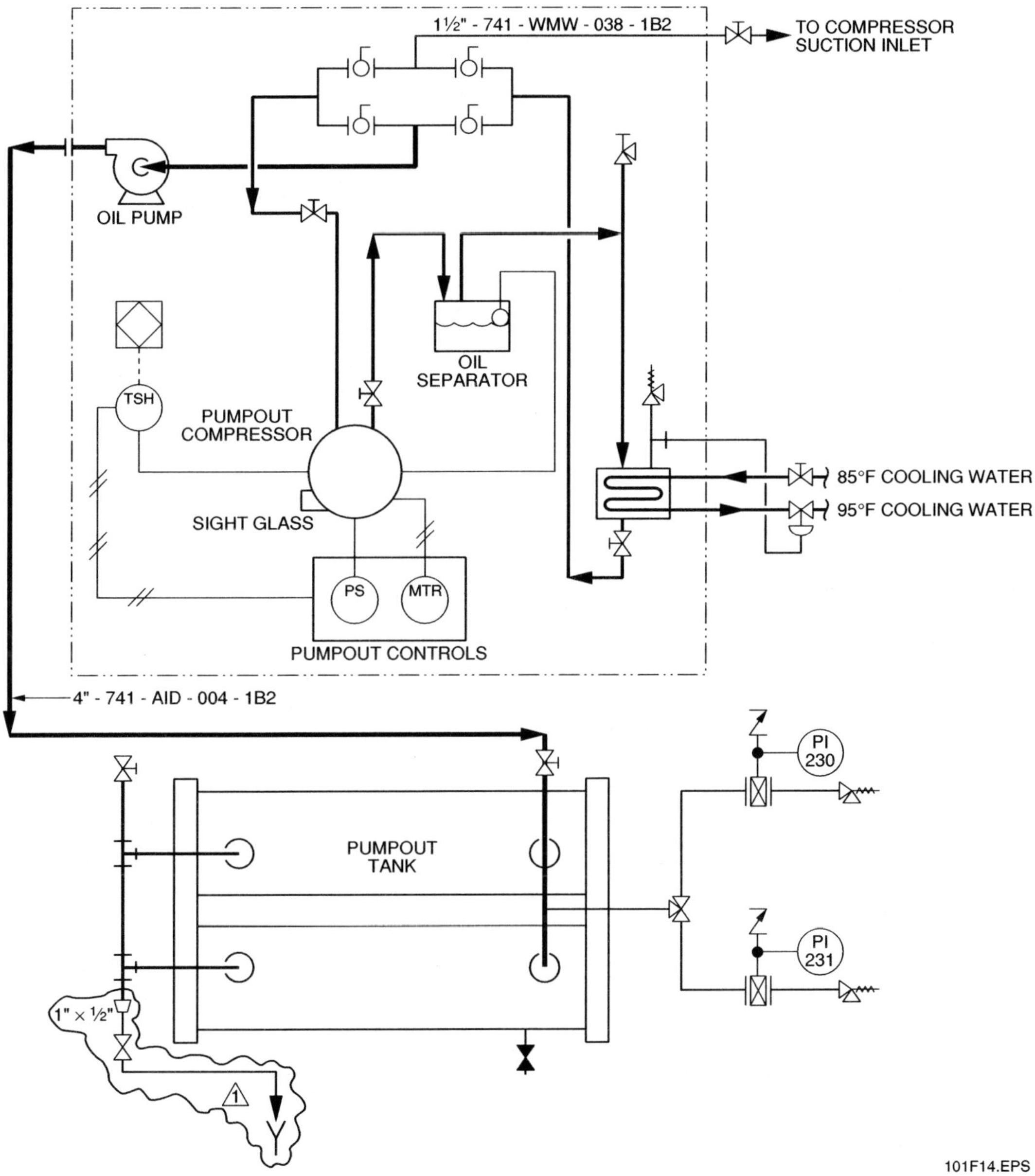

Figure 14 ◈ Sample P&ID.

The P&ID depicts all components of a particular system, including pipe sizes, flange sizes, valve sizes, flow direction, and references to other diagrams. Standardized symbols are used for these components instead of pictures. Industry standards for these symbols have been developed and adopted by organizations, such as the American National Standard Institute (ANSI), the Instrument Society of America (ISA), and the American Society for Mechanical Engineers (ASME), to aid interpretation. The P&ID title block and legend should be examined closely for explanations of the diagram.

3.2.0 Blueprints

Blueprints are duplicates, or copies, of plans for buildings, assemblies, systems, and other constructions. These are working drawings, including specifications for a project that usually includes site plans, floor plans, elevations of buildings, cross sections, and detail drawings. They were originally made by a photographic process that produced white lines against a blue background, hence, the name. However, they are now produced by computer-aided design (CAD) software and generally have black or blue lines on a white background. The blueprint shows the size and location of all structures and equipment in the project to scale and how they are physically put together. *Figure 15* shows a site-plan blueprint.

3.3.0 Strip Maps/Alignment Sheets

Strip maps (see *Figure 16*) are combinations of different types of maps or drawings over the same area such as the pipeline route. The strips of each type of map or drawing are laid out side by side

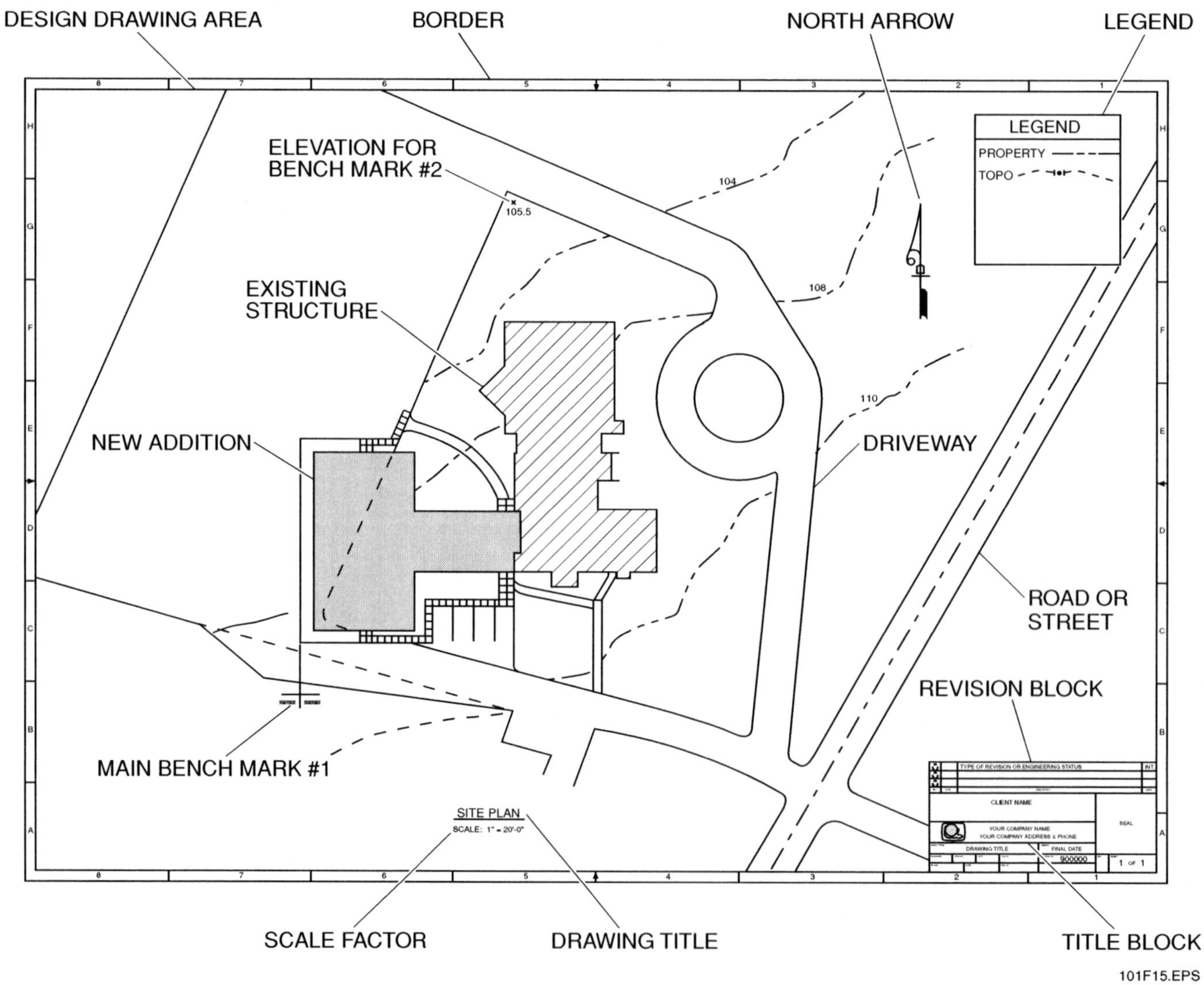

Figure 15 ◆ Site-plan blueprint.

Figure 16 ◆ Sample strip map.

on the same sheet so that they line up along the route. There may be views shown of the ground from an aerial, or plan view, and side, or elevation view, which shows the ground profile along a specific line. The pipeline will be superimposed on all views. Strip maps help to coordinate surface features with topography as well as with other pipeline data at any point on the route.

Alignment sheets are various types of maps, drawings, and diagrams that show different features along the pipeline route. The alignment sheets will show the relationship of the pipeline to true north, geographic features, political boundaries, fences, rivers, property lines, and other pipelines. *Figure 17* shows one type of alignment sheet.

Strip maps and alignment sheets vary greatly from company to company. These names may be used interchangeably or may mean very different things.

4.0.0 ◆ OPERATIONS

Controlling the fast-paced, continuous, and complex operation of the pipeline is a critical job done largely by the scheduler, controller, and dispatcher with the help of other personnel and systems like Supervisory Control and Data Acquisition (SCADA), programmable logic controllers (PLCs), and remote terminal units (RTUs). All control functions are documented by filling out forms or printing out reports.

4.1.0 Scheduling

Before a shipment of product reaches the pipeline and comes under the control of the dispatching office, it has to be accumulated and scheduled with the pipeline scheduler. The scheduler at a pipeline facility typically performs the following functions:

- Takes nominations from shippers
- Creates pumping orders for pipelines to meet shippers' needs
- Validates and reviews operating information for schedule adjustments
- Receives delivery information in time for shipments
- Validates all shipping information for billing
- Utilizes power optimization to save costs and maximize throughput efficiency

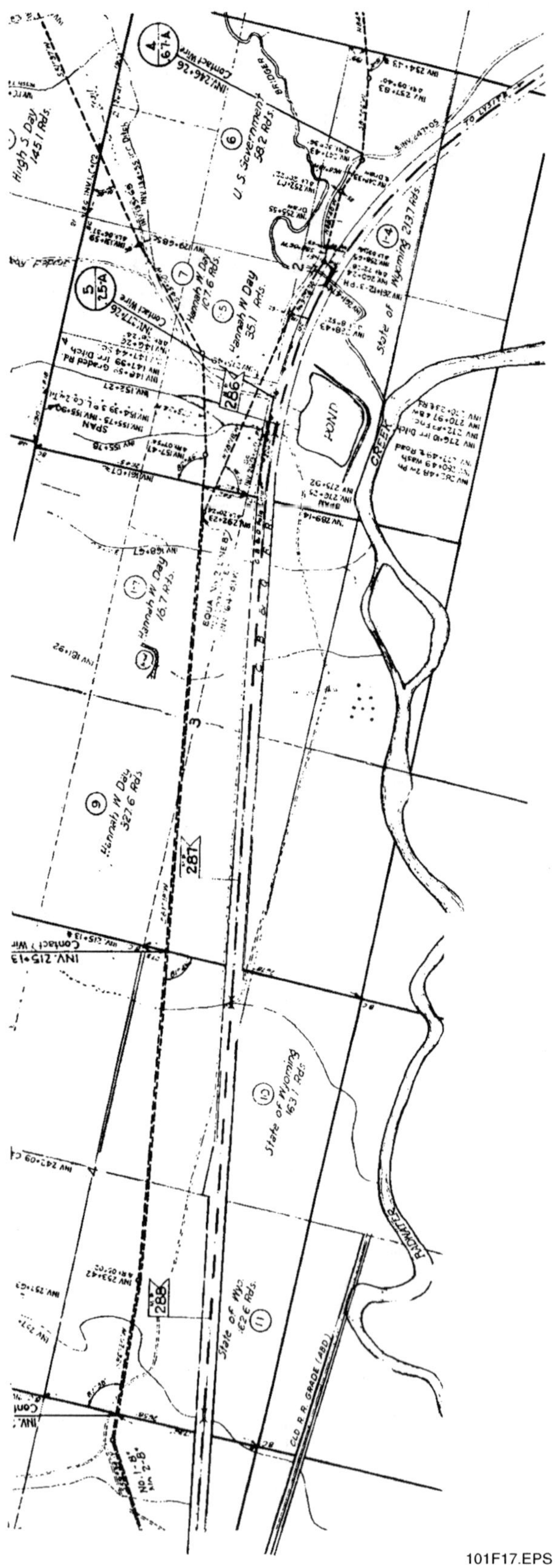

101F17.EPS

Figure 17 ◆ One type of alignment sheet.

4.2.0 Pipeline Operations

The pipeline is monitored closely and inspected for potential equipment failures and line integrity. These problems can arise from bad welds, corrosion spots, faulty valves, third party damage, or shifting of the pipe. When aboveground, the pipe may also be damaged by accidents or sabotage. In areas with faults or history of earthquakes, seismic activity must also be closely monitored. Flow can also be interrupted by obstructions in the pipe, such as stuck pigging devices, foreign objects, and debris.

Computers and automation are used to control and monitor the pipeline operation. Communication is maintained between all stations regarding the status of the operation. Controllers have the ability to shut down the pipeline flow if a problem is detected and to pinpoint the location of the problem. *Figure 18* shows a typical control center layout.

4.3.0 Control Systems

SCADA is the software used to monitor and control the flow of products through the pipeline every day. Wireless telemetry and SCADA systems provide remote alarm, control, and data collection for many types of pipeline operations.

SCADA is installed with the main computer system and hooked up through a server to RTUs and PLCs in the field. Sensors along the pipelines precisely measure operating conditions, flow rates, and the status of control devices, such as flow and pressure regulators and valves. The RTUs collect vital information in the field and transmit the data via telephone, microwave, or satellite circuits to the control center.

4.4.0 Communications

Communications are the systems that keep the pipeline operating. All along the pipeline, everyone must have open communications at all times because everything that occurs on the pipeline affects the entire system.

A variety of communications equipment and devices may be used in a pipeline company's operations. These are set up for continuity during normal or emergency conditions. It is important that communications be reliable and always available. Most companies have redundant, or backup, systems to ensure that this is true. Some of the equipment used includes:

- Dedicated satellites or microwave-based radio communications systems that provide linkage between the control room personnel and various pump stations and valve locations
- Hardwire communications that link the control room and all offices and all remotely monitored and operated facilities, such as pump stations and RTUs
- Portable cellular phones or vehicle phones provided to all key and supervisory personnel
- Locally activated pagers or nationwide pagers provided to key personnel for use in routine operating or emergency events

101F18.EPS

Figure 18 ◆ Typical control center layout.

- Fixed repeaters installed at points within the system that provide internal long distance and local radio communications

In the event of a power loss, backup systems are provided, which may include standby generators for powering communications equipment, batteries, or uninterruptible power systems (UPS). Portable self-contained mobile repeaters may be used at an incident scene and utilized to augment existing radio systems or replace any that may be inoperable. A radio link (base station) may be used with portable units to allow for emergency communications between facilities in the event that telephone and/or microwave transmissions become inoperable. Redundant communications and power systems should be installed at key facilities to allow for control in the event of loss of primary systems.

5.0.0 ◆ FLUIDS AND SYSTEM HYDRAULICS

The operation of any liquid pipeline is built on the principles of hydraulics, or the science of the behavior of fluids at rest or in motion. These principles are in turn built on the properties and characteristics of fluids.

5.1.0 Fluids

A fluid is a substance that flows, and some flow more easily than others. A fluid is also a substance that changes form with a slight force or pressure; it is elastic, however—meaning it will try to return to its previous form when the pressure is removed. Some fluids are more elastic than others. Elasticity is a measure of the fluid's resistance to change in shape.

Fluids include all liquids and gases. Water at ordinary temperature is a liquid fluid; at high temperatures, a gas fluid. Air is a gas fluid. All fluids can be greatly affected by changes in temperature and pressure.

No fluid is perfect and frictionless, and different fluids have different levels of friction, or resistance to flow. This resistance to flow is called *viscosity*.

Some fluids are thicker than others. This is called *density*. Density can be a large factor in the viscosity of a fluid. It is a measure of how tightly packed the fluid molecules are or how much of the fluid's matter exists within a given unit of volume.

Measuring density in a liquid is readily done with a hydrometer because they are not usually compressible. Measuring density in a gas is more difficult because gases can easily be compressed. Compressibility is the ability to be squeezed, or forced into a smaller volume.

Density helps determine the compressibility of a fluid. When a fluid is compressed, it becomes denser. The specific gravity of a substance is derived from its density. Specific gravity is the ratio of the density or mass of a given volume of the substance to the density of the same volume of a standard substance, such as water. The water and the liquid substance must be at the same temperature. For gases, the gas must be at the same temperature and pressure.

American Petroleum Institute (API) gravity is a unit of measure developed specifically for liquid petroleum products. The heavier the substance, the lower the number of its API gravity. Temperature will affect API gravity and should always be corrected to 60°F.

The properties of fluids that are used to describe the fluids are density, viscosity, compressibility, inertia, and elasticity. *Table 2* shows some fluid properties.

Table 2 Fluid Properties

Fluid	API Gravity @ 60°F	Specific Gravity @ 60°F	Vapor Pressure @ 100°F	Absolute Viscosity	Compressibility @ 1000 psi @ 60°F
EP Mix 80/20	214	0.41	550 psi	0.11 cP	7.3%
NGLS	146	0.51	325 psi	0.15 cP	3.8%
Raw Feed Propane	146	0.51	150 psi	0.15 cP	3.8%
Butane	112	0.58	88 psi	0.18 cP	2.0%
Anhydrous Ammonia	97	0.62	200 psi	0.25 cP	1.6%
Gasoline	63	0.66	2 psi	0.70 cP	0.8%
Jet Fuel #1	42	0.82	2 psi	1.4 cP	0.8%
Diesel Fuel #2	33	0.86	1 psi	4.7 cP	0.5%
Crude Oil	50 and below	0.78–0.98	> 1 psi	6–400 cP	> 1.0%
Water	10	1		1.12 cP	0.3%

101T02.EPS

Other properties of fluids, particularly those found in pipelines, include volatility and flammability. Volatility is the tendency of a fluid to readily vaporize. Sudden vaporization can cause a violent reaction or explosion. Flammability is a measure of how readily a fluid catches fire when in contact with heat or sparks. The volatility and flammability of specific pipeline products is described in another section of this module.

5.2.0 Basic Hydraulics

Some laws of hydraulics apply to gases as well as liquids but only under certain conditions. In general, the following sections pertain only to liquid fluids.

Hydraulics is usually divided into two branches: hydrostatics for liquids at rest and hydrodynamics for liquids in motion.

5.2.1 Hydrostatics

Hydrostatic principles are fundamental to the design of many machines and instruments. The design of hydraulic presses and the operation of instruments, such as manometers and barometers, depend on these principles.

A simple hydraulic press consists of a liquid-filled cylinder that contains two pistons, one smaller than the other. When a force is applied to the smaller piston, the force travels through the liquid to the larger piston, producing a force larger than the initial one. The increase in force is proportional to the ratio of the larger piston to the smaller one. *Figure 19* shows a simple hydraulic press.

The laws of hydrostatics are the following:

1. The amount of pressure caused by the weight of a column of liquid is determined by the height of the column.
2. An object placed in a liquid loses an amount of weight equal to the weight of the liquid it displaces.
3. A fluid in a container transmits pressure equally in all directions.

5.2.2 Hydrodynamics

Hydrodynamic principles are basic to hydraulic engineering, to the design of certain machines, and to pipeline engineering. Hydrodynamics define the plans for water systems, canals, and irrigation systems, as well as airplane design and hydraulic machines like water turbines.

Pipelines for water and other fluids have to take into account the pipe size and the fluid's friction on the pipe to determine fluid velocity through the pipe. Liquid flow can be steady or unsteady. An unsteady flow results from changes in the velocity, temperature, or pressure of a liquid or its movements around obstacles. The three basic laws of hydrodynamics apply only to steady-flowing liquids. The laws of hydrodynamics are the following:

1. The velocity of a liquid flowing through a pipe increases as the area of the pipe decreases and vice versa.
2. The pressure of a liquid increases as its velocity decreases and vice versa.
3. The velocity with which a liquid flows through an opening in a container equals the velocity of a body falling from the surface of the liquid to the opening.

Figure 20 shows liquid velocity and pressure relationships.

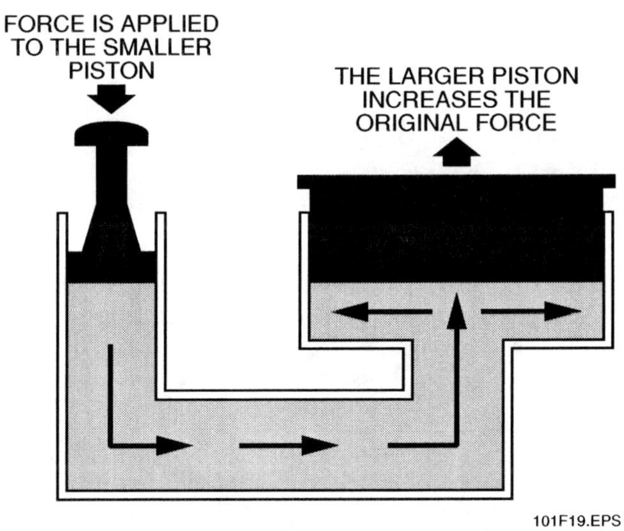

Figure 19 ◆ Simple hydraulic press.

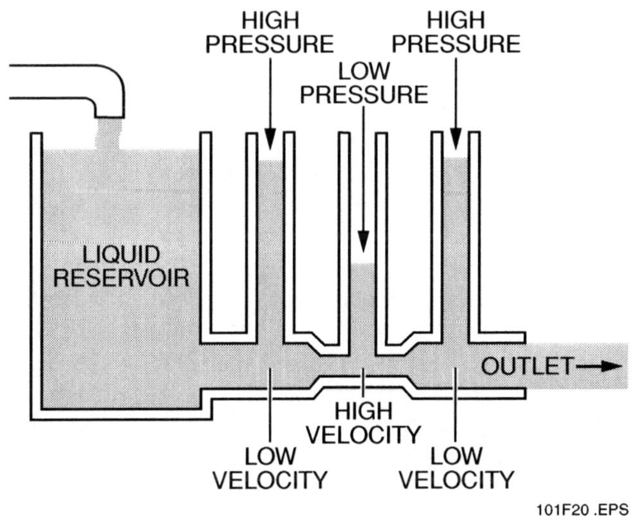

Figure 20 ◆ Liquid velocity and pressure relationships.

5.3.0 Pipeline Hydraulics

Fluid properties are major factors which affect the design of the pipeline. The primary factors in pipeline design are:

- Fluid characteristics
- Desired flow rate
- Maximum operating pressure (MOP)
- Maximum allowable operating pressure (MAOP) (for gas)
- Elevation profile/terrain
- Fittings/valves
- Installation costs
- Environmental concerns

Pressure is referenced in several ways in the pipeline industry. The maximum operating pressure (MOP) is defined as the maximum internal pressure allowed liquid during the operation of a pipeline. The term MOP is typically used more in the liquid pipeline industry and the regulations are referenced in *49 CFR 195*. The MAOP is defined as the maximum internal pressure permitted during the operation of a gas pipeline. The term MAOP is typically used more in the gas pipeline industry and the regulations can be referenced in *49 CFR 192*.

Although fluids are defined by their ability to flow, something has to start the flow and overcome their inertia and friction. A force, such as difference in pressure, gravity, or an external force physically pushing it, can do this. Usually, this is done by pumps. Stopping the flow once started will require another force, such as resistance (friction), or a stationary object, such as a closed valve.

The resistance or friction of the fluid against the pipe will be constantly trying to slow it down. Most pumps work by providing a difference in pressure and must continue providing that force all along the line to overcome the friction. There are drastic differences in the amount of pressure required to deliver the proper volume of different types of products. But the pressure in the system must be kept under the maximum limits.

Friction is also affected by the roughness of the pipe interior, the pipe diameter and length, the product density and viscosity, the flow rate, and obstacles, such as valves and bends. Pipeline design can help reduce some of these, but higher fluid viscosity and density will increase the friction.

When the fluid is traveling uphill, pumps must be used to overcome the fluid's inertia and resistance and the force of gravity. When the flow in the pipe is downhill, gravity may be all that is needed.

Another factor in the flow of the pipeline is the temperature. The higher the temperature, the lower the viscosity and vice versa. When the product is hot, it flows better. Sometimes heaters are included in the pipeline equipment to keep the product at the optimum temperature.

6.0.0 ◆ GASES AND SYSTEM PNEUMATICS

Pneumatics, as it relates to the pipeline industry, is the study of the storage, compression, and transmission of gases. The compression of gas has one primary goal, to deliver the gas at a pressure higher than the original pressure. The original pressure may vary from very low absolute pressures to several thousand pounds. The pressure rise may vary from a few ounces to many thousand pounds, and the volumes handled may vary from a few cubic feet per minute to hundreds or thousands of cubic feet.

There are three variables that have to be calculated and controlled when compressing and moving gas in a pipeline. These variables are pressure (p), volume (v), and temperature (k). These three characteristics of gases have a close relationship with one another and are explained in the ideal gas laws.

6.1.0 The Ideal Gas Laws

Pressure, volume, and temperature of gases have a close relationship with one another. By keeping one of them constant, the other two will have either a direct or inverse proportion. The relationships between pressure, volume, and temperature are explained using the ideal gas laws. These ideal gas laws are Boyle's law, Amonton's (Gay-Lussac's) law, and Charles' law.

6.1.1 Boyle's Law

Boyle's law states that when temperature is kept constant and volume is decreased, pressure will increase. His law gives the relationship between pressure and volume if temperature and amount are held constant.

- If the volume of a container is increased, the pressure decreases.
- If the volume of a container is decreased, the pressure increases.

If the volume is increased, the gas molecules have farther to travel and they will impact the container walls less often. This means the gas pressure will be less because there are fewer molecule impacts per unit of time.

If the volume is decreased, the gas molecules have a shorter distance to travel, thus striking the walls more often. This results in pressure being increased because there are more molecule impacts per unit of time.

6.1.2 Amonton's (Gay-Lussac's) Law

Amonton's law states that when volume is kept constant and temperature is increased, pressure also increases. Amonton's law was discovered by Joseph Louis Gay-Lussac in the early 1800s. His law gives the relationship between pressure and temperature when volume and amount are held constant.

- If the temperature of a container is increased, the pressure increases.
- If the temperature of a container is decreased, the pressure decreases.

If the temperature is increased, the gas molecules will move faster and they will impact the container walls more often. This means the gas pressure inside the container will increase. Gay-Lussac's law is a direct mathematical relationship. This means there are two connected values and when one goes up, the other also increases.

6.1.3 Charles' Law

Charles' law states that when pressure is kept constant and volume is increased, temperature also increases. The pressure that a gas exerts on the walls of its container is determined by the momentum of the atoms and molecules of the gas, which in turn is determined by the temperature. As the temperature increases, the atoms and molecules move faster, and so exert a greater pressure on the walls.

This law gives the relationship between volume and temperature if pressure and amount are held constant.

- If the volume of a container is increased, the temperature increases.
- If the volume of a container is decreased, the temperature decreases.

If the temperature is increased, the gas molecules will move faster and they will impact the container walls more often. This means the gas pressure inside the container will increase, but only for an instant. The greater pressure on the inside of the container walls will push them outward, thus increasing the volume. When this happens, the gas molecules will now have farther to go, thereby lowering the number of impacts and dropping the pressure back to its constant value.

6.2.0 Measuring Energy Gases

Quantities of natural gas are measured in volume units. A cubic foot of natural gas at a temperature of 60°F and an atmospheric pressure of 14.7 pounds per square inch gauge (psig) is the common unit of measure. Gas production from wells and supplies to power plants are measured in thousands or millions of cubic feet (MCF and MMCF). Resources and reserves are calculated in billions of cubic feet (Bcf) or trillions of cubic feet (Tcf). To put this in perspective, a trillion cubic feet is enough to fill a cube with sides 2 miles long.

The amount of energy that is obtained from the burning of a unit volume of natural gas is measured in British thermal units (Btus). One Btu is the amount of heat required to raise the temperature of 1 pound of water from 60° to 61°F at normal atmospheric pressure (14.7 psig). At sea level, it would take about 75 Btus to heat water enough for a cup of tea. A cubic foot of natural gas on the average gives off 1,000 Btus, but the range of values is 500 to 1,500 Btus. Therefore, 1 cubic foot of some natural gas may make only 7 cups of tea, while another would make as many as 20 cups of tea.

Energy content of natural gas varies because natural gas accumulations vary in the amount and types of energy gases they contain: the more noncombustible gases in a natural gas, the lower the Btu value. In addition, how much of any energy gas that is present in a natural gas accumulation—the mix of combustible gases—also influences the Btu value of natural gas. The more carbon atoms in a hydrocarbon gas, the higher its Btu value. To illustrate, methane typically represents more than 80% of energy gases. Methane contains one carbon atom per molecule; burning 1 cubic foot of methane gives off 1,012 Btus. Butane, possessing four carbon atoms, has a Btu value more than three times larger than that of methane. Molecular hydrogen, on the other hand, though combustible, contains no carbon atoms; its Btu value is three times smaller than that of methane.

6.3.0 Commercial Natural Gas Combustible Limits

Commercial natural gas is very flammable when sufficient amounts of oxygen (the oxygen present in air is more than adequate) are present. Mixtures of natural gas and air are explosive only when the mixture is less than 15% or more than 4% natural gas. If the natural gas is less than 4%, it will be too lean a mixture to burn, and over 15%, too rich a mixture to burn. The fact that only a small amount of gas is necessary to create an explosive mixture makes even small leaks in confined spaces very serious.

As mentioned earlier, an odorant is added to commercial natural gas so that it can be easily smelled if there should be a leak. However, it is possible for odorants to be filtered out or partially filtered out of commercial natural gas. This could happen when the natural gas passes through a filtering substance (soil or other). The presence of natural gas should be determined by testing the surrounding air with available instruments such as a combustible gas indicator or another suitable instrument.

6.4.0 Commercial Natural Gas Specific Gravity (Weight)

Specific gravity is a term used in the comparison of the weight of a gas compared to the weight of the same amount of air measured under the same temperature and pressure. Some gases may or may not be heavier than air. Air is judged to be the standard weight of one (1) with commercial natural gas weighing only .6 or 6/10 when compared to air. This means that natural gas weighs less than air so natural gas will rise up or actually float from the ground and fade away into the atmosphere when a leak occurs. Also, commercial natural gas will dissipate rapidly unless confined in a building, vault, or other structure. However, propane has a specific gravity of 1.5, is heavier than air, and will sink or fall to low places in buildings or in the ground. Butane is even heavier, weighing 2.0 or twice the weight of air, and will cling to the ground. It can be extremely dangerous when leaks occur.

Commercial natural gas is lighter than air and will normally rise and travel the path of least resistance. This means that a leak might travel uphill underground until it reaches a blockage and then travel back down the hill because that is the easiest path to follow.

7.0.0 ◆ PIPELINE EQUIPMENT

The pipeline system includes all the pipes, equipment, stations, terminals, and other facilities necessary for operation. The major pipeline equipment consists of:

- Pumps and motors
- Compressors
- Valves
- Actuators/operators
- Tanks
- Piping
- **Manifolds**
- Traps
- Measurement/sampling equipment

7.1.0 Pumps and Motors

Pumps provide the pressure and force to move product. Various types of pumps may be used in the pipeline, including centrifugal, positive displacement, hydraulic, and rotary pumps. Each is used in applications requiring different capacities or other pump characteristics. Power is supplied from many different sources, such as turbines, hydraulics, diesel engines, and electric motors. One pump station may use positive-displacement pumps driven by diesel engines, whereas another station may use centrifugal pumps driven by electric motors.

7.1.1 Centrifugal Pumps

Centrifugal pumps work by rotating and forcing product outward as product flows into the center. Revolving blades or vanes create the suction that pulls product into the pump. The product takes up the space between the vanes and rotates with them. It is then pushed outward until it leaves through the discharge, or exit, pipe. *Figure 21* shows how centrifugal pumps work.

7.1.2 Positive-Displacement Pumps

In a typical positive-displacement pump (see *Figure 22*), a piston slides back and forth in a cylinder. With each movement, it traps part of the product to be moved. This product is then moved toward the discharge end of the pump, where the pressure rises as the product is squeezed by the piston. The pressure forces the product out. There are several different types of positive-displacement pumps, including lift, or suction, pumps and force pumps.

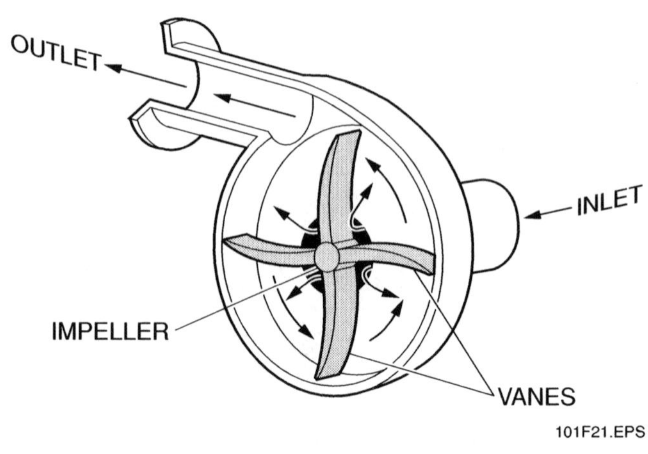

Figure 21 ◆ How the centrifugal pump works.

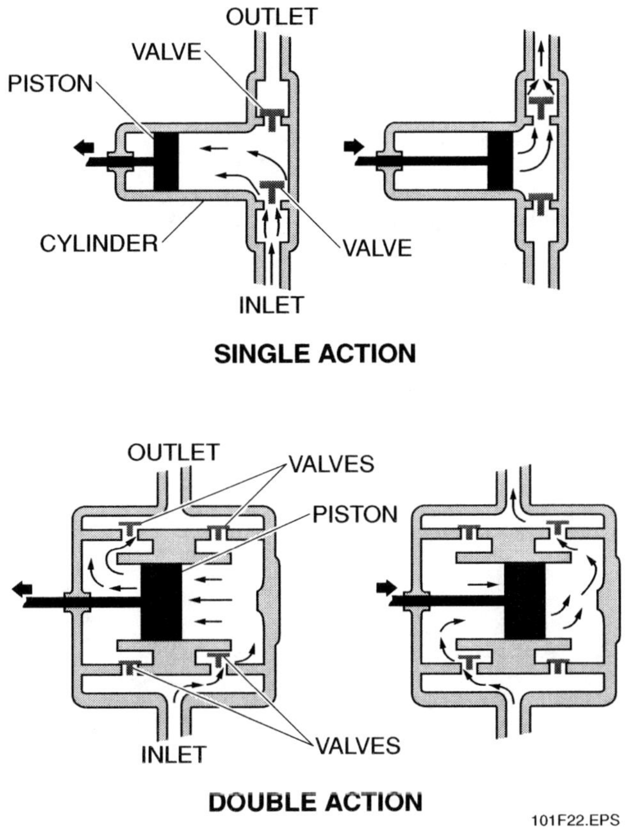

Figure 22 ◆ Two positive-displacement pumps.

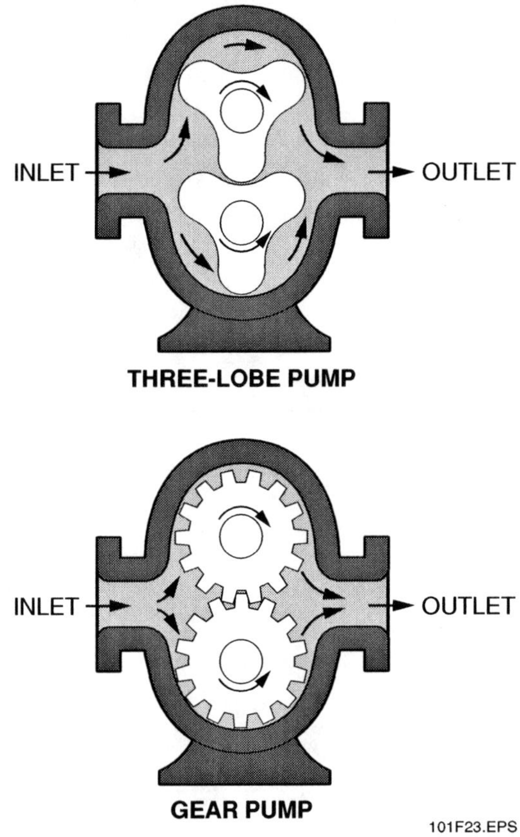

Figure 23 ◆ Two rotary pumps.

7.1.3 Rotary Pumps

Rotary pumps pump air as well as fluids. They have propeller-like blades that rotate on a single shaft. The rotation at the inlet side speeds up the fluid flow in a direction parallel to the shaft, building up pressure between the rows of blades. The pressure moves the fluid to the discharge side of the pump as the blades continue rotating.

A gear pump, which uses gear teeth instead of a piston, is used to pump thick products. The gears rotate in a housing, trapping product between the gear and the casing, moving it to the outlet side. The pressure in the outlet side builds up until it is high enough to discharge the product. Many types of drivers are used to drive these pumps. *Figure 23* shows two rotary pumps.

7.1.4 Drivers

Crude oil, refined liquid products, and gas are pushed through the pipeline grid by powerful pumps and compressors. The size of this machinery depends on the type of pipeline and the products and volumes being transported. Liquids are usually moved through the pipelines by huge pumps, while gases are moved by compressors. These compressors use pressure to push the gas through the pipeline. The pump and compressor drivers may be electric motors with fixed-speed or variable-frequency drive (VFD), diesel or gas engines, or natural-gas-fueled turbines. The choice of driver depends on the power sources available. The most common drive is a fixed-speed electric motor, although variable-frequency drives are becoming more popular. In remote locations, diesel-powered equipment is often used. Diesels are also used as backups during power failures.

7.2.0 Compressors

Compressor stations in a gas line act like pump stations in a fluid line. Compressors boost and maintain the pressure in the line, which keeps the gas flowing. In any compression process, the pressure of gas is increased from one level to another by the addition of energy to the gas being processed.

The centrifugal compressor is classified as dynamic, as opposed to a positive-displacement machine. In a reciprocating compressor, a quantity of gas is drawn into a cylinder, a piston moves, and the trapped gas is compressed by direct volume reduction or positive displacement. By this process, mechanical energy is transformed into pressure energy. *Figure 24* shows a gas compressor.

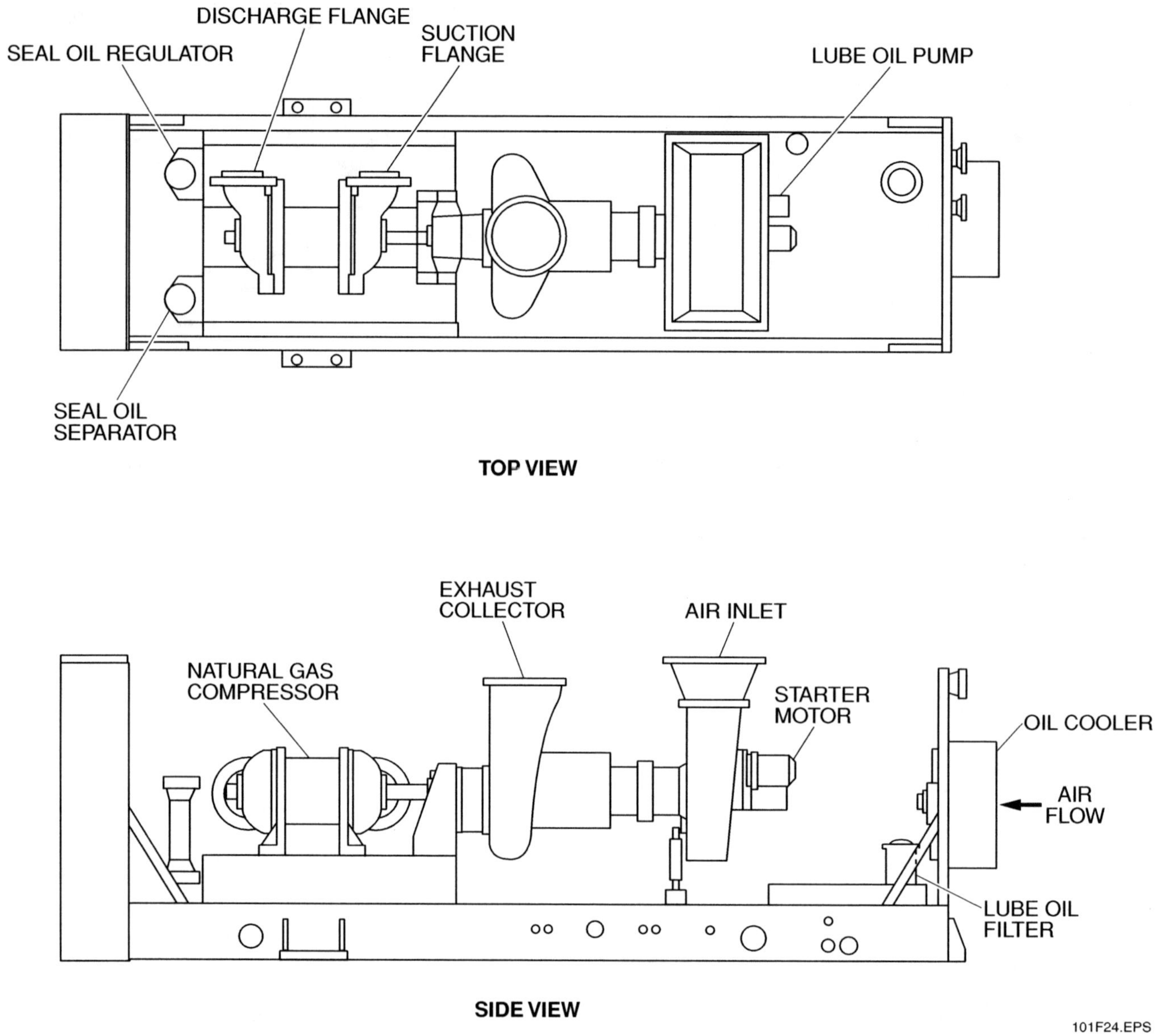

Figure 24 ◆ Gas compressor.

The combination of inlet vane, impeller, and diffuser or diaphragm is referred to as a stage. A single centrifugal stage has a relatively low pressure capability. When larger ratios are desired, additional stages are added in series. The compressor flow can be controlled and varied by changing the pressure differential across the compressor or by changing the power level. Centrifugal compressors apply inertial forces to the gas. The inertial forces are transmitted by an impeller which, by dynamic centrifugal motion, adds velocity energy through acceleration of the gas. The gas flows from the impeller into the diffuser, where the gas decelerates, and velocity energy is transformed to pressure energy.

7.3.0 Valves

Valves are a critical part of pipeline systems. They are used to control the rate of flow in a line, to open or shut off a line completely, or to serve as automatic safety devices. Valves may be identified by type or by function, and can be equipped with locking devices to prevent accidental operation of the valve. Valves may be equipped for manual operation, but most frequently operated valves in the pipeline are equipped with electric-, pneumatic-, or hydraulic-power-driven mechanical operators. Those used extensively include gate, plug, ball, butterfly, check, control, and relief valves.

7.3.1 Gate Valves

In a simple gate valve, a wedge-shaped disc or gate is pressed between inclined seats to close the valve. Operating on the same principle but with different configurations of gates and seats, there are other more complex gate valves. Conduit gate valves have an elongated gate and a port in the lower part for a fluid conduit. *Figure 25* shows a simple gate valve.

7.3.2 Plug Valves

Plug valves (see *Figure 26*) are similar in design to the stopcocks used to control flow from containers. The plug usually has an elongated port that may be smaller in cross-sectional area than the pipe. The valve is operated by giving the plug a quarter turn. Operation may be by a lever fitted to the stem, by a gear system, or by power-driven mechanical operators. Plug valves are used where reliability and fast operation are important. There is considerable sliding contact area between the plug and the valve body, and as size increases, the operating force required also increases. Many plugs have interior channels and grooves in which lubricants can be forced into the contact areas to improve turning and sealing. Plug valves are generally run in pipe in which scrapers are not run.

7.3.3 Ball Valves

Ball valves are an adaptation of a plug valve, using a ball for the plug. The sealing contact area of the ball valve is that of two seal rings. These seals are made practical by use of tough, low-friction, oil-resistant elastomers.

7.3.4 Butterfly Valves

Butterfly valves (see *Figure 27*) control product flow by use of a disc mounted on a stem that can be turned to block or allow flow. This simple lightweight valve is used in parts of the pipeline. Butterfly valves have two valuable features. They are quick-operating and take up minimal space.

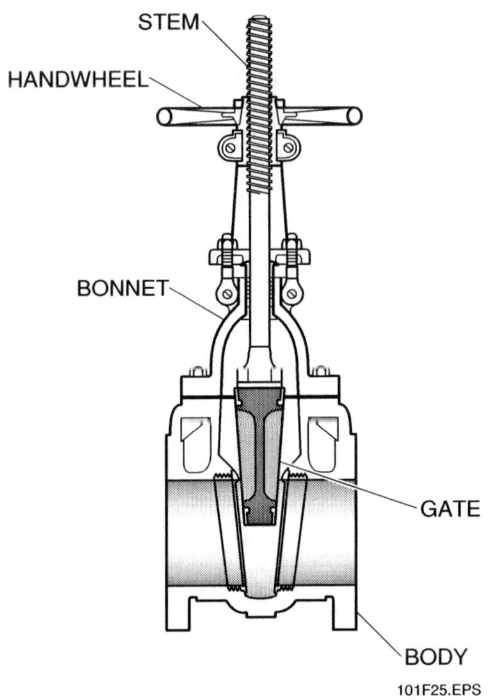

Figure 25 ◆ Simple gate valve.

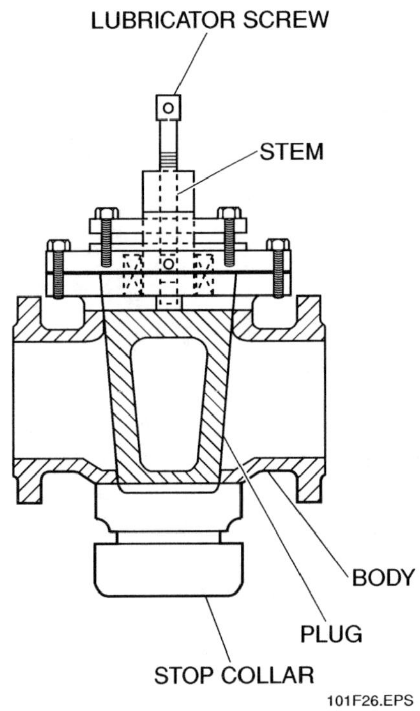

Figure 26 ◆ Plug valve.

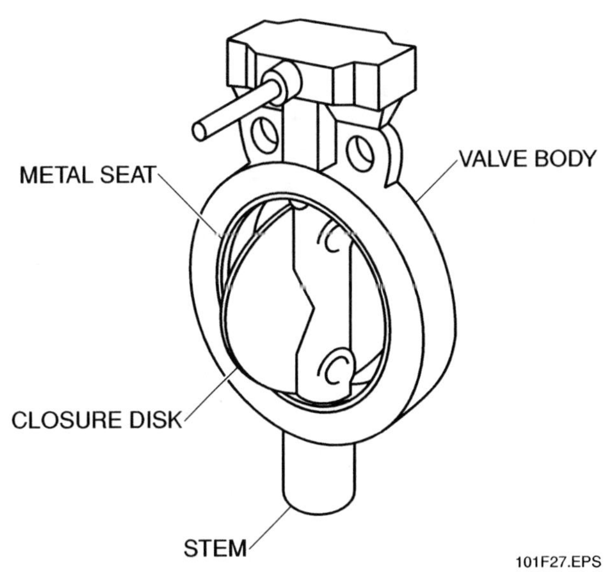

Figure 27 ◆ Butterfly valve.

7.3.5 Check Valves

Check valves (see *Figure 28*) are considered shutoff valves since they permit full flow or prevent flow in one direction. They close automatically when pressure in the flow direction is removed. Check valves can be designed to allow passage of scrapers and are most often used in gathering lines and in station piping. They are often used in conjunction with other shutoff valves.

7.3.6 Control Valves

Automatic control of flow rates and pressures is provided by control valves. They may or may not provide for complete shutoff because flow control is their major function. Port sizes and flow patterns through these valves must be continuously adjusted to provide control by throttling action, increasing their tendency toward damaging vibration, erosion, and corrosion. These valves are used principally at pump stations, although they may be needed at other points in the line.

7.3.7 Relief Valves

Relief valves (see *Figure 29*) may be called safety valves. Their function is to remain closed until pressures rise above design levels. Pressure is released to prevent pressure damage when released products are piped to safe discharge points.

When not opened at set pressure levels, the valves are kept closed by spring pressure, a nitrogen charge, a weight load, or other means. They normally close automatically when the pressure is relieved but may be designed to stay open until reset. They may be operated and monitored remotely.

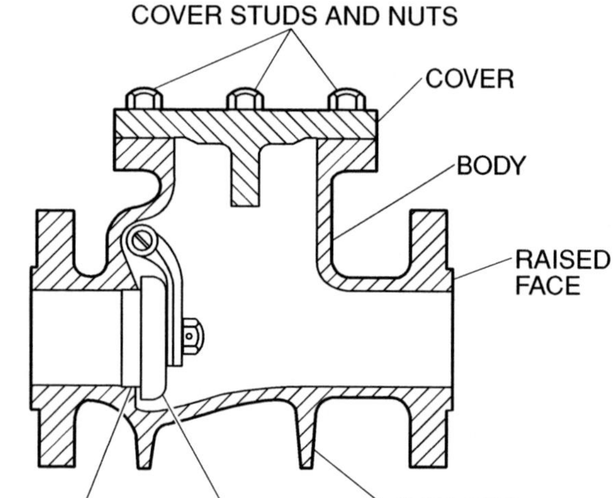

Figure 28 ◆ Check valve.

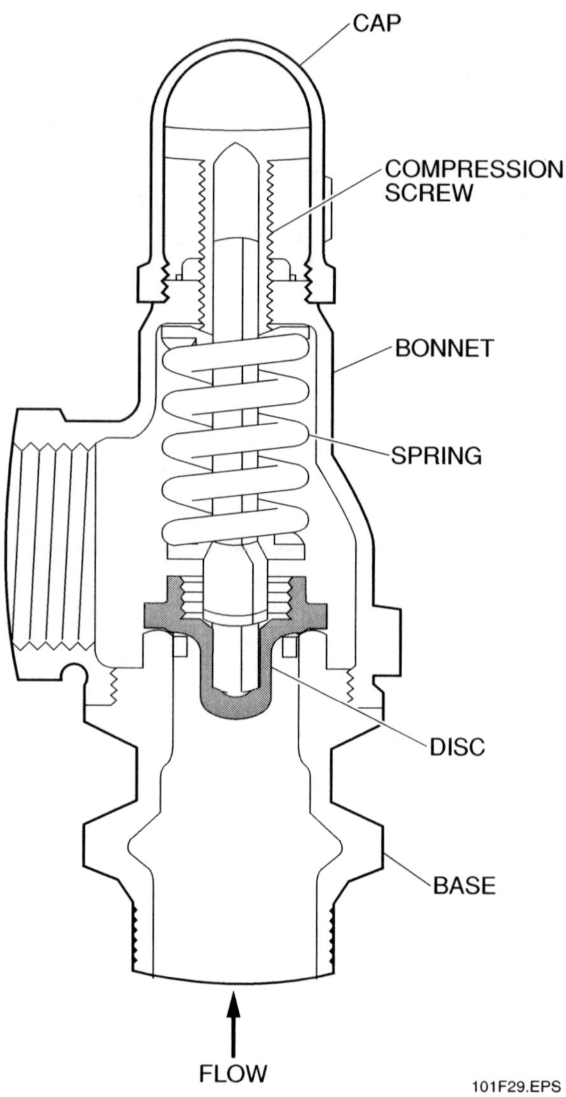

Figure 29 ◆ Relief valve.

7.3.8 Globe Valves

Other valves in the pipeline include globe valves (see *Figure 30*). Globe valves are used for shutoff and throttling, usually in auxiliary systems, such as steam and water systems.

7.3.9 Needle Valves

Needle valves are a type of small, simple, rugged device for throttling and shutting off extremely high pressures compared with other valves. They are used mostly in small auxiliary piping systems.

7.4.0 Actuators/Operators

A valve actuator, commonly called an operator, is a device that opens and closes a valve either manually or automatically. Manual actuators come in the form of handwheels, levers, and wrenches that are attached to the stem. Larger valves that require great effort to turn must be equipped with some type of powered mechanism, including electric, pneumatic, or hydraulic units, mounted on the stem. Some valve actuators use a gearbox with wide reduction ratios. Some manual actuators employ high and low gear ratios to enable personnel to turn the valve.

Valves with automatic actuators are controlled from a console in a control center by remote control, sometimes from many miles away. Such automatic actuators are used when the valve is remote from the main work area, when the frequency of operation would require unreasonable human effort, or if rapid opening and closing of the valve is required. Maintenance crews often carry portable valve actuators to work in outlying installations.

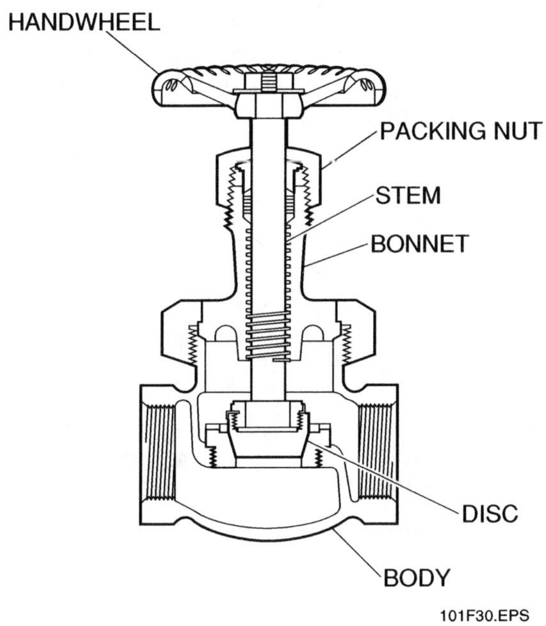

Figure 30 ◆ Globe valve.

7.5.0 Tanks

Tanks are containers used for liquid storage, relief, and breakout. Tanks may exist along the length of the pipeline, at the production site, at pump stations, at terminals, and at delivery points. Although the types and sizes of tanks at each of these sites may vary, they are usually round and made of steel. All atmospheric storage tanks are constructed per *API 650* and repaired per *API 653*.

Tanks must be calibrated for capacity. The dimensions are strapped, or measured, to be used in strapping tables. From these tables, the volume of liquid contained in the tank can be determined. Tank gauges show the tank number and gauge reading from which corresponding volume in barrels can be determined.

Cylindrical tanks may have flat floating roofs, fixed cone-shaped roofs, or both. The first can be used for gasolines and aviation-grade kerosenes, while the second can be used for fuel oils. Some tanks have fixed geodesic domes covering a floating pan. Spheroid tanks are used for volatile products because of their high vapor-pressure capacity.

The floating roof tank (see *Figure 31*) is designed to reduce losses through evaporation of volatile components of crude oil and to reduce the corrosion from sour crude by eliminating the vapor space between the oil and the roof. Seals around the perimeter of the roof prevent vapor loss and rain contamination. The wind girder on the outside rim of the tank adds strength to the shell.

Tanks must also be maintained and repaired per industry standards. Although a mixer helps to keep sediment in suspension in the product, sludge forms in the bottom and must be cleaned out periodically. The tanks are repainted inside and out at specific intervals. Structural inspections and preventive maintenance guard against internal and external damage.

7.6.0 Pressure Vessels

Just as tanks are used for the storage of liquid products, pressure vessels are used along the gas pipeline for the storage of gas products. Liquid vessels are typically cylindrical, but like tanks, can vary in size and shape from site to site, depending on application, capacity, and pressure ratings.

Cylinders and pressure vessels can be fabricated of steel, aluminum, or composite materials. They must be manufactured, inspected, marked, tested, and retested in accordance with United

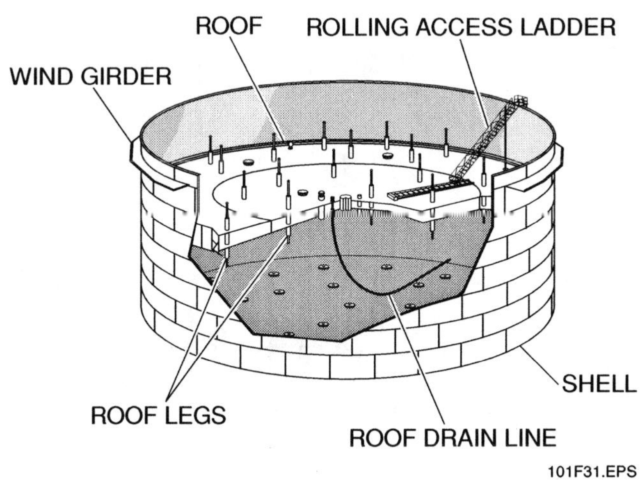

Figure 31 ◆ Open floating roof tank.

States **Department of Transportation (DOT)** regulations and exemptions for the type of service that they are used for. Pressure vessels must also be designed and repaired in compliance with *API Standards 510 and 2510*. These regulations cover the manufacture, marking, maintenance, repair, reconditioning, and testing of seamless pressure vessels, and they prescribe the requirements for transportation of hazardous materials within vessels. *Figure 32* shows a pressure vessel.

7.7.0 Pipe

Line pipe is made of steel through various methods. The main types of pipe used in the oil and gas industry are seamless, electrical resistance welded (ERW), and double submerged arc welded (DSAW). Other piping, less commonly used now, is welded steel pipe.

Seamless pipe is made from a solid length of metal pierced lengthwise by a mandrel with a rounded nose. The manufactured steel tubes and pipes are used in water, gas, and crude oil transmission and distribution lines and as casing pipes in the petroleum industry.

ERW is the most common form of manufacture for pipe in sizes from 2 to 22 inches outside diameter (OD). While seamless pipe is traditionally stronger and more expensive than comparable ERW pipe, ERW technology is improving and being used more and more frequently.

DSAW pipe is pipe that has longitudinal or spiral butt joints. The joints are produced by at least two passes, including at least one on the inside and one on the outside of the pipe. Pressure is not used, and filler metal for the inside and outside welds is obtained from the electrode or electrodes. This pipe is primarily intended for use in the gas and oil transmission industry.

Welded steel pipe is made by bending strips of steel into the form of a tube and welding the longitudinal seam either by electric resistance, by fusion welding, or by heating the tube and pressing the edges together. Two types of older piping that are still found on pipelines are furnace lap-welded and furnace butt-welded piping.

7.8.0 Fittings

Fittings include elbows, tees, reducers, flanges, plugs, and unions. Elbows are connections for directional changes. Tees are used to join three pipe ends. A reducer is used to connect pipes of different sizes. Flanges provide a means of opening and closing piping and attaching pipe ends to mating flanges on valves, tanks, pumps, and other equipment. Plugs and caps close the end of a pipe. A union provides a means of opening and closing piping and permits closing of circuitous piping systems. *Figure 33* shows types of fittings.

Fittings are made in a range of materials based on their service requirements. *Table 3* shows a partial list of materials and their pressure ranges.

Table 3 Materials and Pressures

Material	Pressure
Cast Iron	150 psig
Malleable Iron	250 psig
Cast Steel	150–2,500 psig
Forged Steel	1,000–6,000 psig

101T03.EPS

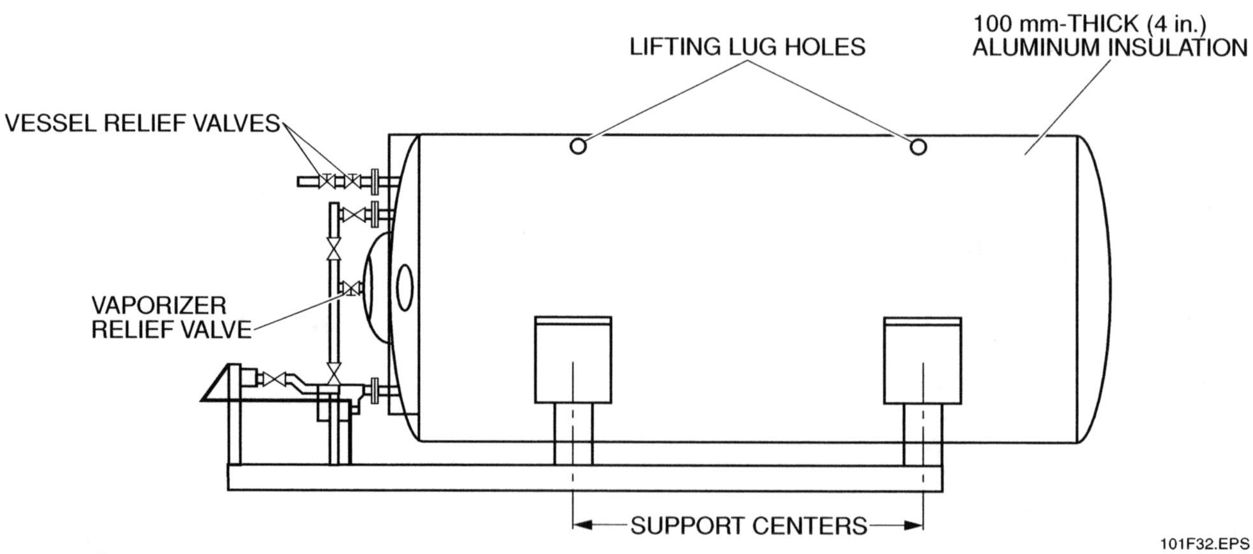

Figure 32 ◆ Pressure vessel.

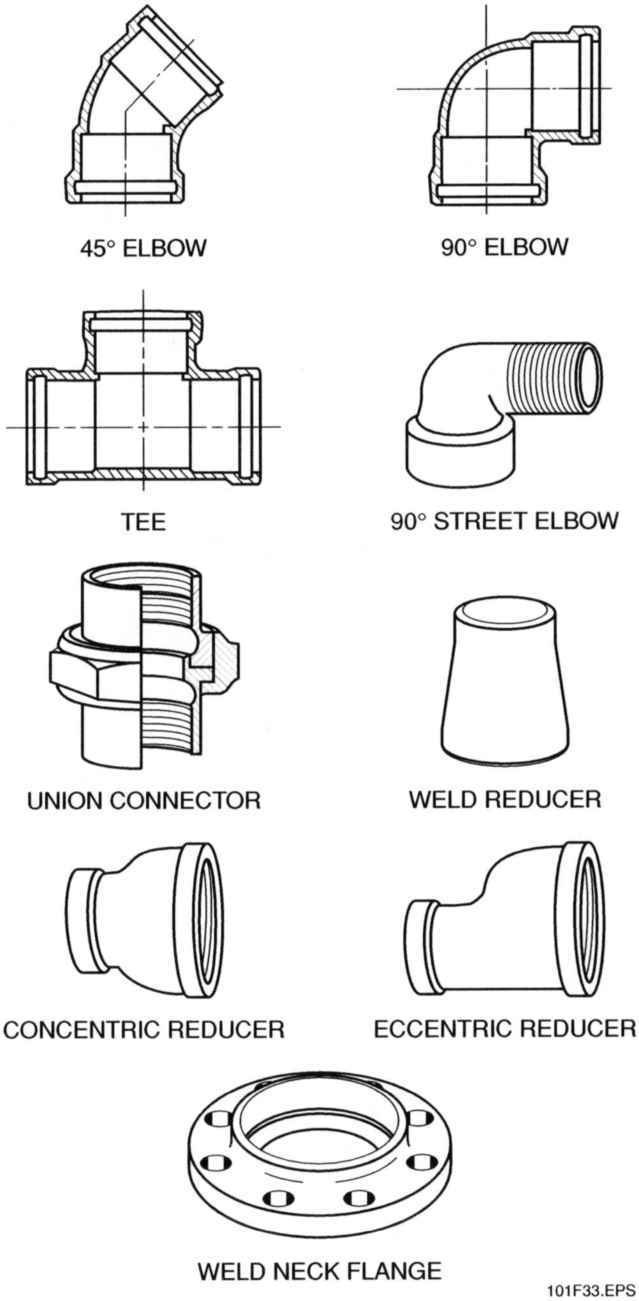

Figure 33 ◆ Types of fittings.

7.9.0 Manifolds

The pipeline uses manifolds to control the flow path of oil, gas, and other petroleum products. Manifolds, with valves and meters, branch off of the line to divert specific volumes of certain products to different locations or other pipelines. *Figure 34* shows a delivery manifold.

The complexity of manifold piping ranges from very simple to very complicated. A booster station manifold may merely allow bypassing of the station. A more extensive manifold at an originating station may connect the main line, field-gathering lines, tanks, and pumps. This manifold may perform any or all of the following operations:

- Pumping product through the main line with tanks inactive
- Receiving product from the field into any tank
- Delivering from the main line into any tank
- Transferring from one tank to another
- Pumping product from a tank into the main line
- Isolating tanks from main-line flow
- Injecting product from any tank into the main line

7.10.0 Traps

Traps in the pipeline are designed to launch or catch scrapers and other pigs that are inserted in the pipeline. Valves and/or manifolds are used to isolate the trap for scraper insertion or removal and then to realign the trap to the pipeline flow. *Figure 35* shows incoming and outgoing scraper/pig traps.

Automation of operation has been extended to the launching of scrapers in some instances. Some traps are arranged so that one or more scrapers may be loaded and then launched at any desired time by remote-controlled valve changes.

7.11.0 Measuring/Sampling Equipment

All volumes entering, leaving, and passing through the pipeline system must be measured and accounted for. Accurate measurement is required for custody transfer, task purposes, leak detection, and delivery schedules.

The primary method of measurement in the pipeline is metering. Meters, whose accuracy is documented with meter provers, measure and record the volume of product passing through. Factors affecting volume are the temperature, pressure, and density of the product being measured. Various instruments are used to measure these factors concurrently with the volume measurement. Product characteristics are also checked periodically with automatic and manual sampling.

7.11.1 Meters

Various meters are used along the pipeline to monitor the volume of fluid or gas. Most product measurements are made by turbine meters or positive-displacement meters, although coriolis-type mass flowmeters, vortex, and ultrasonic meters are also used. Meter readings taken before

Figure 34 ◆ Delivery manifold.

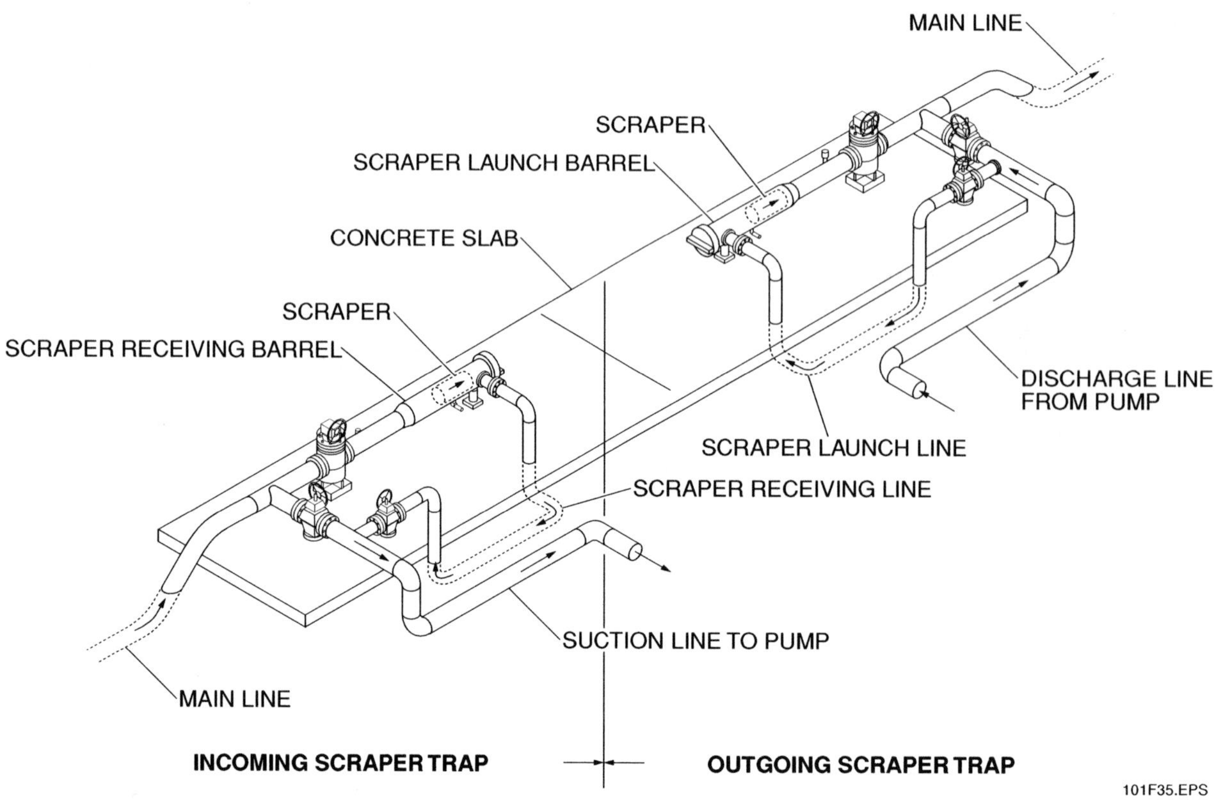

Figure 35 ◆ Incoming and outgoing scraper, or pig, traps.

and after deliveries are used to determine the volume delivered, or **throughput**. Meters are typically equipped with transmitting, remote-reading, and print-out accessories to facilitate the delivery of correct product volumes and for preparation of delivery tickets. Data is often transmitted to remote locations. *Figure 36* shows two types of meters.

Gas is typically measured by orifice, turbine, positive-displacement, ultrasonic, and rotary vane meters. These meters are generally automated to correct gas volumes for moisture content and temperature and to provide remote read-out and printing capability. Strainers are installed upstream of meters to remove any foreign material from the stream before it goes through the meter. The wire mesh in strainers is generally more coarse than filter materials are.

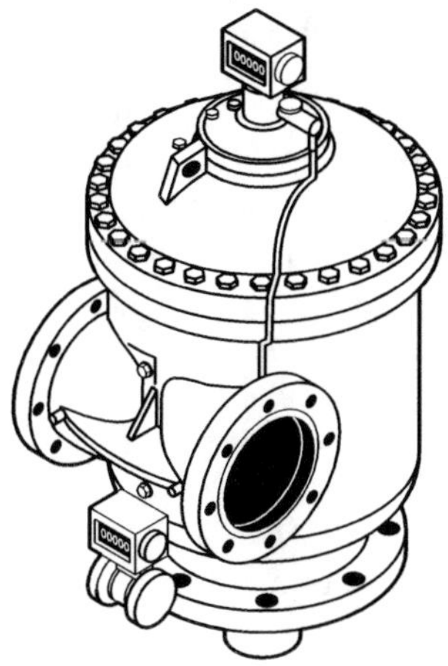

POSITIVE-DISPLACEMENT METER

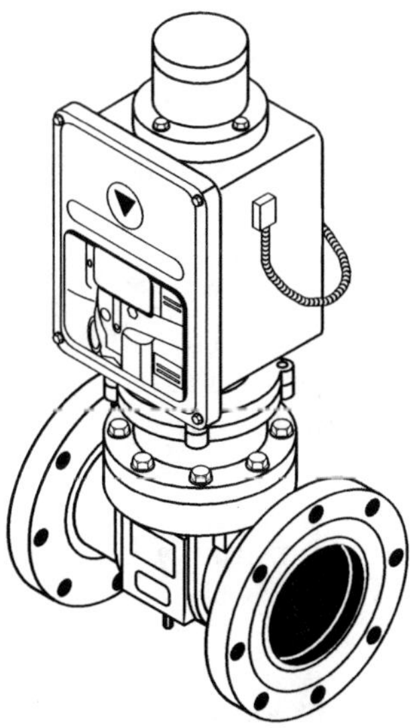

TURBINE METER

101F36.EPS

Figure 36 ◆ Two types of meters.

The orifice plate is the vital component of the orifice meter. It is the part that restricts the flow to develop the difference in pressure that is measured to calculate actual flow. The accuracy of orifice meters is dependent on the quality of the meter that is installed. Expert installation and maintenance are necessary to ensure the continued accuracy of the meter.

Orifice plates are secured between halves of flanges of machined carriers or in the body of a special orifice meter housing constructed so that the plate can be removed for inspection. Some meters allow the plate to be removed without stopping the flow. The orifice meter is made up of the carrier housing or flanges, the orifice plate, gaskets, O-rings, devices for removal, and pressure fittings. Pressure taps are drilled at a location exactly 1 inch on center from either side of the orifice plate. *Figure 37* shows some common orifice plate holders designed for easy removal.

Visual inspection of the housing unit involves close examination of the unit looking for any signs of rust, corrosion, cracks, or any buildup of foreign material around the outside housing joints, suggesting the possibility of a leak. Pressure fittings and sensor connections must also be inspected for leaks or damage of any kind. Plates vary from 2.5 inches to 30 inches in diameter.

7.11.2 Provers

A meter prover is used to calibrate meters for specific conditions of flow and product type. Meters are proved by passing a volume of product through the meter at a known rate and comparing this metered volume with the known volume prover. Two types of provers frequently used are the ball prover and the piston prover. A prover loop is a length of pipe or a calibrated U-shaped pipe with a known volume used to check or prove the accuracy of a meter.

8.0.0 ◆ ELECTRICAL POWER SYSTEMS

The electrical power system used by any pipeline company may be quite large or quite small. Power may be self-generated, or partially so, or may be purchased from another source. For the purposes of an introduction, a basic power system will be described from generation to delivery.

An electrical power system is an assembly of equipment and circuits for generating, transmitting, and distributing electrical energy. The elements of a power system form complex networks that require energy control centers to monitor and regulate their operation. The reliability of the system is critical to pipeline operation. *Figure 38* shows an electrical power system. The principle elements of the system can be divided into the following:

- Power generation and transmission
- Power distribution
- Direct current (DC) power distribution
- Utilization equipment

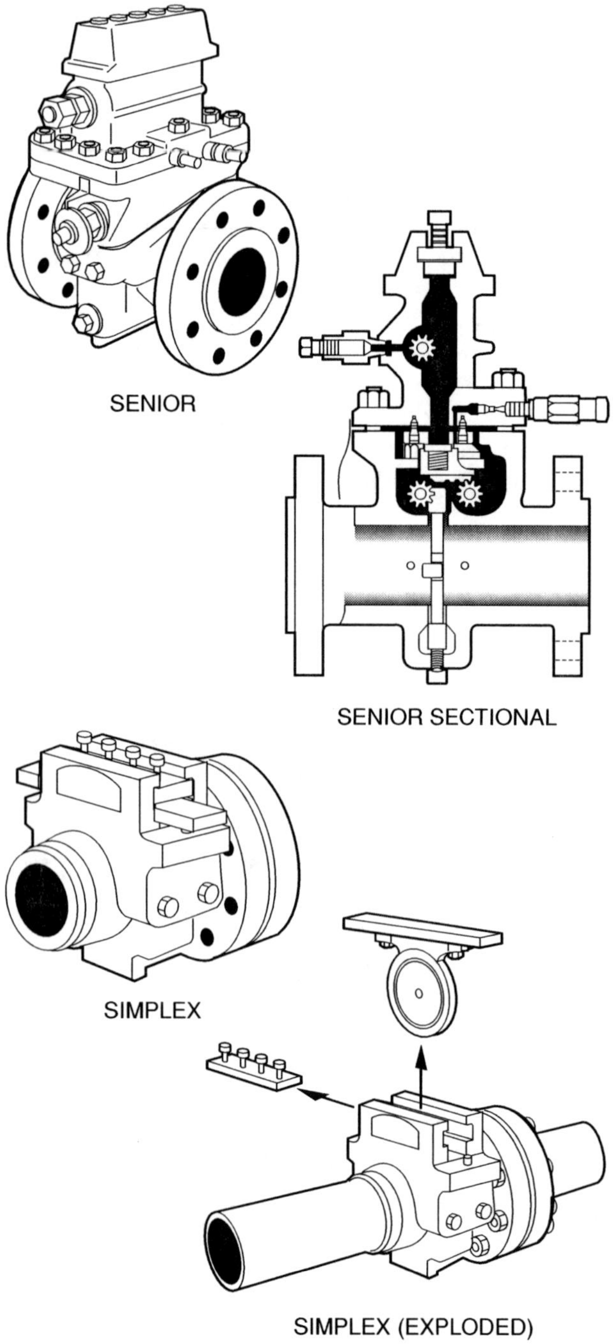

Figure 37 ◆ Common orifice plate holders.

8.1.0 Power Generation and Transmission

Electrical systems require generating stations to produce electric power, and transmission systems to carry it to areas of consumption. Along the way, the voltage of the power delivered must be adapted to the specific needs of consumers. The myriad of uses that consume this energy is the power system load.

8.1.1 Generation

The large quantities of electricity required to supply a system are produced in generating stations, commonly called power plants. These are actually conversion facilities where the chemical or nuclear energy of fuel, such as coal, oil, gas, and uranium, or the kinetic energy of falling water is converted to electricity.

The generators, which supply the electricity, can be driven by steam turbines, hydraulic turbines, gas turbines, or internal combustion engines. Coal, natural gas, and heavy fuel oil are commonly used fuel for steam turbines.

Three-phase 60 hertz (Hz) AC systems are used almost exclusively in the U.S. Consequently, power system generators are wound for three-phase output at a voltage usually limited by design features. The output of modern generating stations is usually stepped up by transformers to the voltage level of transmission circuits used to deliver power to distant load areas.

8.1.2 Transmission

The transmission system carries electric power efficiently and in large amounts from generating stations to consumption areas. Such transmission is also used to interconnect adjacent power systems for mutual assistance in case of emergency.

8.2.0 Power Distribution

Power is reduced and distributed through a series of substations to provide the voltage required in specific applications. A substation is an assembly of equipment through which electrical energy is passed for transmission, distribution, interconnection, transformation, conversion, or switching. Power distribution involves:

- Substations
- Disconnection means
- Transformers
- Cables/conduits
- Switchgear
- Motor control centers

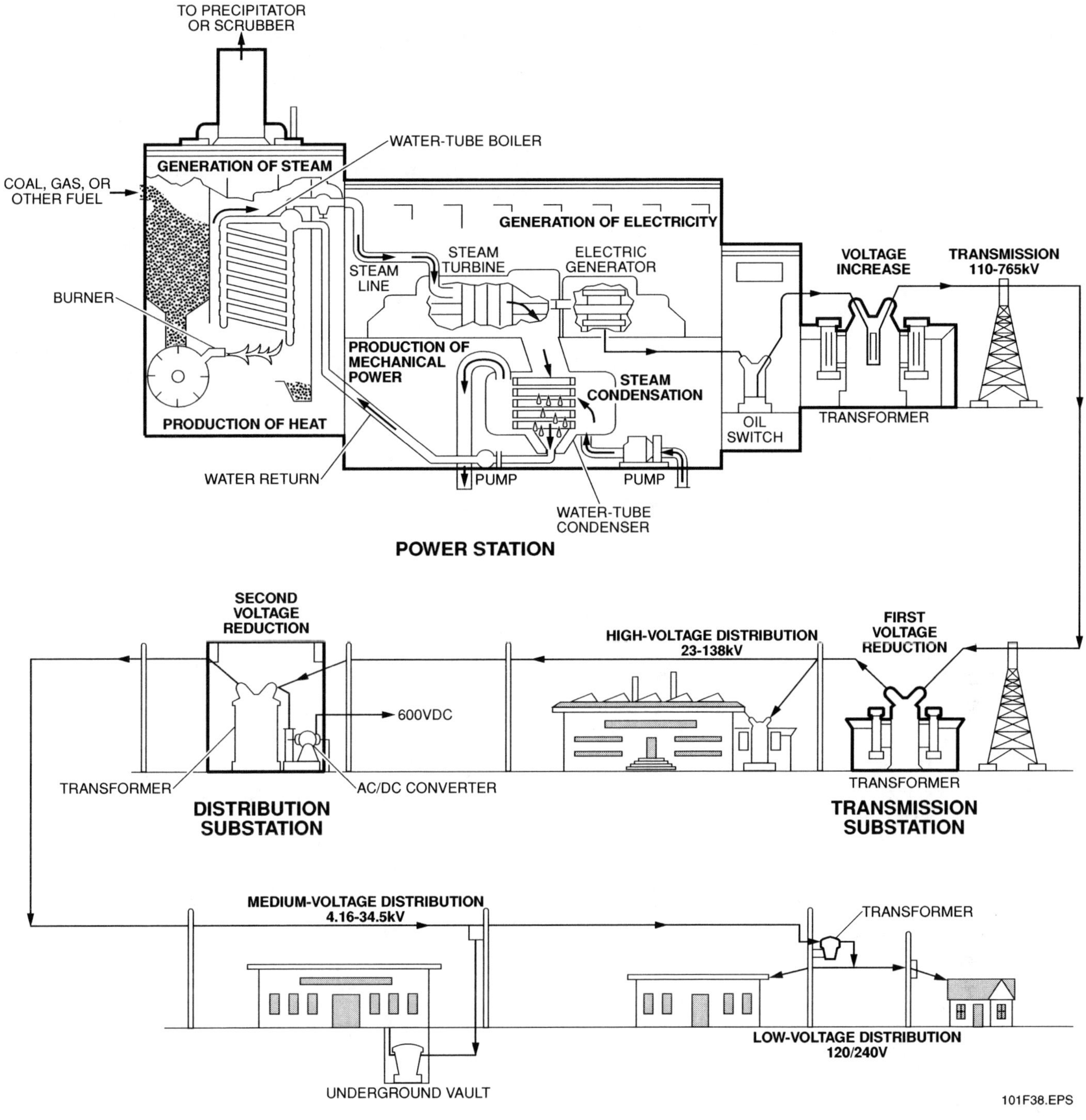

Figure 38 ◆ Electrical power system.

8.2.1 Substations

Substations are used for some or all of the following purposes:

- Connecting generators, transmission or distribution lines, and loads to each other
- Transforming power from one voltage level to another
- Interconnecting alternate sources of power
- Switching for alternate connections and isolating failed or overloaded lines and equipment
- Controlling system voltage and power flow
- Compensating for reactive power
- Suppressing overvoltage
- Detecting faults, monitoring, and recording information, power measurements, and remote communications

Minor distribution or transmission equipment installation is not referred to as a substation. Substations are referred to by the main duty they perform. Broadly speaking, they are classified as transmission substations when associated with high-voltage levels, and distribution substations when associated with low-voltage levels. However, they are also referred to in other ways, such as:

- Transformer substations, when the equipment includes transformers
- Switching substations, when the equipment is mainly for connections and interconnections
- Converter stations, when the main function is conversion from alternating current (AC) to DC and vice versa

8.2.2 Disconnection Means

Disconnect switches are commonly present so that sections or components can be isolated for maintenance. These have isolation and connection capability but may not be capable of interrupting load.

8.2.3 Transformers

Transformers use magnetic core and windings to transform power from one side of the transformer to the other at different voltages. Substation transformers range from small sizes of 1MVA to large sizes of 2,000MVA. Most are insulated and cooled by oil, and adequate precautions must be taken against fire hazard. These include adequate distance from other equipment, fire walls, fire extinguishing means, and pits and drains for containing leaked oil.

Current and potential transformers are for measuring currents and voltages. They are also for providing proportionately low-level currents and voltages for control and protection.

8.2.4 Cables and Conduits

Cables are two or more insulated wires and a grounding wire covered by an outer jacket, or sheath. Cable is usually classified by the type of covering it has, either nonmetallic or metallic. Cable may also be classified according to where it can be used. The *National Electrical Code* has definite guidelines on cables and their allowable uses. Because water can cause power loss or short circuits, cables are classified for either dry, damp, or wet locations. They are also classified for exposure to sunlight or rough use.

All cables are marked to show important properties and uses. These markings show the wire size, number of conductors, cable type, and voltage rating. The information is printed on nonmetallic sheathing and on tags attached to metal sheathing.

Raceway is a general term that refers to a wide variety of circularly and rectangularly enclosed channels used to house electrical wiring. Conduit is a raceway with a circular cross section similar to pipe that contains cables or wires. Conduit is used to provide protection for conductors and route them from one place to another. In addition, conduit makes it easier to replace or add wires to existing structures and, if it is metal, provides a permanent electrical path to ground.

8.2.5 Switchgear

Switchgear includes a wide variety of enclosed circuit breakers, disconnects, busways, and associated items that distribute power from a common point. An appropriate switching arrangement for connections of generators, transformers, lines, and other major equipment is basic to any substation design. Switchgears are use to provide this arrangement.

8.2.6 Motor Control Centers

The motor control center (MCC) contains the isolators for all the electrical equipment associated with a main line pump. Each MCC **cubicle** contains some or all of the following:

- Breaker
- Insulator
- Disconnect
- Isolator
- Reset button
- Hand/auto switch
- Run and stop light indicators
- Opened, closed, and stop light indicators

The isolator is the main isolator for the identified equipment and may be padlocked in the OFF position. The run and stop light indicators display the status of the associated equipment. Opened, closed, and stop light indicators are associated with motor-operated isolation valves.

8.3.0 DC Power Distribution

The DC power system provides the controls, monitor, and indicators of power required for control systems. Normal shutdown and automatic malfunction shutdown functions are also provided in the electrical system.

The DC power supply consists of batteries and chargers. A source of DC power is externally supplied by batteries and a battery charger. The charger and certain other equipment require an external power source. The battery charger will automatically charge the batteries. It does so by changing the batteries to full charge and then provides a float charge that will preserve the charge.

The charger will carry continuous and intermittent DC loads up to 80% of the maximum rated output. The charger has limiting characteristics to avoid exceeding its maximum output rating, and the output is further protected by fuses for short-circuit conditions.

9.0.0 ◆ CORROSION CONTROL

Corrosion is a natural process that, given the right conditions, can attack any metal or alloy. Most pipeline technicians are familiar with corrosion but do not have specific knowledge of the various types of corrosion. Everyone is surely familiar with rust, the most common form of corrosion which results in overall thinning of a metal. Most corrosion failures, however, are not caused by rust or general corrosion but by localized attack—pitting, crevice corrosion, intergranular corrosion, or stress corrosion cracking.

Each year unchecked corrosion can cost the pipeline industry millions of dollars in material damage and downtime. Corrosion can cause catastrophic failures with potentially severe consequences for the environment and serious injury or even death to personnel. It is necessary today to be able to identify potential problems and risks through a corrosion management program. An entire industry has built up around this need.

The largest organization in the world committed to the study of corrosion is the National Association of Corrosion Engineers (NACE). NACE was established in 1943 by eleven corrosion engineers in the pipeline industry. Today, NACE International has more than 50 years of experience in developing corrosion prevention and control standards. NACE is an excellent source of information for further study.

Federal guidelines have been implemented to support safety of both personnel and the environment. The complete text of the corrosion control requirements can be found in *49 CFR Part 192, Subpart I* and *49 CFR Part 195, Subpart H*. Companies are required to have written guidelines and procedures for most corrosion-related activities.

9.1.0 Types and Causes of Corrosion

Many factors contribute to metal corrosion; dissimilar metals that corrode at different rates and are sometimes incompatible, dissimilar soils, dissimilar aeration, bacteria, and stray current are factors that contribute to corrosion.

Pipeline corrosion is most prevalent when the failure of coatings, inhibitors, or cathodic protection occurs in a corrosive environment. These factors do not affect the pipeline equally at all locations, and corrosion does not grow at the same rate throughout a pipeline.

Internal corrosion of oil and gas pipelines is a growing concern to pipeline operators. The causes of internal corrosion of pipelines are chloride, carbon dioxide, hydrogen sulfide, oxygen, and microbiological activity. A recent severe case of internal corrosion caused several failures on a 20-inch oil line after only 2 years of service. The cause of the failure was high carbon dioxide concentrations (9%) in combination with high levels of brine.

Piping and pipeline systems are often great places to grow microbes that can produce corrosive conditions. Microbiologically influenced corrosion (MIC) often works in combination with other forms of corrosion, such as crevice corrosion and under-deposit corrosion, on in the pipe. It can also accelerate corrosion produced by impurities in the system environment, such as hydrogen sulfide and/or carbon dioxide.

For example, MIC can occur in the crevices produced by disbonded coatings on the outside of the pipe that are shielded from cathodic protection. Internally, sites for MIC are where water, debris, and other nutrients accumulate on the pipe wall. Both sites provide the local conditions that promote colonization of microbes, which can make existing corrosion problems worse by producing organic acids or acid-producing gases. There is a long list of microbes that can participate in the corrosion reaction; these include aerobic or anaerobic types (depending on whether they require oxygen or lack of oxygen to survive). They can also make use of nutrients in the environment (sulfates, nitrates, iron) to produce corrodants.

9.2.0 Inspection and Testing

Pipeline integrity programs use tools, technologies, and strategies to ensure that pipeline systems are in a condition to perform safely and reliably. Increasingly sophisticated technology is available. Constant monitoring and inspection is essential to any pipeline integrity program.

9.2.1 Microbiologically Influenced Corrosion (MIC)

A variety of methods is available to detect MIC. Sampling and culturing can be performed for identification of the size of the infection and the active strains of organism present. This information is then used to identify preventative or remedial treatments. Chemical analysis and pH of corrosion products can also give important information relative to the nature of the problem. Evidence taken from the metal surface, called sessile surveys, has a better chance of properly assessing the situation versus sampling of the liquids, called planktonic survey.

9.2.2 Sampling

Injection and sampling systems are fundamental to corrosion control and process control programs. They are applicable to a large variety of processes in the petroleum, chemical, and water treatment industries. Injection systems are used to inject a wide range of chemicals into processes. Such chemicals include biocides, demulsifiers, corrosion inhibitors, oxygen scavengers, glycol and monoethylene glycol, dewaxers, methanol, odorizers, and product additives.

Sampling systems are used to take samples of the process product. Such samples are then analyzed in the laboratory for inhibitor concentration levels, the presence of metal ions, oxygen levels, scale forming compounds, and a wide range of other substances.

9.2.3 Corrosion Coupons

Coupons are specimens of materials that are exposed to tests or real environments to assess the effect of degradation on the material. They can be used inside a pipeline or outside if the pipe is underground. Corrosion coupons represent a widely used and relatively simple form of corrosion monitoring. Usually a coupon is exposed to a corrosive environment to represent an actual component/structure. After exposure for a specified amount of time, the coupons can be weighed and examined microscopically to determine and evaluate the type and average rate of corrosion damage. *Figure 39* shows typical coupons.

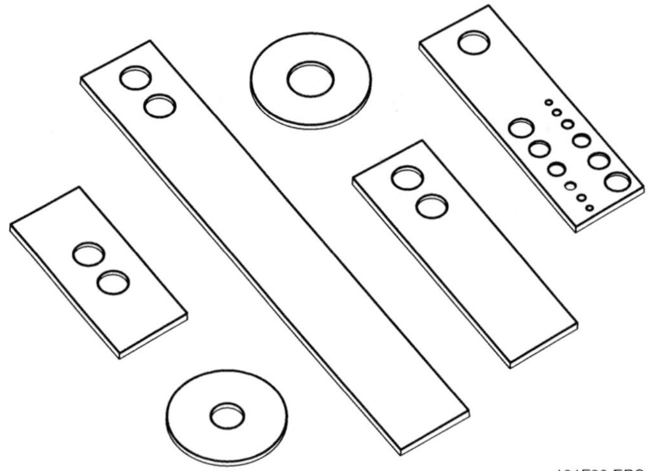
101F39.EPS

Figure 39 ◆ Typical coupons.

9.2.4 In-Line Inspection, or Smart, Pigs

Smart pigs are high-tech inspection tools covered with sensors that gather and record information in onboard computers as they pass through the pipeline. They check for various problems, such as thinning, pitting, and general corrosion. Some smart pigs can detect metal loss in areas around welds. *Figure 40* shows an ultrascan in-line inspection (ILI) pig.

9.3.0 Maintenance and Protection

Monitoring a pipeline for corrosion is essential in preventing failures. If an operator can identify active or growing corrosion defects, then the rate of corrosion can be individually predicted for each defect, allowing an opportunity for corrective measures before the failure occurs.

COURTESY OF PII GROUP LTD
101F40.EPS

Figure 40 ◆ Ultrascan in-line inspection pig.

As soon as production begins in a line, companies begin maintenance programs to reduce the risk of pipeline leaks. Sophisticated electronic control systems monitor changes in pressure, volume, and product flow. A variety of highly technical tools collect information about both the interior and exterior condition of pipelines. In-line inspection (ILI) tools, commonly called smart pigs, assess metal loss, strain, and cracking. This information becomes part of a computer database that is used by experts to identify and plan long-term maintenance programs.

To safeguard the outside surface of the pipe, protective coatings are applied when lines are built. In all cases, as a backup to the protective coating, cathodic protection is installed to help control external corrosion.

9.3.1 Inhibitors

A corrosion inhibitor is a chemical substance that, when added in small concentration to an environment, effectively decreases the corrosion rate. Inhibitors are chemicals that react with a metallic surface or with the environment the surface is exposed to, giving the surface a certain level of protection. Inhibitors often work by adsorbing themselves on the metallic surface, protecting the metallic surface by forming a film. Inhibitors are normally distributed in a solution and injected into a pipeline through a sampling/injection tool. Some are included in a protective coating formulation. Inhibitors slow corrosion processes by one of the following methods:

- Increasing the anodic or cathodic polarization behavior
- Reducing the movement or diffusion of ions to the metallic surface
- Increasing the electrical resistance of the metallic surface

A vapor corrosion inhibitor stops corrosion by releasing protective ions that are attracted to metal. These ions condense on the metal surface, forming a water-repellent layer that contains both positively and negatively charged ions. The layer is maintained and repaired by the constant condensation of new vapor.

Corrosion inhibitors used in deep-water treating of oil and gas wells and deep-water risers differ from the use of land-based inhibitors in several ways, including delivery of the chemical to the point of injection, selection of inhibitor, and monitoring for results.

Each situation must be carefully examined before the proper inhibitor is selected.

9.3.2 Cathodic Protection

Cathodic protection effectively protects underground or submerged metallic structures through the use of a negative potential applied by an external source to the pipeline or structure. Usually, once the pipeline has been made sufficiently negative, it will resist environmental (soil or moisture) corrosion. The method is typically applied to underground pipelines, storage tanks, the interior of water storage tanks, ocean pilings, and electrical transmission towers. Buried steel structures will revert back to their natural state as an iron oxide without proper protection.

Cathodic protection is the most widely applied electrochemical corrosion control technique that reduces the corrosion of a metal surface by making that surface the cathode of an electrochemical cell. Cathodic protection is accomplished by creating and controlling the action of large corrosion cells by connecting the negative terminals of DC power sources to the pipeline and the positive terminals to groundbed anodes (see *Figure 41*).

10.0.0 ◆ REGULATIONS, PROCEDURES, AND DOCUMENTATION

Each pipeline company has published operations and maintenance (O&M) and emergency procedures. These procedures are based on federal, state, and local regulations and on industry standards. Regulations have the effect of laws and are enforced by government agencies. Standards are generally accepted practices gained through experience. All pipeline employees are expected to become familiar with all procedures that are relevant to their job.

10.1.0 Regulations

The U.S. Department of Transportation (DOT) Office of Pipeline Safety also handles pipeline safety and the quality of pipeline design, construction, and operation. This office administers two sets of regulations, *Code of Federal Regulations (CFR) 49 Part 192* and *CFR 49 Part 195*, that address transportation of natural gas and other gases and liquids by pipeline. The pipeline industry is greatly affected by federal regulations. The Federal Energy Regulatory Commission (FERC) controls the natural gas and crude oil and products lines in addition to the regulations of other federal agencies. The Occupational Safety and Health Administration (OSHA) oversees every facet of worker safety and health, from the wearing of personal protective equipment to provision of hazard information to availability of emergency equipment.

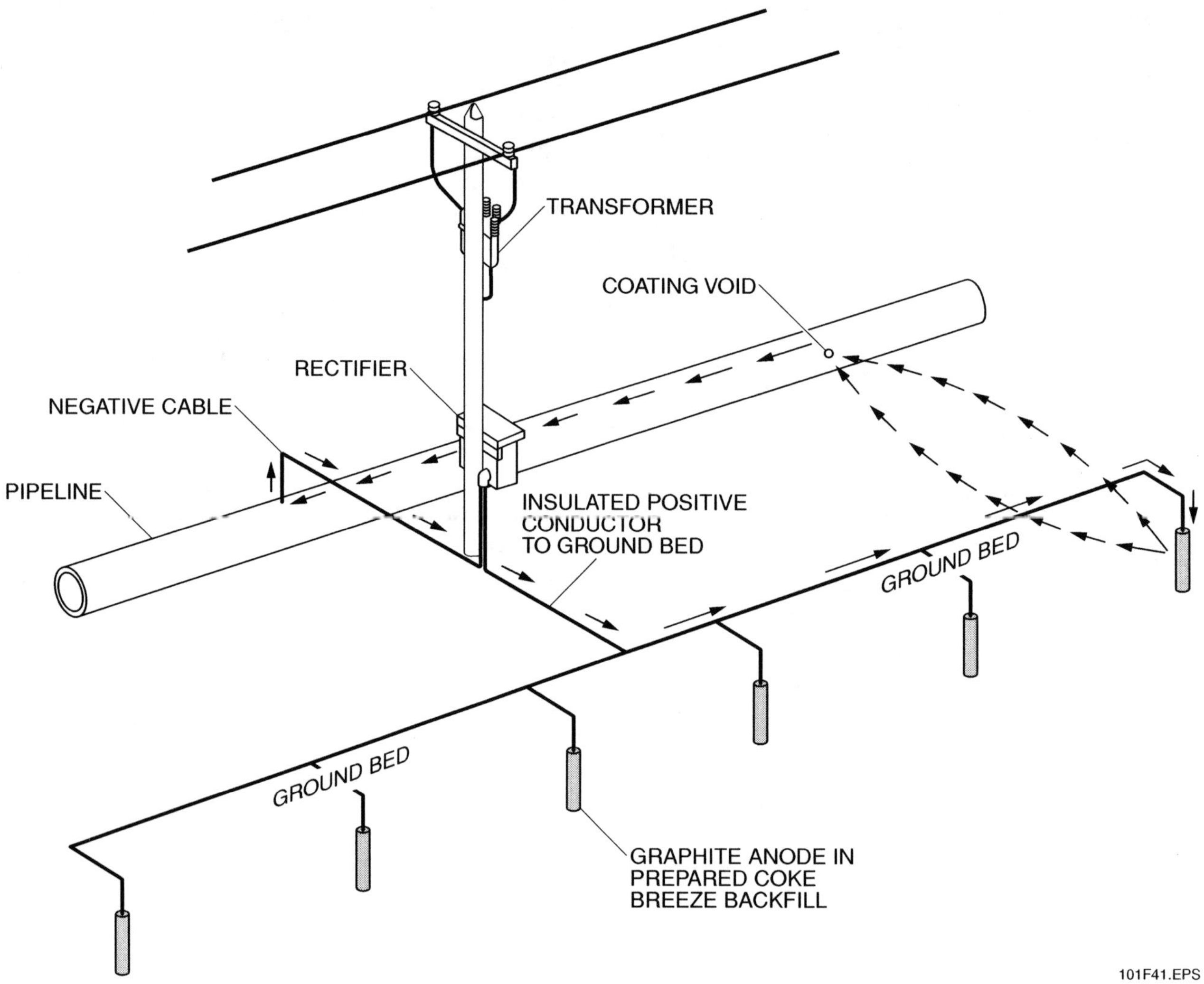

Figure 41 ◆ Cathodic protection arrangement.

State agencies can adopt federal regulations or enforce more stringent regulations than federal because each state has developed their own form of regulation regarding pipelines. Pipelines operate on an interstate or intrastate basis but not both. Interstate pipelines are located through more than one state and fall under federal jurisdiction and classification. Intrastate pipelines are located in a single state, and different states may classify the pipeline differently. Procedures should be relevant to the states involved.

10.2.0 Abnormal Operating Conditions

Abnormal operating conditions (AOCs) are specified by federal law in *49 CFR Part 195.503 and CFR 192.803*. The Operator Qualification Rule requires that an individual recognize and respond to any AOC that could happen while performing a task. An AOC refers to any condition identified by the individual that may indicate failure of a component, or deviation from normal operations that may indicate a condition exceeding design limits or that may result in a hazard to persons, property, or the environment. The following list identifies the AOCs that can occur when operating a pipeline system:

- Abnormal facility condition
- Activation of a safety device
- Communications failure
- Control system failure
- Power interruption
- Fire, explosion, or natural disaster
- Pipeline system damage
- Unexpected hazardous liquid or carbon dioxide encountered
- Unexplained flow rate deviation
- Unexplained pressure deviations
- Unexplained status change
- Other company specific circumstances

10.3.0 Company Policies and Procedures

Company policy and procedure manuals typically include:

- Operating procedures manual
- Maintenance procedures manual
- Safety procedures manual
- Emergency procedures manual

Each procedure will typically include procedure details, purpose, responsibilities, and other supporting information. There are many different requirements depending on the company and the location of the pipeline.

10.4.0 Qualifications

With the introduction of the new Operator Qualification Rule in *49 CFR 195.503*, pipeline operators are now required to have a written qualification program. Required elements of the qualification program include:

- Identification of covered tasks
- Qualification in the performance of covered tasks through operator evaluations
- Evaluation measures for individuals following a DOT-reportable accident or for other reasonable causes
- Communication processes
- Identification of intervals at which reevaluations will occur

10.5.0 Documentation

Documentation is an important part of the pipeline operation from origin to end point. The DOT requires that many operations and maintenance procedures be documented to record their correct and timely performance. Problems must be documented to track down the cause and to prevent recurrence.

Some documents are forms that must be completed by operations personnel and technicians. Others are forms, reports, and summaries that are printed out periodically through the control system. Some of the documentation required for operations includes:

- Accident/incident reports
- Operations logs
- Work orders
- Daily events/event logs
- Personnel qualifications
- Inspection and repairs

10.5.1 Accident/Incident Reports

Accident/incident reports are required, as specified in the Office Pipeline System (OPS) regulations, whenever there is a reportable incident along the pipeline. An incident is defined as an unexpected situation that causes, or could be expected to cause, an operational problem, harm to personnel, or damage to the environment. Reportable situations are defined in the regulations.

10.5.2 Operations Logs

Operations logs record the daily events of operation. These are usually summarized on reports printed from the control system; they include all activities communicated to and from the control room as associated with the pipeline.

10.5.3 Work Orders

Work orders and/or work permits are issued when specific maintenance or other tasks need to be performed. Each work order or permit has its own number and is recorded on a work order or permit log along with a description of the job and other details of the assignment.

These forms are required to ensure systematic maintenance of the pipeline, to assist with scheduling, and to record work done. Work orders and schedules for tasks done at specific time intervals can be triggered by the computer system. Not all pipeline companies use a work order system.

10.5.4 Daily Events and Event Logs

Daily events are recorded in the event log as specific tasks or activities that are performed during the day's operations. These might include repairs, inspections, AOCs, investigations, maintenance, or many other events.

10.5.5 Personnel Qualifications

The OPS regulation provides for formal documentation of the qualifications of personnel responsible for pipeline operations and maintenance in each company. The regulation was designed to reduce the risk of accidents on pipeline facilities caused by human error. Each company must have a written program specific to its own system for the qualification of individuals, including employees, contractors, and subcontractors who work on that system. These qualifications often cover about fifty different safety-related tasks, usually by position, such as pipeline controllers, and corrosion, maintenance, and operations personnel. The rule requires that a number of specific elements be present in all operator plans, including:

- Personnel must demonstrate the ability to recognize and respond to abnormal operating conditions.
- Companies must keep records that demonstrate compliance to the OPS.

10.5.6 Inspection and Repairs

When repairs are made, the pipe specifications, defect description, stress levels, and action taken must be documented and maintained in case of an audit. Other information, in accordance with the requirements of regulatory authorities, must also be prepared. Some documents are forms that must be filled in by operations and maintenance personnel. Others are forms, reports, and summaries that are printed out periodically through the control system.

Federal guidelines require that all inspections, treatments, and repairs be documented. Some of the documentation required for corrosion control includes:

- Cathodic protection system inspections
- In-line inspection results
- Pipeline coating inspections
- Incident reports
- Maintenance reports

Record-keeping is a huge and important task. Tremendous amounts of inspection and monitoring data are generated over the life of a pipeline, including:

- Development of written pipeline integrity procedures and coordination with operation, maintenance, contingency, and emergency response procedures
- Identification and implementation of repairs and modifications
- Program evaluation

Many software programs are now available to make record-keeping easier and to help companies stay in compliance with regulations. Failure to maintain good documentation can result in significant penalties for your company.

Summary

The term *pipeline* has many meanings. In the simplest terms, it applies to the pipe and fittings used to construct facilities for transporting commodities. In a broader sense, it applies to all of the facilities in addition to the pipe and fittings that are needed to transport the commodities from the point of origin to the point of delivery. Pipeline facilities include the means of measuring, controlling, and accounting for quantities and qualities of product delivered from the custody of the original transporter to the shipper.

In order to efficiently operate and maintain the pipeline facilities, personnel must understand the basic principles of pipeline hydraulics when working with liquid pipelines, and pipeline pneumatics when working with gas pipelines. Personnel must also familiarize themselves with basic pipeline equipment and its proper use.

Electrical power systems supply power to the pipeline equipment and facilities. Electrical and instrumentation technicians maintain and repair this equipment to ensure that the pipeline remains in service. Another major task of pipeline personnel is to monitor corrosion and practice measures to prevent or minimize the presence of corrosion on the pipeline and equipment.

Each pipeline company has published operations and maintenance and emergency procedures. These procedures are based on federal, state, and local regulations and on industry standards. All pipeline personnel must familiarize themselves with these standards in order to operate the pipeline in a manner that will ensure the safety of the employees and the general public and protect the environment.

Review Questions

1. Oil and gas are formed from _____.
   a. decaying organic matter
   b. buried rocks
   c. salt domes
   d. groundwater

2. Different types of pumps or valves are used to cap the well, depending on the _____ of the fluid being withdrawn.
   a. temperature
   b. destination
   c. properties
   d. amount

3. Lines from the oil fields are called _____ lines.
   a. field
   b. main
   c. distribution
   d. gathering

4. The pipe must be _____ tested to verify the pipeline integrity.
   a. strength-
   b. X-ray-
   c. hydrostatically
   d. sonar-

5. Different petroleum products can be shipped consecutively in the same pipeline. This is called _____.
   a. bunching
   b. mixing
   c. batching
   d. blending

6. Large volumes of natural gas are typically stored in _____.
   a. underground storage fields
   b. surge tanks
   c. relief tanks
   d. open roof tanks

7. Oil containing a large amount of sulfur compounds is called a _____.
   a. sweet crude
   b. sour crude
   c. paraffin
   d. light crude

8. There is a set of _____ for each hazardous substance, listing its properties and hazards.
   a. chemical descriptions
   b. OSHA labels
   c. product papers
   d. MSDSs

9. The marine environment is particularly sensitive to _____.
   a. high pressure
   b. overproduction
   c. oil spills
   d. vapor accumulation

10. A(n) _____ depicts the functional layout of a piping system and its components as clearly and accurately as possible.
    a. P&ID
    b. alignment sheet
    c. strip map
    d. blueprint

11. _____ is the software used to monitor and control the flow of millions of barrels of oil and products through the pipeline.
    a. RTUS
    b. SCADA
    c. SEACA
    d. PLCS

12. The science of the behavior of fluids at rest or in motion is referred to as _____.
    a. pneumatics
    b. hydraulics
    c. compressibility
    d. viscosity

13. Boyle's law states that when temperature is kept constant and volume is decreased, pressure will _____.
    a. decrease
    b. increase
    c. stay the same
    d. not be affected

14. In a typical positive-displacement pump, a _____ slides back and forth in a cylinder.
    a. rotor
    b. vane
    c. piston
    d. blade

15. _____ are used to control the rate of flow in a line, to open or shut off a line, or to serve as automatic safety devices.

a. Motors
b. Valves
c. Pumps
d. Manifolds

16. Tanks must be calibrated for _____.

a. strength
b. sealing ability
c. capacity
d. wall thickness

17. _____ are used to join three pipe ends.

a. Flanges
b. Unions
c. Elbows
d. Tees

18. The pipeline uses _____ to control the flow path of oil and petroleum products.

a. manifolds
b. pumps
c. tanks
d. vents

19. Various _____ are used along the pipeline to monitor the volume of fluid.

a. gauges
b. tests
c. meters
d. provers

20. Communications failure is one example of an _____.

a. emergency
b. AOC
c. accident
d. interruption

GLOSSARY

Trade Terms Introduced in This Module

Anomaly: A defect in the pipeline or pipeline coating caused by corrosion, wear, or other external means.

Barrel: The common measure of petroleum products: one barrel equals 42 U.S. gallons or 0.15899 cubic meters or 9,702 cubic inches.

Cathodic protection: A means of preventing the destructive electrochemical process of corrosion of metal by using the metal as the cathode of a cell with a sacrificial anode.

Corrosion: The gradual destruction, dissolving, or wearing away of a substance, usually a metal or an alloy, caused by a chemical or electrochemical reaction with its environment. Corrosion can be a problem in different locations along the pipeline, both externally and internally.

Crude oil: Unrefined liquid petroleum. Crude oil ranges in color from yellow to black, and may have a paraffin, asphalt, or mixed base. Crude oils may be referred to as heavy or light according to API gravity, with the lighter oils having the higher gravities.

Cubicle: An enclosure for high-voltage equipment.

Department of Transportation (DOT): The federal agency that sets standards for, and regulates, transportation.

Hydrostatic testing: The most common final quality check on a complete pipeline. Water is pumped into the line and pressured up to a specified level. The pipe is checked for leaks or ruptures.

Main line: The primary pipe between distant points; it is also called the trunk line.

Manifold: An accessory system of piping to a main piping system that serves to divide the flow into parts, to combine several flows into one, or to reroute a flow to several possible locations.

Material safety data sheet (MSDS): A form containing information on a potentially hazardous material, including names, contents, test results, characteristics, potential hazards, and handling procedures.

Natural gas: A highly compressible, highly expandable mixture of hydrocarbons with a low specific gravity. It occurs naturally in a gaseous form. Besides hydrocarbon gases, it may contain nitrogen, helium, carbon dioxide, hydrogen sulfide, and water vapor. The gases are variable in form and may be found as gases or liquids under various conditions of pressure and temperature. Natural gas liquids include butane, propane, and natural gasoline.

Natural gas liquids (NGLs): The portions of reservoir gas that are liquefied at the surface in lease separators, field facilities, or gas plants.

Petroleum: Unrefined liquid hydrocarbons, including *natural gas liquids (NGLs)*.

Pig: A scraping tool with blades or brushes that is forced through a pipeline to clean out accumulations of wax, scale, and debris from the pipe walls; it is also called a line scraper or go-devil.

Product: Any fluid, refined or not, that is transported through a pipeline.

Refined products: Manufactured petroleum products generated from crude oil.

Refinery: System of process units used to convert crude petroleum into fuels, lubricants, and other petroleum-derived products.

Seismic activity: The phenomena of earth movements, earthquakes, and earth vibrations, including those that are artificially induced.

Tank: A metal, plastic, or wooden container used to store a liquid. Three types include mud tanks for drilling, production tanks, and storage tanks.

Throughput: The volume of product that goes through a section of equipment during a specified amount of time.

The NCCER makes every effort to keep these textbooks up-to-date and free of technical errors. We appreciate your help in this process. If you have an idea for improving this textbook, or if you find an error, a typographical mistake, or an inaccuracy in NCCER's Contren™ textbooks, please write us, using this form or a photocopy. Be sure to include the exact module number, page number, a detailed description, and the correction, if applicable. Your input will be brought to the attention of the Technical Review Committee. Thank you for your assistance.

Instructors – If you found that additional materials were necessary in order to teach this module effectively, please let us know so that we may include them in the Equipment/Materials list in the Instructor's Guide.

Write: Curriculum Revision and Development Department
National Center for Construction Education and Research
P.O. Box 141104, Gainesville, FL 32614-1104

Fax: 352-334-0932

E-mail: curriculum@nccer.org

Craft ______ Module Name ______

Copyright Date ______ Module Number ______ Page Number(s) ______

Description ______

______

______

______

(Optional) Correction ______

______

______

(Optional) Your Name and Address ______

______

______

Module 66102-02

Liquid Pipeline General Abnormal Operating Conditions

Introduction to the Pipeline Operator Qualification Rule, API, and NCCER

It could be said that petroleum products keep America running. Petroleum products like oil and natural gas are the principal sources of energy in the United States and are used in many manufacturing and chemical processes. Petroleum products are primarily transported by pipeline. There are more than a million miles of liquid and natural gas pipeline in this country, ranging in diameter from two inches to four feet. These pipelines bring crude oil and natural gas from production fields to processing plants and refineries—and ultimately to the consumer.

Pipeline Operator Qualification

Petroleum products are considered hazardous, so pipeline personnel must take extra precautions and follow specific procedures when working with them. To help ensure the safe handling and transportation of petroleum products, the United States Department of Transportation (DOT) mandates safe practices as federal regulations.

One such regulation, the Operator Qualification (OQ) rule, requires pipeline operators to develop a written qualification plan to evaluate their personnel and their contract personnel (see *49 CFR Part 192 Subpart N—Gas,* and *49 CFR Part 195 Subpart G—Liquid*). The evaluation hinges on the operator's ability to perform certain covered tasks and to be able to recognize and respond to abnormal operating conditions that may be encountered while performing these covered tasks.

Covered Tasks

A pipeline performance task is deemed *covered* if the activity passes the following four-part test:

1. Is it performed on a pipeline facility?
2. Is it an operations or maintenance task?
3. Is it performed as a requirement of *49 CFR Part 195 (Liquid)* or *49 CFR Part 192 (Gas)*?
4. Does it affect the operation or integrity of the pipeline facility?

If the answer to all four questions is *yes*, then the activity is considered *covered* and must figure into the evaluation process of pipeline personnel.

American Petroleum Institute

The OQ rule and its related covered tasks provided the catalyst to develop a unified pipeline training and assessment program—so that pipeline operators and owners have ready access to the assessment and training tools they need. This effort was the direct result of the vision and leadership provided by The American Petroleum Institute (API) and API members.

Efforts to unify the industry are nothing new to API. Indeed, the impetus for forming the Institute in 1919 was the need to standardize engineering specifications for drilling and production equipment. Today, the most pressing issues revolve around public perceptions and government policies toward the industry. Speaking with one voice on these issues has become as essential as having interchangeable parts in the field. API's members determine the petroleum industry's positions on public policy as well as the standards that govern its day-to-day operations worldwide.

Consortium on Operator Qualification

API assembled a focus group whose primary purpose was to identify covered tasks and provide compliance recommendations to pipeline operators.

The Consortium on Operator Qualification (COOQ) became the guiding force for the identification and development of covered tasks for the liquid pipeline industry. Their work product, API publication 1161, *Guidance Document for the Qualification of Liquid Pipeline Personnel,* formed the basis for the resulting Pipeliner Training and Assessment Program.

The covered task list created in API publication 1161 has now evolved to include gas covered tasks and is maintained by NCCER. The most current covered task list can be downloaded at www.nccer.org.

National Center for Construction Education and Research

With the guidance and support of API, its members, and the Consortium on Operator Qualification, the National Center for Construction Education and Research (NCCER) was able to design and implement the Pipeliner Training and Assessment Program.

NCCER is a not-for-profit 501(c)(3) education foundation established in 1995 by the world's largest and most progressive construction and maintenance contractors and national contractor associations. NCCER was founded to address the

severe workforce shortage facing the construction and maintenance industries and to develop a standardized training process and curricula. Today, NCCER is supported by hundreds of leading construction and maintenance companies, manufacturers, and national associations and developed more than 35 construction and maintenance titles.

NCCER also maintains a National Registry, which contains the records of trainees and instructors who have received NCCER training. All completion information for every trainee is sent to the NCCER and filed within the Registry. The Registry can then confirm training and skills for workers as they move from company to company, state to state, or even within their own company.

Standardized Curriculum (NCCER's Contren™ Learning Series)

NCCER curricula are developed using the expertise of industry subject matter experts (SMEs). The use of SMEs as an integral part of every new development or revision ensures that the curriculum is current, industry-driven, and indicative of industry practices. The development and/or revision process is facilitated from start to finish by NCCER project managers, who work closely with the SMEs and a select team of technical writer(s) on the development and presentation of the curriculum.

Pipeline: Five Training Paths

The pipeline curriculum focuses on essential skills and competencies within five skill areas:

- Corrosion Control
- Electrical and Instrumentation
- Maintenance
- Mechanical
- Operations (Field and Control Center)

SMEs from the pipeline industry have determined the skills and competencies needed based on API's covered tasks and the consensus of the group/industry. Specific company information is generally not incorporated into the curriculum unless the SMEs determine that the information is the industry standard.

Training Modules and Covered Tasks

The smallest unit of instruction in the curriculum is a *module*. Each curriculum is composed of modules (the equivalent of stand-alone chapters) that are logically sequenced into complete *levels* of instruction. Every module is task-oriented and requires the successful completion of a written (knowledge-based) examination and a performance (hands-on) test. Modules may be individually selected and/or combined for efficient task training. Modules in the pipeline curricula, to the extent possible, will align with API-developed covered tasks. While not all modules will relate to covered tasks, all covered tasks will relate to at least one module.

Benefits of Training to the Pipeline Marketplace

Applying NCCER's standardized training process allows pipeline owners and contractors in the United States to establish the benchmark for acceptable skill and competency levels within the pipeline industry. This process will assist training organizations in developing and/or improving standardized pipeline training programs that effectively match their resources and operating conditions while responding to regulatory demands and to the needs of the industry. The process will also recognize training organizations that meet the minimum standards of competence, warranting regulatory, public, and professional confidence. This in turn will foster national unity within the pipeline industry for the mutual benefit of individual technicians, owners, contractors, and ultimately pipeline-users and society.

Additional Resources

To review the Pipeliner Training and Assessment Program Implementation and Resource Guide and the latest version of pipeline covered tasks, visit NCCER's Web site at www.nccer.org, or contact:

National Center for Construction Education and Research
PO Box 141104
Gainesville, FL 32614-1104
(888) 622-3720
Web site: www.nccer.org

For more information on API, please contact:

American Petroleum Institute
1220 L Street, Northwest
Washington, D.C. 20005
(202) 682-8000
Web site: www.api.org

To order *API Pub 1161*, contact Global Engineering Documents by calling 1-800-854-7149 or visiting their Web site at www.global.his.com.

Additional sources of information on the pipeline operator qualification regulation include:

DOT – Office of Pipeline Safety
www.ops.dot.gov

Transportation Safety Institute
www.tsi.dot.gov

COURSE MAP

This course map shows each curriculum path in the Pipeliner Training program. The suggested training order begins on the left with the Pipeline Core modules and proceeds to the right. Which curriculum path you follow depends on your areas of specialty and interest. On a given curriculum path, skills levels increase as you advance to the right on the course map, although some are single-level curricula. The local Training Program Sponsor may adjust the training order.

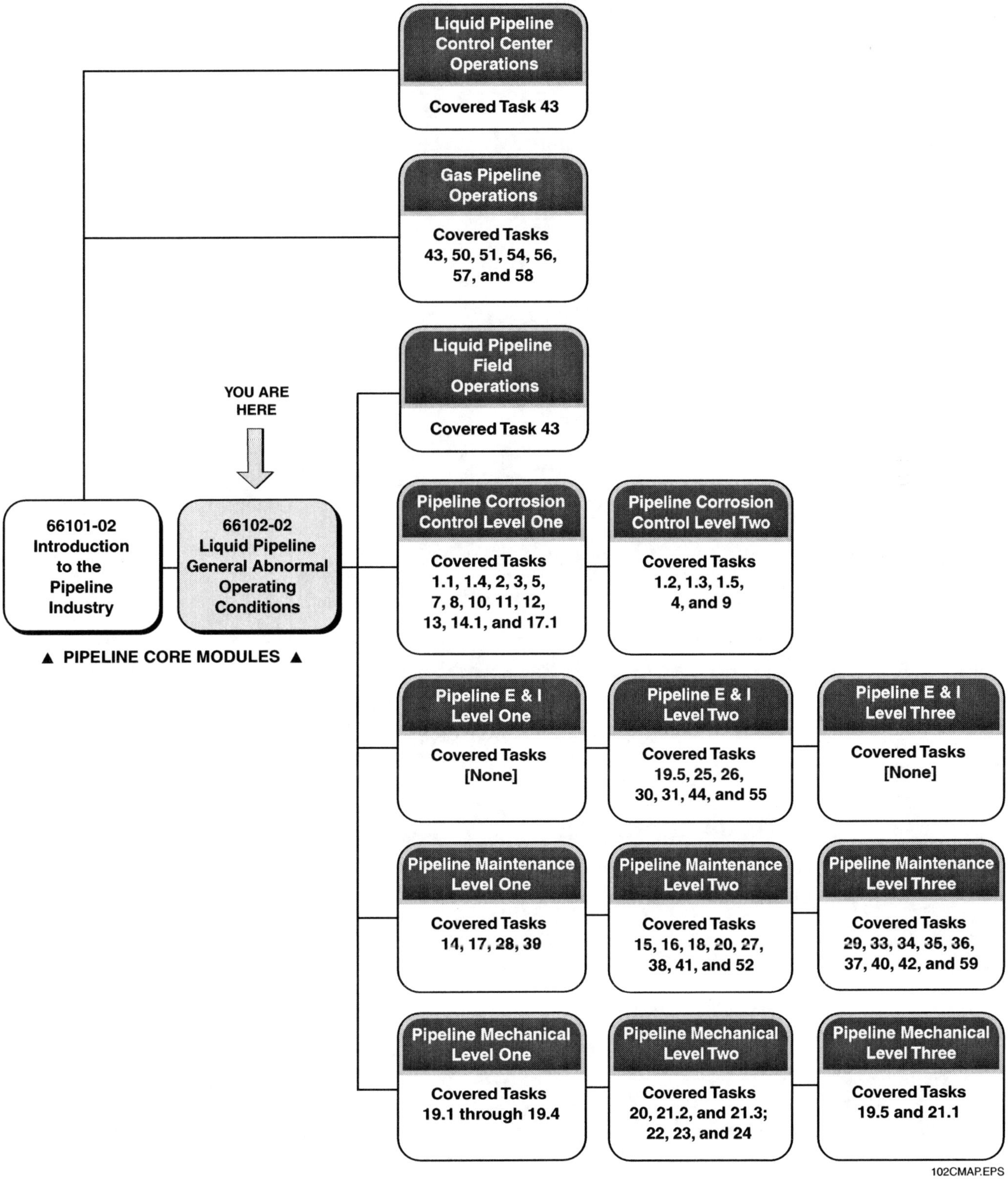

MODULE 66102-02 CONTENTS

Figures

Table

Liquid Pipeline General Abnormal Operating Conditions

Objectives

When you have completed this module, you will be able to do the following:

1. Recognize and react to abnormal facility conditions.
2. Recognize and react to activation of a safety device.
3. Recognize and react to communications failures.
4. Recognize and react to power interruptions.
5. Respond appropriately to fire, explosions, and natural disasters.
6. Recognize and react to pipeline system damage.
7. Recognize and react to unexpected hazardous liquid or carbon dioxide (CO_2) leaks.
8. Recognize and react to unexplained pressure deviations.

Prerequisites

Before you begin this module, it is recommended that you successfully complete the following: Core Curriculum; Pipeline Core, Module 66101-02.

Required Trainee Materials

1. Pencil and paper
2. *Code of Federal Regulations, Title 49, Part 195*

1.0.0 ◆ INTRODUCTION

Abnormal operating conditions (AOCs) do not happen frequently on the pipeline. Companies have programs, procedures, training, safety equipment, warning devices, and fail-safe devices on practically every facet of operations. Computers and electronic controls are used to operate much of the pipeline and eliminate many problems and errors before they occur. However, the operator has the responsibility for making certain that when equipment does fail, the pipeline is kept in a safe condition. The operator must make use of training, skills, and good judgment until solutions are found to maintain safe operation.

Time is of the essence in such conditions. Personnel are expected to take decisive action to overcome the abnormal condition, maintaining the safety of pipeline operations. Personnel must follow company procedures. As regulation *49 CFR 195.402* states, each company must ensure its employees maintain a thorough knowledge of its operations, maintenance, and emergency manuals.

If it is necessary to deviate from company policy for the safety and protection of personnel, the public, the environment, or equipment, then the operator is expected to handle the situation. The operator goes through several logical procedures in making decisions and performing actions. Although not every procedure is included in every action to secure the safety and operation of the pipeline, the list of actions might be divided as follows:

- Recognize the AOC.
- Investigate the cause of the alert.
- Perform the actions planned.
- Make the appropriate notifications.

1.1.0 Definition

An AOC is defined by federal pipeline safety law *49 U.S.C. 60102* and pipeline safety regulations *49 CFR 192.605* and *49 CFR 195.402*. AOCs are also specified by federal law in *49 CFR 195.503*. This module will concentrate on AOCs related to *49 CFR 195 – Transportation of Hazardous Liquids by Pipeline*.

An AOC is any condition identified by the individual that may indicate failure of a component or deviation from normal operations. It may also be a condition exceeding design limits or resulting in hazards to persons, property, or the environment.

The Operator Qualification rule requires that an individual be able to recognize and respond to any AOC that could happen while performing a covered task. *Table 1* shows a layout of abnormal operating conditions.

Table 1 Layout of Abnormal Operating Conditions (1 of 2)

Abnormal operating condition (Recognize and react to...)	Example	Reaction
Abnormal facility condition	During the course of performing a covered task, the individual could encounter one of the following: • Exposed pipe • Missing or damaged pipeline supports • Incorrect ANSI pressure rating of a component or material type • Disbonded coatings • Unauthorized equipment on the right-of-way • Sunken floating tank roof	Determine the extent of damage, perform repairs if properly trained to do so, make the appropriate notifications as required by company policy, and complete any required documentation.
Activation of a safety device	During the course of a maintenance or operating activity, the following safety devices could activate: • Pressure relief valve relieves to tank/sump/atmosphere • Emergency or abnormal shutdown • High-pressure shutdown • Hazardous atmosphere alarm indicating the presence of an LEL(20%) or HEL(40%) explosive atmosphere • High sump level or shutdown alarm • High tank shutdown alarm These devices are typically designed to operate and reduce/eliminate a hazardous situation.	Contact the operations or control center and make the appropriate company notifications, minimize the situation if possible, and complete any required documentation.
Communications failure	During the course of operations, the operator could have a loss of the following communication devices: • Radio communications • Telephone • SCADA communication • Satellite	Immediately contact the control center, make the appropriate notifications as required by company policy, and complete any required documentation.
Power interruption	During the course of operating a pipeline component, the individual could experience a loss of commercial or local AC power due to the following: • Utility power failure due to equipment problems or storms • Station or terminal power problems (for example, motor overload, breaker trips)	Inform the power company of the outage if applicable, make the appropriate notifications as required by company policy, and complete any required documentation.

102T01A.EPS

Table 1 Layout of Abnormal Operating Conditions (2 of 2)

Abnormal operating condition (Recognize and react to...)	Example	Reaction
Fire, explosion, or natural disaster	During the course of a maintenance or operations activity, the individual could discover a fire or explosion. This fire or explosion could be caused by the individual during the performance of a task due to things such as: • Ignition of a hazardous vapor • Unintentional detonation Fire and explosion may also result from natural disasters, such as: • Lightning strike • Flooding • Landslide • Tornado	Notify local 911 emergency personnel; limit access to area as much as possible; make the appropriate notifications as required by company policy, and complete any required documentation.
Pipeline system damage	During the course of a maintenance or operations activity, the individual could encounter or cause damage to the pipeline such as: • Damaged or inoperable pipeline components (pipe, valves, pumps, tank gauges) from external force • Gouges/dents/pitting • Damaged pipe support • Wrinkles/buckle/cracks • Tank shell damage • Sunken roof, leaking pontoon • Cracked or significantly corroded flange or bolts • Damage due to flooding or land movement	Anticipate the proper response for unexpected hazardous liquid or carbon dioxide encounter and for fire/explosion if appropriate; determine the extent of the damage; make the appropriate notifications as required by company policy, and complete any required documentation.
Unexpected hazardous liquid or carbon dioxide encountered or unintended release	During the course of maintaining or operating the pipeline facility, the individual could encounter an unexpected presence of hazardous liquid or carbon dioxide, or could cause an unintended release during the performance of a task. This does not include the controlled transfer, sampling, venting, draining, or handling of hazardous liquid or carbon dioxide. The individual could be expected to respond to: • Hydrocarbon release • Vapors	Notify local 911 emergency personnel and/or appropriate persons/authorities according to company procedure, and complete any required documentation.
Unexplained pressure deviations	During the course of maintaining or operating the pipeline facility, the individual could observe the following pressure deviations: • Increased pressure • Decreased pressure • High-pressure limit • Low-pressure limit • Loss of line pressure	Follow troubleshooting procedures, perform shutdown of the facility if necessary, make the appropriate notifications as required by company policy, and complete any required documentation.

102T01B.EPS

2.0.0 ◆ RESPOND AND REACT TO ABNORMAL FACILITY CONDITIONS

Abnormal facility conditions relate to problems involving the physical structure of the pipeline. Examples of abnormal facility conditions include exposed pipe, missing or damaged pipeline supports, evidence of corrosion, missing signs/markers, equipment stalled over the line, hazardous atmosphere alarms, or resistance temperature device (RTD) alarms. Incorrect data caused by equipment malfunction, old or incorrect maps or survey records, as well as the incorrect interpretation of the data also create AOCs.

In almost all incidences of noted abnormal facility conditions, personnel should determine the extent of damage if applicable, perform repairs within a reasonable time if properly trained to do so, make the appropriate notifications as required by company policy, and complete the required documentation.

2.1.0 Respond to High Pump Case Temperature/Pressure Alarm

Main-line pumps are fitted with RTDs that give temperature readings for the pump case and product temperature. The devices are preset to signal an alarm in the control center when the temperature reaches a critical level and will shut the pump down automatically if some action cannot be taken to lower the temperature. When a pump case temperature reaches such a critical level and automatically shuts down, it is considered an AOC. *Figure 1* shows pump case temperature RTDs.

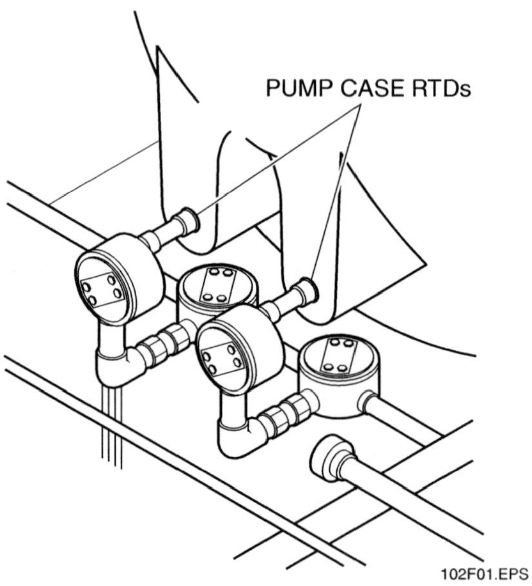

Figure 1 ◆ Pump case temperature RTDs.

Most main-line pumps are set to shut down on low pressure. Companies fit pumps with differential pressure switches and temperature switches to protect the equipment from damage caused by low pressure, no pressure, or no flow and to avoid exceeding **maximum operating pressure (MOP)**.

Pumps also have high discharge pressure alarms that will shut down the pump when the discharge pressure exceeds a preset limit. Many times, this indicates a downstream line obstruction that is not allowing the product to flow properly from the pump.

Another type of alarm that can be received from a pumping unit is a seal failure alarm. If responding to a seal failure alarm, operations and/or maintenance personnel should investigate the validity of the alarm, check the station sump level, and notify the appropriate personnel. This may not necessarily result in an AOC, but it would require control center personnel to document the alarm and notify the appropriate personnel to check and possibly replace the seal. The control center operator should never attempt to restart the pump before investigating the cause of the alarm and responding to the alarm per company policy.

2.2.0 Respond to Damaged Pipeline Supports

Damaged pipeline supports (see *Figure 2*) can be serious problems on the pipeline but may not require immediately shutting down the system. A report detailing the damage is normally submitted by maintenance or inspection personnel who are traversing the pipeline. Action is determined by the extent of the damage.

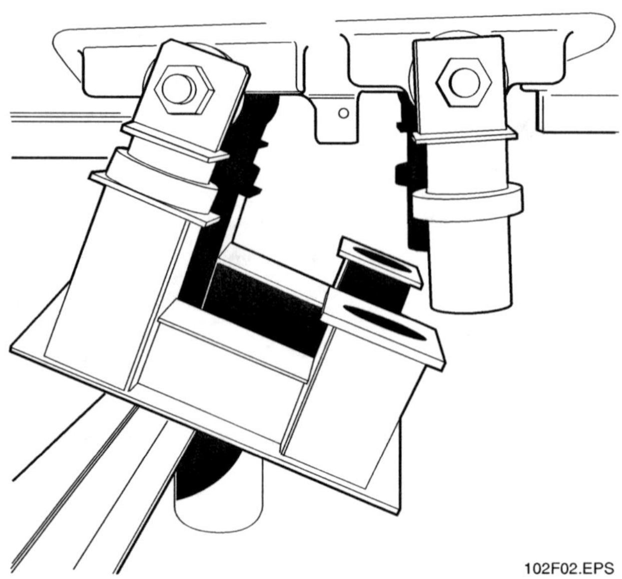

Figure 2 ◆ Damaged pipeline support.

If the support is damaged, the pipeline may have to be shut down. If the damage is not critical, it may be possible to continue operation under heightened awareness. Such a decision is a judgement call on the part of the operator in collaboration with qualified maintenance or inspection personnel. The operator must submit a report to appropriate personnel so that repairs can be scheduled.

2.2.1 Reaction

If the damage to the support is critical or indicative of a possible release, the individual making the discovery must contact the control center or the appropriate company personnel immediately if the pipe is in danger of excess movement. The operator and/or call taker should then follow company policy precisely to investigate the condition and dispatch qualified personnel to the scene. Until the arrival of qualified personnel, the individual at the scene should stay a safe distance away from the site, evacuate the area if necessary, and make the appropriate notifications as required by company policy.

2.3.0 Documentation

The documentation and records required following an AOC depend upon the size of the incident and whether a release occurred. Any material defect or physical damage that impairs the serviceability of a pipeline requires the operator to complete a safety-related condition report and the designated company official to submit the report to government agencies (*CFR 195.55*). *Figure 3* shows a typical form for a safety-related condition report.

> **NOTE**
> A safety-related condition report is only required if the condition is not corrected within a given amount of time, or occurs within a certain distance of residences, streams, or other public areas, or if the condition is indicative of general corrosion.

3.0.0 ◆ RESPOND AND REACT TO ACTIVATION OF A SAFETY DEVICE

Safety devices are installed to protect the safety of pipeline operations and equipment. Any time a safety device activates, an alarm is sent to the control center. Safety devices include pressure relief systems, pressure limiting devices, high pump case pressure alarms, overfill protection devices, and main-line pressure shut-down alarms.

Pumps also have high discharge pressure alarms that will shut down the pump when the discharge pressure exceeds a preset limit for the pipeline.

> **CAUTION**
> An individual's actions during the performance of a covered task could cause a safety device to activate. The activation represents a deviation from normal operations, resulting in an AOC.

3.1.0 Respond to the Activation of a Pressure Relief System

Activation of **relief devices** along the pipeline system is an abnormal occurrence. Relief valves can discharge into tanks or vessels that are designed and constructed in compliance with *API Standard 650* or *API 2510* for pressure vessels. Relief valves (see *Figure 4*) can be installed on pipeline or pressure **breakout tanks**.The purpose of a relief valve is to protect the pipeline or pressure tank from overpressure. These tanks can also be used to temporarily absorb **surge pressure** from the main-line and manifold pressure relief valves to lower pressure on the line. An abnormal operating condition exists when a pressure relief occurs.

3.1.1 Reaction

Each relief event should be recognized and documented by operations personnel. If relief becomes excessive, operations personnel must immediately begin to investigate and make adjustments to ensure that the pressure of the breakout tank or pipeline does not exceed 110% of the MOP.

If the alarm is signaled and the pressure remains high, control center operations personnel must take immediate action to reduce pressure. A pressure reduction is usually achieved by shutting down the upstream pumping station, closing a block valve, or shutting down the pipeline system.

3.1.2 Documentation

Any malfunction or operating error that causes the pressure of a pipeline to rise above 110% of its maximum operating pressure may require the operator to complete a safety-related condition report and the designated company official to submit the report to government agencies (*CFR 195.55*).

Safety-Related Condition Report

Date: ________________ Government Letter #: ________

APSC FILE #: ______________

Information Resources Manager
Office of Pipeline Safety
Research and Special Programs Administration
U.S. Department of Transportation
Room 2335
400 Seventh Street SW
Washington, D.C. 20590

Safety-Related Condition Report

49 CFR 195.55
Subject:

Operator:

Date of Report:

Person Submitting Report: Senior Vice President
Business Unit

Person Determining Condition:
Name:

Title:

Telephone:

Date Condition Discovered:

Date Condition Determined:

Location of Condition:

Description of Condition:

Corrective Action:

Sincerely,

Sr. Vice President, Business Unit

102F03.EPS

Figure 3 ◆ Safety-related condition report form.

102F04.EPS

Figure 4 ◆ Relief valve.

When filing a safety-related condition report, management must include a description of the condition, including the circumstances leading to its discovery, any significant effects of the condition on safety, and the name of the commodity being transported or stored. Management must also describe the corrective action taken (including reduction of pressure or shutdown) and the planned follow-up or future corrective action, including the anticipated schedule for starting and concluding such action.

3.2.0 Respond to Unintended Valve Closure or Shutdown

The unintended closure of a main-line block valve, remote block valve, manual block valve, pump station suction or discharge valve, and certain terminal valves (manifold/tank) will result in an automatic shutdown of the pipeline. In these cases, the operator should inspect the equipment to determine why closure occurred and reset/open/close the valve when it is safe to do so. Such occurrences may not cause an AOC, but if the event causes the pipeline to exceed design limits in any way, its classification changes to abnormal.

3.3.0 Testing Safety Devices

Safety devices provided at each pump station must be tested under conditions approximating actual operations (*CFR 195.262*). If, for example, the control room calls to check the **lower explosive limit (LEL)** alarm at a pumping station, it must first shut down the station and then check the LEL alarm at the station.

If, while testing a safety device, an alarm indicates a malfunction not related to the device being tested, operations and/or maintenance personnel should contact control center operations, make appropriate company notifications, and complete the required documentation.

> **WARNING!**
> If the hazardous atmosphere alarm indicates the presence of a LEL of 20%, evacuate the area.

4.0.0 ◆ RECOGNIZE AND REACT TO COMMUNICATIONS FAILURES

Failure of communications usually refers to a loss of contact between the control center **supervisory control and data acquisition (SCADA)** system and remotely controlled operating components on the pipeline. Essentially, operations personnel lose the ability to remotely monitor and control the pipeline facility. Communication failures may include:

- Intermittent facility outages
- Loss of the SCADA system

Sometimes the communications with various components can be verified, but the affected control system will not work. In these cases, the problem involves other components of the control system that may be mechanical, electrical, or hydraulic.

The failure to communicate with a remote control valve operator is considered a communications AOC because the valve cannot be opened or closed by the operator. The valve could also have a mechanical failure that prevents it from being operated. Sometimes electrical components or connections burn out, and if this happens on some pump drivers, they would not be able to respond in order to start up or shut down. Control of equipment is lost in this situation and could happen in many others.

4.1.0 Reaction

If it is discovered that there is a loss of a communications device, personnel should immediately contact the control center to verify the condition. Operations and/or other appropriate personnel should then determine the exact cause and take decisive action to control the condition, while always following company policies and procedures. In some instances, operations personnel might be able to bypass the offending component until appropriate maintenance can be performed. In other cases, the situation may warrant a shutdown.

4.2.0 Notifications

If the component can be operated manually, the problem can be alleviated until repairs can be made. Operations must monitor pressure, however, and if 110% of MOP is exceeded, the pipeline must be shut down without discussion. If the pressure remains within limits or the component is not one that will affect the operation directly, the condition must still be monitored closely if the pipeline is to stay in operation. Even after normal conditions have been restored, operations personnel must continue to monitor the condition to make certain that repairs have been successful.

5.0.0 ◆ RECOGNIZE AND REACT TO POWER INTERRUPTIONS

A loss of power impacts the abilities of the control center or facility operations to safely control and monitor the pipeline facility. Virtually every component of the pipeline is in some way dependent upon electricity. If power goes off, electronic indicators will be lost. The SCADA system may be off. Pumps will shut down. Remote control valves will not operate, making control of the system difficult.

The pipeline may have a problem if the power failure is of a local nature because only a part of the system will shut down without electricity. Company policy and procedures must be followed. Area-wide power failure may also cut telephone communications and isolate the affected station. At the receiving terminal, a shipment that is in process may not have storage and cannot be shipped further without pumps. A tank being filled that cannot be switched may reach a high level or possibly overflow.

When power failure occurs, some systems can be protected by direct current (DC) backup power supplies, but most stations, terminals, and equipment are operated on alternating current (AC) or are equipped with uninterruptible power systems (UPS), a battery backup that converts DC to AC. The UPS or back-up generators supply emergency AC power to allow minimum operating capabilities or at least the ability to safely shut down the facility until the power is restored.

5.1.0 Reaction

Reaction to power failure is determined by what part and how much of the pipeline is affected. If only a single component or station is involved, it may be possible to adjust line pressure and flow to keep the pipeline in operation.

Suppose power interruption occurs while operating a valve. In this scenario, the crew/individual should be prepared to operate the valve manually, either by hand or by another type of operator.

Suppose a pipeline system with multiple stations is flowing with all units running. Power is lost at a station that is needed to keep pressure below MOP. In this situation, the control center operator would have to shut down the pipeline system to avoid overpressurizing the system. The terminal operator must be sure to switch off any pumps that were running so that when power returns, they will not start up. Yard personnel must be stationed outside so valves can be operated manually. If a shutdown is not required, the shipment will continue to arrive, and the schedule must be followed and the tasks completed manually.

6.0.0 ◆ RESPOND APPROPRIATELY TO FIRE/EXPLOSION

Some pipelines and facilities are fitted with detection devices that send an alarm when certain conditions occur. Such equipment includes flame detectors, fire detectors, and seismic detectors. Reaction to these types of alarms may require evacuation of facilities.

Flame detectors (see *Figure 5*) are thermal detectors that have normally open contacts. They work by closing their contacts when exposed to rapidly rising temperature, as would be experienced in the presence of open flames. When the contacts close, an alarm is signaled on the console in the control center and on the fire control panel in the station.

Fire alarms are activated by ultraviolet (UV) detectors or by manual pull-switch. Fire detectors are UV detectors that are set off by the light given off by open flame. Their response is almost instantaneous, and they signal fires in the very beginning stages. UV detectors are connected to a firefighting system and immediately trigger a dump of a fire suppression material into the area.

> **NOTE**
> Verify that a hazardous alarm test has been performed before performing any work, such as welding or cutting, opening electrical boxes, or operating non-explosion-proof tools in hazardous classified areas, that may produce an ignition source.

102F05.EPS

Figure 5 ◆ Flame detector.

6.1.0 Reaction

Reaction to a flame detector or fire detector alarm must be immediate and decisive. Personnel must first verify that it is not a false alarm. A false alarm or an expected alarm due to maintenance activities requires no further action unless, in the judgment of operations, the system needs to be investigated. Alarms often appear on the control center control panel or SCADA screen.

Confirmation of the alarm gives the exact location of an active fire. Company policy dictates procedures for such emergency action and it must be followed. Procedures are site-specific and may allow for some flexibility in responding. However, maintenance and operations personnel should not need to consult a manual in case of an alarm. All personnel must be adequately trained and familiar with emergency procedures in advance. It is important to take necessary action, including a shutdown of the pipeline if required to control the emergency.

NOTE

According to *CFR 195.403*, operators shall establish and conduct a continuing training program to instruct emergency personnel in the proper use of firefighting procedures and equipment, in the identification of hazardous conditions likely to cause emergencies, in the steps necessary to control accidental releases, and in the characteristics of hazardous liquids or carbon dioxide transported.

The safety of personnel is always top priority. Protection of property is always second. Attention must be totally focused on the emergency. Response must be implemented relative to the type of fire in progress.

If the fire is a small or incipient (beginning) fire, personnel should use the available fire extinguishing equipment to control the situation. Personnel should respond in the way they have been trained. Such fires are usually the result of faulty electrical equipment or a smaller item such as a trash can fire. Responders should make certain to use the proper type of fire extinguisher for the type of fire.

If the fire is larger than the immediate workforce can contain, operations personnel should minimize the situation, notify appropriate personnel and agencies, and complete the required documentation. For example, during an electrical storm, if lightning strikes a full tank at the facility and the roof seal catches on fire, operations should follow company policy and contact the fire department immediately to contain the fire.

6.2.0 Respond Appropriately to Natural Disasters

Natural disasters include landslides, earthquakes, flooding, or other dangerous weather patterns that could be precursors to soil and pipeline movement. The impact of moving soil may result in forces sufficient enough to cause pipeline failure or product release.

Additional stresses are imposed on piping by soil movement or subsidence. Rains resulting in saturated soil conditions may also result in soil movement in some areas. Pipeline personnel should be alert to weather patterns and conditions that could cause soil movement and have personnel available for patrolling, preventive maintenance activities, and emergency response actions such as shutdown, isolation, and containment in case of a pipeline failure.

Patrols should be designed to effectively evaluate right-of-way (ROW) conditions in areas susceptible to soil movement that might jeopardize the operation of the pipeline. Personnel should evaluate areas of potential soil movement and, as appropriate, monitor soil subsidence and take preventive actions to ensure the safe operation of the pipeline.

Finally, pipeline personnel should evaluate the accessibility of pipeline facilities, such as valves, needed to isolate pipeline segments or sections of pipelines that might be jeopardized by soil subsidence.

6.3.0 Perform Emergency Shutdown

Events occurring on the pipeline that present immediate hazard to persons or property require an emergency shutdown. An event might be caused by human error, mechanical failure, electrical failure, or an accident on the pipeline. It might also be brought about by threats from weather or civil disobedience. Regardless of the reason, the line must be shut down quickly and safely. Each company has a standard operating procedure for such an action. (In case of a pump shutdown from low pressure, there may be some time to discuss the move with appropriate management if the line has enough flexibility to withstand the stress.)

However, once the need is confirmed, decisive action must be taken immediately.

> **NOTE**
>
> In emergency conditions such as a possible release, the pipeline operator has full responsibility and authority to make shutdown decisions and take appropriate action. These actions are governed by the company release policy above all other considerations.

In the case of a fire, explosion, or release, discussion time is obviously not an option. The pipeline must be isolated by operations personnel by safely shutting down the originating station and continuing downstream in a safe manner until all pumps are offline. Operations personnel will then close all block valves to complete the shutdown. The responsible operations and maintenance personnel should be notified.

Threats from natural disaster or civil disobedience may come without warning. If an individual witnesses these kinds of threats to the pipeline, the supervisor or control center should be notified immediately. Such affected facilities should be isolated. Once facilities are isolated, pipeline shutdown may follow.

6.4.0 Evacuate the Facility

Safety of personnel and the public is the first priority in any event. Regardless of the emergency, if there is a threat of injury or death, all affected persons must be evacuated. The operator must evaluate every emergency to determine possible danger to the people involved. Events that could result in the need for an evacuation are fire, explosion, release of hydrocarbon or other hazardous vapors, seismic activity, and other natural phenomena.

The responsibility to maintain the safety of personnel and the public cannot be taken lightly. If the determination is made that a hazardous atmosphere or condition exists, the command to evacuate must be given. In the case of a release of product, the concentration of vapors downwind should be tested repeatedly to determine if there is a threat of explosion. If a hazardous atmosphere is traveling toward a populated area, the area must be warned and possibly evacuated.

6.4.1 Company Procedures

Companies have standard procedures in place to guide evacuations. By law, Emergency Response Plans must be in place and evaluated annually. Facility layouts with evacuation maps are required to be posted in every area or department. Safe areas are designated for all personnel. Once the command to evacuate is given, designated persons are required to call a roll of each employee name and check off who is present in the safe area. Every person must be accounted for.

6.4.2 General Public

In case of a release of product, the atmosphere downwind must be tested repeatedly. A vapor cloud traveling toward a populated area requires that the people be warned and possibly evacuated to a safe area. That area must be determined by the direction in which the hazardous vapors are traveling. Law enforcement and fire and rescue personnel must be solicited to aid in directing people away in a route that will give them the most protection.

In a major release, company records must be consulted so that potentially affected isolated people can be contacted in person. The path of the vapors or cloud should be plotted continuously so that if there are people, animals, or roads that the vapors will cross, the evacuation can be expanded. All roads leading to the area must be blocked until the danger has passed. All means should be employed to keep the product from reaching sewers, underground caverns, or waterways.

7.0.0 ◆ RECOGNIZE AND REACT TO PIPELINE DAMAGE

Pipeline damage occasionally occurs. Damaged and inoperable components, such as valves, pipe, or pumps, should be reported. One case may be a pipeline being hit or struck by excavation equipment. In this case, personnel may see gouges, scrapes, dents, or even buckling, but some signs are less obvious. A sheen on the water that covers submerged pipelines could be an indication of valve damage. A frost ball or vapor cloud that is discovered along the right-of-way may indicate a pipeline release of **highly volatile liquid (HVL)**. Pipelines in some areas have been shot by hunters accidentally or by other people intentionally. Vandalism is also a potential cause of damage.

7.1.0 Reaction

The first reaction should be to determine the extent of the damage. Some of these incidents are serious enough to require an emergency shutdown. A shutdown would more likely be involved if the pipeline were struck or damaged in such a way as to involve a major release and/or explosion. If the pipeline were struck and there were no release, then the abnormal operating condition would indicate a pipeline anomaly.

The least desirable way to detect a release is by receiving notification from a third party, yet this is the most important call an operator can receive. Conditions of damaged pipe are frequently brought to the company's attention through a telephone contact from outside the company. Frequently, the call comes from an aerial patrol of the pipeline ROW or an individual landowner who contacts the One-Call notification system to report a release, damage, or other such problem.

Personnel should always anticipate the proper response to potential emergency situations, make the appropriate notifications per company policy, and complete the required documentation.

Examples of times that a pipeline would have to be shut down due to damage are if an excavator operating a bulldozer gouges the pipeline causing a release, or if a field operator loses control of a vehicle and dents a large section of the pipeline or runs into a valve. In these situations, control center operations should be notified immediately. In some cases when a known release is involved, the control center operator may be able to drain a portion of the pipeline into a delivery system to lessen the amount of product that could be released from the pipeline.

If qualified and trained to do so, personnel should perform an initial site assessment, determine the location of the release, isolate it, and contact 911 emergency personnel if necessary. Regardless, control center operations should dispatch qualified personnel to assess the conditions. If the investigation reveals an immediate hazard, the line must be shut down in a safe manner as rapidly as possible. On multiple pump station systems, the pump units must be shut down in the order of upstream units first and then downstream units.

7.2.0 Abnormal Movement or Loading of an In-Service Pipeline

Great care must be taken during the process of moving in-service pipelines. Additional stress could be placed on the pipeline and could lead to pipe failure. When in-service pipe is lowered, the ambient temperature is also a critical factor. As the weather changes, the pipe may expand or contract and expand.

8.0.0 ◆ RECOGNIZE AND REACT TO UNEXPECTED/UNINTENDED RELEASE OR ENCOUNTER OF HAZARDOUS LIQUID

An operator may suddenly observe product being released or sprayed from a manifold, valve, or ruptured piece of equipment. A pump could start accidentally during a repair process, resulting in an increase in pressure in the pipeline. An isolation device might fail. The pipe may be of insufficient wall thickness (corrosion). There could also be an occasion when product vapor is discovered to be exceedingly strong in an unexpected area or a pool of product is observed during ROW inspections.

Such discoveries may signal the presence of a release. A major release in the field, terminals, or tank farms demonstrates many of the same characteristics. Decisive action must be taken to control the event as quickly as possible. If it is not controlled quickly, explosion, fire, and injury is imminent, as well as pollution of the environment and loss of product.

8.1.0 Reaction

Reactions to product release are determined by the type of facility or location involved. If the hazard is discovered in a building or enclosed space, it must be handled somewhat differently than if it is discovered outside.

8.1.1 Enclosed Areas

If the hazard is in a building or enclosure, personnel must get to a safe location immediately. Manually operated doors and windows should be opened to allow vapors to dissipate. The source of the product release must be determined quickly.

The nearest cutoff valve should be closed. If a major line or equipment is involved and can be operated from the control center, it is better for personnel to move to a safe location and call the control center operator to have it closed remotely.

No electrical equipment should be operated in the area. If power cannot safely be cut off at the switchgear, operations personnel must call the electrical supplier and have power cut off from outside the facility. Personnel should only be allowed to enter the area using appropriate personal protective equipment (PPE).

NOTE

To identify the hazards associated with the chemical being released, consult the material safety data sheets (MSDS).

8.1.2 Outdoor Areas

After first assuring the protection of the public, report the release to control center operations and/or appropriate personnel. The source of product being released must be discovered and cut off. Sources of ignition must be avoided and the power cut off. Personnel must remove themselves to a safe location. Vapor associated with the release must be allowed to dissipate before personnel attempt to enter the area. If an odor of gasoline has been detected even a few hundred feet from the source, control center operations must be contacted to shut down the line. If the leak is from equipment at a remote station, drivers should approach the equipment from the upwind side of the station.

WARNING!

If you are witnessing a release and there is a downed person in the area of the release, notify the dispatch control center to call 911 and clear the area.

NOTE

Process Safety Management (PSM) stipulates that employers must ensure that contract employees are trained in the work practices necessary to perform their job safely.

Personnel should only be allowed to enter the area using appropriate PPE. Repairs must be carried out with the permits and documentation as required. Product on the ground will require vacuum equipment and/or other means of cleanup.

8.1.3 Response

Responses to both enclosed and outdoor releases would include some or all of the following:

- Open all windows and doors.
- Find the source of the release.
- Use appropriate PPE.
- Avoid opening switchgear doors.
- Operate no electrical switches or equipment including lights.
- Exit any area where meters or smells indicate vapors are increasing.
- Have the power company cut power to shut down any pumps running if required before the problem is solved. (Go to an outside telephone to call the power company.)

8.2.0 Notifications

Appropriate personnel must be notified if conditions advance to dangerous levels. In case of a pipeline shutdown or other major action, appropriate personnel must be notified as soon as possible. If it is determined that product has escaped or if there has been an injury or major property damage, government agencies must receive appropriate reports.

An accident report is required for each failure in a pipeline system in which there is a release of hazardous liquid or carbon dioxide resulting in any of the following (see *CFR 195.50*):

- Any emergency that includes fire, explosion, or injury
- Releases of five gallons or more of HVLs
- All releases on water
- Releases that exceed five or more barrels of product
- Death of any person
- Bodily harm
- Estimated property damage exceeding $50,000

The accident report should be filed with the National Response Center (NRC) no later than 30 days after the discovery of the incident.

9.0.0 ◆ RECOGNIZE AND REACT TO UNEXPECTED CARBON DIOXIDE ENCOUNTER

Carbon dioxide is a colorless, odorless, gaseous compound used as a propellant in many domestic applications, such as fire extinguishing equipment, and is a by-product of combustion. Being heavier than air, it can collect in low-lying areas where it may displace oxygen; exposure can result in suffocation and death. A leak detector can detect a hazardous atmosphere, but without it, personnel can be overcome without realizing the problem. Carbon dioxide is often a by-product of incomplete burning and can be a deadly hazard if personnel are not aware of its possible existence. Some pipeline operations pump carbon dioxide commercially, and in such operations, the hazards are increased.

9.1.0 Reaction

In case of a fire, the operator must make certain that no one is allowed to enter the building or area after the fire is out unless the proper gas detection meter is carried along with appropriate PPE. The meter must be calibrated for detection of carbon dioxide immediately before entry. Of special concern until the atmosphere has been tested for carbon dioxide would be low-lying areas or lower rooms and compartments of buildings involved.

The operator must make certain that the first personnel to enter such questionable atmospheres are wearing self-contained breathing apparatus (SCBA). Cases have been noted in which co-workers have discovered associates lying in such areas unconscious. Rushing to their aid, the co-workers were also overcome by carbon dioxide.

In such emergencies, personnel who are trained and qualified in the use of SCBAs and are trained in rescue must enter first. Other workers must be trained to understand the danger and to stay a safe distance away until the deadly atmosphere has been cleared. In the case of a building or compartment, signs must be posted at the doors warning of the dangerous atmosphere. The operator must make certain that all such areas are plainly marked so that personnel will not enter.

10.0.0 ◆ INJURIES AS A RESULT OF UNEXPECTED/UNINTENDED RELEASE OR ENCOUNTER

It is the operator's duty to call for emergency rescue and medical assistance as soon as the incident is known. The operator must make certain that any injured or unconscious personnel are provided medical attention as quickly as possible.

Company procedure must be strictly followed in such incidents. In some cases, if there are physical injuries, the person must only be moved by trained and qualified personnel. In most companies, properly trained personnel are allowed to apply cardiopulmonary resuscitation (CPR) and first aid until an ambulance arrives.

10.1.0 Notifications

The supervisor must make certain that the incident and the injury are properly recorded and documented. In the case of personnel being injured, reports must be made to the Occupational Safety and Health Administration (OSHA) as required by the federal standard. The death of an employee requires notification of the NRC immediately and OSHA within a specified number of hours. A written accident report must be submitted to the NRC as soon as possible but no longer than 30 days from the time of the incident.

When the problem has been solved, the operator must complete the required documentation. Included in the list of documentation that must be completed are:

- Incident Investigation and Analysis Report
- Incident Report
- Event Notification forms
- Control center logs
- Station logs (if involved)
- Records of recommendations for corrective actions and training
- Documentation of such corrections and training after completion

11.0.0 ◆ RECOGNIZE AND REACT TO UNEXPLAINED PRESSURE DEVIATIONS

Unexplained pressure deviations can include an absence of pressure, a sudden increase in pressure, or a sudden decrease in pressure. Deviations in pressure indicate a serious condition on the pipeline. Occurrences such as unintended valve closures cause AOCs that will endanger the pipeline. The condition must be recognized, and reaction must be quick and decisive. If not, an emergency condition could result very quickly.

11.1.0 Reaction

If a change in pressure occurs that is different from the expected pressure, personnel should immediately investigate the change. If the cause is not determined, the operating lines should be shut down, referring to the abnormal condition shutdown procedure. Personnel should then proceed with emergency response notification.

Summary

An AOC is a condition that may indicate a malfunction of a component or deviation from normal operations, which may exceed design limits or result in a hazard to persons, property, or the environment. As required by law, companies have standard operating procedures in place to give personnel direction as to how to alleviate these conditions. Time is of the essence in such conditions. Operations personnel should be able to recognize and react to these conditions, using knowledge of policies and procedures as well as their own skill and good judgment.

In the event of a fire, explosion, or other disaster, operations personnel must be prepared to aid in emergency operations. Whether the AOC is the result of a valve closed inadvertently or a major release of product, personnel must react quickly to ensure the protection of the public and to preserve the safety and efficiency of the pipeline.

Review Questions

1. An example of an abnormal facility condition would be _____.
   a. a high-pressure shutdown
   b. unauthorized equipment on the ROW
   c. transformer failure
   d. a lightning strike

2. The first reaction once pipeline damage has been discovered is to _____.
   a. shut down the pipeline facility
   b. contact the control center
   c. evacuate the surrounding area
   d. determine the extent of the damage

3. Main-line pumps have high discharge pressure alarms that will shut down the pump when the discharge pressure _____.
   a. exceeds 100% of the MOP
   b. exceeds a preset limit
   c. is too low
   d. is at null

4. If 110% of maximum operating pressure is being exceeded, operations personnel must _____.
   a. shut down the pipeline
   b. continue to monitor the situation closely
   c. file an accident report form
   d. contact OSHA

5. _____ failures are somewhat interlinked with communications failures because the SCADA system may be involved.
   a. Power
   b. Equipment
   c. Control system
   d. Pump

6. In the case of a release, pipeline operations personnel should first _____.
   a. shut down the originating station, continuing until all pumps are offline
   b. engage the emergency response team
   c. close all block valves to shut down the system
   d. debate the situation with their co-workers

7. An uninterruptible power system _____.
   a. is a device that converts alternating current (AC) to direct current (DC)
   b. provides DC backup in case of power failure
   c. breaks down electric power from a high-voltage source to appropriate levels for distribution
   d. provides protection against primary AC power failure

8. An accident report is required for a failure in a pipeline system in which there is a release of product resulting in _____.
   a. releases of carbon dioxide
   b. releases that exceed five barrels of product
   c. estimated property damage exceeding $10,000
   d. releases of three gallons of HVLs or more

9. Carbon dioxide can collect in low-lying areas and displace oxygen because _____.
   a. it is a by-product of combustion
   b. it is heavier than air
   c. it is 1.5 times less dense than air
   d. it is produced by respiration

10. If an odor of _____ has been detected even a few hundred feet from the source, control center operations must be contacted to shut down the line.
    a. carbon dioxide
    b. propane
    c. gasoline
    d. carbon monoxide

GLOSSARY

Trade Terms Introduced in this Module

Abnormal operating conditions (AOCs): Conditions within a pipeline system that exceed operating design limits as specified in *49 CFR 195*.

Breakout tank: A DOT term used to identify a tank that relieves surges in a jurisdictional hazardous pipeline system or receives and stores hazardous liquid transported by a pipeline for subsequent reinjection and continued transportation by the jurisdictional pipeline. The breakout tank is a jurisdictional tank.

Highly volatile liquid (HVL): A hazardous liquid that will form a vapor cloud when released to the atmosphere and has a vapor pressure exceeding 40 pounds per square inch absolute at 100°F.

Lower explosive limit (LEL): The minimum concentration of vapor in the air below which propagation of a flame does not occur in the presence of an ignition source.

Maximum operating pressure (MOP): The maximum internal gauge pressure at which a hazardous liquid or carbon dioxide pipeline system or segment of such may be operated. The highest level of pressure allowed for a particular pipeline as set by the Department of Transportation to regulate the range of safe operations for that pipeline. Broadly speaking, it is set at 80 percent of the original design pressure, considering all components, or the highest pressure the pipeline was operated at during the last five years.

Relief device: Device for diverting a stream of product into a tank to take pressure off the pipeline during an overpressure condition.

Supervisory Control and Data Acquisition (SCADA): Computer system that is used to monitor and control the operation of a pipeline.

Surge pressure: Pressure produced by a change in velocity of the moving stream that results from shutting down a pump station or pumping unit, closure of a valve, or any other change in velocity of the moving stream.

The NCCER makes every effort to keep these textbooks up-to-date and free of technical errors. We appreciate your help in this process. If you have an idea for improving this textbook, or if you find an error, a typographical mistake, or an inaccuracy in NCCER's Contren™ textbooks, please write us, using this form or a photocopy. Be sure to include the exact module number, page number, a detailed description, and the correction, if applicable. Your input will be brought to the attention of the Technical Review Committee. Thank you for your assistance.

Instructors – If you found that additional materials were necessary in order to teach this module effectively, please let us know so that we may include them in the Equipment/Materials list in the Instructor's Guide.

Write: Curriculum Revision and Development Department
National Center for Construction Education and Research
P.O. Box 141104, Gainesville, FL 32614-1104

Fax: 352-334-0932

E-mail: curriculum@nccer.org

Craft ______________________ Module Name ______________________

Copyright Date ____________ Module Number ____________ Page Number(s) ____________

Description

____

____

____

____

(Optional) Correction

____

____

____

(Optional) Your Name and Address

____

____

____

Module 63103-02

Pipeline Mechanic Hand and Power Tools

Introduction to the Pipeline Operator Qualification Rule, API, and NCCER

It could be said that petroleum products keep America running. Petroleum products like oil and natural gas are the principal sources of energy in the United States and are used in many manufacturing and chemical processes. Petroleum products are primarily transported by pipeline. There are more than a million miles of liquid and natural gas pipeline in this country, ranging in diameter from two inches to four feet. These pipelines bring crude oil and natural gas from production fields to processing plants and refineries—and ultimately to the consumer.

Pipeline Operator Qualification

Petroleum products are considered hazardous, so pipeline personnel must take extra precautions and follow specific procedures when working with them. To help ensure the safe handling and transportation of petroleum products, the United States Department of Transportation (DOT) mandates safe practices as federal regulations.

One such regulation, the Operator Qualification (OQ) rule, requires pipeline operators to develop a written qualification plan to evaluate their personnel and their contract personnel (see *49 CFR Part 192 Subpart N—Gas*, and *49 CFR Part 195 Subpart G—Liquid*). The evaluation hinges on the operator's ability to perform certain covered tasks and to be able to recognize and respond to abnormal operating conditions that may be encountered while performing these covered tasks.

Covered Tasks

A pipeline performance task is deemed *covered* if the activity passes the following four-part test:

1. Is it performed on a pipeline facility?
2. Is it an operations or maintenance task?
3. Is it performed as a requirement of *49 CFR Part 195 (Liquid)* or *49 CFR Part 192 (Gas)*?
4. Does it affect the operation or integrity of the pipeline facility?

If the answer to all four questions is *yes*, then the activity is considered *covered* and must figure into the evaluation process of pipeline personnel.

American Petroleum Institute

The OQ rule and its related covered tasks provided the catalyst to develop a unified pipeline training and assessment program—so that pipeline operators and owners have ready access to the assessment and training tools they need. This effort was the direct result of the vision and leadership provided by The American Petroleum Institute (API) and API members.

Efforts to unify the industry are nothing new to API. Indeed, the impetus for forming the Institute in 1919 was the need to standardize engineering specifications for drilling and production equipment. Today, the most pressing issues revolve around public perceptions and government policies toward the industry. Speaking with one voice on these issues has become as essential as having interchangeable parts in the field. API's members determine the petroleum industry's positions on public policy as well as the standards that govern its day-to-day operations worldwide.

Consortium on Operator Qualification

API assembled a focus group whose primary purpose was to identify covered tasks and provide compliance recommendations to pipeline operators.

The Consortium on Operator Qualification (COOQ) became the guiding force for the identification and development of covered tasks for the liquid pipeline industry. Their work product, API publication 1161, *Guidance Document for the Qualification of Liquid Pipeline Personnel*, formed the basis for the resulting Pipeliner Training and Assessment Program.

The covered task list created in API publication 1161 has now evolved to include gas covered tasks and is maintained by NCCER. The most current covered task list can be downloaded at www.nccer.org.

National Center for Construction Education and Research

With the guidance and support of API, its members, and the Consortium on Operator Qualification, the National Center for Construction Education and Research (NCCER) was able to design and implement the Pipeliner Training and Assessment Program.

NCCER is a not-for-profit 501(c)(3) education foundation established in 1995 by the world's largest and most progressive construction and maintenance contractors and national contractor associations. NCCER was founded to address the

severe workforce shortage facing the construction and maintenance industries and to develop a standardized training process and curricula. Today, NCCER is supported by hundreds of leading construction and maintenance companies, manufacturers, and national associations and developed more than 35 construction and maintenance titles.

NCCER also maintains a National Registry, which contains the records of trainees and instructors who have received NCCER training. All completion information for every trainee is sent to the NCCER and filed within the Registry. The Registry can then confirm training and skills for workers as they move from company to company, state to state, or even within their own company.

Standardized Curriculum (NCCER's Contren™ Learning Series)

NCCER curricula are developed using the expertise of industry subject matter experts (SMEs). The use of SMEs as an integral part of every new development or revision ensures that the curriculum is current, industry-driven, and indicative of industry practices. The development and/or revision process is facilitated from start to finish by NCCER project managers, who work closely with the SMEs and a select team of technical writer(s) on the development and presentation of the curriculum.

Pipeline: Five Training Paths

The pipeline curriculum focuses on essential skills and competencies within five skill areas:

- Corrosion Control
- Electrical and Instrumentation
- Maintenance
- Mechanical
- Operations (Field and Control Center)

SMEs from the pipeline industry have determined the skills and competencies needed based on API's covered tasks and the consensus of the group/industry. Specific company information is generally not incorporated into the curriculum unless the SMEs determine that the information is the industry standard.

Training Modules and Covered Tasks

The smallest unit of instruction in the curriculum is a *module*. Each curriculum is composed of modules (the equivalent of stand-alone chapters) that are logically sequenced into complete *levels* of instruction. Every module is task-oriented and requires the successful completion of a written (knowledge-based) examination and a performance (hands-on) test. Modules may be individually selected and/or combined for efficient task training. Modules in the pipeline curricula, to the extent possible, will align with API-developed covered tasks. While not all modules will relate to covered tasks, all covered tasks will relate to at least one module.

Benefits of Training to the Pipeline Marketplace

Applying NCCER's standardized training process allows pipeline owners and contractors in the United States to establish the benchmark for acceptable skill and competency levels within the pipeline industry. This process will assist training organizations in developing and/or improving standardized pipeline training programs that effectively match their resources and operating conditions while responding to regulatory demands and to the needs of the industry. The process will also recognize training organizations that meet the minimum standards of competence, warranting regulatory, public, and professional confidence. This in turn will foster national unity within the pipeline industry for the mutual benefit of individual technicians, owners, contractors, and ultimately pipeline-users and society.

Additional Resources

To review the Pipeliner Training and Assessment Program Implementation and Resource Guide and the latest version of pipeline covered tasks, visit NCCER's Web site at www.nccer.org, or contact:

National Center for Construction Education and Research
PO Box 141104
Gainesville, FL 32614-1104
(888) 622-3720
Web site: www.nccer.org

For more information on API, please contact:

American Petroleum Institute
1220 L Street, Northwest
Washington, D.C. 20005
(202) 682-8000
Web site: www.api.org

To order *API Pub 1161*, contact Global Engineering Documents by calling 1-800-854-7149 or visiting their Web site at www.global.his.com.

Additional sources of information on the pipeline operator qualification regulation include:

DOT – Office of Pipeline Safety
www.ops.dot.gov

Transportation Safety Institute
www.tsi.dot.gov

COURSE MAP

This course map shows all of the modules in the first level of the Pipeline Mechanical curriculum. The suggested training order begins at the bottom and proceeds up. Skill levels increase as you advance on the course map. The local Training Program Sponsor may adjust the training order.

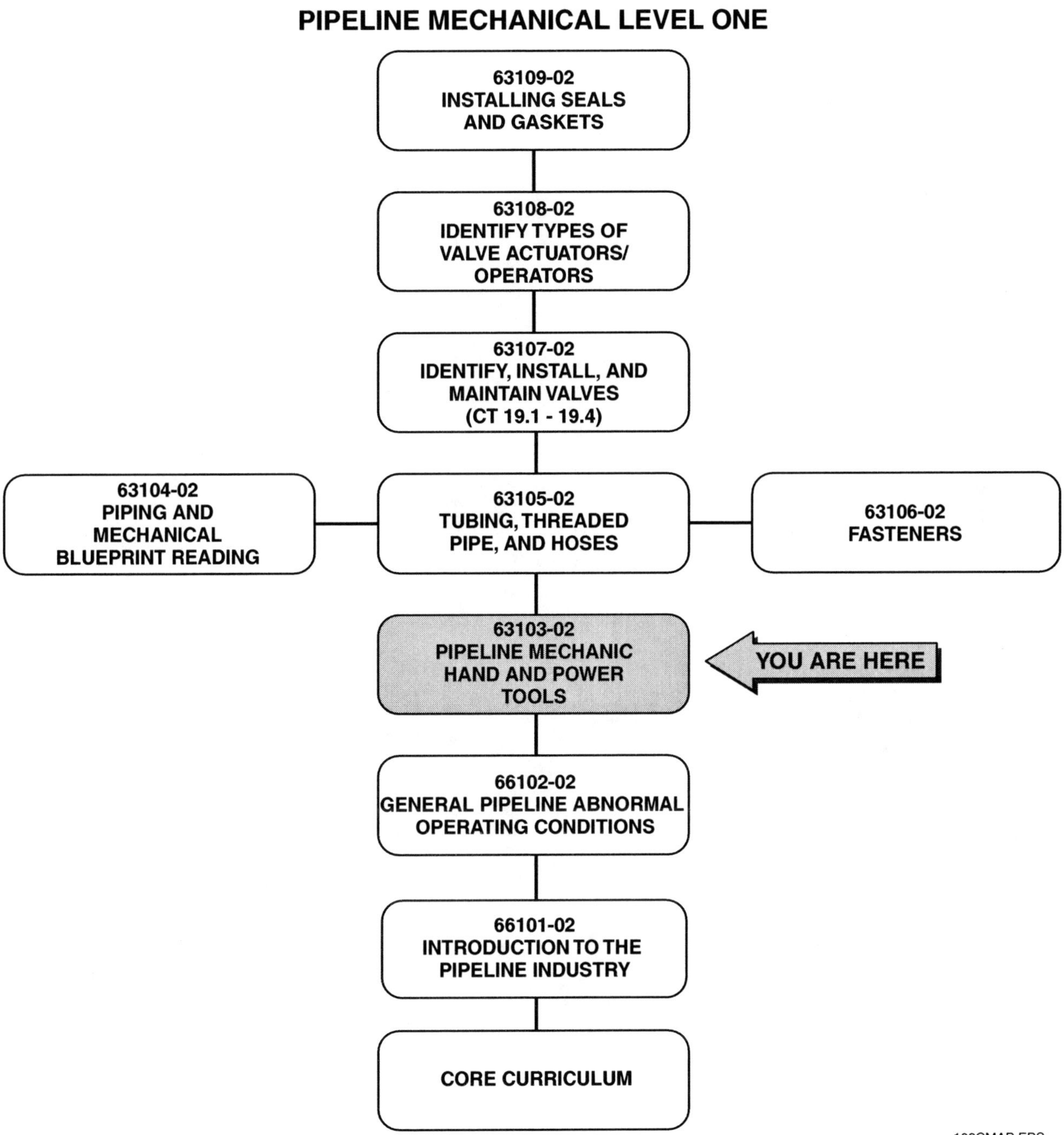

103CMAP.EPS

MODULE 63103-02 CONTENTS

Figures

Tables

MODULE 63103-02

Pipeline Mechanic Hand and Power Tools

Objectives

When you have completed this module, you will be able to do the following:

1. Safely use and care for pipeline mechanic hand tools.
2. Safely use and care for drill presses, hydraulic presses, and pipe threading machines.
3. Safely use and care for selected sheet metal tools.
4. Identify and explain surface grinders and belt sanders.
5. Identify and explain Woodruff key seaters and key broaches.
6. Safely use and care for bearing heaters and drills and perform precision drilling.

Prerequisites

Before you begin this module, it is recommended that you successfully complete the following: Core Curriculum; Pipeline Core, Modules 66101 and 66102.

Required Trainee Materials

1. Pencil and paper
2. Appropriate personal protective equipment

1.0.0 ◆ INTRODUCTION

Pipeline mechanics use hand tools for practically every job that they perform. There are hand tools that have multiple uses and those that are specially made for a specific purpose. As a pipeline mechanic, you must be able to recognize, select, inspect, maintain, and safely use hand tools to perform a job safely and efficiently.

The widespread use of power tools by pipeline mechanics has greatly increased individual production. Power tools are available for cutting, grinding, and shaping all types of materials. Pipeline mechanics must know how to safely operate these power tools for the specific jobs for which the tools are designed.

This module introduces the trainee to the different types of hand and power tools and procedures for selecting, using, caring for, and maintaining these tools.

2.0.0 ◆ TOOL SAFETY

Follow the safety guidelines outlined below when using hand and power tools.

2.1.0 Hand Tool Safety

Although hand tools do not present the immediate hazards of power tools, improper use of hand tools can cause serious personal injury. Safely using hand tools requires following some basic safety rules. Some hand tools are more hazardous than others and require additional personal protection and care when using them. For safe use of hand tools, follow the basic rules listed below:

- Always choose the proper tool for the job.
- Never use a tool for anything other than its intended purpose. For example, do not use a screwdriver as a prybar.
- Never use dull or broken tools. Dull tools require greater force to do the job. A sharp tool is a safe tool.
- Always keep your hands and fingers away from the sharp edges of cutting tools.

- Always work away from your body when using cutting tools.
- Always ensure that a tool is in good condition and that handles are tightly fastened to the tool.
- Always keep tools clean and free of rust.
- Always wear eye protection, such as safety glasses or goggles, when using chisels, punches, or other tools that may produce flying particles and debris.
- Always wear gloves when using tools that require them.
- Always use tools in a safe and proper manner. An improperly held wrench can slip, causing injury to the hand or knuckles.
- When working on gas pipeline components, always use brass nonsparking tools.

2.2.0 Power Tool Safety

If not used properly, power tools can cause severe injuries and even death. The following safety precautions should be used when using power tools.

Before use:

- Always keep the work area clean. Many accidents are caused by materials or tools carelessly spread around the work area.
- Always wear proper clothing. Do not wear loose-fitting clothing or jewelry when working with or around power tools. Keep long hair tied back behind the shoulders. There is always the danger of loose clothing and hair getting caught the moving parts of the tool.
- Wear safety glasses when using power tools. Most power tools cause material to be thrown, which could severely damage the eyes.
- Look for the job safety analysis (JSA) sticker on the tool. The JSA explains the basic safe operating procedures that must be followed when operating the tool.
- Inspect power tools before each use. Make sure that the tool and power cord are not damaged and that all the safety guards are in place.
- Always select explosion-proof power tools when working on gas pipeline components.
- Avoid accidental starts. Make sure that the power switch is in the OFF position before plugging in the power cord.
- Inspect power cords to ensure that they are in good condition before use.

During use:

- Use tools only for their intended purposes, and use the proper tool for each job. Know the capabilities of the tools, and work within those limits. If a tool is not suited for a particular job, select a tool that can perform the job.
- Never force a tool beyond its capabilities. Forcing a tool to work beyond the limits of its design not only wears out the tool prematurely, but also places you in an unsafe situation.
- Do not alter or change a power tool in any way.
- Always maintain proper footing and balance when using a power tool. Do not overreach. Any fall can be dangerous, but a fall while using a power tool can be much worse.
- Never carry a power tool by the power cord or an air tool by the air line.
- Keep power cords out of situations in which they can be frayed, cut, or damaged. Always disconnect a power cord by grasping the plug and pulling. Never yank on a cord to disconnect it.
- Do not use electrical power tools in wet or damp locations. There is always a danger of electrical shock, even if the tool is properly grounded.

After use:

- Disconnect the power source before performing maintenance on the tool or changing accessories.
- Respect your tools and always follow the recommended operations and maintenance procedures. A well-maintained tool is a safe tool.
- Allow only qualified technicians to repair power tools.
- Store tools when they are not being used. Allowing power tools to lie around the work area when they are not in use increases the possibility that the tools may be damaged and become unsafe for use.

3.0.0 ◆ USING AND CARING FOR PIPELINE MECHANIC HAND TOOLS

Hand tools are necessary for pipeline mechanics to perform their work. Good hand tools can be very expensive to buy. If you must provide your own hand tools, it can mean a substantial cost out of your pocket. If your employer furnishes your hand tools, this expense adds to the cost of doing business and affects the company's ability to do work in a competitive manner. You can see that properly using and caring for these tools is important. This section explains the use and care of the following tools:

- Pipe wrenches
- Strap/chain wrenches
- Spanner wrenches
- Taper plug gauges
- Allen wrenches

- Pipe and tubing cutters
- Honing stones
- Putty knives/scrapers
- Drift pins
- Diagonal cutters
- Tin snips
- Taps and **dies**
- Thread gauges
- Scribers
- Tension meters
- Sheave gauges
- Cylinder hones
- Gear pullers
- Packing pullers
- Reamers
- Inspection mirrors
- Retaining ring pliers
- Plasti-gauges
- Taper gauges
- Extractors

3.1.0 Pipe Wrenches

Pipe wrenches come in many shapes and sizes to satisfy different needs. The most common is the straight pipe wrench. Another type is the offset pipe wrench, which is used where the amount of swing is limited.

The jaws of the pipe wrench have teeth that grip pipe or other round objects. The lower jaw is fixed in the handle, while the upper jaw floats in the wrench housing. The opening between the jaws is adjusted by turning the knurled nut on the shank end of the upper jaw. *Figure 1* shows a straight pipe wrench and an offset pipe wrench.

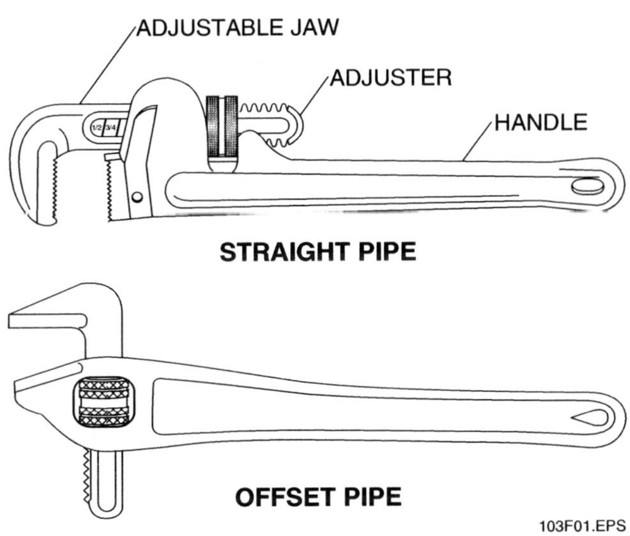

Figure 1 ◆ Straight pipe wrench and offset pipe wrench.

Follow these steps to select, inspect, use, and maintain pipe wrenches:

Step 1 Select a pipe wrench with a jaw opening wide enough to fit the pipe being turned and a handle long enough to provide the amount of pressure necessary to loosen or tighten the pipe.

> **NOTE**
> Although the jaw opening is adjustable to some extent, there are pipe wrenches for different sizes of pipe. Choose a wrench that will fit the pipe without adjusting it all the way out.

Step 2 Inspect the selected wrench to ensure that the teeth are in good condition, the threads on the upper jaw shank are not damaged or excessively worn, and the housing is not cracked or otherwise damaged.

Step 3 Oil the threads at the adjusting nut lightly.

Step 4 Adjust the jaw opening to fit the pipe.

> **NOTE**
> The jaws should be adjusted to fit snugly on the pipe. They will tighten when pressure is applied.

Step 5 Position the wrench on the pipe.

> **NOTE**
> The wrench should be positioned so that the pipe is in the center of the jaws. The pipe should not be allowed to contact the jaw shank, as this will prevent the wrench from getting a tight grip on the pipe. *Figure 2* shows positioning the wrench on the pipe.

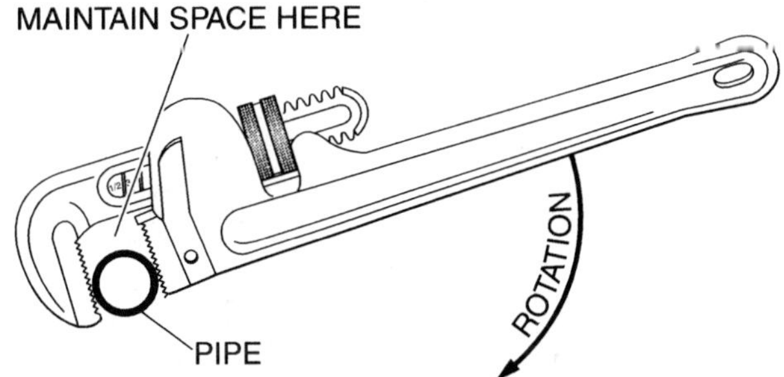

Figure 2 ◆ Positioning the wrench on the pipe.

Step 6 Grip the wrench handle near the end, and slowly apply pressure until the jaws get a firm grip on the pipe.

> **NOTE**
> In some cases it may be necessary to use another pipe wrench to hold the fitting or equipment that the pipe is screwed into, to keep it from turning. In this case, you would turn the second wrench backward from the first wrench to hold the work.

Step 7 Continue to apply pressure until the pipe turns.

> **WARNING!**
> If the pipe cannot be turned with the selected wrench, exchange it for a larger wrench. Never use a cheater bar or any other extension on the wrench handle, because this could break the wrench and cause personal injury.

Step 8 Rotate the wrench as far as clearances will permit or as far as is convenient.

Step 9 Release pressure, return the wrench to the starting position, establish a new grip, and turn the pipe again. Repeat this process until the pipe has been given as many turns as needed.

Step 10 Clean the wrench thoroughly.

Step 11 Oil the wrench lightly to prevent rust.

Step 12 Store the wrench in its proper place.

3.2.0 Strap/Chain Wrenches

Strap wrenches and chain wrenches are variations of pipe wrenches and are used to turn and hold pipe and fittings (see *Figure 3*). They may also be used in situations where a pipe wrench will not work. For instance, when working on a pipe that is close to a wall, there need only be enough room between the pipe and the wall to pass the chain or strap through in order to use a chain or strap wrench. Chain wrenches come in various sizes to handle pipe up to 18 inches in **diameter**. Strap wrench sizes go up to 12 inches. Strap wrenches are often used on pipe or round material that is too hard for pipe or chain wrenches to get a grip on as well as finished surfaces that pipe wrenches or chain wrenches would mar.

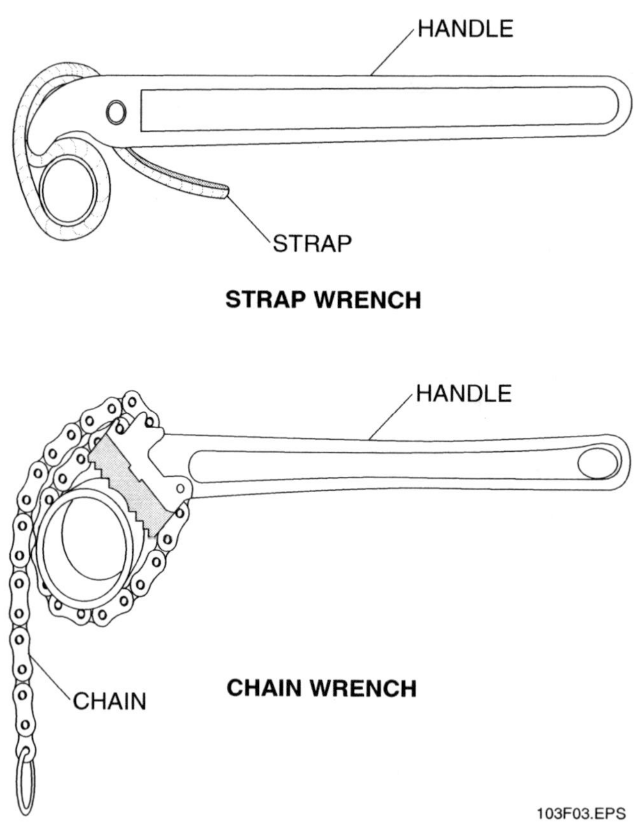

Figure 3 ◆ Chain wrench and strap wrench.

Chain wrenches and strap wrenches are used in much the same manner. Follow these steps to select, inspect, use, and maintain chain wrenches:

Step 1 Select a chain wrench of adequate size to fit around the pipe.

Step 2 Inspect the wrench to ensure that the chain is not broken or excessively worn and that the handle is not cracked or broken.

Step 3 Position the wrench on the pipe, wrap the chain around the pipe, and hook the chain into the hooks on the handle.

> **NOTE**
> The chain will fit snugly, but not tightly, around the pipe. The chain will tighten on the pipe when pressure is applied.

Step 4 Apply pressure on the wrench handle in the direction the pipe is intended to turn until the chain tightens and the wrench grips the pipe.

Step 5 Rotate the wrench as far as clearances will permit or as far as is convenient.

Step 6 Release pressure, return the wrench to the starting position, establish a new grip, and turn the pipe again. Repeat this process until the pipe has been given as many turns as needed.

Step 7 Clean the wrench thoroughly.

Step 8 Oil the wrench lightly to prevent rust.

> **CAUTION**
> Do not oil the strap of a strap wrench. This will make the strap slick and defeat its gripping ability.

Step 9 Store the wrench in its proper place.

3.3.0 Spanner Wrenches

Spanner wrenches are specially made wrenches that have holes, projections, or hooks on one or both ends of the head. They are usually either furnished with the tool or equipment on which they are used on or are offered as service items for the equipment. Refer to a parts list or service bulletin to identify a spanner for a particular application. *Figure 4* shows some common spanner wrenches.

Spanner wrenches are relatively easy to use, and all spanners are used in a similar manner. Follow these steps to select, inspect, use, and maintain a spanner wrench:

Step 1 Select a spanner wrench that is suitable for the job to be done. It must be the correct size or be adjustable.

Step 2 Inspect the wrench to ensure that the pins are not bent, excessively worn, or otherwise damaged and that the handle is not bent or damaged.

Step 3 Clean the holes of the tool or equipment part being worked on thoroughly to remove dirt, grease, or other obstructions that would prevent the wrench from fitting into the holes.

Step 4 Insert the pins of the wrench into the holes provided in the tool or equipment part being worked on.

> **NOTE**
> Be sure that the pins fit snugly into the holes and that the holes are not disfigured beyond use.

Step 5 Hold the wrench in the holes with one hand, and apply pressure to the wrench with the other.

> **CAUTION**
> Be sure that you are turning the wrench in the proper direction. Although most nuts are loosened by turning them in a counterclockwise direction, some nuts have left-hand threads and are loosened by turning them clockwise.

Step 6 Turn the wrench as far as clearances will allow or as far as is convenient.

Step 7 Remove the wrench from the holes, return to the starting position, and reinsert the wrench. Continue turning the nut until it has been rotated as many turns as needed.

Step 8 Clean the wrench thoroughly, using a rag and solvent if needed.

Step 9 Oil the wrench lightly to prevent rust.

Step 10 Store the wrench in its proper place.

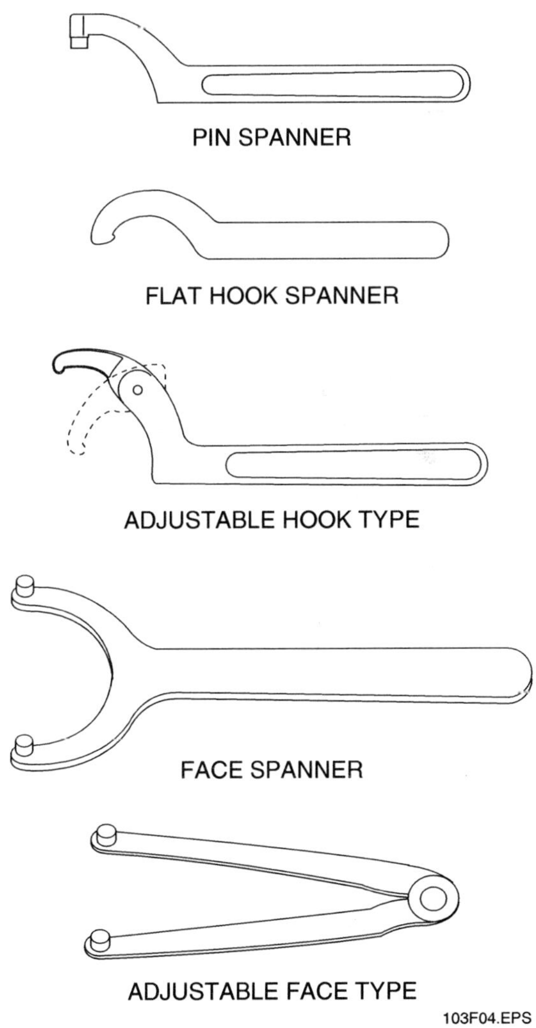

Figure 4 ◆ Common spanner wrenches.

3.4.0 Taper Plug Gauges

A taper plug gauge consists of a solid shaft-shaped bar with a tapered section that is used to measure the inside diameters of tapered holes (see *Figure 5*). Taper plug gauges come in various sizes to measure different size holes. There is a gauge line on the tapered part of the gauge. To use a taper gauge, the gauge is inserted in a tapered hole and read at the gauge line. The hole is the right diameter if the gauge fits exactly up to the gauge line. If the gauge does not go all the way up to the gauge line, the hole is too small. If the gauge goes in past the gauge line, the hole is too large. Plug gauges are precision instruments and are very fragile. When not in use, the gauge should be wrapped in an oily cloth to prevent rust and stored in a place where other tools and materials will not be thrown or dropped on them. Once a gauge is damaged, it is of no further use.

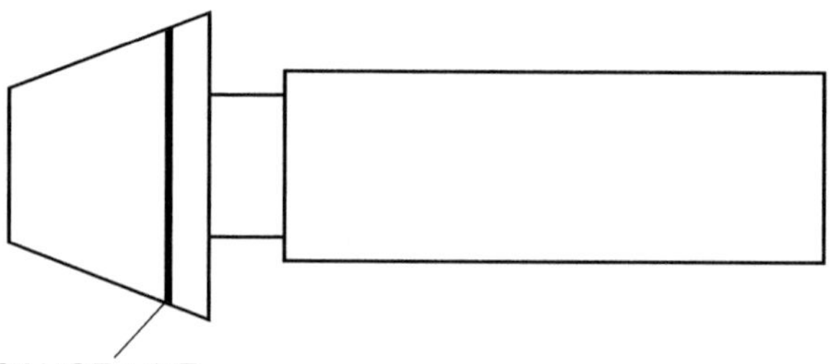

Figure 5 ◆ Taper plug gauge.

3.5.0 Allen Wrenches

The Allen wrench, or hex wrench, (see *Figure 6*) is an L-shaped, six-sided wrench made from tool steel. This wrench is used to install and remove recessed head Allen screws. Allen wrenches come in various sizes, up to ¾ inch.

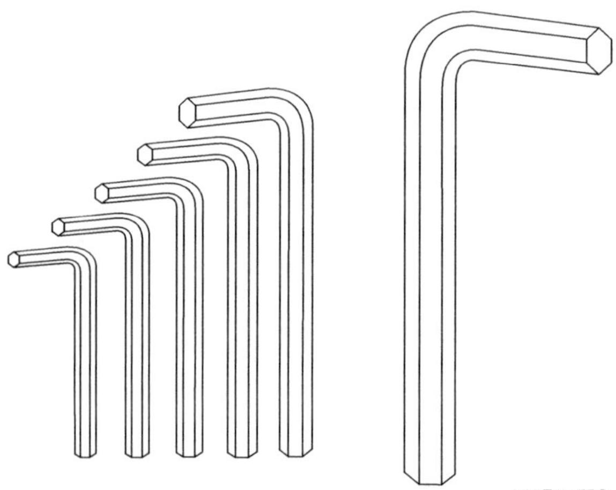

Figure 6 ◆ Set of Allen wrenches.

Follow these steps to select, inspect, use, and maintain Allen wrenches:

Step 1 Select an Allen wrench of the exact size of the screw to be turned.

> **CAUTION**
> The Allen wrench must be the correct size to prevent rounding or spreading the screw or damaging the wrench.

> **NOTE**
> Usually the wrench size will be one and one-half times smaller than the screw size.

Step 2 Inspect the wrench to ensure that the corners and end are not rounded and that the wrench is not bent or twisted.

Step 3 Clean the hole of the screw to remove any dirt, grease, or other material that would prevent the screw from seating in the bottom of the hole.

Step 4 Insert the wrench into the screw and ensure that it fits all the way in.

Step 5 Hold the wrench in the hole with one hand, and apply pressure with the other.

> **CAUTION**
> Be sure that you are turning the wrench in the proper direction. Although most screws are loosened by turning them in a counterclockwise direction, some screws have left-hand threads and are loosened by turning them clockwise.

Step 6 Turn the wrench as far as clearances will allow or as far as is convenient.

Step 7 Remove the wrench from the hole, return to the starting position, and reinsert the wrench. Continue turning the screw until it has been rotated as many turns as needed.

Step 8 Clean the wrench thoroughly, using a rag and solvent if needed.

Step 9 Oil the wrench lightly to prevent rust.

Step 10 Store the wrench in its proper place.

3.6.0 Pipe and Tubing Cutters

Pipe cutters are used to cut steel, brass, copper, and iron pipe. A pipe cutter has a special alloy-steel cutting wheel and two pressure rollers. The cutting wheel is mounted so that it is stationary in the cutter frame. The pressure rollers are mounted on a sliding frame and are adjusted and tightened by turning the handle. To cut a pipe, the cutter is placed on the pipe, pressure is applied to the pressure rollers, and the cutter is rotated around the pipe until it is cut.

Tubing cutters closely resemble pipe cutters and are used in the same way. Tubing cutters are used to cut iron, brass, steel, copper, and aluminum tubing with thinner walls. There are various sizes of pipe and tubing cutters for different sizes of pipe and tubing (*Figure 7*).

Pipe cutters and tubing cutters are used in essentially the same way. Follow these steps to select, inspect, use, and maintain pipe and tubing cutters:

Step 1 Select the proper type and size cutter for the pipe to be cut.

> **NOTE**
>
> The cutter must be of the proper size to fit over the pipe and heavy enough to cut the wall thickness of the pipe being cut. Pipe with heavy walls cannot be cut with this type of cutter.

Step 2 Inspect the cutter wheel to ensure that it is not nicked, bent, or dulled.

Step 3 Inspect the pressure rollers to ensure that they are not damaged and that there is no buildup of foreign materials on the rollers.

Step 4 Inspect the cutter frame to ensure that it is not bent, cracked, or otherwise damaged.

Step 5 Place the cutter over the pipe at the point where it is to be cut, and tighten the pressure rollers until there is light pressure on the pipe.

Step 6 Grip the cutter by the handle, and rotate it around the pipe one complete turn.

Step 7 Tighten the pressure rollers a small amount, about one to one and one-half turns of the handle, and rotate the cutter another turn around the pipe.

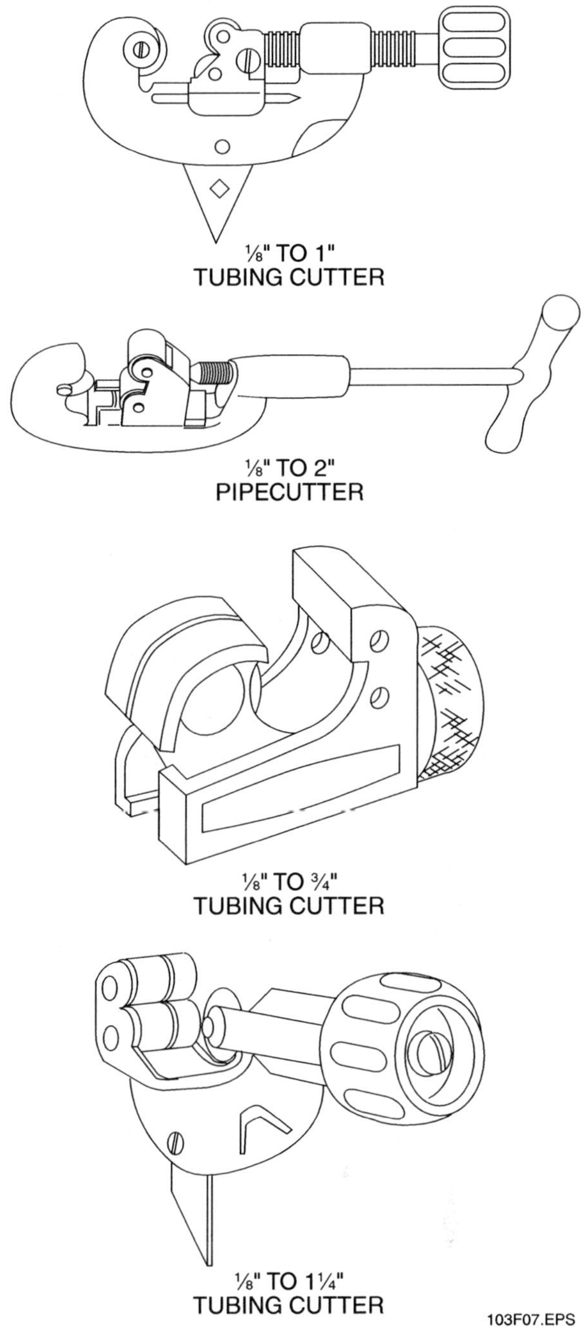

Figure 7 ◆ Different sizes and types of pipe and tubing cutters.

> **CAUTION**
>
> Do not overtighten the cutter. The cutting wheel could be damaged or the cutter broken. Be sure that the cutter follows in the same groove cut by the first turn on the second and all consecutive turns of the cutter. If the cutter is moving out of the groove, you may be applying too much pressure, or the cutter may be defective.

Step 8 Continue this procedure until the cut is finished.

Step 9 Clean the cutter thoroughly to remove any dirt, grease, or buildup.

Step 10 Oil the cutter lightly to prevent rust.

Step 11 Store the cutter in its proper place.

3.7.0 Honing Stones

Honing stones belong to one of two groups: natural or artificial. Artificial stones are usually made of silicone, aluminum oxide, or carbide. Some natural stones are treated with oil during manufacture and are often called oilstones. Natural stones have very fine grains and are good for putting sharp edges on knives and fine cutting tools.

Honing stones are available in various grains: coarse, medium, fine, and very fine. It is important to use the proper stone for any honing operation. Use a coarse stone to sharpen large, dull, or nicked tools. Use a medium stone to sharpen tools that do not require a finished edge. Use a fine stone to sharpen a knife or tools requiring a very sharp, razor-like edge. Some stones will have one type of grain on one side and another type of grain on the other side.

Honing stones are fragile and must be stored in boxes or kept in special racks to protect them from breakage. Never leave a stone on a bench or table where it may be knocked off or where heavy objects may fall on it. Do not store honing stones in a hot place.

Honing stones come in many shapes and sizes. *Figure 8* shows several types of honing stones.

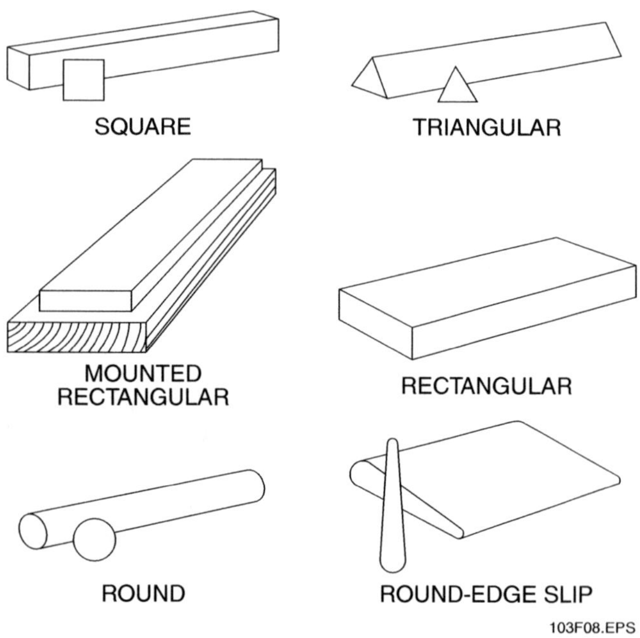

Figure 8 ◆ Honing stones.

Honing stones are used to sharpen a variety of tools. This procedure explains sharpening a wood chisel. Follow these steps to select, inspect, use, and maintain a honing stone, using a chisel as an example tool to be sharpened:

Step 1 Select a honing stone with a coarse grain on one side and a fine grain on the other.

Step 2 Inspect the stone to ensure that it is not cracked or broken.

Step 3 Coat the stone with a thin layer of machine oil to prevent glazing.

NOTE

Glazing is a condition that occurs when steel that is ground off a tool clogs the stone.

Step 4 Make sure that the stone is firmly held in its holder and cannot move.

Step 5 Hold the chisel in one hand with the bevel flat against the coarse side of the stone.

Step 6 Steady the chisel and hold it down against the stone with the fingers of the other hand.

Step 7 Work the chisel back and forth parallel to the stone, using smooth, even strokes.

CAUTION

Work the chisel over the entire surface of the stone to prevent wearing a hollow in the center of the stone. Do not rock the blade. Keep the angle of the blade constant during sharpening.

Step 8 Turn the chisel over, placing the flat side against the stone, and make a few strokes on the fine grit side of the stone to remove any feather edge from the chisel.

CAUTION

Be sure to hold the chisel perfectly flat when whetting the flat side of the blade. Raising the chisel even slightly will produce a bevel on the flat side of the blade. This requires grinding the blade to remove the bevel and then sharpening the blade again.

Step 9 Turn the chisel over and whet the bevel side a few strokes to remove any remaining feather edge or **burr**.

Step 10 Clean and oil the chisel (or other tool you are sharpening), and put it in its proper place.

Step 11 Clean the stone with a rag, using ammonia or dry cleaning solvent if needed, to remove any glazing.

Step 12 Cover the stone with a light coat of machine oil.

Step 13 Put the stone in its protective box, and store it in its proper place.

3.8.0 Putty Knives/Scrapers

Putty knives are used to scrape off gasket, glue, sealer, and other substances when cleaning pipe flanges and other machinery and equipment (see *Figure 9*). Putty knives have a wide, square blade and are available in many different lengths and widths.

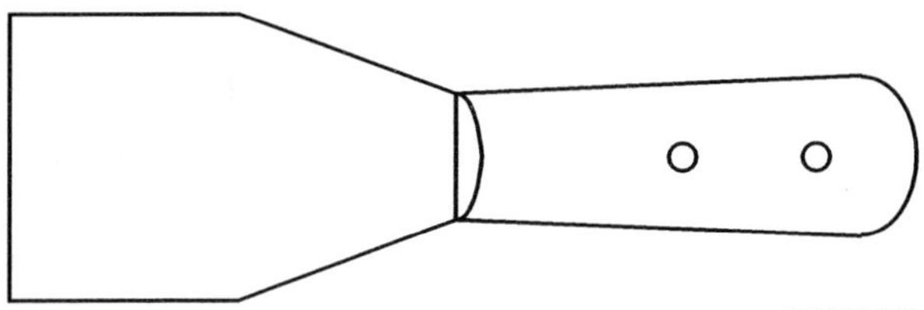

Figure 9 ◆ Common putty knife.

This procedure explains scraping off a gasket with a putty knife. Follow these steps to select, inspect, use, and maintain a putty knife:

Step 1 Select a putty knife of the proper length and width to do the job.

> **WARNING!**
> Do not choose a knife that is larger than needed and cannot be handled safely.

Step 2 Inspect the putty knife to ensure that the blade is not bent or dull and that the handle is in good condition and firmly attached to the blade.

> **WARNING!**
> Do not try to chip the gasket with the putty knife. This could nick the surface being scraped and dull the knife. Keep the knife blade flat on the surface. Do not try to scrape with the side edge of the blade. Keep your hands away from the cutting edge of the blade, and work away from your body to prevent personal injury.

Step 3 Hold the knife in one hand at an angle to the work, and scrape the gasket off using smooth, even strokes.

Step 4 Continue to scrape the surface until all of the gasket is removed.

Step 5 Clean the knife thoroughly with a rag, using solvent if needed.

Step 6 Oil the knife lightly to prevent rust.

Step 7 Store the knife in its proper place.

3.9.0 Drift Pins

A drift pin is a round steel piece that is tapered on one end. A barrel pin is a drift pin that is tapered on both ends. Drift pins and barrel pins are used to **align** bolts holes when connecting flanges, structural steel, and other holes in mating surfaces. Drift pins and barrel pins come in various sizes to fit different size bolt holes. *Figure 10* shows a common drift pin and barrel pin.

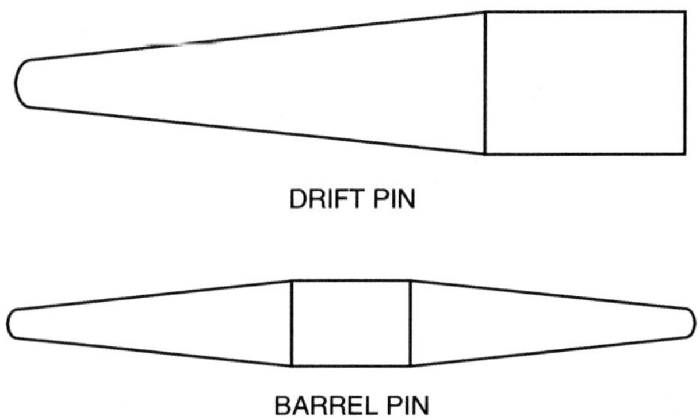

Figure 10 ◆ Common drift pin and barrel pin.

Follow these steps to select inspect, use, and maintain a drift pin:

Step 1 Select a pin of the proper size for the holes being aligned.

> **NOTE**
> The proper size drift or barrel pin will fit all the way through both holes being aligned and still be large enough and strong enough to pull the mating pieces together.

Step 2 Inspect the pin to ensure that it is not bent or have burrs that could cut your hands.

Step 3 Insert the pin into the holes of the mating pieces, and use it to pull the holes into alignment.

Step 4 Clamp the mating pieces together, using a C-clamp if possible, to hold the parts in alignment while the pin is removed and a bolt is installed in its place.

NOTE

On pieces with more than one hole, the drift pin can be left in until bolts can be installed in the other holes. Then the drift pin is removed, and a bolt is installed in its place.

Step 5 Clean the pin, using a rag and solvent if needed.

Step 6 Cover the pin with a light coat of oil to prevent rust.

Step 7 Store the pin in its proper place.

3.10.0 Diagonal Cutters

Diagonal cutters are small pliers used to cut wire, cotter pins, and other items made of light material that are inaccessible with other types of cutters (see *Figure 11*). Because the cutting edges are diagonally offset approximately 15 degrees, they are also used to cut small objects close to a surface.

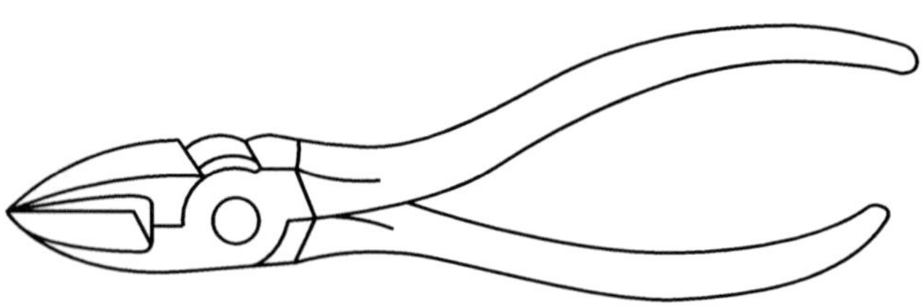
103F11.EPS

Figure 11 ◆ Diagonal cutter.

Follow these steps to select, inspect, use, and maintain a diagonal cutter:

Step 1 Select a diagonal cutter of the proper size for the material to be cut.

Step 2 Inspect the cutter to ensure that the cutting edges are not nicked or dull.

Step 3 Position the cutter to cut the material in the desired location.

Step 4 Squeeze the handles of the cutter to cut the material.

Step 5 Clean the cutter.

Step 6 Store the cutter in its proper place.

3.11.0 Tin Snips

Tin snips work like scissors. They are used to cut light-gauge sheet metal where the shape, position, or location prevents the use of power metal-cutting tools. Tin snips are made in a variety of configurations for specific jobs.

There are several types of snips, including those that cut straight and circular (*Figure 12*). Both are available in right-hand and left-hand styles.

Some of the most commonly used snips are aviation snips. The handles of aviation snips are color-coded to indicate whether they cut a curve to the right, a curve to left, or a straight line. Red-handled aviation snips cut a curve to the left. Green-handled aviation snips cut a curve to the right. Yellow-handled aviation snips cut a straight line.

Tin snips are used to cut many shapes from sheet metal. This procedure explains cutting a hole in a piece of sheet metal. Follow these steps to select, inspect, use, and maintain tin snips:

Step 1 Select a snip that is of adequate capacity to cut the sheet metal and of the style needed to make the cut (right cut or left cut).

Step 2 Inspect the snips to ensure that they are not bent or damaged, that the blades are not sprung, gaped, or dull, and that the handles are in good condition.

NOTE

The blades of tin snips can be easily damaged if they are used improperly. They should never be used to cut wire, rod, or material that is beyond their capacity. The blades must be sharp to produce a good cut.

Step 3 Ensure that the cut line on the sheet metal is distinct and easily seen.

Step 4 Drill or cut a hole, using a hammer and chisel, to make a starter hole in the center of the intended cut.

Step 5 Slip the blade of the snips into the starter hole and start the cut, using the snips like a pair of scissors and allowing the cut to curve as you go.

WARNING!

You should wear gloves when cutting sheet metal. Cutting the metal leaves a sharp edge and may also create burrs that can seriously cut your hands.

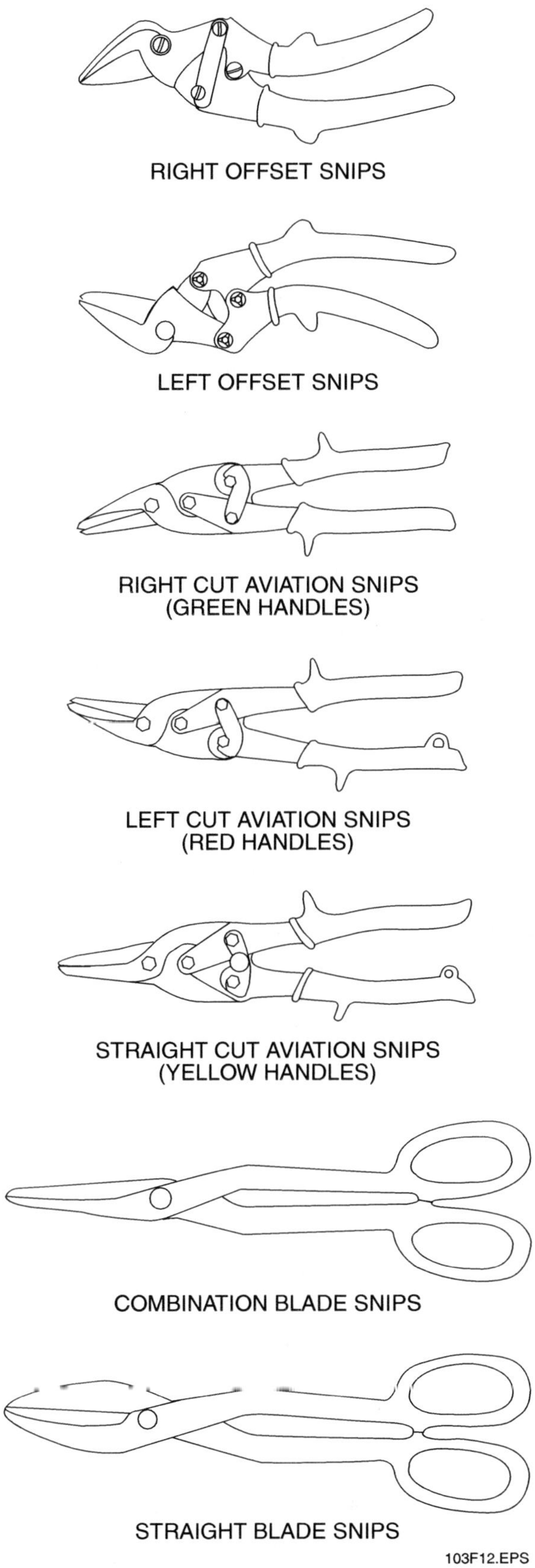

Figure 12 ◆ Common tin snips.

Step 6 Continue to cut in a circular motion from the starter hole until you reach the cut line (see *Figure 13*).

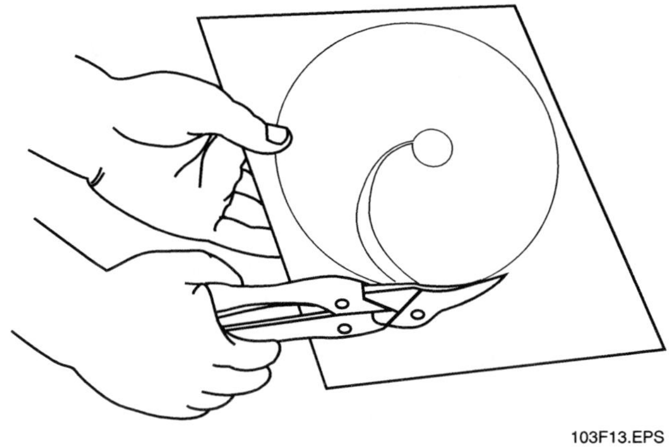

Figure 13 ◆ Starting a circle cut.

Step 7 Cut along the cut line around the circle until the cut is complete.

> **NOTE**
> You should not completely close the snips during the cut. Close the snips approximately halfway on each cut; slip the snips further into the cut, and cut halfway again. Closing the snips completely will cause burrs on the metal that must be filed off later.

Step 8 Clean the snips, using a rag.

Step 9 Coat the blades of the snips lightly with oil.

Step 10 Store the snips in their proper place.

3.12.0 Taps and Dies

Taps and dies are used to cut threads in materials such as metal, plastic, and hard rubber. Taps are used to cut internal threads in materials, and dies are used to cut external threads on bolts and rods.

Taps are used to thread holes that go all the way through a piece of material and holes that end inside the material (blind holes). The most common types of taps are (*Figure 14*):

- *Pipe taps* – Used where extremely tight fits are necessary, as with pipe fittings. The diameter of the tap from end to end of the threaded portion is tapered at the rate of ¾ inch per foot.

- *Taper taps* – Used when starting a tapping operation or when tapping through holes. They have a **chamfer** length of 8 to 10 threads.
- *Plug taps* – Designed to be used after the taper tap. They have a chamfer length of 3 to 5 threads.
- *Bottoming taps* – Always used after the taper and plug taps have already been used. They are used for threading the bottom of a blind hole and have a chamfer length of only 1 to 1½ threads.

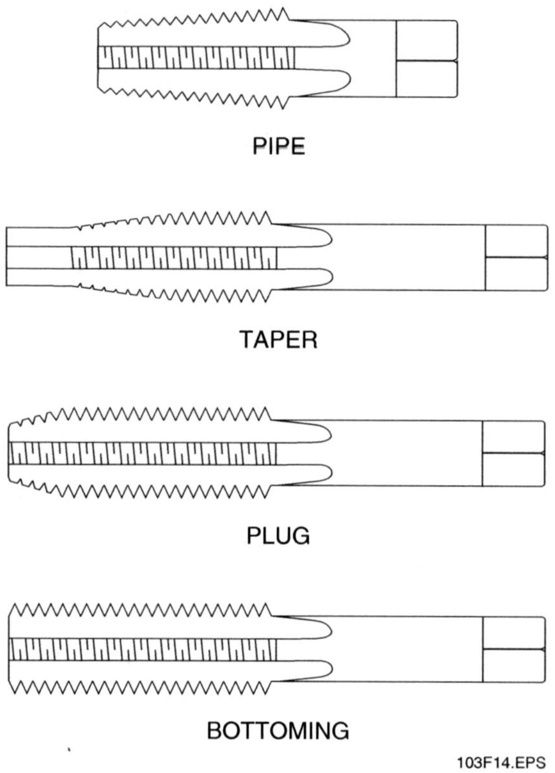

Figure 14 ◆ Types of taps.

Dies are made in a variety of shapes and sizes and are solid or adjustable. The following are common types of dies:

- *Pipe dies* – Will only cut American Standard Pipe Threads.
- *Rethreading dies (die nuts)* – Used for repairing damaged or rusty threads on bolts or screws. They are usually hexagon-shaped and can be turned using any type of wrench that will fit.
- *Screw adjusting dies* – Adjusted by a screw in the die.
- *Open adjusting dies* – Adjusted by three screws in the die holder.
- *Two-piece rectangular dies* – Adjusted by set screws and held in ordinary or ratchet die-stocks.

Taps and dies can be obtained in complete sets that include various taps and dies, as well as die-stocks, tap wrenches, guides, and the screwdrivers and wrenches necessary to loosen and tighten adjusting screws. *Figure 15* shows a common tap and die set.

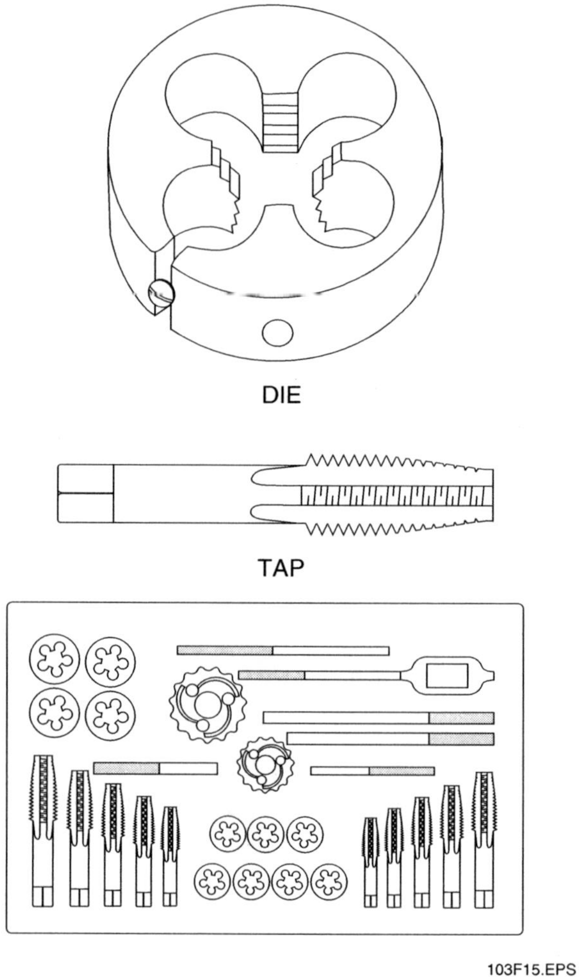

Figure 15 ◆ Common tap and die set.

This procedure explains threading a blind hole. Follow these steps to select, inspect, use, and maintain a tap:

Step 1 Select a tap of the proper size for the hole being tapped and the proper type for the tapping operation.

> **NOTE**
> All three types of dies are used when tapping a blind hole. A taper die is used to start the tapping operation.

Step 2 Inspect the tap to ensure that there are no broken or nicked teeth and that the tap is sharp and clean of any shavings or chips.

Step 3 Place the end of the tap in the hole, and turn it while applying light pressure on the tap wrench to start the tap.

Step 4 Check the tap using a square to ensure that it is square to the surface of the work.

> **CAUTION**
> The tap must be square to the surface of the work to ensure that the hole is tapped straight. Check the tap in several places using the square.

Step 5 Continue to turn the tap, applying light pressure until the tap feeds itself. The pressure can then be relaxed.

> **CAUTION**
> The tap must be kept lubricated during the tapping process to prevent the tap from galling or breaking. Lubricate the tap regularly during the entire tapping process.

Step 6 Stop the forward cutting, and back the tap out one-quarter turn to clear away chips.

> **CAUTION**
> During the tapping process, the tap should be turned forward one-half turn, then backed out one-quarter turn to prevent chips from clogging the flutes of the tap.

Step 7 Continue the tapping process of turning in one-half turn and then backing out one-quarter turn until the tap touches the bottom of the hole.

Step 8 Back the tap out of the hole.

Step 9 Select a plug tap for the second phase of the tapping process.

Step 10 Tap the hole using the plug tap, following the same procedure as with the taper tap.

Step 11 Back the plug tap out of the hole.

Step 12 Select a bottom tap for the final phase of the tapping process.

Step 13 Tap the hole using the bottom tap.

> **CAUTION**
> The procedure for the bottom tap is the same as for the taper and plug taps, except that the tap must be backed all the way out every two or three turns and the chips cleaned out to prevent them from collecting in the bottom of the hole.

Step 14 Continue the process of turning the tap forward two or three turns, backing it out, and cleaning the hole until the bottom tap touches the bottom of the hole.

Step 15 Clean the taps thoroughly to remove all chips and shavings.

Step 16 Coat the tools lightly with oil to prevent rust.

Step 17 Return the tools to their protective case, and store them in their proper place.

3.13.0 Thread Gauges

Thread gauges, or screw **pitch** gauges, are used to measure the pitch of threaded fasteners. Thread gauges can be used to measure both external and internal threads. A thread gauge is a thin, flat metal leaf that has teeth that correspond to standard thread sections. Thread gauges usually come in sets of several sizes, held together by a screw. *Figure 16* shows a thread gauge that measures internal and external threads.

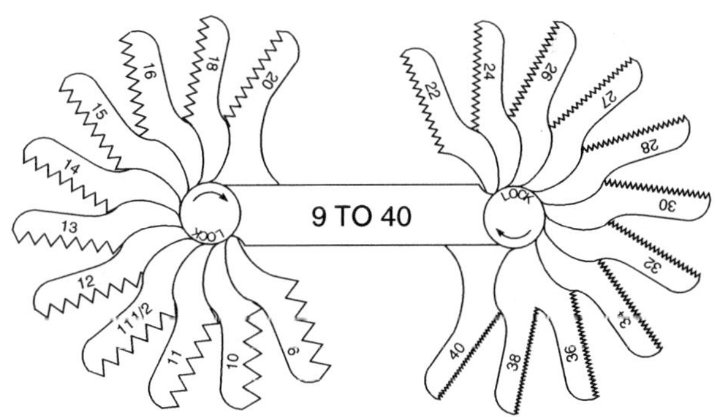

103F16.EPS

Figure 16 ◆ Thread gauge that measures internal and external threads.

Follow these steps to select, inspect, use, and maintain thread gauges:

Step 1 Select a complete set of gauges.

Step 2 Inspect the gauges to ensure that they are not damaged, dirty, or rusty.

Step 3 Hold one of the gauges against the threads being measured, and compare the fit.

Step 4 Try various sizes of gauges until you find an exact fit.

NOTE
The correct size will fit the threads perfectly. The number on the gauge indicates the number of threads per inch. Once this value is determined, you can use it to select nuts, bolts, taps, and dies of the proper pitch.

Step 5 Clean the gauges thoroughly.

Step 6 Coat the gauges lightly with oil to prevent rust.

Step 7 Store the gauges in their proper place.

3.14.0 Scribers

Scribers are sharp, pointed, pencil-like tools used to mark lines on sheet or plate metal (see *Figure 17*). They are usually made of 3/16-inch round tool steel and vary in length from 6 to 12 inches, depending on the handle style. A blue or purple dye, called *layout dye*, is frequently used to make scriber marks more visible. Layout dye is painted on the surface and allowed to dry before marking.

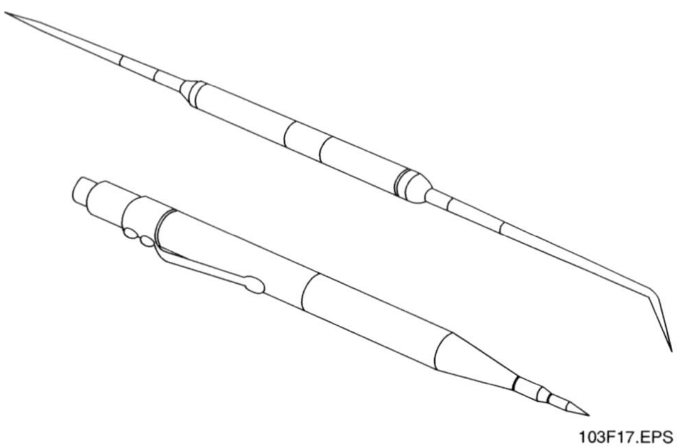

Figure 17 ◆ Two styles of scribers.

Follow these steps to select, inspect, use, and maintain a scriber:

Step 1 Select a scriber of a convenient length for the job.

Step 2 Inspect the scriber to ensure that the points are not bent or dull, and sharpen the points of the scriber if needed.

NOTE
If the points of the scriber are not sharp, it will not produce a distinct mark. Sharpen the points on a honing stone if needed.

Step 3 Hold the scriber at a 45-degree angle to the surface of the material. Apply moderate pressure and pull it toward you, following a straightedge or pattern to mark the surface.

CAUTION
Hold the point of the scribe tightly against the straightedge or pattern so that your marks will be accurate.

Step 4 Clean the scriber.

Step 5 Store the scriber in its proper place.

3.15.0 Tension Meters

A tension meter is used to measure the tension on belts between two pulleys. It is a precision instrument and must be cared for as you would any precision instrument. It should be kept clean and should be stored in a place where it will not get damaged by other tools or materials being thrown on top of it. To use a tension meter, press it against the belt about midway between the pulleys. Read the tension on the scale on the side of the meter. *Figure 18* shows using a tension meter.

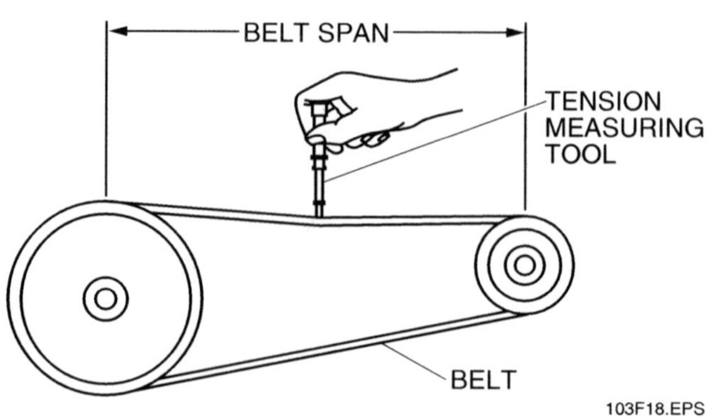

Figure 18 ◆ Using a tension meter.

3.16.0 Sheave Gauges

A sheave gauge is used to measure the side wall wear of a belt sheave. It is used when changing V-belts or when trouble is detected in a V-belt drive. When measuring sheave grooves, ensure that the proper gauge is used. The gauge must be correct for the type and size of belt used and for the pitch diameter of the sheave being checked. *Figure 19* shows a sheave being measured using a sheave gauge.

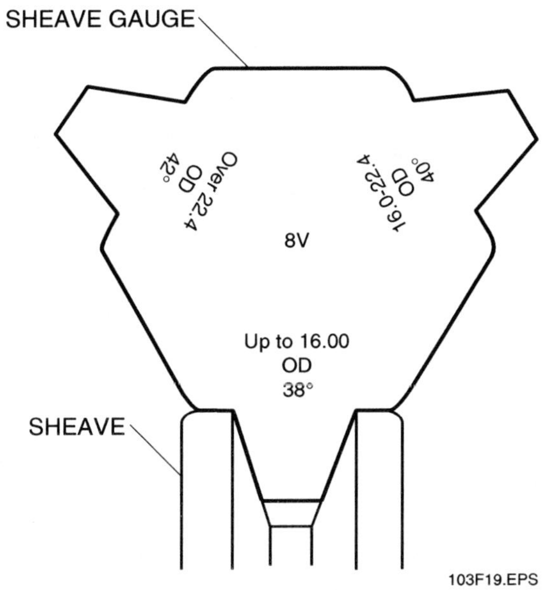

Figure 19 ◆ A sheave being measured using a sheave gauge.

Follow these steps to select, inspect, use, and maintain a sheave gauge:

Step 1 Select a gauge that is correct for the type and size of V-belts being used and for the correct pitch diameter of the sheave being checked.

Step 2 Inspect the gauge to ensure that it is not bent, worn, or otherwise damaged.

Step 3 Insert the gauge into the groove of the sheave, and observe the fit.

CAUTION

It is important that the correct groove dimensions and angles be maintained on sheaves. Refer to the manufacturer's dimension data to ensure that the sheave is within tolerances.

Step 4 Clean the gauge thoroughly.

Step 5 Coat the gauge lightly with oil to prevent rust.

Step 6 Store the gauge in its proper place.

3.17.0 Cylinder Hones

A cylinder hone is used to dress the walls of cylinders when rebuilding and repairing engines, pumps, compressors, and other equipment with cylinders. To hone a cylinder, the hone is turned in the cylinder by an electric drill or motor. The hone is equipped with abrasive stones that cut small amounts of metal away from the cylinder wall as the hone turns. If large amounts of metal must be removed to true the cylinder, the cylinder must be bored. The amount of cylinder wear and the type of rings to be installed usually determine whether the cylinder should be honed or bored. Honing is a precise procedure, and the hone must be held rigid and square to the cylinder to produce a good job. *Figure 20* shows a hone set up to hone a cylinder block.

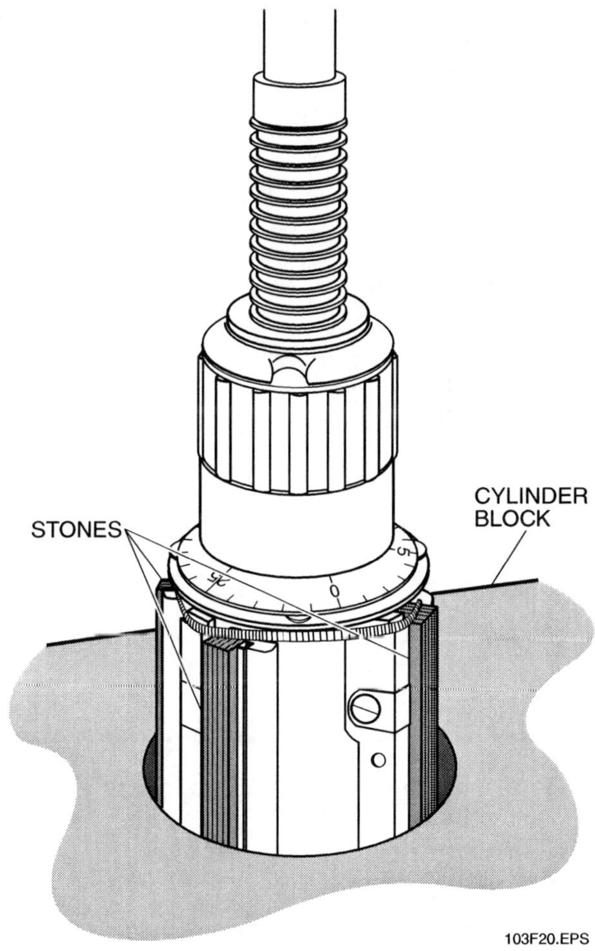

Figure 20 ◆ Hone setup to hone a cylinder block.

Follow these steps to select, inspect, use, and maintain a cylinder hone:

Step 1 Select a hone of the proper size for the cylinder being honed.

Step 2 Inspect the hone to ensure that it is not bent and that the stones are not chipped, broken, or excessively worn.

Step 3 Install the hone in a drill, and set up the honing operation.

> **CAUTION**
> On cylinders that are small enough, it is best to set up the honing operation on a drill press or in a lathe. On larger equipment, a jig must be set up to hold the drill and hone rigid and true to the cylinder. If the hone is not held true to the cylinder, the cylinder can be ruined.

Step 4 Move the hone to the center of the cylinder.

Step 5 Tighten the adjusting knob by hand or by using a small screwdriver to apply light pressure to the stones.

Step 6 Lubricate the stones according to the manufacturer's recommendations.

Step 7 Adjust the speed of the drill.

> **CAUTION**
> The hone should operate at 300 to 700 rpm to avoid overheating and damaging the stones or the cylinder.

Step 8 Start the drill.

> **WARNING!**
> Be sure to wear eye protection to prevent the possibility of injury to your eyes by flying debris.

Step 9 Move the hone up and down a few strokes in the lower half of the cylinder until the bottom of the cylinder begins to increase in diameter.

Step 10 Increase the stroke toward the top of the cylinder gradually until the hone travels the full length of the cylinder bore.

> **CAUTION**
> Do not extend the hone more than ¾ inch past either end of the cylinder to prevent damaging the stones.

Step 11 Stop the drill and readjust the pressure on the stones as the diameter of the cylinder increases and the pressure on the stones decreases.

Step 12 Change to a fine, or finishing, stone if coarse stones were used on the first part of this procedure, and hone the cylinder again to remove all rough hone marks.

Step 13 Extract the hone from the drill when honing is complete, and clean the drill thoroughly.

Step 14 Coat the hone and stones lightly with oil.

Step 15 Store the hone in its proper place.

Step 16 Take down the setup, store all tools, and clean up the work area.

3.18.0 Gear Pullers

Many precision gears and bearings are press-fitted in place and are difficult to remove. Attempting to remove them with a hammer or by prying them off may damage the gear or bearing and the shaft. Gear pullers are made to grip the gears or bearings in a way that will not damage them. Gear pullers apply force to the gear or bearing in a straight and even motion. Pullers are usually made of tool steel and have two or more interchangeable, reversible jaws and a pressure bolt mounted in the middle of a yoke. Gear pullers come in various styles and sizes for different applications (see *Figure 21*).

Follow these steps to select, inspect, use, and maintain gear pullers:

Step 1 Select a gear puller with the correct jaw length and spread width for the gear or bearing being pulled.

Step 2 Inspect the puller to ensure that the jaws are not bent or excessively worn and that the pressure bolt threads are not stripped.

Step 3 Ensure that all set screws or locking screws have been loosened or removed from the gear or bearing.

Step 4 Inspect the shaft to ensure that there are no nicks, burrs, or foreign objects that would interfere with the gear or bearing's sliding off the shaft.

Step 5 Position the puller on the gear or bearing, and screw the pressure screw in until it presses against the shaft.

Step 6 Check the puller to ensure that the jaws are properly positioned on the gear or bearing.

CAUTION

The puller jaws must be positioned so that they will provide a straight, even pull on the gear or bearing and not damage it.

Step 7 Turn the pressure bolt slowly clockwise to begin applying pressure to the gear or bearing.

WARNING!

Always wear eye protection when using a puller because particles could break off and injure your eyes.

Step 8 Continue to tighten the pressure bolt until the gear or bearing is removed from the shaft.

Step 9 Clean the puller thoroughly.

Step 10 Coat the puller lightly with oil to prevent rust.

Step 11 Store the puller in its proper place.

3.19.0 Packing Pullers

A packing puller is used to remove packing material from valves, pumps, and some motors. There are several types of packing pullers (*Figure 22*). Some pullers resemble a pick with a curved point. Other types are flexible and have a corkscrew-like end on them. Special care must be taken when using a packing puller to prevent scarring the valve stem, pump shaft, or the stuffing box.

103F22.EPS

Figure 22 ◆ Packing pullers.

Follow these steps to select, inspect, use, and maintain a packing puller:

Step 1 Select a corkscrew-type puller of the proper size and length for the job.

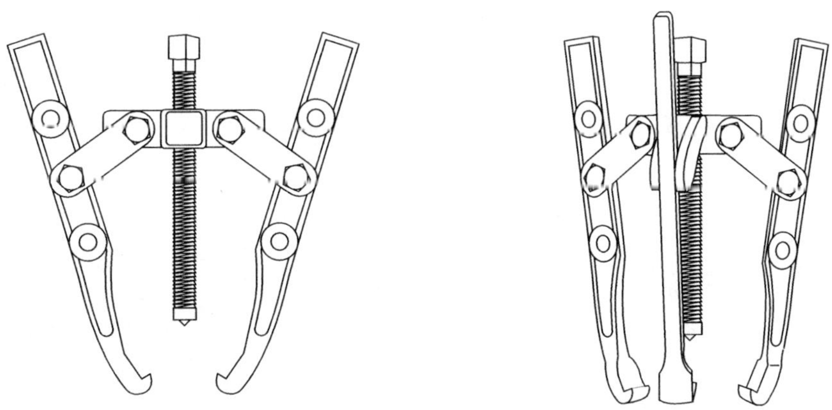

103F21.EPS

Figure 21 ◆ Common gear pullers.

Step 2 Inspect the puller to ensure that it is sharp and not bent out of shape.

Step 3 Insert the puller into the stuffing box, turning it one or two turns clockwise to screw it into the packing.

Step 4 Pull on the puller carefully to remove the packing from the box.

NOTE
Usually the packing will break up into small pieces when it is removed.

Step 5 Repeat Step 4 until all the packing has been removed from the stuffing box.

NOTE
After the packing has been removed, there will usually be some packing pieces left that the corkscrew puller cannot pick up. Choose a pick-type puller and remove the packing pieces.

Step 6 Clean the puller thoroughly.

Step 7 Coat the puller lightly with oil to prevent rust.

Step 8 Store the puller in their proper place.

3.20.0 Reamers

Reamers are used to enlarge and true holes. A reamer consists of three parts: the body, the shank, and the cutting blades. The shank has a square end that allows the reamer to be turned with a wrench. The blades are made of hardened steel and are supported by the body of the reamer. Reamers are available in standard sizes, as well as in size variations of 0.001 inch for special work.

The hardened blades of a reamer are brittle and fragile. Special care should be taken when using and storing the reamer to prevent chipping the blades. *Figure 23* shows some common reamers.

Follow these steps to select, inspect, use, and maintain a reamer:

Step 1 Select a reamer of the proper size for the hole to be reamed.

Step 2 Inspect the reamer to ensure that it is not bent and that the blades are not chipped or nicked.

Step 3 Select a wrench of the proper size to turn the reamer.

Step 4 Position the reamer in the hole to be reamed.

Step 5 Squirt a small amount of oil into the hole around the reamer.

Step 6 Hold the reamer square to the surface of the piece being worked on, and turn the reamer clockwise to start reaming.

CAUTION
Always turn the reamer in the cutting direction (clockwise). Never turn it backwards because this could chip or dull the blades.

Step 7 Continue to turn the reamer until the hole is properly reamed. Remove the reamer from the hole.

Step 8 Clean the reamer thoroughly to remove shavings and debris.

Step 9 Wrap the reamer in an oily cloth, and store it in its proper place.

CAUTION
The reamer should be stored in a place where other tools will not be thrown or dropped on it, because this can damage the blades.

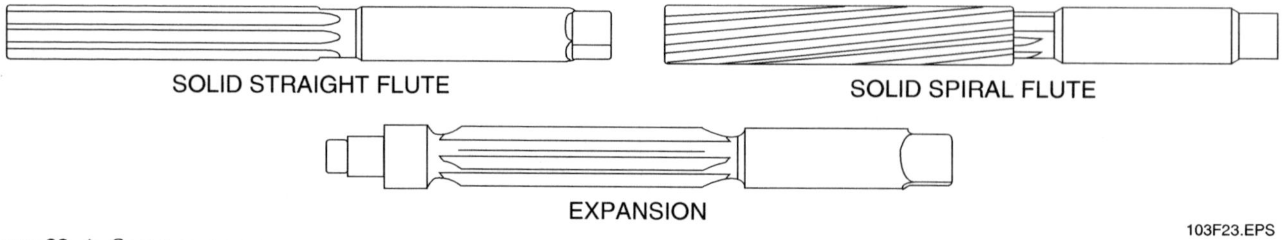

Figure 23 ◆ Some common reamers.

3.21.0 Inspection Mirrors

An inspection mirror (*Figure 24*) is used for close inspection of areas that cannot be seen directly. Inspection mirrors come in a variety of sizes and may be rectangular or round. The mirror is attached to the end of rod and may be fixed or adjustable. By using a flashlight and holding the mirror in the proper position, you can inspect most areas. Some of the newer mirrors have built-in lights for more convenience.

Inspection mirrors should be kept clean and must be stored in a place where they will not be damaged.

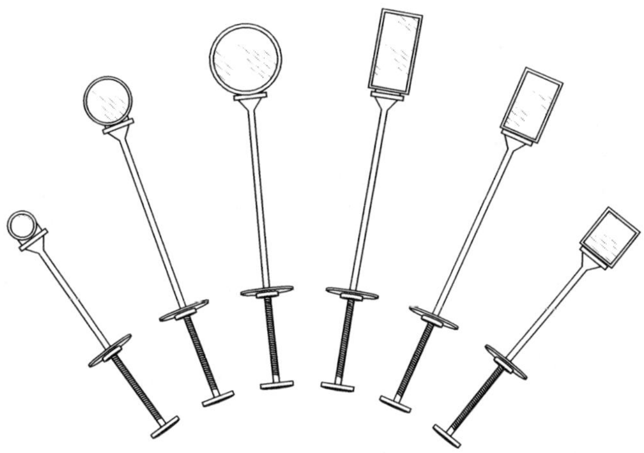

103F24.EPS

Figure 24 ◆ Adjustable inspection mirrors.

3.22.0 Retaining Ring Pliers

Retaining ring pliers are used to remove and install retaining rings from shafts, rods, and spindles. They come in various sizes and shapes for different uses (see *Figure 25*). Some retaining ring pliers come with interchangeable jaws. Retaining ring pliers are made to use on retaining rings only and should not be used for any other purpose. They are fragile and should be used with care and stored in a safe place to prevent damage. To use retaining ring pliers, the points of the pliers are inserted into the holes of the ring. The handle is squeezed to open the ring, and the ring is removed or installed.

3.23.0 Plasti-Gauges

A plasti-gauge is used to measure the amount of oil clearance in a rod bearing (*Figure 26*). It is a plastic thread that flattens when pressure is applied to it.

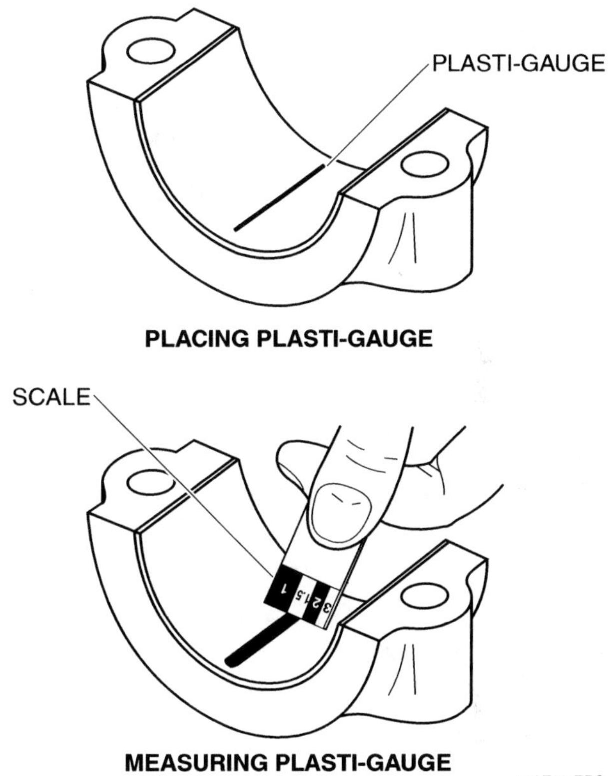

103F26.EPS

Figure 26 ◆ Using a plasti-gauge.

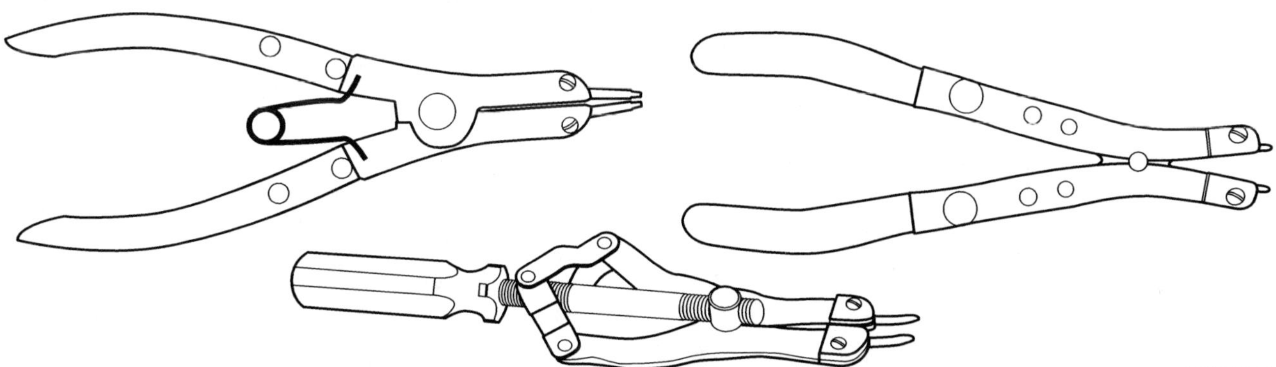

103F25.EPS

Figure 25 ◆ Several types of retaining ring pliers.

To use a plasti-gauge, a strip of the material is placed on the inside of the rod bearing cap, and the cap is installed and tightened on the crank. The cap is then removed, and the plasti-gauge is measured with the scale that is provided to determine the amount of flattening, or oil clearance, of the rod bearing. It is important not to turn the crank while the plasti-gauge is in place, because this further flattens the material and spoils the clearance measurement. The plasti-gauge should be stored in a cool place where it will not be crushed or disfigured in any way. shows using a plasti-gauge.

3.24.0 Taper Gauges

Taper gauges are used for measuring the inside diameters of pipe and tubing, measuring slot widths and hole sizes, and setting calipers. They are made of spring-tempered steel and have thin, tapered leaves that are graduated to measure diameters and widths from $\frac{1}{16}$ to $1\frac{1}{16}$ inch in **graduations** of 64ths of an inch. Taper gauges usually come in sets (see *Figure 27*).

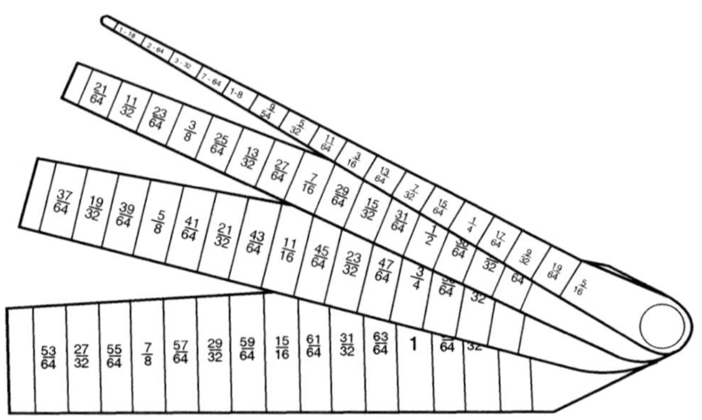

103F27.EPS

Figure 27 ◆ Set of taper gauges.

Follow these steps to select, inspect, use, and maintain taper gauges:

Step 1 Select a complete set of taper gauges.

Step 2 Inspect the gauges to ensure that they are not bent or worn.

Step 3 Find the gauge that will fit in the space to be measured.

Step 4 Hold the gauge square to the space being measured, and slide the gauge into the space.

Step 5 Read the gauge at the point where it touches the edge of the piece being measured.

Step 6 Clean the gauge thoroughly.

Step 7 Coat the gauge lightly with oil to prevent rust.

Step 8 Put the gauge in its protective box, and store it in its proper place.

CAUTION

Taper gauges are fragile and must be stored in a place where other tools and materials will not be thrown or dropped on them and in an atmosphere where they will not rust.

3.25.0 Extractors

If a screw or tap breaks off in a workpiece, it is necessary to remove it. If the screw or tap protrudes above the surface of the work, removing it may be as simple as using a pair of vise-grip pliers to grip and remove it. If the screw or tap is broken near or below the surface of the work, you will need an extractor to remove it. Two common types of extractors are:

- Spiral screw extractors
- Tap extractors

3.25.1 Spiral Screw Extractors

Spiral screw extractors, or easy-outs, are reverse-threaded to remove right-hand threaded screws or bolts. They are made in various sizes to use on many size screws. The easy-out manufacturer provides recommendations as to what size easy-out should be used to remove a given size screw and what size drill should be used with the easy-out. You should follow the manufacturer's recommendations because an easy-out that is too small may not be strong enough to turn the screw. The easy-out may break off in the screw, compounding the problem of removing the screw. Follow these steps to select, inspect, use, and maintain an easy-out:

Step 1 Select an easy-out of the proper size, according to the manufacturer's recommendations.

Step 2 Inspect the easy-out to ensure that the threads are not damaged.

Step 3 Select a drill that is the proper size for the easy-out, according to the manufacturer's recommendations.

Step 4 Apply penetrating oil to the broken screw, and allow it to soak.

> **NOTE**
> It is best to allow the penetrating oil to soak overnight if possible, adding oil periodically. Allow it to soak as long as time will allow.

Step 5 Prick the exact center of the screw, using a center punch.

> **NOTE**
> Pricking the center of the screw will help to ensure that the hole will be drilled in the center of the screw.

Step 6 Drill a hole in the center of the screw.

> **CAUTION**
> Take care not to drill all the way through the screw and into the workpiece, because this could damage the workpiece.

Step 7 Insert the easy-out into the drilled hole, and begin turning it in a counterclockwise direction until it gets a bite.

Step 8 Continue to slowly turn the easy-out until the screw is removed (see *Figure 28*).

Step 9 Clean the easy-out thoroughly.

Step 10 Oil the easy-out lightly, and store it in its proper place.

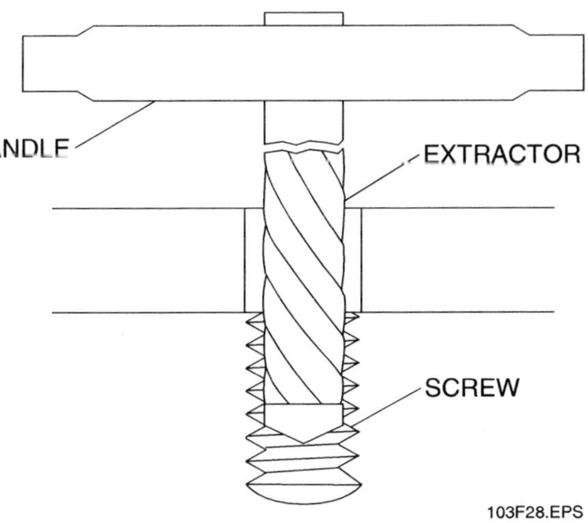

Figure 28 ◆ Using an easy-out.

3.25.2 Tap Extractors

Tap extractors are specially made to remove broken taps from a workpiece. The tap extractor has four prongs that fit into the flutes of the broken tap to turn it. There is a special size tap extractor for each size tap. Follow these steps to select, inspect, use, and maintain tap extractors:

Step 1 Select an extractor of the proper size for the tap being removed.

Step 2 Inspect the tap to ensure that it is not damaged.

Step 3 Insert the prongs of the extractor into the flutes of the tap.

Step 4 Push the lower collar of the extractor down against the surface of the workpiece to hold the prongs securely against the body of the tap.

Step 5 Turn the tap back and forth slowly, using a tap handle on the extractor to loosen the tap.

Step 6 Turn the tap counterclockwise to screw the tap out of the hole.

Step 7 Clean the extractor thoroughly.

Step 8 Oil the extractor lightly to prevent rust.

Step 9 Store the extractor in its proper place.

4.0.0 ◆ DRILL PRESSES

A drill press consists of a stationary mounted drill, a drive motor, an adjustable table, and the necessary controls, all mounted on a column with a base that is fastened to the floor (see *Figure 29*). The drill spindle is raised and lowered using a hand lever. The table can be adjusted to any height on the column to allow drilling operations on various sizes of material. The drill press usually has some type of vise arrangement that fits on the table to hold the work being drilled. Special jigs can be fabricated to hold odd-shaped objects on the table.

> **WARNING!**
> Keep hands away from the drill spindle. Never hold the workpiece being drilled by hand. Some type of mechanical clamping device must be used to secure the workpiece to the table. Never wear gloves while using a drill press, as they may get caught into the drill spindle. If shavings begin to pile up, use a piece of wood rather than your hand to knock the shavings free.

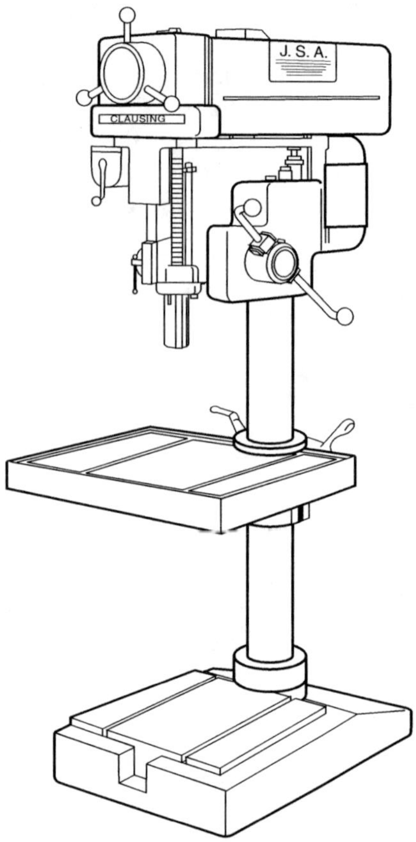

Figure 29 ◆ Drill press.

The spindle of a drill press can be locked in any vertical position for router or shaper work. The spindle is set at the desired position using the hand lever and locked in position by tightening the quill lever (see *Figure 30*).

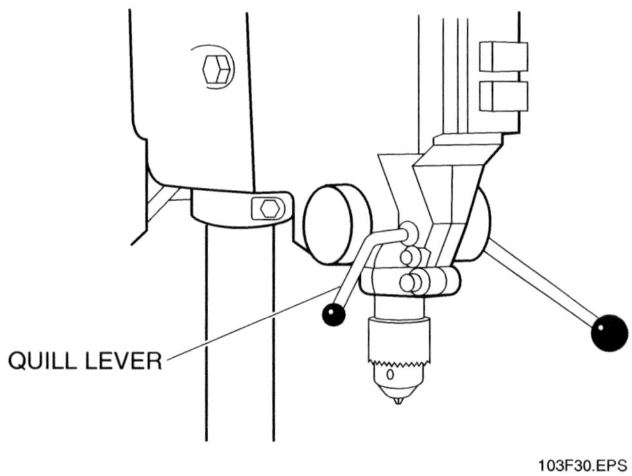

Figure 30 ◆ Quill lever.

Many drill presses are variable-speed, belt-driven machines. The speed of the drill is adjusted by positioning the belt on the appropriate pulley (see *Figure 31*).

> **WARNING!**
> To prevent accidental startup and personal injury, be sure that the drill press is unplugged from the power source before changing the belts.

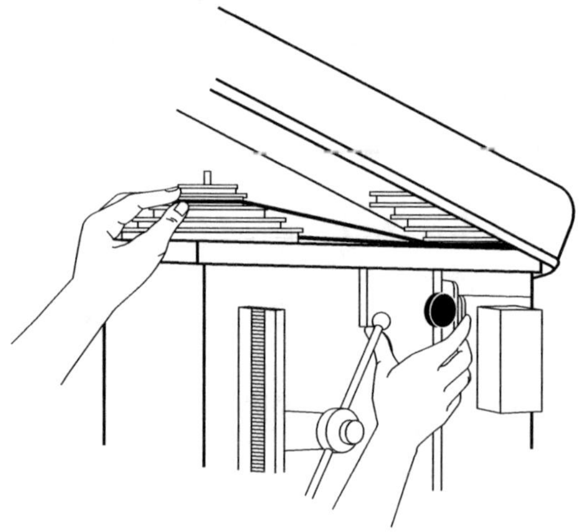

Figure 31 ◆ Positioning a belt on pulleys.

Other drill presses are equipped with a variable-speed drive (see *Figure 32*). To vary the speed of the drill, turn the pilot wheel until the pointer indicates the desired speed. Always check the manufacturer's instructions before attempting to change the speed of the drill.

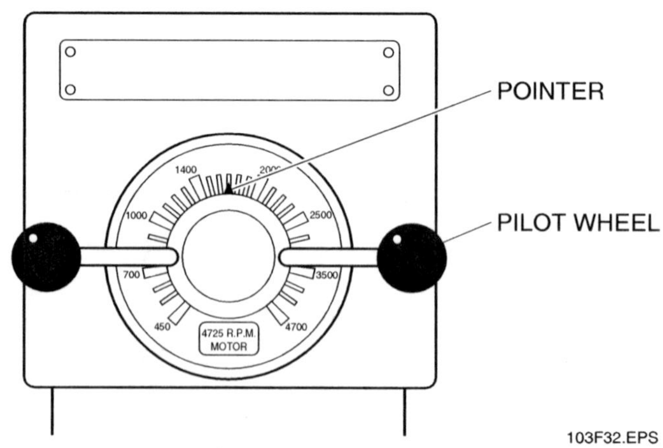

Figure 32 ◆ Variable-speed drive control.

5.0.0 ◆ HYDRAULIC PRESSES

A hydraulic press can be used for bending, straightening, pressing, assembling, and other pressing operations. Hydraulic presses can be vertical or horizontal, but the vertical type is the most common (see *Figure 33*). Shop presses vary in size from 5- to 60-ton capacities and may be hand-pumped or have an electric pump to power the cylinder.

A hydraulic press can produce a great amount of pressure. If an excessive amount of pressure is applied to a part, the part could reach the breaking point without any visible indication that too much pressure is being applied. The only way to determine the amount of pressure being exerted by the press is by watching the pressure gauge.

To prevent damage to the part or personal injury, observe the following precautions when working with a hydraulic press:

- Remove all burrs from the workpiece and clean the workpiece before placing it in the press.
- Ensure that the workpiece is properly supported.
- Use a lubricant between mating parts to prevent **seizing** when pressing parts together.
- Monitor the pressure gauge to ensure that you do not apply excessive pressure.
- Ensure that the workpiece is properly positioned in the press.
- Place a piece of brass or other material that is softer than the workpiece between the ram and the workpiece to prevent marring the workpiece.

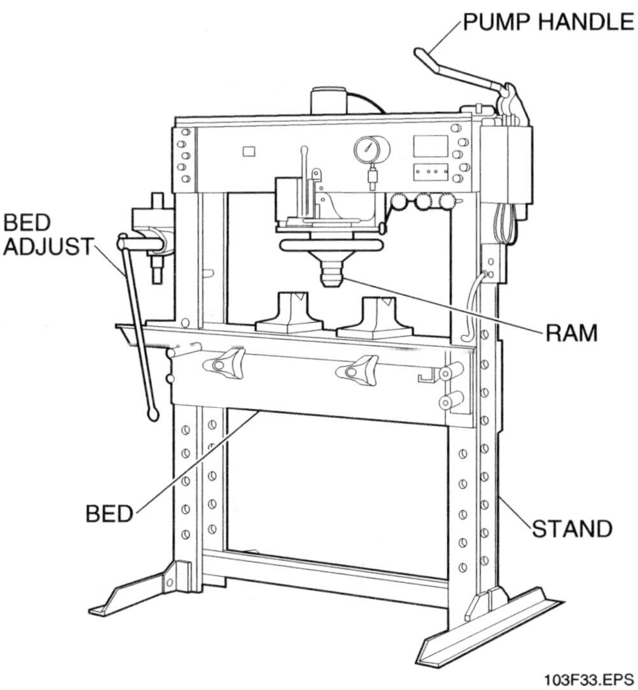

Figure 33 ◆ Hydraulic press.

6.0.0 ◆ PIPE THREADING MACHINES

Pipe threading machines are multiple-purpose power tools that center, hold, and rotate pipe, conduit, or bolt stock while cutting, reaming, and threading operations are performed (see *Figure 34*). All pipe threading machines are designed with a gear drive; **chucks**, which hold and turn the pipe; and a three-position selector switch consisting of forward, off, and reverse drive positions. Depending on the size and complexity of the machine, pipe threading machines can be mounted on **tripods**, tables, or special stands.

Pipe threading machines come in a variety of models, from the basic gear-drive type with front and rear pipe-holding chucks to the all-purpose model with built-in cutters, oilers, reamers, and a pipe threading die.

By following a few simple guidelines, you can expect a long life of service from a threading machine. Use the following guidelines to keep the threading machine in proper working condition:

- Never reverse the rotation of the machine while the machine is running. Always turn off the power and wait until the drive has stopped rotating before changing the direction of the drive mechanism.
- At the start of each day, clean the chuck jaws, using a stiff brush to remove rust, scale, chips, pipe coating, or other foreign matter. Apply lubricating oil to the machine bed ways, the cutter rollers, and the feed screw.
- Always use sharp dies, which produce smoother threads and require less motor power than dull dies.

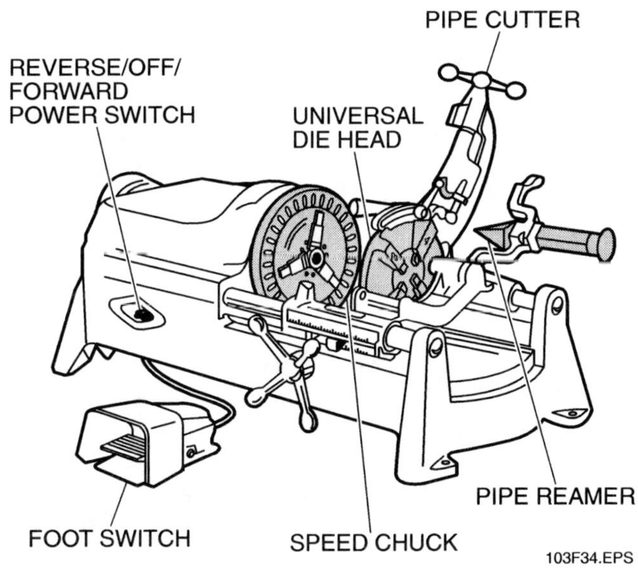

Figure 34 ◆ Pipe threading machine.

- Refer to and follow the manufacturer's instructions on lubrication schedules for bearings and gears.
- Ensure that there is always plenty of cutting oil in the machine. Make sure that the die is flooded with cutting oil at all times while threading pipe. The cutting oil should be periodically cleaned to remove accumulated sludge, chips, and other foreign matter. Replace the cutting oil when it becomes dirty and contaminated.
- Remove the oil filter periodically, and clean it with solvent. Blow the filter clean with compressed air. Do not operate the machine without the oil filter.
- Keep the chuck jaws in good condition, and replace the jaw inserts as necessary when they become worn.

7.0.0 ◆ NIBBLERS

Nibblers cut metal by punching out small notches as the metal is moved over the bench-mounted nibbler or as the portable nibbler is applied to the sheet metal. Nibblers are used for cutting irregular shapes and do not leave a finished edge. The width of the cut is usually more than ⅛ inch. Nibblers are usually electrically powered and can be either bench-mounted or portable (see *Figure 35*).

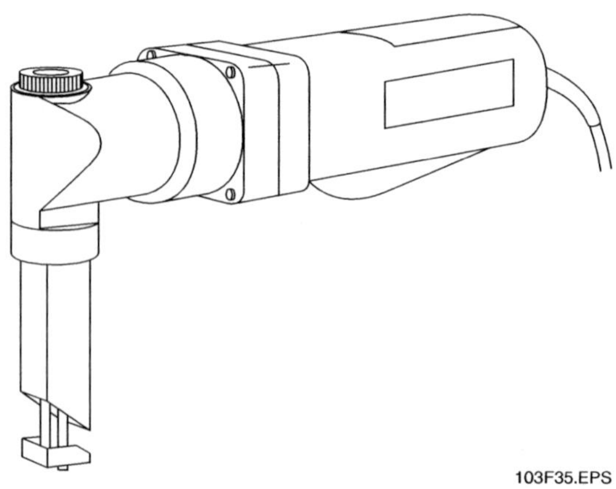

Figure 35 ◆ Portable nibbler.

8.0.0 ◆ PRESS BRAKES

The press brake is a production machine generally used in large-volume manufacturing for forming metal. It forms the metal by pressing it into specially formed dies. Through a series of pressings and by changing the upper and lower dies, the press brake can be used for forming either light sheet metal or heavy metal plate into a variety of production shapes and fittings.

Press brakes are either hydraulically or electrically operated and exert much more force on a workpiece than is possible with a manual hand brake. They are extremely dangerous unless proper safety precautions are observed. Manufacturers of press brakes provide several types of safety devices for the operator, such as both-hands-required controls located above the work area or pull-back restraints that automatically yank the operator's hands back from the danger zone if accidental pressing occurs. Press brakes are available in many sizes, from 4 to 35 feet in length, and they are capable of many different sheet metal forming operations.

9.0.0 ◆ SURFACE GRINDERS

Surface grinders are used to produce flat, smooth surfaces on parts (see *Figure 36*). The surface grinder grinds away surface material until the required smoothness and size tolerances are achieved. There are many types of surface grinders.

The major parts of this type of surface grinder are the following:

- Base
- Table
- Wheel head assembly
- Coolant system
- Wheel

9.1.0 Base

The surface grinder base has a column at the back that supports the wheel head. The saddle that supports the table is mounted on the base at the front of the machine.

9.2.0 Table

The table is precision-fitted to the saddle. It moves left and right (**traverse travel**) under the grinding wheel. This type of table is known as a *reciprocating table*. The table has adjustable reversing dogs that can be set to limit table travel. The table feed can be manually controlled using a handwheel, but it also has an automatic power feed.

The table usually has T-slots, which are used with bolts to hold the workpiece or a vise in position on the table. Magnetic chucks, which hold the workpiece without the use of bolts or a vise, are also built in or attached to the table.

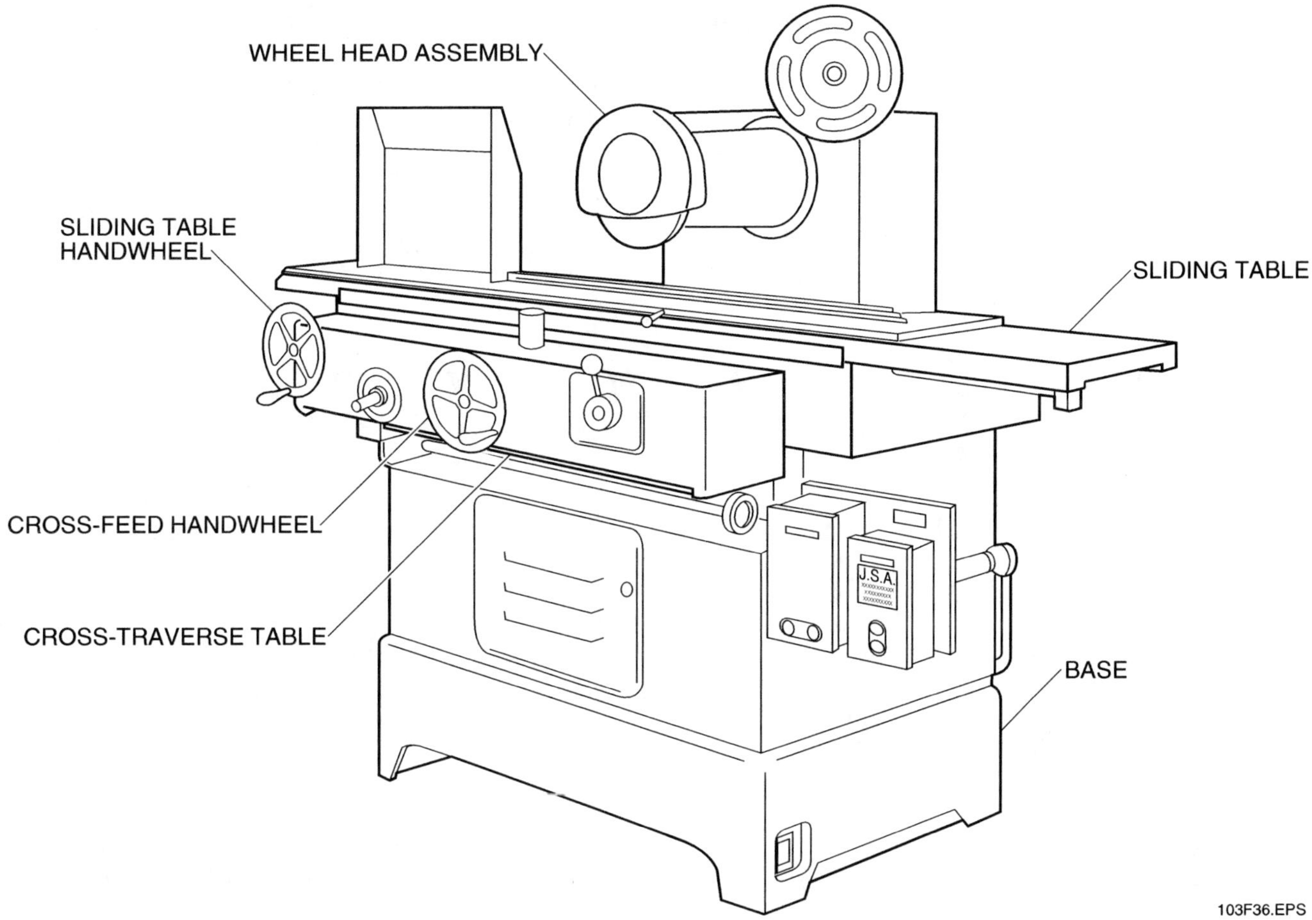

Figure 36 ◆ Surface grinder.

9.3.0 Wheel Head Assembly

The wheel head assembly is precision-fitted to the column and moves up and down. The up and down movement of the wheel head is controlled manually by a handwheel. The grinding wheel is mounted vertically in the wheel head. The wheel acts against a surface that is mounted on a horizontal table. This is why this type of grinder is sometimes called a *horizontal grinder*. The grinding wheel is driven by an electric motor mounted in the head. The wheel grinds the workpiece as it feeds and reciprocates under the wheel.

9.4.0 Coolant System

Grinders have a coolant reservoir and a pump that provide coolant to regulate the temperature of the workpiece being ground.

9.5.0 Wheel

Several types of wheels are used on a surface grinder. For most common operations, a plain wheel is used. The plain wheel cuts on the edge of a workpiece. For grinding shoulders, a recessed wheel may be used. Wheels with formed edges are used to cut Vs and other shapes.

10.0.0 ◆ BELT SANDERS

A belt sander has many uses, including sanding, sizing, and shaping objects. Depending on the type of sanding belt used, the sander can be used on metal, wood, plastic, fiberglass, and other types of materials. *Figure 37* shows a common type of belt sander.

The belt sander has a tilting table, which can be tilted from 0 to 45 degrees. The table is held in position by star-grip table locks on both sides of the table. The table also has a groove for a miter gauge.

The abrasive sanding belt can be used vertically, horizontally, or at an angle. The table can remain in a level position while the platen is tilted to different angles. When using a belt sander with the head in the horizontal position, the backstop must be used to keep the work from being carried along with the belt (see *Figure 38*).

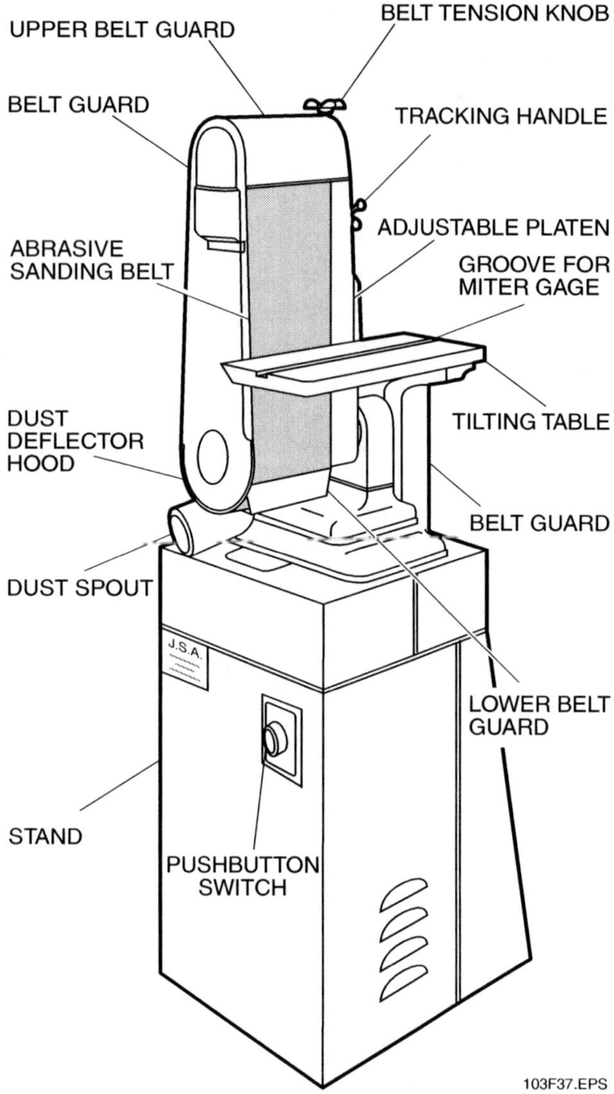

Figure 37 ◆ Common type of belt sander.

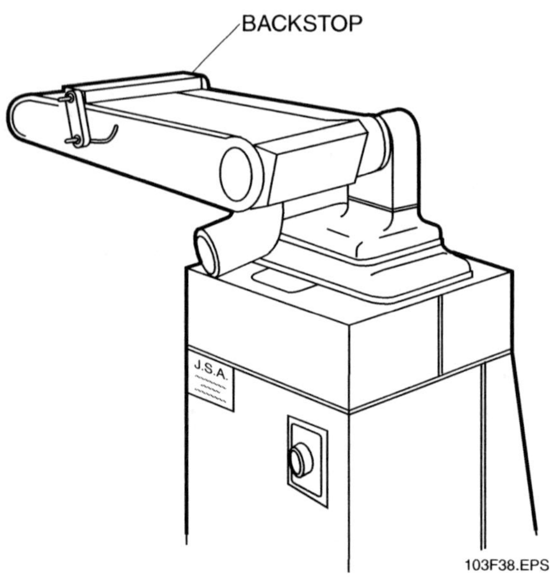

Figure 38 ◆ Belt sander with backstop.

The belt sander can also be used for contour finishing when the upper belt guard is removed. The idler drum can then be used to contour finished pieces.

The abrasive sanding belt rides on a bed with tracking lines, which keep the belt on track. By rotating the tracking handle while the sander is running, the edge of the belt can be lined on the tracking lines. When adjusting the tracking, make sure that the belt is running from top to bottom.

11.0.0 ◆ WOODRUFF KEY SEATERS

A Woodruff key is a half-moon key that requires a special type of slot in the shaft of the equipment. The Woodruff key seater is used to cut this slot. The Woodruff key seater has deep flutes cut across the surface of the teeth. It is slightly thicker at the crest of the teeth than at the center of the cutter to provide clearance between the sides of the slot and the cutter. Key seaters smaller than 2 inches are made in one piece, with the cutter and the shank made together. Cutters 2 inches and larger have a hole in the center for arbor mounting. *Figure 39* shows the two types of Woodruff key seaters.

To use a Woodruff key seater, the cutter is held in an arbor or in a drill chuck mounted in the spindle of a milling machine. The shaft is held in a vise or other jig on the milling machine table and is moved up to the cutter until the desired key depth is obtained. The cutter must be the same width and diameter of the key as the slot is being cut.

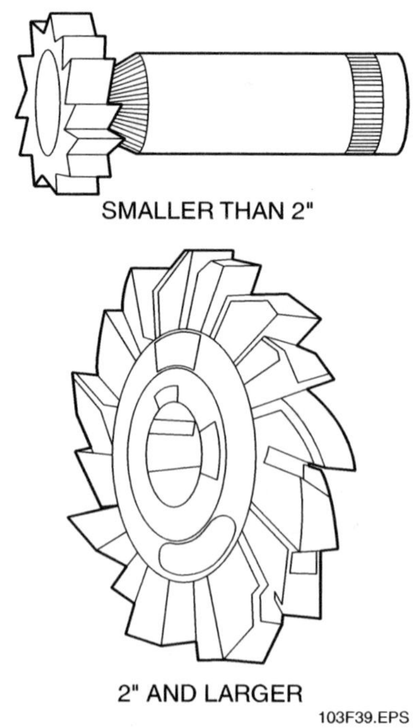

Figure 39 ◆ Woodruff key seaters.

12.0.0 ◆ KEY BROACHES

A key broach is a hardened, tapered tool with metal cutting teeth used to cut holes. The teeth are small at the small end of the broach and get progressively larger toward the large end of the broach. The last few teeth are the correct size for the hole being cut. Key broaches come in shapes to cut round, square, triangular, and hexagonal holes. They can also be used to cut internal keyways and gears.

A key broach is an expensive and fragile tool. It can be damaged beyond use if not used properly and stored in a safe place. It should be stored where no other tools or materials will chip or dull the cutting edges.

A mechanical or hydraulic press is usually used to push the broach through the part being cut. A considerable amount of pressure is required to broach a hole. The setup for broaching should be rigid and secure. Follow these steps to broach in a workpiece:

Step 1 Mount the broach in the chuck of the press.

Step 2 Position the workpiece in a vise or other setup on the press table.

Step 3 Bring the broach down to almost touch the workpiece.

Step 4 Adjust the setup until the broach is properly positioned to cut the desired hole.

Step 5 Lubricate the broach with broaching oil.

Step 6 Operate the press to force the broach through the workpiece.

CAUTION

Pushing the broaching through the workpiece with a slow, even pressure will produce the best results without damaging the broach.

Step 7 Release the broach from the chuck, and let it fall through the hole.

CAUTION

Do not pull the broach back through the hole, because this may damage the cutting edges. Be sure to catch the broach with a soft cushion of some type. Do not let it drop onto the floor, because this could damage it.

Step 8 Clean the broach thoroughly.

Step 9 Oil the broach lightly to prevent rust.

Step 10 Store the broach in a safe place.

13.0.0 ◆ BEARING HEATERS

Interference fit, or shrink fit, bearings have an inside dimension that is smaller than the outside dimension of the shaft on which they fit. In an interference installation, the bearing is heated to expand the inner ring so that it will fit onto the larger shaft. Once the bearing is installed and allowed to cool, it will shrink to fit tightly on the shaft.

Bearings must be heated slowly and evenly to prevent damage. You should never heat a bearing with a torch or other direct heating method, because this would heat the bearing unevenly and damage the bearing beyond use. A bearing should never be heated to a temperature above 250°F. Higher temperatures cause changes in the metal of the bearing, and the bearing would not be suitable for use. Use an approved method of heating bearings, such as one of the following:

- Heat lamps
- Electric ovens
- Induction heaters

13.1.0 Heat Lamps

A heat, or infrared, lamp can be used to heat small bearings (see *Figure 40*). It is a slow method, but it has the advantage of allowing the bearing to be heated either wrapped or unwrapped.

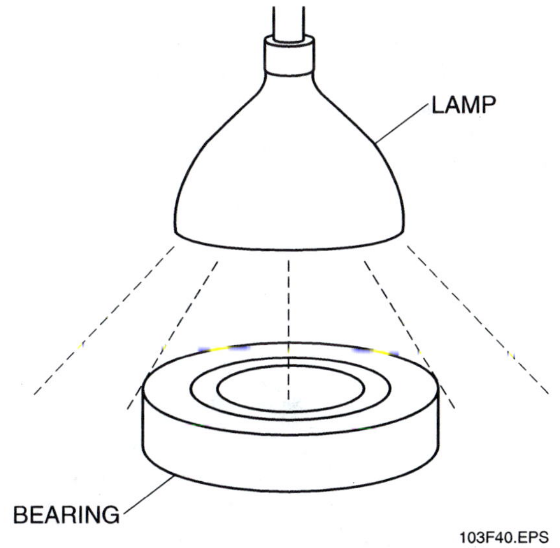

Figure 40 ◆ Heat lamp.

13.2.0 Electric Ovens

A small electric oven can also be used to heat bearings (see *Figure 41*). This is a relatively safe way to heat bearings because the temperature controls can be set to a specific temperature. The bearing is placed in the oven and heated to the required temperature.

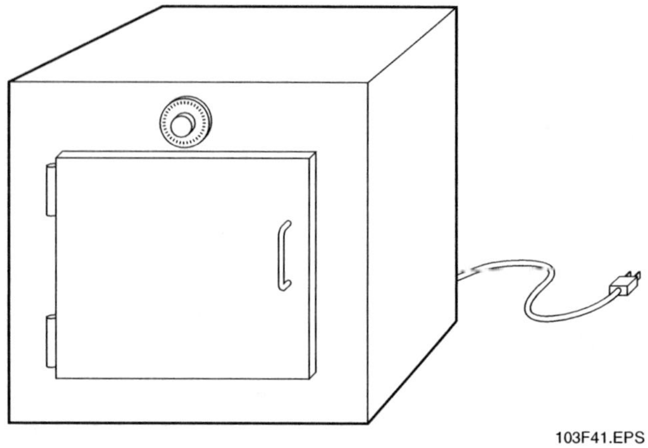

Figure 41 ◆ Electric oven.

Follow these steps to use an electric oven:

Step 1 Inspect the oven and clean it, using an appropriate solvent.

Step 2 Check the calibration of the thermostat.

Step 3 Determine the temperature at which to heat the part.

Step 4 Place the part gently onto the oven rack.

Step 5 Allow the part to heat until it reaches the desired temperature.

> **CAUTION**
> Excessive temperature discolors the part. Excessive discoloration of the metal indicates damage.

Step 6 Measure the diameter of the part to ensure that it is the proper size.

> **NOTE**
> The part can be measured with a snap gauge, an inside micrometer, temperature sticks, or outside or inside calipers.

> **WARNING!**
> Be sure to wear heat-resistant gloves when handling hot parts.

Step 7 Remove the part from the oven manually.

Step 8 Turn off the oven.

> **WARNING!**
> The part remains hot for a while after it is removed from the oven. Ensure that all personnel are aware of the hazard.

> **NOTE**
> The part must be protected from contamination after it is installed.

13.3.0 Induction Heaters

Induction heaters are the most commonly used bearing heaters (see *Figure 42*). They are safe, fast, and easy to use. With an induction heater, the bearing is heated from the inside out. The bearing is placed on the bar of the heater, and 110 volts of electricity passes through the bar. The heater is set at the desired temperature for a preset amount of time, and the bearing is heated. Induction heaters can also be used to heat bushings, shrink rings, pulleys, and gears.

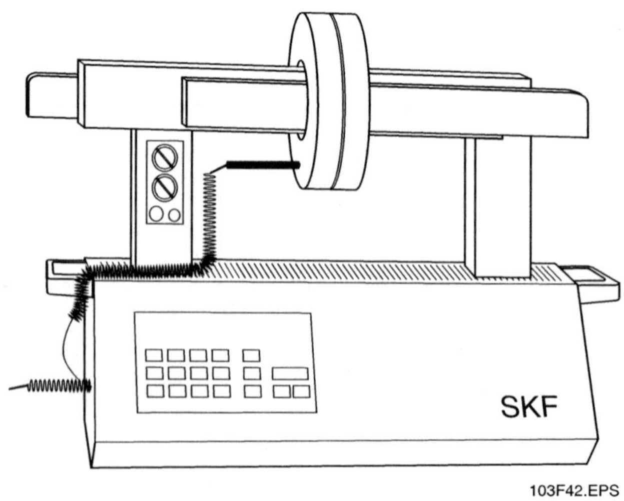

Figure 42 ◆ Induction heater.

Follow these steps to use an induction heater:

> **WARNING!**
> Be sure to wear Kevlar® or cotton gloves when handling hot parts.

Step 1 Clean and inspect the bar.

Step 2 Determine the temperature at which the part is to be heated and the appropriate **temp stick** to use.

Step 3 Inspect the part to be sure that it is not damaged.

Step 4 Place the part on the bar.

Step 5 Turn on the heater.

> **NOTE**
> Rotate the part if necessary.

Step 6 Check the temperature of the part frequently, using the temp stick until the desired temperature is reached.

Step 7 Turn the demagnetizer switch to the ON position to demagnetize the part while it is on the bar.

Step 8 Turn the demagnetizer switch to the OFF position.

Step 9 Turn off the heater.

Step 10 Remove the part from the bar, and measure it if necessary.

Step 11 Install the part.

14.0.0 ◆ DRILLS

Drills are used in many operations in the shop and the field. Drills, especially larger drills, are expensive. Most drills can be resharpened many times in order to save on replacement costs. They must be sharpened accurately if they are to drill properly. Accurately sharpening drills takes practice and patience.

There are many types of drill points for different applications (see *Figure 43*). Drill points are ground at different angles and shapes for different applications.

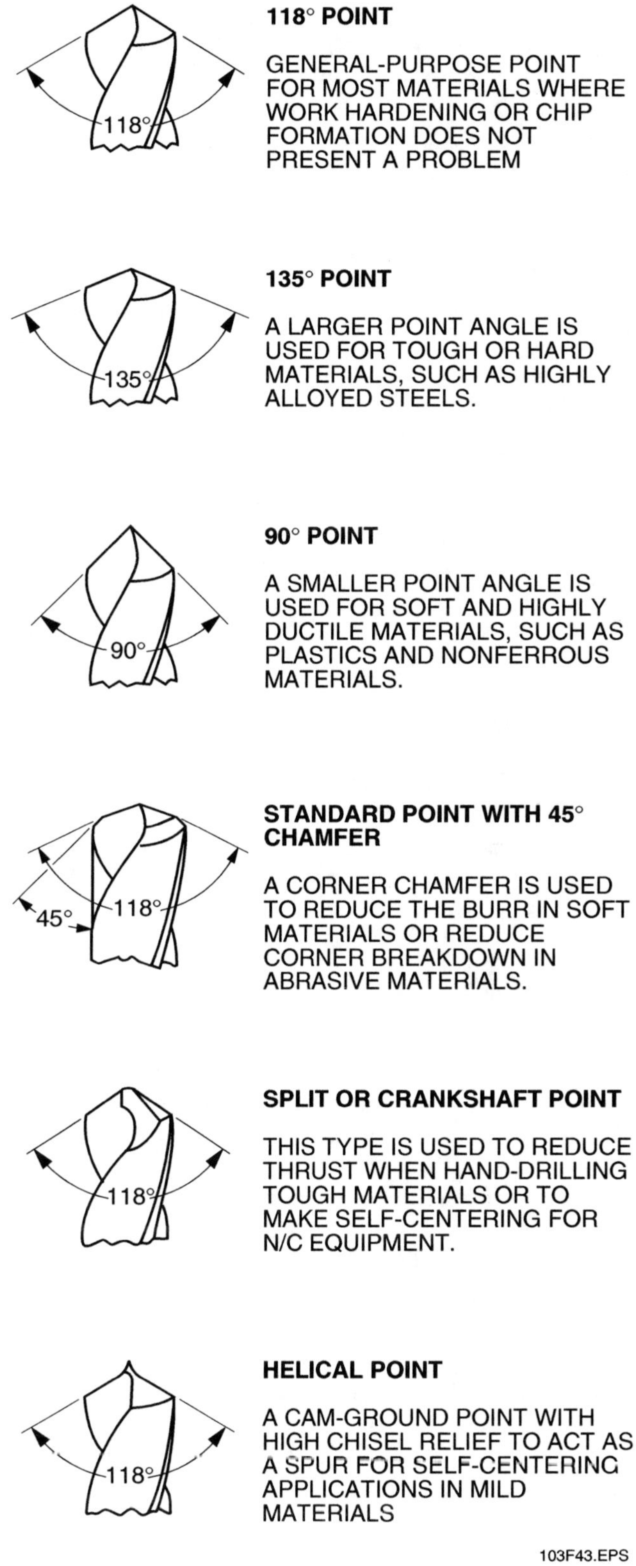

Figure 43 ◆ Types of drill points.

Table 1 lists the suggested relief angles on drill points for general purpose use. Use this table as a guide for repointing and lip-relief angles. For harder materials, reduce these values by 2 percent. For soft and nonferrous materials, increase these values by 2 percent.

Table 1 Lip Relief Angles

Drill Diameter Range	Lip-Relief Angle
#80 to #61 (0.0135 to 0.0390)	20°
#60 to #31 (0.0400 to 0.1200)	17°
⅛ to ¼ (0.1250 to 0.2500)	14°
F to ½ (0.2570 to 0.5000)	12°
33/64 to 1 (0.5156 to 1.0000)	10°
1 1/64 to 1¾ (1.0156 to 1.7500)	8°
1 49/64 to 3½ (1.7656 to 3.500)	6°

103T01.EPS

14.1.0 Web Thinning

Web thinning is the grinding away of metal inside the web of the drill. It is a method of sharpening a drill. Web thinning of drill points is necessary for some types of drilling operations. *Figure 44* shows proper web thinning.

14.2.0 Point Grinding

As furnished from the manufacturer, most twist drills are machine-point ground. This ensures uniform chip load and wear, which in turn results in maximum drill life. Most drill failures can be directly attributed to improper point grinding. It is important that great care be taken when regrinding drill points. *Figure 45* shows proper and improper drill point grinding.

14.3.0 Troubleshooting Drills

Many times, problems encountered in cutting operations can be directly attributed to the drill being used. *Table 2* lists drilling problems and probable causes.

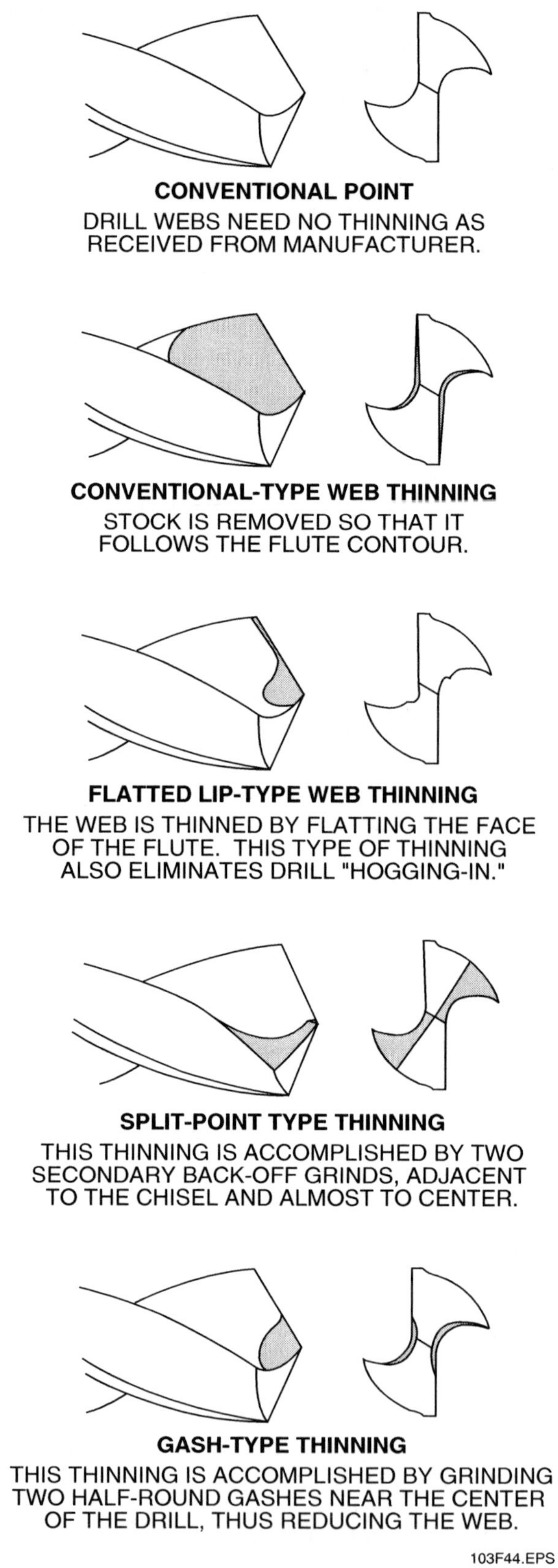

Figure 44 ◆ Proper web thinning.

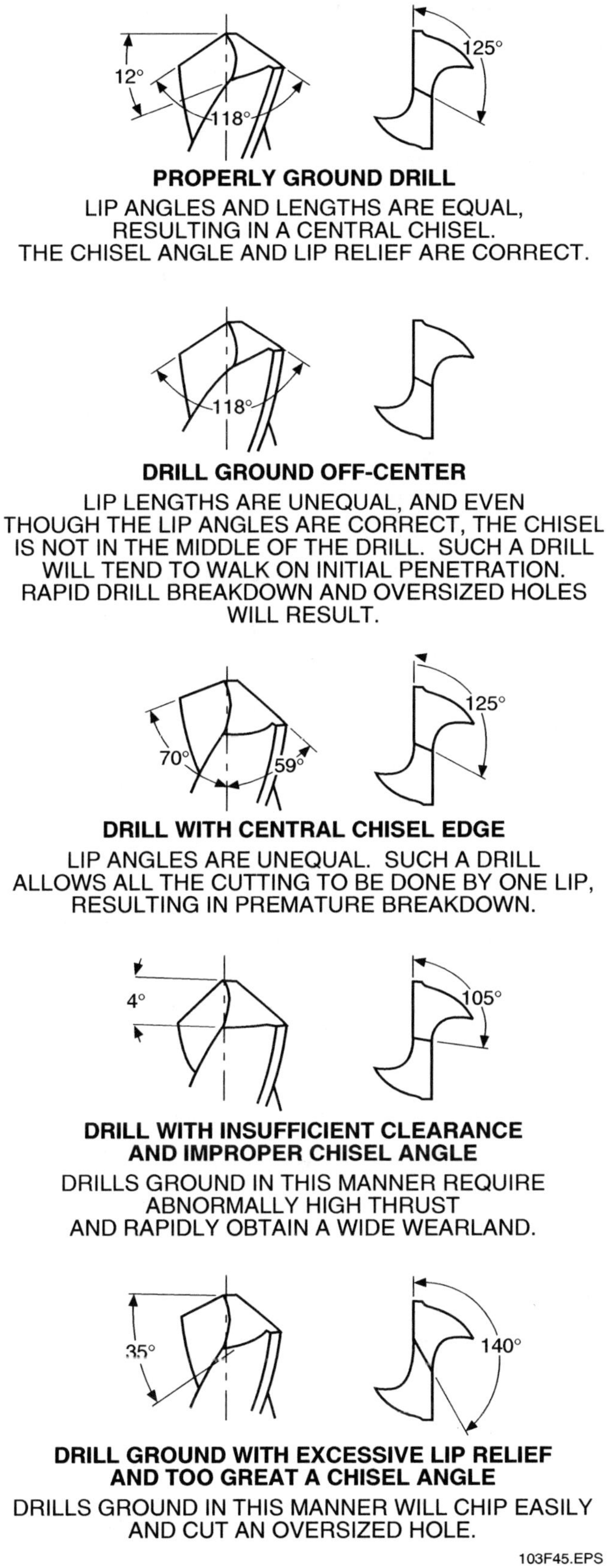

Figure 45 ◆ Proper and improper drill point grinding.

Table 2 Drilling Problems and Probable Causes

Problem	Probable Cause
Outer corner breakdown	• Cutting speed too high • Hard spots in material • No cutting compound at drill point • Flutes clogged with chips
Cutting lips chip	• Too much feed
Checks or cracks in cutting lips	• Drill overheated or cooled too quickly while sharpening or drilling
Margin chips	• Jig bushing too big
Drill breaks	• Point improperly ground • Feed too heavy • Spring or backlash in drill press, fixture, or work • Drill is dull • Flutes clogged
Tang breaks	• Imperfect fit between taper shank and socket, caused by dirt, chips, or burred or badly worn sockets
Drill breaks while drilling brass or wood	• Flutes clogged with chips • Drilling with improper type drill
Drill splits up center	• Lip relief too small • Too much feed
Drill will not enter work	• Drill is dull • Lip relief too small • Web is too heavy
Hole rough	• Point is dull or improperly ground • No cutting compound at drill point • Feed too great • Fixture not rigid
Hole oversized	• Unequal angle or length of cutting edges • Spindle is loose
Chip shape changes while drilling	• Drill becoming dull • Cutting lips chipped
Large chips coming out of one flute, small chips coming out of the other	• Points improperly ground, one lip doing all the cutting

103T02.EPS

15.0.0 ◆ PRECISION DRILLING

Drills of various types are used to drill holes in a variety of materials. Many times, drilling does not need to be extremely accurate. However, fabricating precision parts requires extreme care to drill the holes perfectly straight and the proper size and depth. The following sections explain precision drilling using a hand drill, a magnetic drill, and a drill press.

15.1.0 Hand Drill

Although absolute precision drilling is impossible using a hand-held drill, you can produce a fairly accurate hole if the drill is used properly. Follow these steps to drill with a hand drill:

Step 1 Ensure that the drill bit has a properly sharpened point.

> **NOTE**
> If the bit point is not sharpened properly, the bit may walk away from the intended hole location when pressure is applied. Even new bits are sometimes improperly sharpened and should be checked before use.

Step 2 Prick-mark the intended hole location, using a center punch, to guide the bit.

Step 3 Apply cutting oil to help the bit cut smoothly.

Step 4 Ensure that the drill is square with the surface being drilled.

> **NOTE**
> Use a small square to check the alignment of the drill to ensure that it is square. If the drill is not square with the surface being drilled, the hole will not be straight and the fastener will not fit properly when installed. Some drill motors have built-in levels to ensure squareness.

Step 5 Hold the drill firmly, and apply enough pressure to start the bit cutting.

Step 6 Increase pressure once the bit has started cutting.

> **CAUTION**
> Too much pressure on the bit can cause overheating and galling of the bit. Always use plenty of cutting oil and apply only enough pressure to keep the bit cutting smoothly.

Step 7 Decrease pressure just before the bit cuts all of the way through the part to prevent exit damage.

> **NOTE**
> If you are cutting a hole to a certain depth, use a gauge or stop of some kind on the drill to prevent cutting too deep.

15.2.0 Magnetic Drill

A magnetic drill is much better than a hand drill for precision drilling. It is held firmly in place by a powerful magnet and, when set up properly, will drill a precision hole. Follow these steps to drill, using a magnetic drill:

Step 1 Ensure that the area where the drill is to be set up is clean and free of oil or any objects that would prevent the drill from sitting flat on the surface.

Step 2 Ensure that the drill bit has a properly sharpened point.

Step 3 Prick-mark the intended hole location, using a center punch, to guide the bit.

Step 4 Align the drill so that the bit is perfectly aligned with the prick mark.

Step 5 Turn the magnet on, and ensure that it is holding firmly.

Step 6 Check the bit with a square to ensure that it is square with the surface to be drilled.

Step 7 Apply cutting oil at the point of the bit.

Step 8 Turn on the drill motor.

Step 9 Apply light pressure and begin drilling.

Step 10 Increase pressure once the bit has started cutting.

> **CAUTION**
> Too much pressure on the bit can cause overheating and galling of the bit. Always use plenty of cutting oil and apply only enough pressure to keep the bit cutting smoothly.

Step 11 Decrease pressure just before the bit cuts all of the way through the part to prevent exit damage.

> **NOTE**
> If you are cutting a hole to a certain depth, use a gauge or stop of some kind on the drill to prevent cutting too deep.

15.3.0 Drill Press

Of the three methods mentioned, the drill press is the best method for precision drilling. The drill press is stationary and rigid and provides excellent stability for drilling a true and accurate hole. Follow these steps to drill, using a drill press:

Step 1 Ensure that the drill bit has a properly sharpened point.

Step 2 Prick-mark the intended hole location in the workpiece, using a center punch, to guide the bit.

Step 3 Align the workpiece so that the bit is perfectly aligned with the prick mark.

Step 4 Clamp the workpiece firmly on the bed of the drill press.

Step 5 Check the bit with a square to ensure that it is square with the surface to be drilled.

Step 6 Apply cutting oil at the point of the bit.

Step 7 Turn on the drill motor.

Step 8 Apply light pressure and begin drilling.

Step 9 Increase pressure once the bit has started cutting.

> **CAUTION**
> Too much pressure on the bit can cause overheating and galling of the bit. Always use plenty of cutting oil and apply only enough pressure to keep the bit cutting smoothly.

Step 10 Decrease pressure just before the bit cuts all the way through the part, to prevent exit damage.

> **NOTE**
> If you are cutting a hole to a certain depth, use the drill press depth gauge to set the depth of the hole to prevent cutting too deep.

Summary

Hand tools are used in some way for practically every job that pipeline mechanics do. Tools are expensive to replace and should be used and maintained in a manner that will ensure a long tool service life. Tools should be matched to the job being done. Always consider safety, and use tools in the safest manner possible.

The use of power tools has greatly increased the individual production of the pipeline mechanic. Power tools are available for cutting, grinding, and shaping all types of materials. Pipeline mechanics must know how to safely operate these power tools for the specific jobs for which the tools are designed.

Review Questions

1. Inspect power tools _____.
   a. at the beginning of each shift
   b. once each day
   c. before each use
   d. according to the plant maintenance schedule

2. On finished surfaces, a(n) _____ wrench is often used to prevent marring.
   a. straight
   b. offset
   c. spanner
   d. strap

3. To measure the inside diameter of tapered holes, a _____ is used.
   a. honing stone
   b. taper plug gauge
   c. drift pin
   d. snap gauge

4. Thin-wall iron, brass, steel, copper, and aluminum tubing are cut with a _____.
   a. pipe cutter
   b. soil pipe cutter
   c. tubing cutter
   d. hacksaw

5. When scraping a gasket, keep the putty knife _____ the surface.
   a. at a 45-degree angle to
   b. at a 90-degree angle to
   c. perpendicular to
   d. at an effective working angle to

6. Light-gauge sheet metal is cut with _____ where the shape, position, or location prevents the use of power metal-cutting tools.
   a. "old men"
   b. offset cutters
   c. tin snips
   d. diagonal cutters

7. _____ dies are used for repairing damaged or rusty threads on bolts or screws.
   a. Screw adjusting
   b. Rethreading
   c. Pipe
   d. Open adjusting

8. To dress the walls of cylinders when rebuilding and repairing engines, pumps, and compressors, _____ are used.
   a. reamers
   b. honing stones
   c. cylinder hones
   d. taps and dies

9. A packing puller is used to remove packing material from _____.
   a. all motors
   b. piping
   c. valves, pumps, and some motors
   d. cylinders

10. Nibblers are used for cutting _____.
    a. regular shapes with a finished edge
    b. regular shapes without a finished edge
    c. irregular shapes with a finished edge
    d. irregular shapes without a finished edge

11. A production machine used in large-volume manufacturing for forming metal is the _____.
    a. press brake
    b. nibbler
    c. surface grinder
    d. ball end mill

12. A hardened, tapered tool with metal cutting teeth used to cut holes is the _____.
    a. key broach
    b. bushing
    c. Woodruff key seater
    d. tailstock

13. The most commonly used bearing heaters are _____.
    a. heat lamps
    b. electric ovens
    c. induction heaters
    d. oil baths

14. To check the alignment of a hand drill, use a _____.
    a. tape measure
    b. height gauge
    c. depth gauge
    d. small square

15. The best tool to use for precision drilling is the _____.
    a. magnetic drill
    b. hand drill
    c. drill press
    d. drill motor

GLOSSARY

Trade Terms Introduced in This Module

Align: To line up two or more parts.

Burr: A sharp, ragged edge produced by cutting sheet metal, metallic piping, and nonmetallic piping.

Chamfer (length): (1) To cut off the edge or corner of; bevel. (2) To bevel a sharp edge on a machined part.

Chuck: The part of a machine that holds work pieces or tools for machine cutting jobs.

Diameter: The width of a circle. The measurement from side to side.

Die: A tool used to make male threads on a pipe or bolt.

Graduation: Marks on a measuring instrument that indicate size.

Pitch: Number of threads per inch on bolts, screws, and threaded rods.

Seizing: The fusing or freezing of one part to another because of excessive temperature or pressure.

Temp stick: A crayon that, when applied to a surface, indicates when the surface exceeds a given value by changing color. Also called a *tempilstick*.

Traverse travel: Moving right and left.

Tripod: A three-legged stand used to support materials and equipment.

CONTREN™ LEARNING SERIES — USER UPDATES

The NCCER makes every effort to keep these textbooks up-to-date and free of technical errors. We appreciate your help in this process. If you have an idea for improving this textbook, or if you find an error, a typographical mistake, or an inaccuracy in NCCER's Contren™ textbooks, please write us, using this form or a photocopy. Be sure to include the exact module number, page number, a detailed description, and the correction, if applicable. Your input will be brought to the attention of the Technical Review Committee. Thank you for your assistance.

Instructors – If you found that additional materials were necessary in order to teach this module effectively, please let us know so that we may include them in the Equipment/Materials list in the Instructor's Guide.

Write: Curriculum Revision and Development Department
National Center for Construction Education and Research
P.O. Box 141104, Gainesville, FL 32614-1104

Fax: 352-334-0932

E-mail: curriculum@nccer.org

Craft ____________________ Module Name ____________________

Copyright Date __________ Module Number __________ Page Number(s) __________

Description

____

____

____

____

(Optional) Correction

____

____

____

(Optional) Your Name and Address

____

____

____

Module 63104-02

Piping and Mechanical Blueprint Reading

Introduction to the Pipeline Operator Qualification Rule, API, and NCCER

It could be said that petroleum products keep America running. Petroleum products like oil and natural gas are the principal sources of energy in the United States and are used in many manufacturing and chemical processes. Petroleum products are primarily transported by pipeline. There are more than a million miles of liquid and natural gas pipeline in this country, ranging in diameter from two inches to four feet. These pipelines bring crude oil and natural gas from production fields to processing plants and refineries—and ultimately to the consumer.

Pipeline Operator Qualification

Petroleum products are considered hazardous, so pipeline personnel must take extra precautions and follow specific procedures when working with them. To help ensure the safe handling and transportation of petroleum products, the United States Department of Transportation (DOT) mandates safe practices as federal regulations.

One such regulation, the Operator Qualification (OQ) rule, requires pipeline operators to develop a written qualification plan to evaluate their personnel and their contract personnel (see *49 CFR Part 192 Subpart N—Gas*, and *49 CFR Part 195 Subpart G—Liquid*). The evaluation hinges on the operator's ability to perform certain covered tasks and to be able to recognize and respond to abnormal operating conditions that may be encountered while performing these covered tasks.

Covered Tasks

A pipeline performance task is deemed *covered* if the activity passes the following four-part test:

1. Is it performed on a pipeline facility?
2. Is it an operations or maintenance task?
3. Is it performed as a requirement of *49 CFR Part 195 (Liquid)* or *49 CFR Part 192 (Gas)*?
4. Does it affect the operation or integrity of the pipeline facility?

If the answer to all four questions is *yes*, then the activity is considered *covered* and must figure into the evaluation process of pipeline personnel.

American Petroleum Institute

The OQ rule and its related covered tasks provided the catalyst to develop a unified pipeline training and assessment program—so that pipeline operators and owners have ready access to the assessment and training tools they need. This effort was the direct result of the vision and leadership provided by The American Petroleum Institute (API) and API members.

Efforts to unify the industry are nothing new to API. Indeed, the impetus for forming the Institute in 1919 was the need to standardize engineering specifications for drilling and production equipment. Today, the most pressing issues revolve around public perceptions and government policies toward the industry. Speaking with one voice on these issues has become as essential as having interchangeable parts in the field. API's members determine the petroleum industry's positions on public policy as well as the standards that govern its day-to-day operations worldwide.

Consortium on Operator Qualification

API assembled a focus group whose primary purpose was to identify covered tasks and provide compliance recommendations to pipeline operators.

The Consortium on Operator Qualification (COOQ) became the guiding force for the identification and development of covered tasks for the liquid pipeline industry. Their work product, API publication 1161, *Guidance Document for the Qualification of Liquid Pipeline Personnel*, formed the basis for the resulting Pipeliner Training and Assessment Program.

The covered task list created in API publication 1161 has now evolved to include gas covered tasks and is maintained by NCCER. The most current covered task list can be downloaded at www.nccer.org.

National Center for Construction Education and Research

With the guidance and support of API, its members, and the Consortium on Operator Qualification, the National Center for Construction Education and Research (NCCER) was able to design and implement the Pipeliner Training and Assessment Program.

NCCER is a not-for-profit 501(c)(3) education foundation established in 1995 by the world's largest and most progressive construction and maintenance contractors and national contractor associations. NCCER was founded to address the

severe workforce shortage facing the construction and maintenance industries and to develop a standardized training process and curricula. Today, NCCER is supported by hundreds of leading construction and maintenance companies, manufacturers, and national associations and developed more than 35 construction and maintenance titles.

NCCER also maintains a National Registry, which contains the records of trainees and instructors who have received NCCER training. All completion information for every trainee is sent to the NCCER and filed within the Registry. The Registry can then confirm training and skills for workers as they move from company to company, state to state, or even within their own company.

Standardized Curriculum (NCCER's Contren™ Learning Series)

NCCER curricula are developed using the expertise of industry subject matter experts (SMEs). The use of SMEs as an integral part of every new development or revision ensures that the curriculum is current, industry-driven, and indicative of industry practices. The development and/or revision process is facilitated from start to finish by NCCER project managers, who work closely with the SMEs and a select team of technical writer(s) on the development and presentation of the curriculum.

Pipeline: Five Training Paths

The pipeline curriculum focuses on essential skills and competencies within five skill areas:

- Corrosion Control
- Electrical and Instrumentation
- Maintenance
- Mechanical
- Operations (Field and Control Center)

SMEs from the pipeline industry have determined the skills and competencies needed based on API's covered tasks and the consensus of the group/industry. Specific company information is generally not incorporated into the curriculum unless the SMEs determine that the information is the industry standard.

Training Modules and Covered Tasks

The smallest unit of instruction in the curriculum is a *module*. Each curriculum is composed of modules (the equivalent of stand-alone chapters) that are logically sequenced into complete *levels* of instruction. Every module is task-oriented and requires the successful completion of a written (knowledge-based) examination and a performance (hands-on) test. Modules may be individually selected and/or combined for efficient task training. Modules in the pipeline curricula, to the extent possible, will align with API-developed covered tasks. While not all modules will relate to covered tasks, all covered tasks will relate to at least one module.

Benefits of Training to the Pipeline Marketplace

Applying NCCER's standardized training process allows pipeline owners and contractors in the United States to establish the benchmark for acceptable skill and competency levels within the pipeline industry. This process will assist training organizations in developing and/or improving standardized pipeline training programs that effectively match their resources and operating conditions while responding to regulatory demands and to the needs of the industry. The process will also recognize training organizations that meet the minimum standards of competence, warranting regulatory, public, and professional confidence. This in turn will foster national unity within the pipeline industry for the mutual benefit of individual technicians, owners, contractors, and ultimately pipeline-users and society.

Additional Resources

To review the Pipeliner Training and Assessment Program Implementation and Resource Guide and the latest version of pipeline covered tasks, visit NCCER's Web site at www.nccer.org, or contact:

National Center for Construction Education and Research
PO Box 141104
Gainesville, FL 32614-1104
(888) 622-3720
Web site: www.nccer.org

For more information on API, please contact:

American Petroleum Institute
1220 L Street, Northwest
Washington, D.C. 20005
(202) 682-8000
Web site: www.api.org

To order *API Pub 1161*, contact Global Engineering Documents by calling 1-800-854-7149 or visiting their Web site at www.global.his.com.

Additional sources of information on the pipeline operator qualification regulation include:

DOT – Office of Pipeline Safety
www.ops.dot.gov

Transportation Safety Institute
www.tsi.dot.gov

COURSE MAP

This course map shows all of the modules in the first level of the Pipeline Mechanical curriculum. The suggested training order begins at the bottom and proceeds up. Skill levels increase as you advance on the course map. The local Training Program Sponsor may adjust the training order.

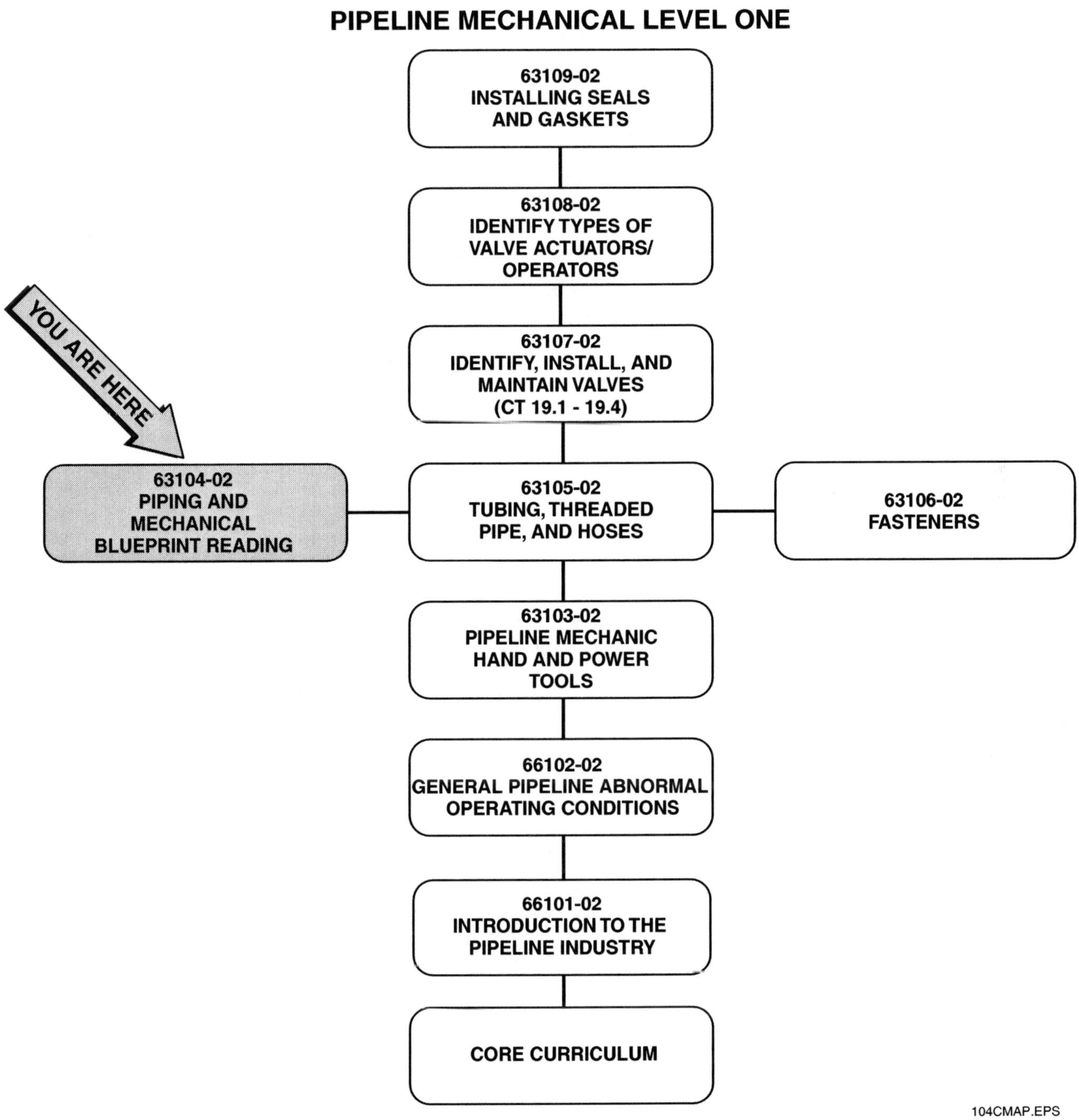

MODULE 63104-02 CONTENTS

Figures

Tables

MODULE 63104-02

Piping and Mechanical Blueprint Reading

Objectives

When you have completed this module, you will be able to do the following:

1. Identify and explain types and parts of drawings.
2. Read and interpret P&IDs, plan views, section views, and isometric drawings.
3. Read and interpret machine drawings.

Prerequisites

Before you begin this module, it is recommended that you successfully complete the following: Core Curriculum; Pipeline Core, Modules 66101 and 66102; Pipeline Mechanical Level One, Module 63103.

Required Trainee Materials

1. Pencil and paper
2. Appropriate personal protective equipment

1.0.0 ◆ INTRODUCTION

This module identifies the types of drawings that you will need to be familiar with in order to locate and maintain equipment in the piping system. **Piping and instrumentation diagrams (P&IDs)**, piping arrangement drawings, **isometric drawings (ISO)**, and mechanical **drawing packages** are explained in more depth with explanations on how to read and interpret the drawings. Control points, coordinates, elevations, and material take-outs are also explained.

2.0.0 ◆ PARTS OF DRAWINGS

Blueprints contain a great deal of information in a limited space. To do this, designers and engineers use certain conventions and standard ways of drawing objects, including the following:

- **Title blocks**
- Scales and measurements
- Symbols and abbreviations
- Notes
- **Revision** blocks
- Control points
- Coordinates
- Elevations
- Line lists

2.1.0 Title Blocks

The title block contains information used to identify the project or any part of the project. It is usually designed by the engineering division and uses a coding system that designates all drawings used on that project. The following information is found in the title block:

- *Company name* – Includes the logo, name, and address of the design firm.
- *Project title* – The project name and location.
- *Drawing title* – Explains what is drawn on that sheet. If it is a detail drawing, the part is identified. If the sheet contains an assembly drawing, this block is left blank.
- *Professional stamp* – A registered seal of approval used by the engineer or architect.

- *Design supervisor* – The name or initials of the person who designed the project and the date on which it was signed by the design supervisor.
- *Drawn* – The name or initials of the person who drafted the drawing and the date on which it was drawn.
- *Checked* – The name or initials of the person who reviewed the drawing and the date on which it was reviewed.
- *Approved* – The name or initials of the person who approved the drawing, usually the project engineer, and the date on which it was approved.
- *Owner's approval* – The signature of the owner or a representative and the date on which the owner approved the drawing.
- *Scale* – Shows the relationship between the size of the drawing and the size of the actual object. If the drawing is the same size as the object, the scale is full size.
- *Project number* – The code number assigned to the project.
- *Sheet number* – Indicates the sequence of this sheet in the series of drawings.
- *Revision number (Rev.)* – Indicates the current drawing. The number in this block can be referenced to the revision block for information about each revision.
- *Key plan* – Shown if one plan drawing does not include the entire system being designed. The key plan shows where this drawing fits into the overall geographic location. The shaded area shows the location of the drawing. The key plan may be shown on the plan view for reference.

The title block is usually located on the lower right corner of each drawing. *Figure 1* shows a typical title block.

2.2.0 Scales and Measurements

Scales on drawings are usually written in inches or in feet and inches. Measurements of less than 1 foot are written as 0 feet and the number of inches. The scale of a drawing is written in the title block. It shows the size of the drawing as compared to the actual size of the object represented.

The scale of a drawing varies according to the size of the object being drawn and the size of the paper used. The drawing scale is usually indicated by a ratio between the size of the object and the size of the drawing. For example, a scale of 1 inch = 1 foot means that for every inch shown on the drawing, the actual size of the object is 1 foot. The standard scale of piping drawings is ⅜ inch = 1 foot. The letters U.N. in a scale stand for "unless noted otherwise on drawing." This indicates that some parts of the drawing may use a different scale.

Graphic scales on drawings show distances that correspond with a unit of length on the ground or with the object represented in the drawing (see *Figure 2*). They are usually placed in or near the title block of the drawing.

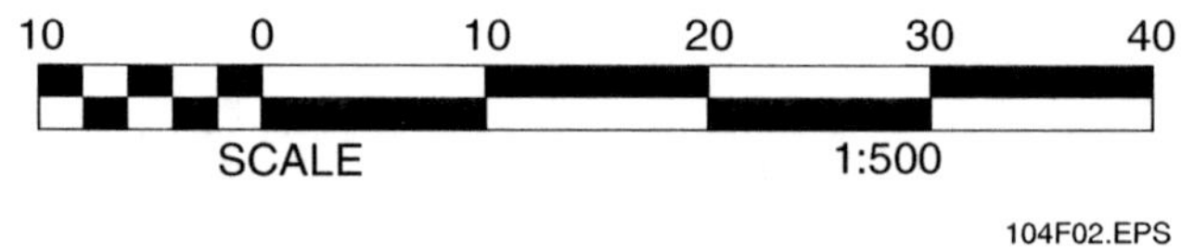

Figure 2 ◆ Graphic scale.

2.3.0 Symbols and Abbreviations

Symbols and abbreviations are used on drawings to show information in a small area. Always check the symbols and abbreviations used in the drawing with the legend that should be included with the drawings.

2.3.1 Symbols

Symbols are used to show objects, such as valves, fittings, flanges, and connections, that would require too much detail if drawn accurately. These objects are represented on drawings by small, simple line drawings. Symbols can vary from job to job (see *Figure 3*).

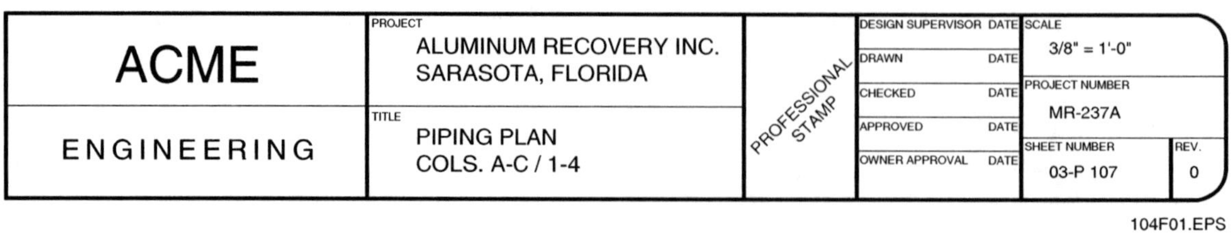

Figure 1 ◆ Typical title block.

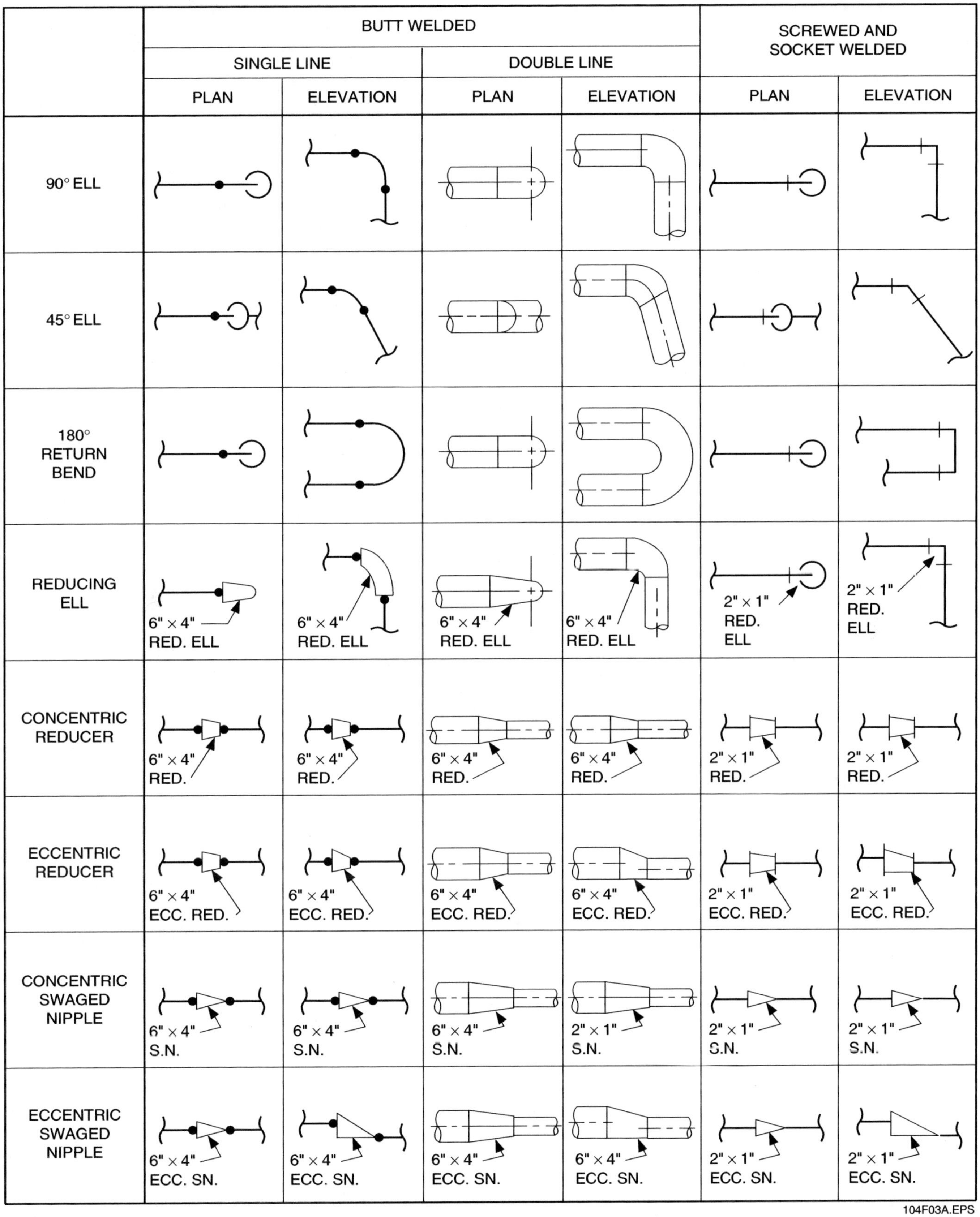

Figure 3 ◆ Commonly used piping symbols (1 of 4).

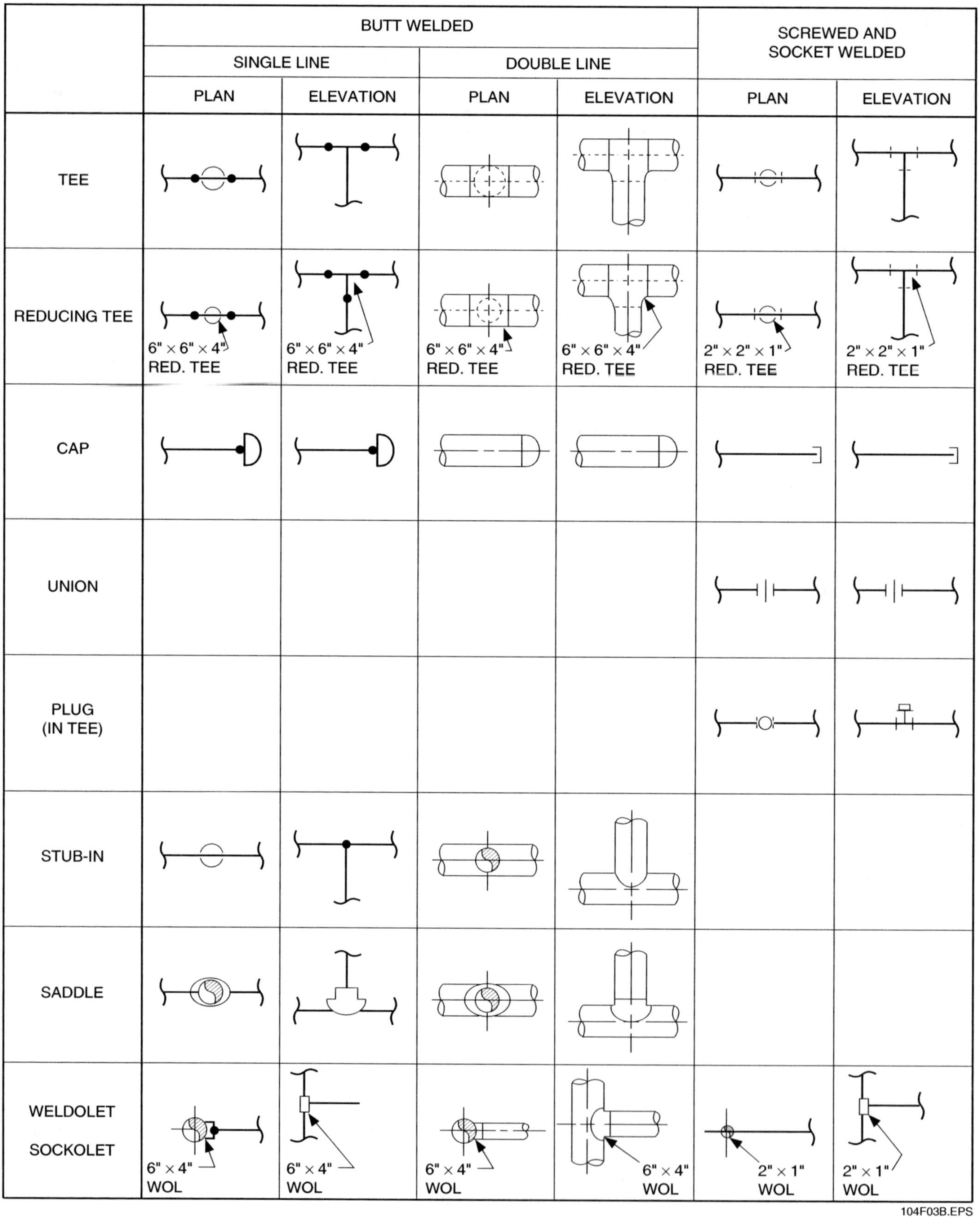

Figure 3 ◆ Commonly used piping symbols (2 of 4).

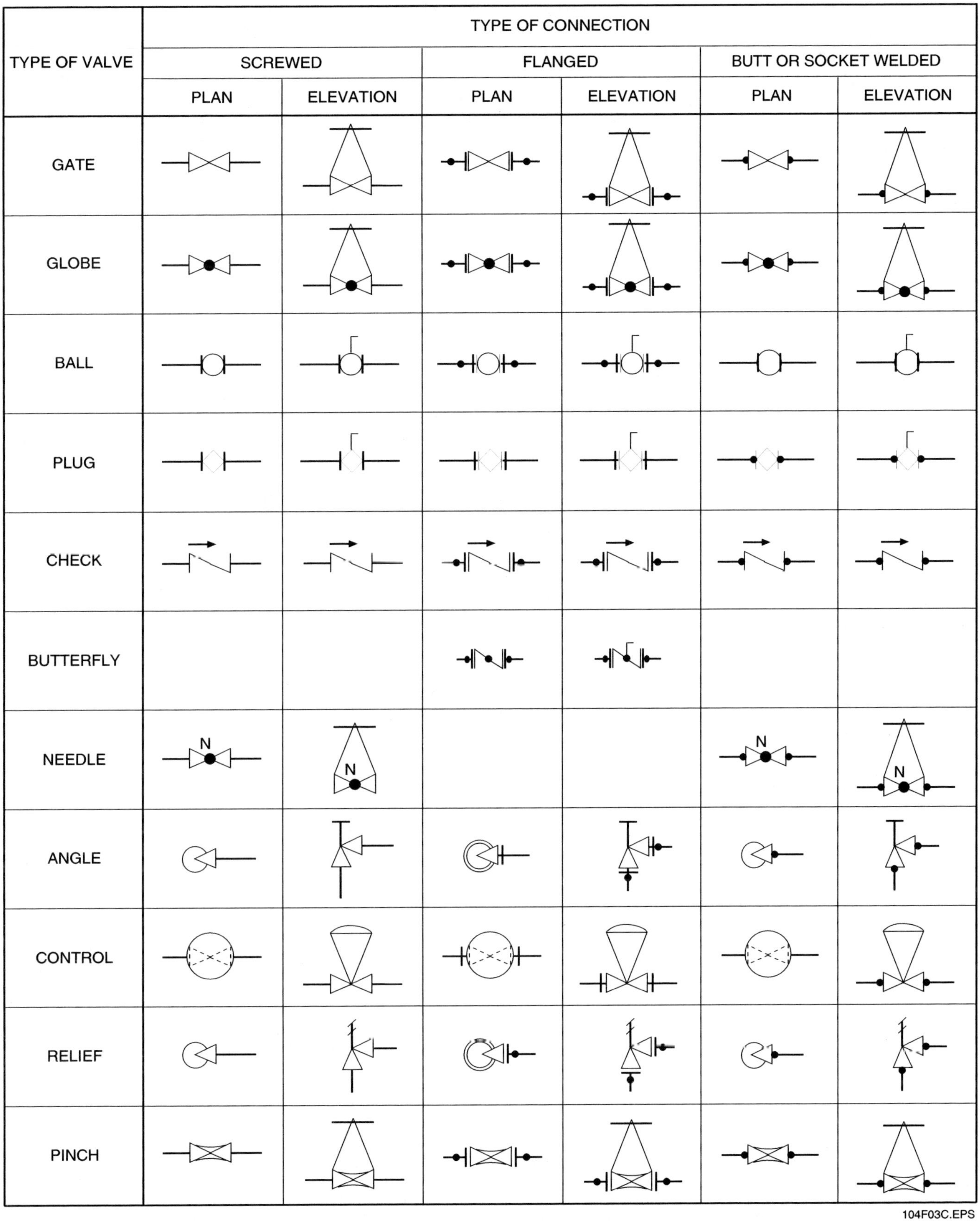

Figure 3 ◆ Commonly used piping symbols (3 of 4).

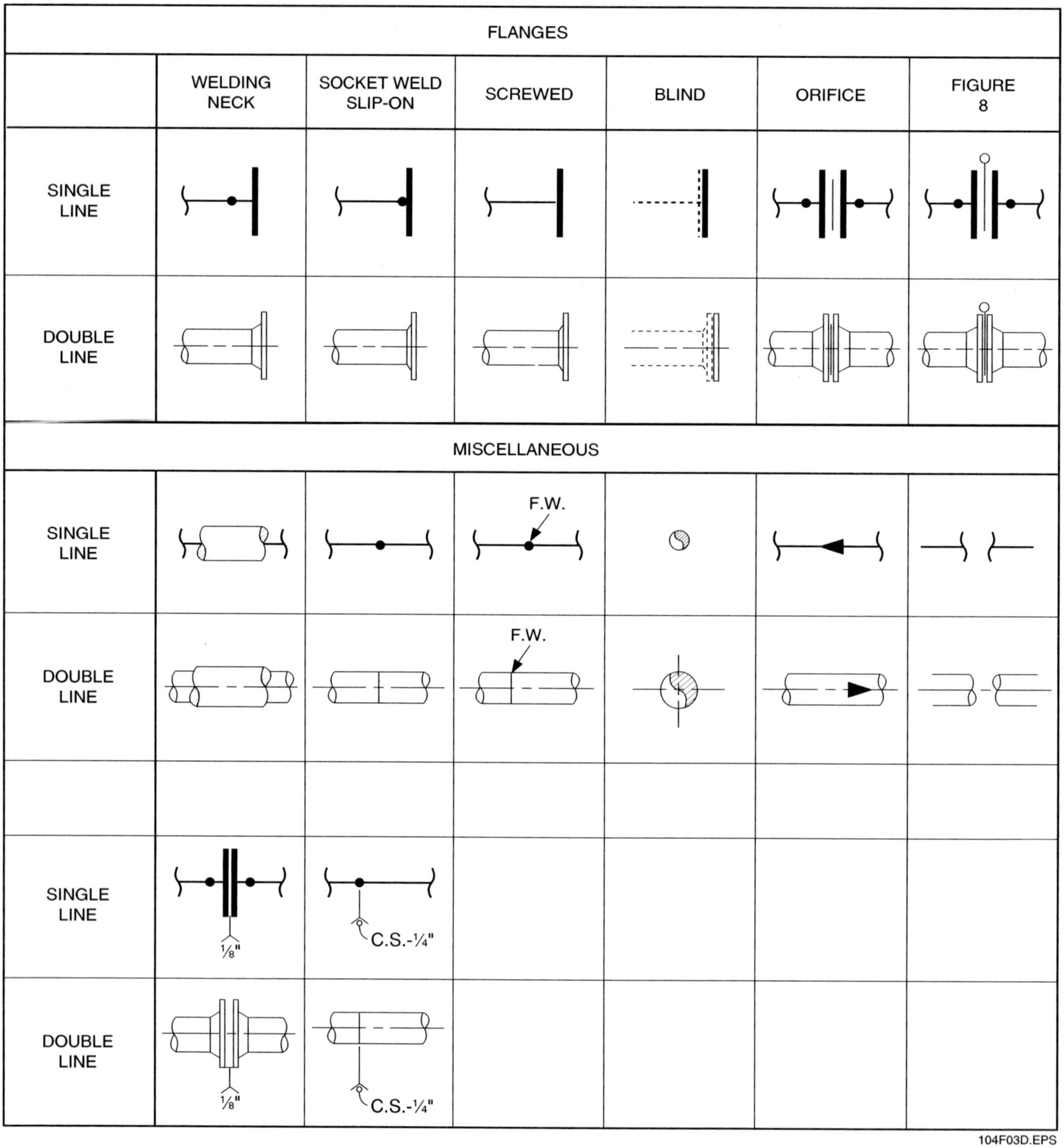

Figure 3 ◆ Commonly used piping symbols (4 of 4).

2.3.2 Abbreviations

Many words and phrases are commonly used by pipeline mechanics when referring to drawings. To avoid writing out each word or phrase every time it is used, standard abbreviations have been developed. The mechanic should recognize these abbreviations and the meaning of each. *Table 1* lists commonly used drawing abbreviations.

Table 1 Commonly Used Drawing Abbreviations (1 of 2)

Adapter	ADPT
Air preheating	A
American Iron and Steel Institute	AISI
American National Standards Institute	ANSI
American Petroleum Institute	API
American Society for Testing Materials	ASTM
American Society of Mechanical Engineers	ASME
Ash removal water, sluice or jet	AW
Aspirating air	AA
Auxiliary steam	AS
Bench mark	BM
Beveled	B
Beveled end	BE
Beveled, both ends	BBE
Beveled, large end	BLE
Beveled, one end	BOE
Beveled, small end	BSE
Bill of material	B/M
Blind flange	BF
Blowoff lines	BO
Bottom of pipe	BOP
Butt weld	BW
Carbon steel or cold spring	CS
Cast iron	CI
Ceiling	CLG
Chain operated	CO
Chemical feed	CF
Circulating water	CW
Cold reheat steam	CR
Compressed air	CA
Concentric or concrete	CONC
Condensate	C
Condenser air removal	AR
Continue, continuation	CONT
Coupling	CPLG
Detail	DET
Diameter	DIA
Dimension	DIM
Discharge	DISCH
Double extra-strong	XX STRG
Drain	DR
Drain funnel	DF
Drawing	DWG
Drip leg or dummy leg	DL
Ductile iron	DI
Dust collector	DC
Each	EA
Eccentric	ECC
Elbolet	EOL
Elbow	ELB
Electric resistance weld	ERW
Elevation	EL
Equipment	EQUIP
Evaporator vapor	EV
Exhaust steam	E
Expansion	EXP
Expansion joint	EXP JT
Extraction steam	ES
Fabrication (dimension)	FAB
Face of flange	FOF
Faced and drilled	F&D
Factory mutual	FM
Far side	FS
Feed pump balancing line	FB
Feed pump discharge	FD
Feed pump recirculating	FR
Feed pump suction	FS
Female	F
Female	FM
Female pipe thread	FPT
Field support on forged steel	FS
Field weld	FW
Figure	FIG
Fillet weld	W
Finish floor	F/F
Finish floor	FIN FL
Finish grade	FIN GR
Fitting	FTG
Fitting makeup	FMU
Fitting to fitting	FTF
Flange	FLG
Flat face	FF
Flat on bottom	FOB
Flat on top	FOT
Floor drain	FD
Foundation	FDN
Fuel gas	FG
Fuel oil	FO
Gauge	GA
Galvanized	GALV
Gasket	GSKT
Grating	GRTG
Hanger	HGR
Hanger rod	HR
Hardware	HDW
Header	HDR
Heat traced, heat tracing	HT
Heater drains	HD
Heating system	HS
Heating, ventilating, and air conditioning	HVAC
Hexagon	HEX
Hexagon head	HEX HD
High point	HPT
High pressure	HP
Horizontal	HORIZ
Hot reheat steam	HR
Hydraulic	HYDR
Increaser	INCR
Input/output	I/O
Inside diameter	ID
Insulation	INS
Invert elevation	IE
Iron pipe size	IPS
Isometric	ISO
Issued for construction	IFC

104T01A.EPS

Table 1 Commonly Used Drawing Abbreviations (2 of 2)

Lap weld	LW	Safety valve vents	SV
Large male	LM	Sanitary	SAN
Length	LG	Saturated steam	SS
Long radius	LR	Schedule	SCH
Long tangent	LT	Screwed	SCRD
Long weld neck	LWN	Seamless	SMLS
Low pressure	LP	Section	SECT
Low-pressure drains	DR	Service and cooling water	SW
Low-pressure steam	LPS	Sheet	SH
Lubricating oil	LO	Short radius	SR
		Slip-on	SO
Main steam	MS	Socket weld	SW
Main system blowouts	BL	Sockolet	SOL
Makeup water	MU	Stainless steel	SS
Male	M	Standard weight	STD WT
Malleable iron	MI	Steel	STL
Manufacture	MFR	Suction	SUCT
Manufacturer's Standard Society	MSS	Superheater drains	SD
		Swage	SWG
National Pipe Thread	NPT	Swaged nipple	SN
Nipolet	NOL		
Nipple	NIP	Temperature or temporary	TEMP
Nominal	NOM	That is	I.E.
Not to scale	NTS	Thick	THK
Nozzle	NOZ	Thousand	M
		Thread, threaded	THRD
Outside battery limits	OSBL	Threaded	T
Outside diameter	OD	Threaded end	TE
Outside screw and yoke	OS&Y	Threaded, both ends	TBE
Overflow	OF	Threaded, large ends	TLE
		Threaded, one end	TOE
Pipe support	PS	Threaded, small end	TSE
Pipe tap	PT	Threadolet	TOL
Piping and instrumentation diagram	P&ID	Top of pipe or top of platform	TOP
Plain	P	Top of steel or top of support	TOS
Plain end	PE	Treated water	TW
Plain, both ends	PBE	Turbine	TURB
Plain, one end	POE	Typical	TYP
Point of intersection	PI		
Point of tangent	PT	Underwriters Laboratory	UL
Pounds per square inch	PSI		
Process flow diagram	PFD	Vacuum	VAC
Purchase order	PO	Vacuum cleaning	VC
		Vents	V
Radius	RAD	Vitrified tile	VT
Raised face	RF		
Raised face slip-on	RFSO	Wall thickness or weight	WT
Raised face smooth finish	RFSF	Weld neck	WN
Raised face weld neck	RFWN	Weldolet	WOL
Raw water	RW	Well water	WW
Reducer, reducing	RED	Wide flange	WF
Relief valve	RV		
Ring-type joint	RTJ		
Rod hanger or right hand	RH		

104T01B.EPS

2.4.0 Notes

Notes are provided in addition to drawings to explain items that cannot be easily drawn (*Figure 4*). These items include general requirements, quality assurance criteria, calibration tolerances, and similar information. Notes may be located in a corner of a blueprint, or they may be listed separately and included with a set of blueprints.

NOTES:
1. FOR PIPING GENERAL NOTES AND REFERENCE DRAWINGS SEE DWG 5439-3.
2. ALL PUMPS ARE LOCATED TO CENTER OF DISCHARGE UNLESS OTHERWISE NOTED.

104F04.EPS

Figure 4 ◆ Typical notes.

Notes can be divided into two categories: general notes that apply to many parts of the structure, and local notes that explain a specific part of the construction and are keyed to that point with a line.

2.5.0 Revision Blocks

The revision block is usually located in the lower center of the drawing and is used to record any changes or revisions to the drawing. All revisions are noted in this block. They are dated and are identified by a letter or number and a description of the revision. A revised drawing is shown by the addition of a letter or number next to the original drawing number in the title block. *Figure 5* shows a typical revision block.

PROJ	NO	REVISION	RVSD	CHKD	APPD	DATE
3483	01	RELEASED FOR CONSTRUCTION		APD	NWS	JULY 92
3483	02	DELETED PART OF LINE 12037		APD	NWS	AUG 92
3483	03	△ ADDED WELDING SYMBOL		APD	NWS	AUG 92

104F05.EPS

Figure 5 ◆ Typical revision block.

2.6.0 Control Points

Column line control points are used to locate the work horizontally. The nearest column line on a drawing is shown, and the piping is then **dimensioned** from this column. This column is designated using the grid line system. The grid line system designates every column inside a building or structure. On a grid line drawing, an imaginary grid is made using the column lines. The areas within the column lines are known as *bays*. *Figure 6* shows a grid line drawing.

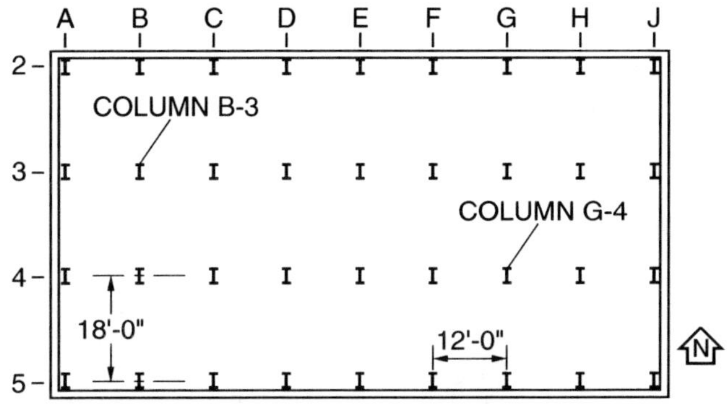

104F06.EPS

Figure 6 ◆ Grid line drawing.

The numbering and lettering of the system normally begins in the upper left- or right-hand corner of the system, which is the north side of the building or structure. The numbers are normally assigned to the vertical rows, and the letters are assigned to the horizontal rows. The letters I, O, and Q and the numbers 1 and 0 have been eliminated from the system to avoid confusion. This system allows each beam in the building to be given a reference number or control point, such as B-3. Normally, the engineer labels each beam in a building for ease in identification. The use of column lines permits anyone who is familiar with the system to locate any point exactly. When column lines are used with elevations, the locations can be specified three-dimensionally.

2.7.0 Coordinates

Coordinates are a system of numbers used to define the geographic position of a point, line, piece of equipment, or other object on the plant site. A coordinate is the distance from a given point to a base point, which is commonly called a *bench mark*. Coordinates are written as a letter and number combination. The letter designates the direction of the coordinate, normally north or east, in relation to the bench mark, and the number designates the distance in feet and inches. Coordinates are used to locate major equipment on drawings.

A property plan indicates the position of the originating point or coordinate bench mark. From this point, a grid system is developed and is used by all design groups to coordinate their drawings.

The north-south axis and the east-west axis intersect through this originating point, and every point on the drawing can be referenced by the originating point. For example, a point that is 30 feet east of the north-south axis through the originating point has an east coordinate of E. 30' – 0". If this same point were 1 foot north of the east-west axis through the originating point, it would have a north coordinate of N. 1' – 0". Put together, the two coordinates pinpoint the location of the point as E. 30' – 0" and N. 1' – 0". *Figure 7* shows coordinate origin.

2.8.0 Elevations

Elevations locate piping vertically in relation to ground elevation. Elevations may be referenced to the center line of the pipe, to the bottom of a pipe, to the face of a flange, or to any other convenient point. The drawing must indicate which point the reference is made to and must always be consistent. Engineers normally mark elevations on the structural columns, using a reference line. These elevations are written in feet and inches. To determine the exact elevation at which a pipeline must be set, locate the elevation reference point marked by the engineer.

Piping elevations are specified in different ways. Sometimes, the elevation is given as an above-the-finished-floor elevation, in which case you would measure from the finished floor to the desired height to establish the equipment elevation. In other cases, the elevation must be determined from an established elevation bench mark.

An elevation bench mark may be inside or outside the building. Usually, the bench mark is located on building columns in various places throughout the building. To determine a rough elevation from a bench mark on a column, you can measure from the bench mark to the floor, using the floor as a measuring point. For a more accurate measurement, more precise methods must be used.

Elevations can be transferred from the bench mark to the bottom of the pipe or face of the flange, using a string level, tubing water level, or an optical level. The most precise tool is the optical level.

2.9.0 Line Lists

The line list, or line table, provides the following information about piping lines:

- **Line numbers** in numerical order
- Class or specification of each line and size
- Whether or not each line is insulated
- What each line transports (liquid or vapor)
- Starting and termination points
- Design pressure and temperature
- Operating pressure and temperature
- Testing pressure

The piping engineering department usually fills out the line list and selects the piping specifications that can be used with the system. *Figure 8* shows an example of a line list.

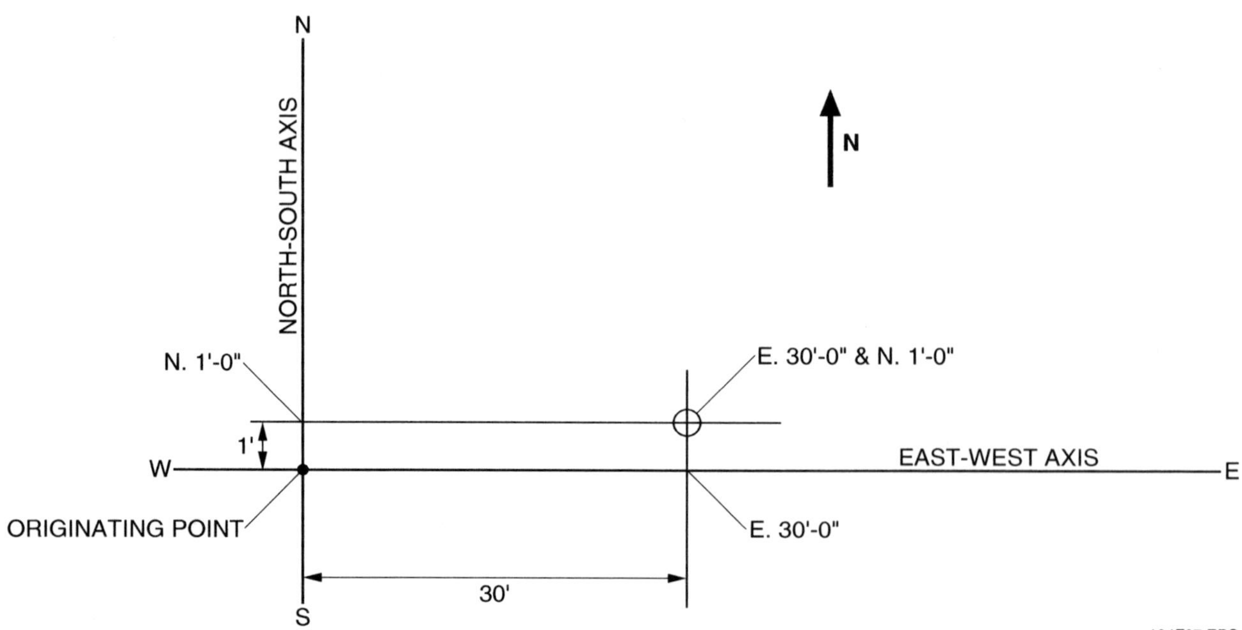

Figure 7 ◆ Coordinate origin.

LINE LIST

JOB NO. ____________
SHEET NO. _____ REV. _____
DATE _______ BY _______

REV	FLOW DIAG. NO.	LINE NO.	LINE CLASS	LINE SIZE	INSU	COMMODITY	VAP. / LIQ.	DESCRIPTION: ORIGIN (FROM)	DESCRIPTION: TERMINUS (TO)	DES psig	OPER. psig	DES °F	OPER. °F	REINF. PAD	POST HEAT TREAT
△		1	C	6"	IH	HYDRO-CARBON FEED	V L	BATTERY LIMITS	T-1	300	255	250	210	NO	NO
△		2	C	—	IH	HYDRO-CARBON	V L	T-1 TRIM MATERIAL		300	250	275	245	NO	NO
△		3	C	16"	IH	HYDRO-CARBON	L	T-1	E-2	300	250	275	230	NO	NO
△		4	C	20"	IH	HYDRO-CARBON	V	E-2	T-1	300	250	275	245	NO	NO
△		5	C	14"	NO	HYDRO-CARBON	V	T-1	E-1	300	245	150	125	NO	NO
△		6	C	10"	NO	HYDRO-CARBON	L	E-1	V-1	300	240	150	118	NO	NO
△		7	C	—	NO	HYDRO-CARBON	L V	V-1 TRIM MATERIAL		300	240	150	118	NO	NO
△		8	C	1-1/2"	NO	HYDRO-CARBON	L	V-1	OILY WATER SEWER	300	240	150	118	NO	NO
△		9	C	2"	NO	HC FLARE	V	V-1	PC - 1 AT LINE 15 A - 2"	300	240	150	118	NO	NO
△		10	C	10"	NO	PROPANE	L	V-1	P - 1 A & B	300	240	150	118	NO	NO
△		11	C	8"	NO	PROPANE	L	P - 1 A & B	T - 1 REFLUX & TO LC - 2 @ LINE 12 A - 4"	310	280	150	118	NO	NO
△		12	A	4"	NO	PROPANE	L	LINE 11C - 4"	BATTERY LIMITS (TO STORAGE)	245	220	140	118	NO	NO
△		13	C	6"	IH	HYDRO-CARBON	L	LINE 3C - 16"	BATTERY LIMITS (TO STORAGE)	300	250	275	230	NO	NO
△		14	A	8"	NO	HYDRO-CARBON	V	RV - 1	ATMOSPHERE	—	ATM.	—	AMB.	NO	NO
△		15	A	2"	NO	HC FLARE	V	PC - 1 AT LINE 9C - 2"	BATTERY LIMITS (FLARE)	100	40	150	118	NO	NO

104F08.EPS

Figure 8 ◆ Example of a line list.

3.0.0 ◆ TYPES OF DRAWINGS

Construction drawings show as much construction detail as possible. The details are presented graphically. Most construction drawings consist of orthographic views and are presented as plans and elevations. The following types of drawings are normally used by pipeline mechanics:

- Plot plans
- Drawing indexes
- Equipment arrangement drawings
- Piping and instrumentation diagrams (P&IDs)
- Isometric drawings
- Equipment drawings
- Pipe support drawings and detail sheets

3.1.0 Plot Plans

A plot plan shows the location of a building on a piece of property and usually indicates where the water and sewer mains are installed (*Figure 9*). The plot plan includes information on sections of the land where nothing can be built, such as the following:

- *Setback* – This is the minimum distance that must be maintained between the street and the structure.
- *Easements* – These are places where utilities can be installed.
- *Side yards* – These areas provide access to rear yards, reduce the possibility of fire transferring from one building to another, and promote ventilation.

3.2.0 Drawing Indexes

The drawing index, provided on the first page of a complete set of drawings, is a list of all drawings included in the set. It is developed from the plot plan.

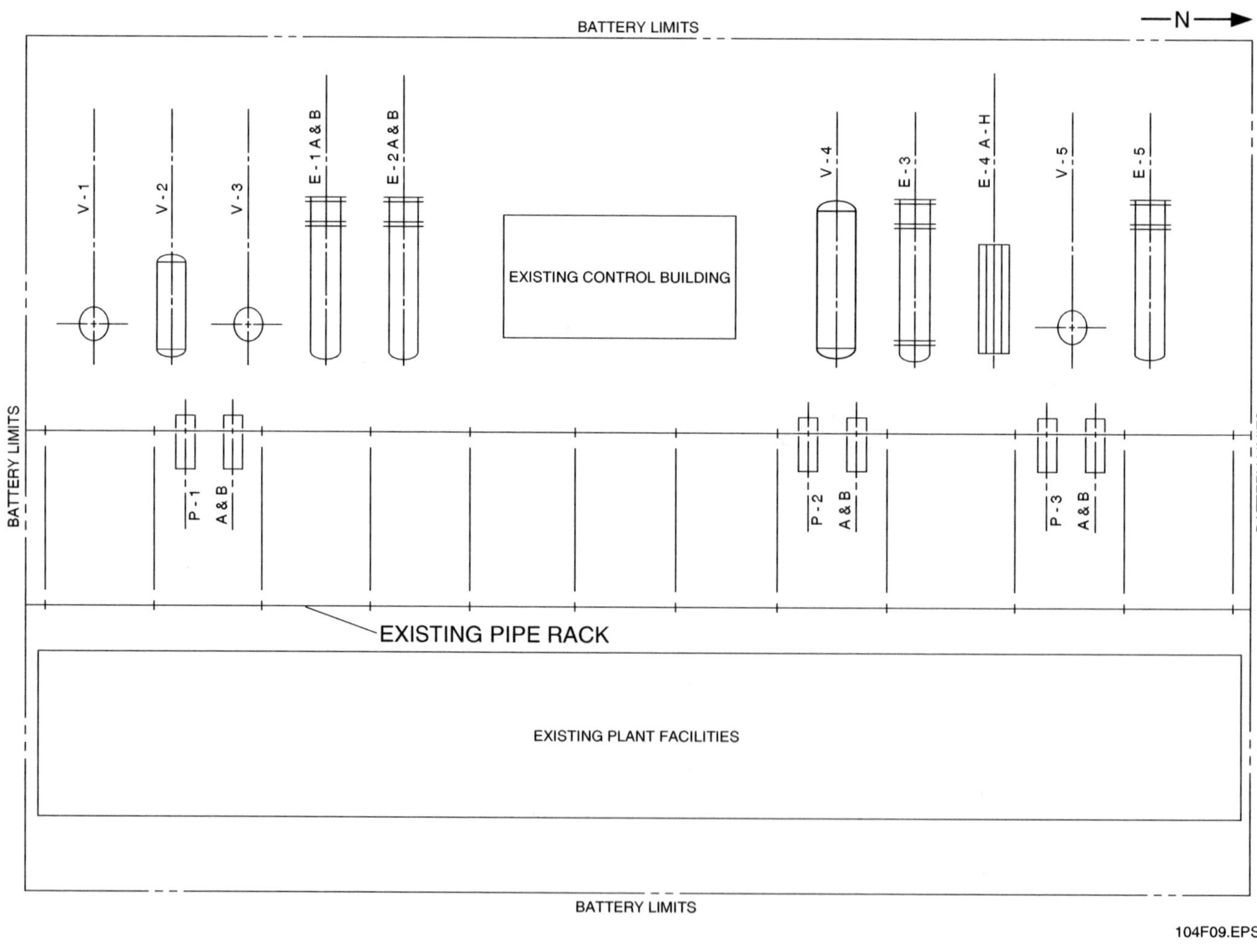

Figure 9 ◆ Simplified plot plan.

The drawings are usually organized into the following major categories:

- Civil (C)
- Architectural (A)
- Interior/graphics (D)
- Structural (S)
- Mechanical equipment (M)
- Piping and plumbing (P)
- Electrical (E)

3.3.0 Equipment Arrangement Drawings

Equipment arrangement drawings show how the equipment is arranged and the locations of all equipment used with the system. Equipment arrangement drawings can be either plan or section views. Equipment on these drawings is usually drawn to correspond to the basic shape of the item represented. These representations are common to most companies.

3.3.1 Plan Views

A plan view is a view of an object or area from above, looking down on the system. The plan view shows the exact locations of the equipment in relation to established reference points, objects, or surfaces (see *Figure 10*). These reference points may be building columns, walls, bench marks, or other equipment.

Plan views are usually laid out with north at the top of the drawing. If the need for space makes it necessary to rotate the drawing, it is rotated so that north is to the left side of the sheet. A plan view normally includes a key plan that references where the drawing fits into the overall system.

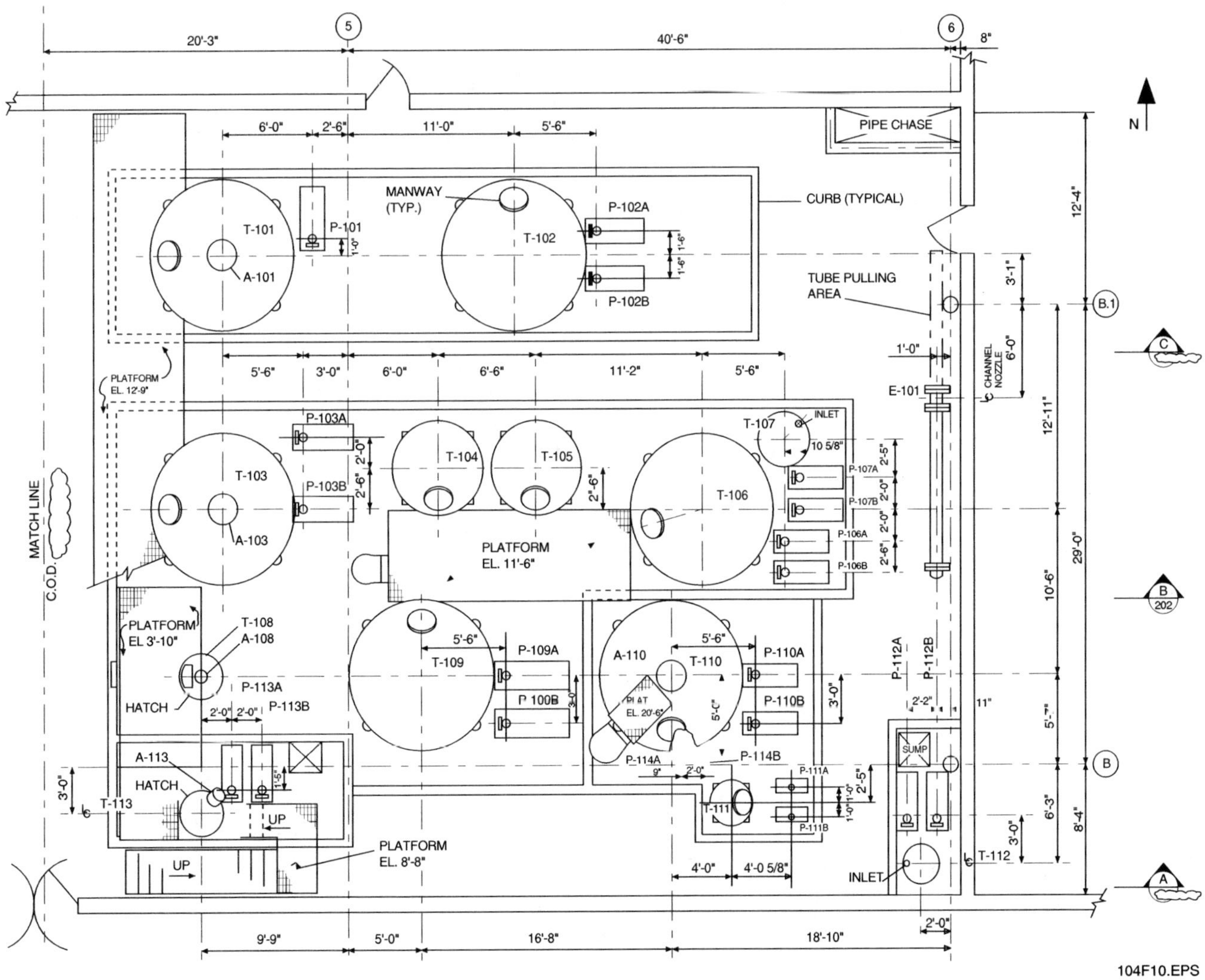

Figure 10 ◆ Equipment arrangement plan view.

3.3.2 Section Views

It is not possible to show all the necessary information in plans and elevations, so a section view must be used. A section view shows the profile of objects as if an imaginary line is cut through the equipment or facility (see *Figure 11*). Most section views are vertical, but horizontal or plan sections may be included. In section views, the parts are usually labeled.

The section view shows how the equipment or parts fit together in the system. From this drawing, the height, width, and location of the equipment can be seen. This is accomplished by drawing what the component looks like on the inside. Section drawings are usually drawn on a larger scale than floor plans and elevations.

3.4.0 Piping Arrangement Drawings

Piping arrangement drawings show how the equipment is arranged and the locations of all equipment used with the system. Piping arrangement drawings can be either plan or section views. Equipment on these drawings is usually drawn to correspond to the basic shape of the item represented. These representations are common to most companies.

A plan view is a view of an object or area from above, looking down on the system. The plan view shows the exact locations of the equipment in relation to established reference points, objects, or surfaces (see *Figure 12*). These reference points may be building columns, walls, bench marks, or other equipment.

Plan views are usually laid out with plant north at the top of the drawing. If the need for space makes it necessary to rotate the drawing, it is rotated so that north is to the left side of the sheet.

Sections, or sectional views, are projections of the vertical plane. Section views are detailed drawings showing a side view of a particular part of a plan drawing. A section view shows what the internal characteristics look like on the cutting plane. A cutting plane may not be continuous, and a horizontal cutting plane of a building may be staggered to include specific information.

Because of the way section views are drawn, they provide visual information not included on elevations or plan views. Sections are sometimes a cutaway view of an object to show internal characteristics that may not be obvious on other views. Many symbols are used on drawings to designate section cutting planes. *Figure 13* shows common symbols used for section views. *Figure 14* shows section views taken from *Figure 12*.

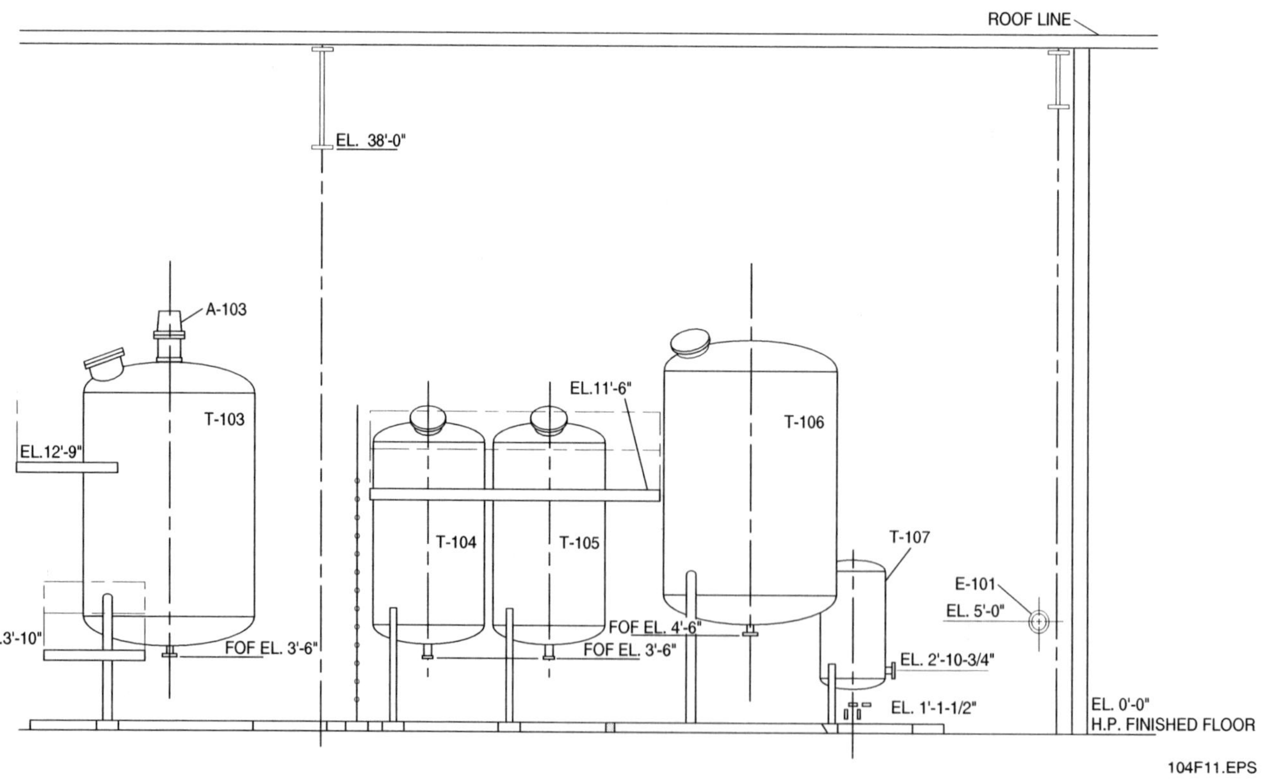

Figure 11 ◆ Equipment arrangement section view.

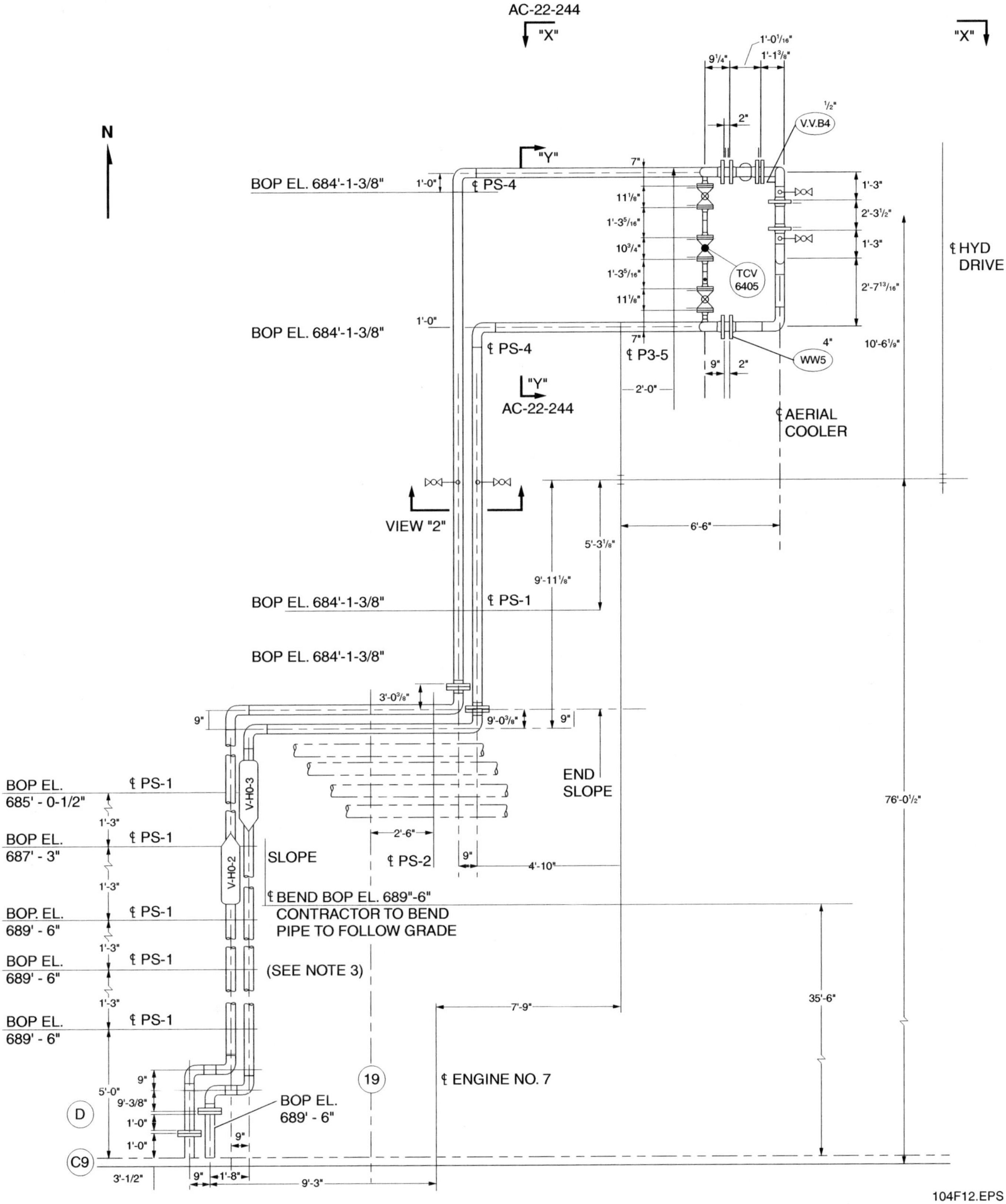

Figure 12 ◆ Piping arrangement plan view.

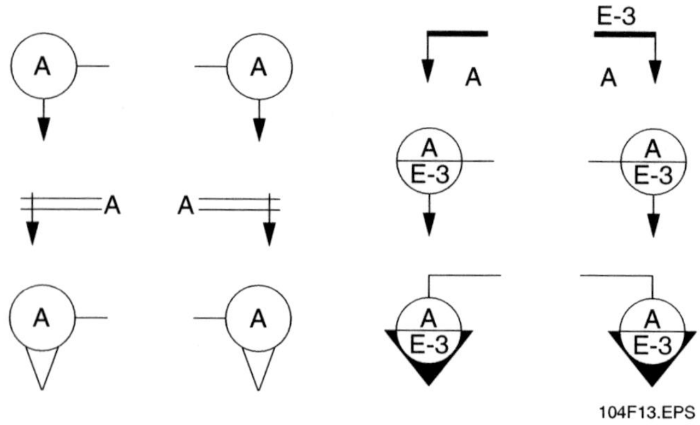

Figure 13 ◆ Common symbols used for section views.

Piping arrangement drawings use control points, coordinates, and elevations to indicate the exact location at which a pipeline, floor penetration, or piece of equipment is to be placed in a facility. Elevations refer to the vertical placement of an object; control points and coordinates refer to the horizontal locations. When identifying the location of an object in a building or area, first locate the horizontal location and then establish the elevation.

3.5.0 Piping and Instrumentation Diagrams

Piping and instrumentation diagrams, or P&IDs, are schematic diagrams of a complete system or systems (see *Figure 15*).

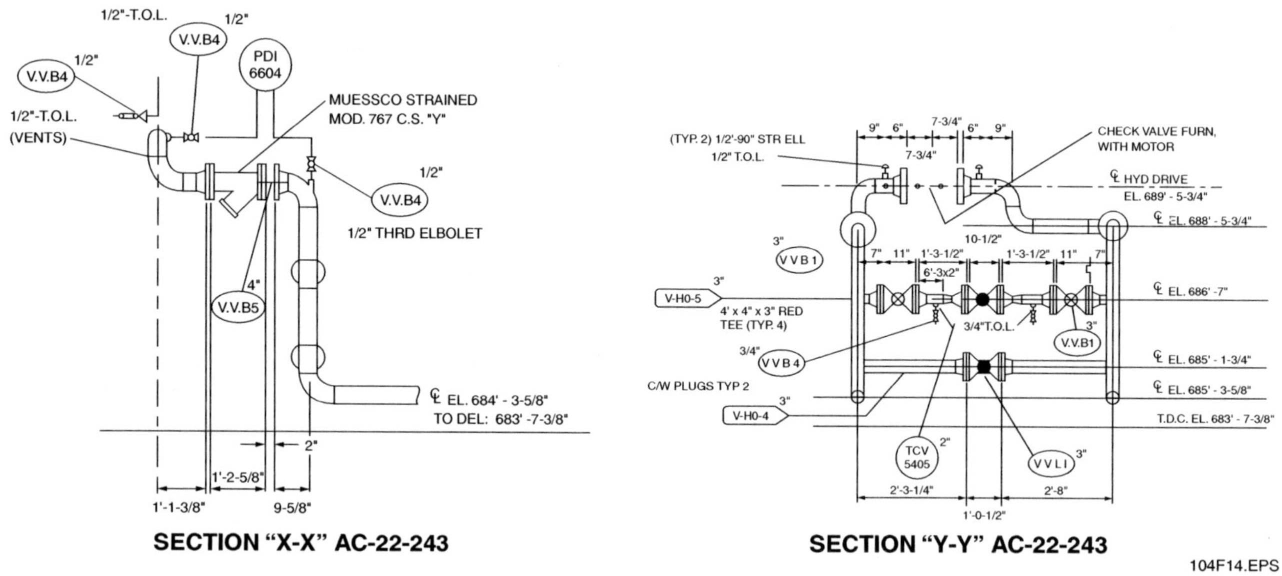

Figure 14 ◆ Section views taken from *Figure 12*.

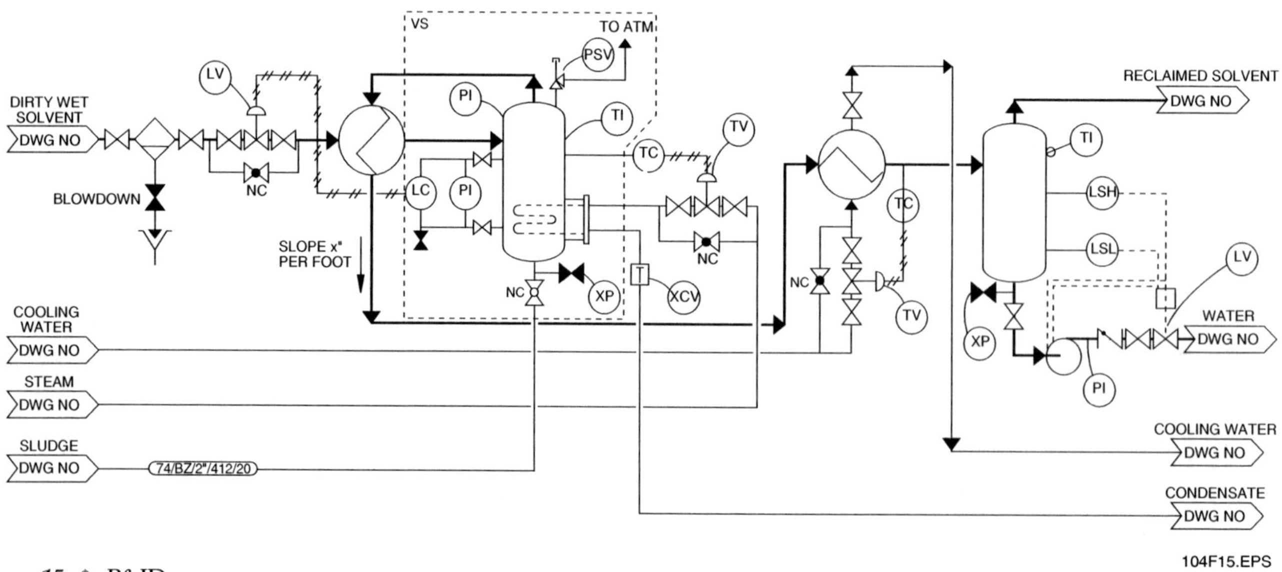

Figure 15 ◆ P&ID.

P&IDs show the process flow. They also show all equipment, pipelines, valves, instruments, and controls necessary for the operation of that system. Because their purpose is only to provide a representation of the work to be done, P&IDs are not drawn to scale or dimensioned. Pieces of equipment on P&IDs are usually drawn to correspond with the actual shape of the object. Valves and other parts are usually indicated with symbols. P&IDs will be explained in more detail later in this module.

3.6.0 Isometric Drawings

An isometric drawing, or ISO, combines the plan and elevation views of a system or part of a system into one drawing. Usually, only one pipeline is shown on an isometric drawing, although for simple or duplicate pipelines, more than one line may be shown. An isometric pipe drawing shows the details needed to fabricate and install a section of a piping system (see *Figure 16*). An ISO usually includes a bill of material, which lists the types of pipe, flanges, valves, bolts, and gaskets required for the pipe run.

Isometric piping drawings use the same symbols as P&IDs or any other piping drawings. The ISO has arrows that show the direction of flow. To read an ISO, begin at the end of the pipe run nearest the source and work in the direction of flow. Trace the line from the first fitting, and identify all the fittings and components by their symbols and abbreviations. Also, note the dimensions from one fitting or component to the other. The dimensions for valves and other components are shown on the ISO. The takeout for fittings and flanges must be calculated from the dimensions given on the ISO. The ISO also provides pipeline elevations for installation purposes.

3.7.0 Equipment Drawings

Equipment, or vendor, drawings are supplied by equipment manufacturers. Equipment drawings include the fabrication, erection, and setting drawings that are necessary to install a piece of equipment. They may also include manufacturer's standard drawings or catalog cuts, performance charts, brochures, and other data that illustrates the items. Samples of some materials are also required.

Pipeline mechanical technicians use equipment drawings to install equipment and to ensure that all the necessary prepiping is in its required location. Equipment drawings are required for almost every product that is fabricated away from the building site.

Equipment drawings are usually prepared after the contract is awarded and construction is under way. They are prepared by the equipment vendors or fabricators and are approved by the contractor and architect.

Equipment drawings are either stamped *certified* or *noncertified* by the manufacturer. If the drawings are stamped *certified*, the manufacturer guarantees that the piece of equipment is made exactly like the dimensions shown in the drawing. If the piece of equipment is different from what the drawing shows, the manufacturer is held liable for the equipment. If the drawing is stamped *noncertified*, the equipment is not guaranteed to the specifications shown in the drawing. Noncertified drawings and equipment are much less expensive than certified drawings and equipment.

3.8.0 Pipe Support Drawings and Detail Sheets

Before a piping system is installed, engineers analyze the system and determine the type, size, and placement of all hangers and supports in the system. Piping drawings are then made, showing the placement of each type of hanger and support. The technician must be able to read and interpret these drawings to install the hangers and supports in the proper place.

Symbols for pipe supports vary from job to job. On piping drawings, the hanger or support has a reference number that refers to the specific detail sheets used on that job. Check the detail sheets to determine the exact type of hanger to use. Check your company's standards and specifications to determine the reference prefixes used on your job site.

Each engineering company also uses pipe hanger and support symbols that are specific to a job site (*Figure 17*). These symbols appear on plan and isometric drawings and are called out by the reference number as explained above.

Refer to the detail sheets of a pipe hanger symbol to determine the details of a particular hanger. The detail sheets include all of the information needed to fabricate and install a hanger. Notice the symbol in the upper right-hand corner of *Figure 17* that designates a clevis hanger with reference number 5HR. Before installing this hanger, refer to the detail sheet for 5HR (*Figure 18*).

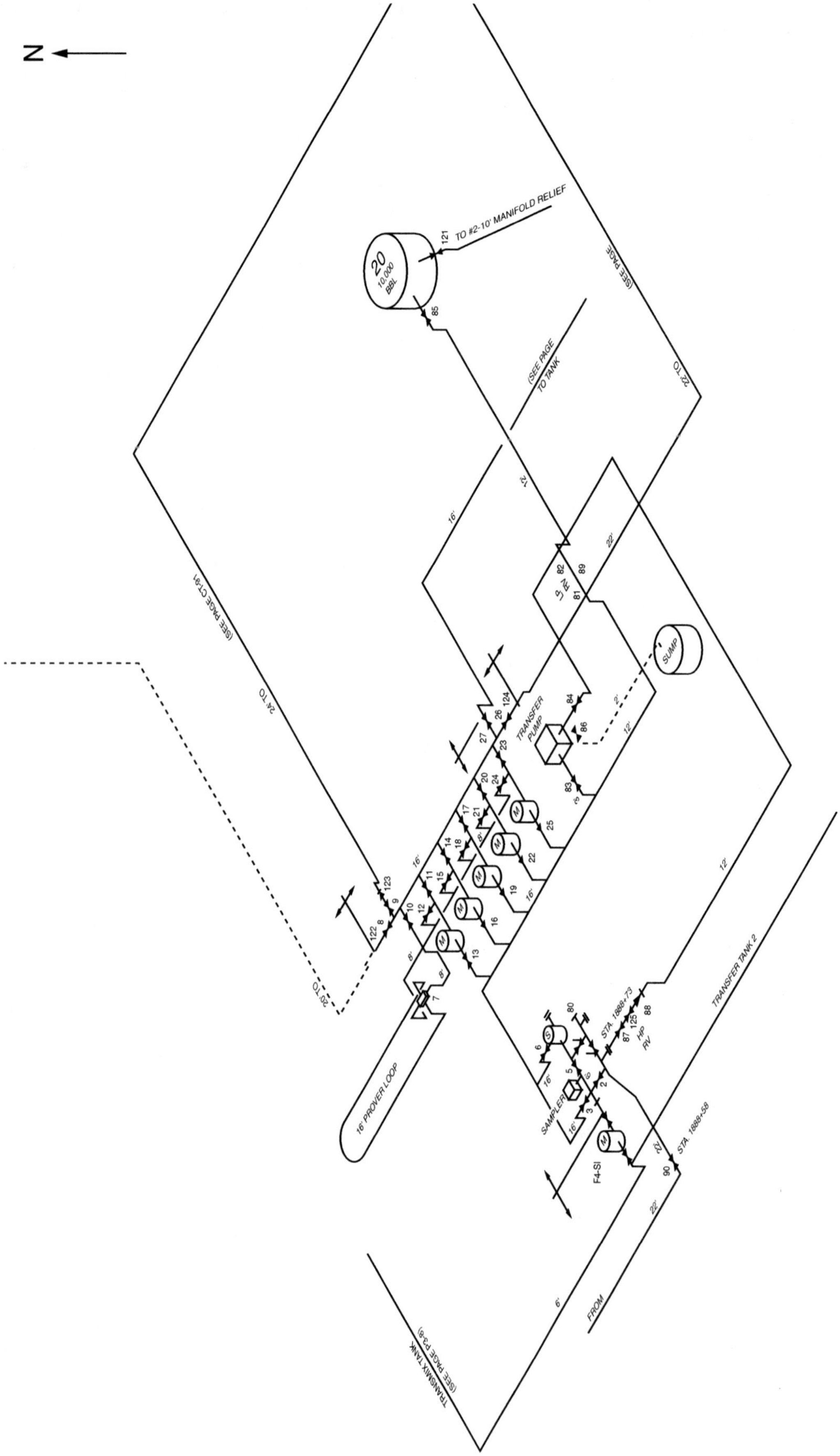

104F16.EPS

Figure 16 ◆ Piping isometric drawing.

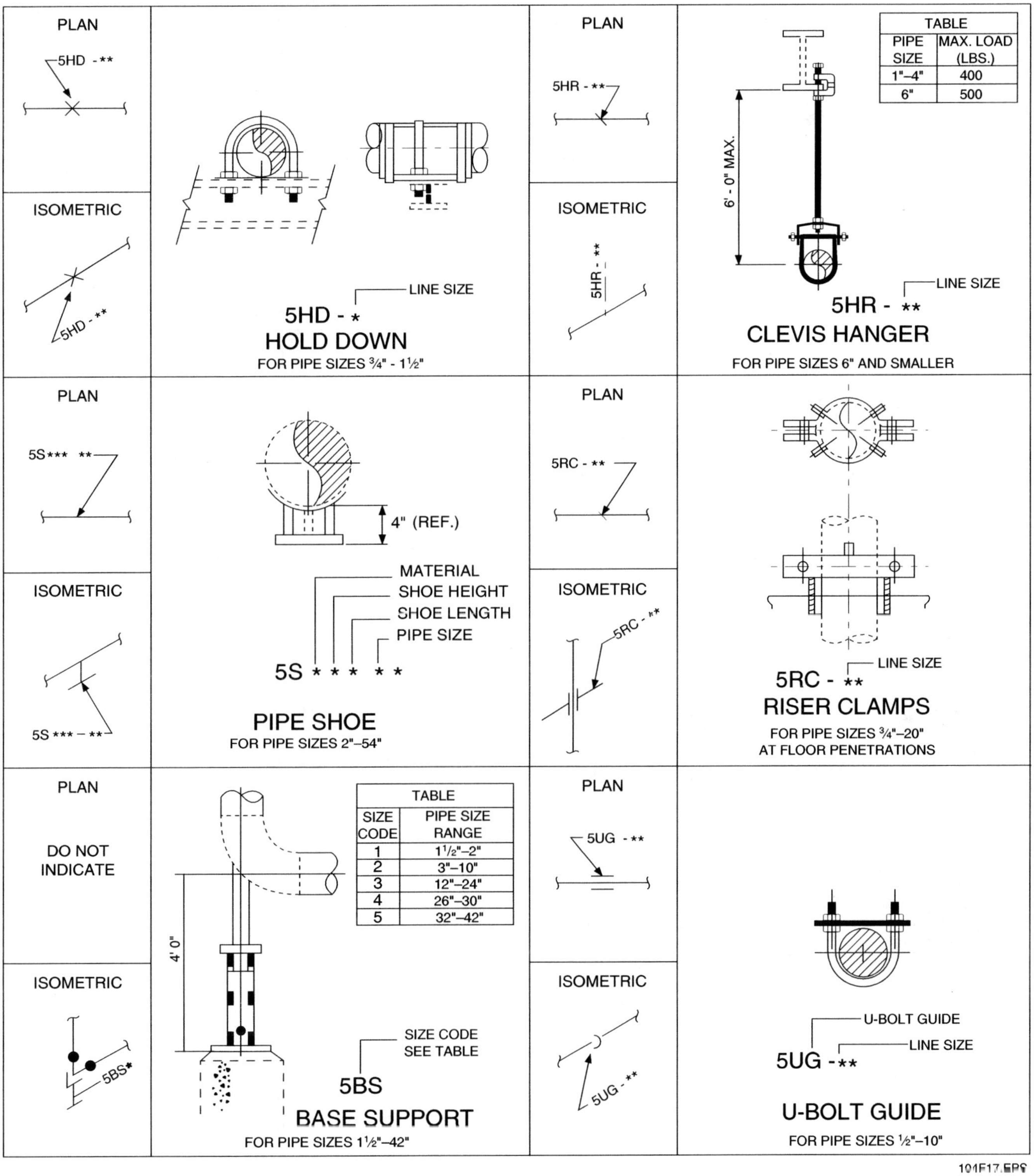

TABLE

PIPE SIZE	MAX. LOAD (LBS.)
1"–4"	400
6"	500

TABLE

SIZE CODE	PIPE SIZE RANGE
1	1½"–2"
2	3"–10"
3	12"–24"
4	26"–30"
5	32"–42"

Figure 17 ◆ Example of pipe support drawings and symbols.

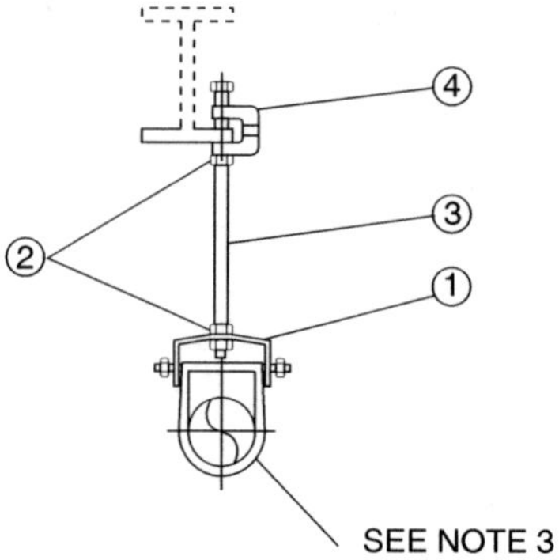

CLEVIS HANGER SELECTION CHART		
PIPE SIZE	ROD SIZE	CLEVIS LOAD (LBS.)
1"	3/8"	400
1 1/4"	3/8"	400
1 1/2"	3/8"	400
2"	3/8"	400
2 1/2"	1/2"	400
3"	1/2"	400
3 1/2"	1/2"	400
4"	5/8"	440
5"	5/8"	440
6"	3/4"	500

ASSEMBLY TAG NOS. ARE AS FOLLOWS:

5HR - **

HANGER ROD DESIGNATION (5HR) — LINE SIZE (**)

FABRICATION NOTES FOR METALLIC APPLICATIONS

1. PARTS EQUAL TO POWER PIPING COMPONENTS MAY BE USED.
2. FIELD TO CUT TO LENGTH REQUIRED AND TO THREAD BOTH ENDS.
3. FOR STAINLESS STEEL PIPE (1/2" - 6") USE 16-GAUGE STAINLESS STEEL BEARING PLATE.

TABLE 1			
LINE SIZE	CLAMP FIGURE NO.	BEAM BRACKET & BOLT SIZE	ROD DIA. & WELDLESS EYE& HEX NUT SIZE
1	222	1/2"	3/8"
2	222	1/2"	3/8"
3	222	1/2"	1/2"
4	222	3/4"	1/2"
6	222	3/4"	3/4"

MARK NO.	DESCRIPTION	TYPE	QUAN. REQ'D
①	ADJUSTABLE CLEVIS HANGER, POWER PIPING FIG. NO. 11 (SEE NOTE 1)	5HR	1
②	HEX NUT SAME SIZE AS ROD DIAMETER, POWER PIPING FIG. NO. 61 (SEE NOTE 1)	5HR	3
③	6'-0" LONG CARBON STEEL ROD, ASTM-A36, NOT THREADED, DIAMETER PER TABLE 1 (SEE NOTE 2)	5HR	1
④	CLAMP WITH RETAINING CLIP	5HR	1

ENGINEERING COMPANY NAME	HANGER RODS 5HR - **
	* = LINE SIZE
	SPECIFICATION NUMBER 50201 — SHEET 1 OF 1

104F18.EPS

Figure 18 ◆ Detail sheet for 5HR.

4.0.0 ◆ READING PIPING AND INSTRUMENTATION DIAGRAMS

A P&ID depicts the functional layout of a fluid system and its piping, valves, and instrumentation as clearly and accurately as possible. It is accurate to the extent that all components are connected to each other as shown in relation to flow path orientation. A P&ID does not represent the actual physical layout of equipment; that is, a valve which may appear to be right at the discharge of a pump can actually be located some distance away or even on another floor. Many times a P&ID will use a broken line around groups of equipment to indicate that they are all in the same building. Refer to the P&IDs in the drawing package when reviewing this section.

The P&ID depicts all components of a particular system including pipe sizes, flange sizes, valve sizes, flow direction, and references to other diagrams. Standardized symbols are used for these components instead of pictures. Industry standards for these symbols have been developed and adopted by organizations such as the American National Standard Institute (ANSI), Instrument Society of America (ISA), and American Society for Mechanical Engineers (ASME) to aid interpretation. The P&ID title block and legend should be examined closely for explanations of the diagram.

Although you may remember many of the common symbols and abbreviations used on a P&ID, it is impossible to memorize all the symbols. There will also be minor differences in the symbols and abbreviations used on different contractors' drawings. A legend showing the symbols and abbreviations and the meaning of each often appears in a space on the working diagram or on a separate diagram.

P&IDs provide the following information about a system:

- Facility piping
- Piping components
- Pipeline facility
- Instrumentation

4.1.0 Facility Piping

P&IDs show the entire piping system and provide information about each pipeline through the use of lines, line numbers, and match lines.

4.1.1 Lines

The lines on a P&ID show the process flow through a piping system from one place or piece of equipment to another. Each type of line on a P&ID is used for a special purpose (see *Figure 19*). The way that a line is drawn provides the pipefitter with information about that line.

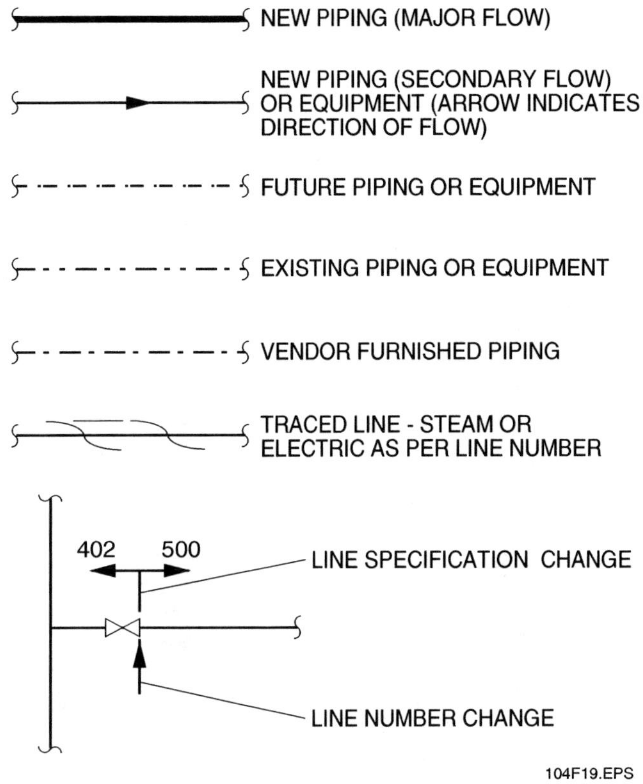

Figure 19 ◆ Types of lines used on P&IDs.

4.1.2 Line Numbers

Line numbers are used on P&IDs to indicate the size of pipe, service of line, piping material specification group number, number of the pipe in a particular service, and insulation required. In other P&IDs or other types of drawings of the same system, line numbers are used to identify lines and to transfer pipelines from one drawing to another.

Line numbers are normally enclosed in an oval-shaped symbol with an arrow at one end. The arrow indicates the direction of flow through the pipeline and should be used as often as necessary to maintain continuity. The line number can be shown directly on the line or above the line, with an arrow pointing to the line (see *Figure 20*).

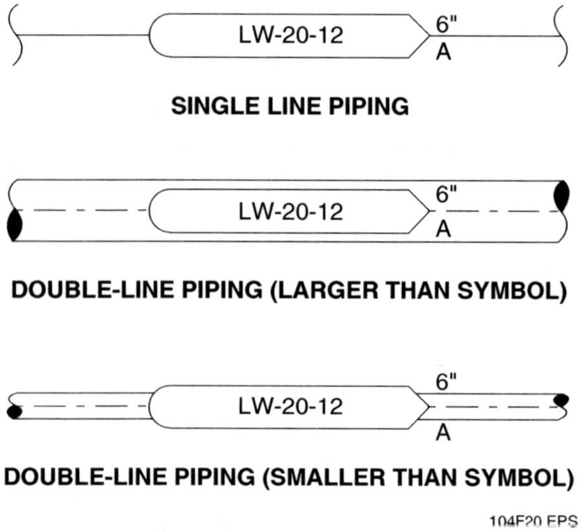

Figure 20 ◆ Typical line numbers.

In *Figure 20*, 6" indicates the nominal pipe size; LW is the fluid symbol and represents white liquor in this case; the number 20 indicates the piping material specification; the number 12 indicates the particular line number in the white liquor system; and the letter A indicates the insulation group numbers from the specifications. If no insulation letter is given, insulation is not required.

Line numbers vary between different engineering companies, and you must always check the legend of the P&ID to determine the system for reading the line numbers.

4.1.3 Match Lines

Reading a P&ID is a simple matter of tracing a system from one location or piece of equipment to another and interpreting the symbols and abbreviations on the diagram. Sometimes, an entire system is shown on a single sheet, which makes it very easy to trace the system. Larger systems may require the use of two or more sheets to show the entire system, and the system must be traced from one P&ID to another. Match lines help you do this.

A numbered match line is provided at the end of the P&ID to show where the system leaves one sheet and starts again on another sheet. The system is traced to the numbered match line on one sheet and continues at the same number on the match line of the next sheet. The system always continues on the next P&ID in the numbering sequence of P&ID sheets unless otherwise noted on the match line. If the system continues on a P&ID sheet other than the next sheet in sequence, a note to the effect appears in a box on the system line at the match line (see *Figure 21*).

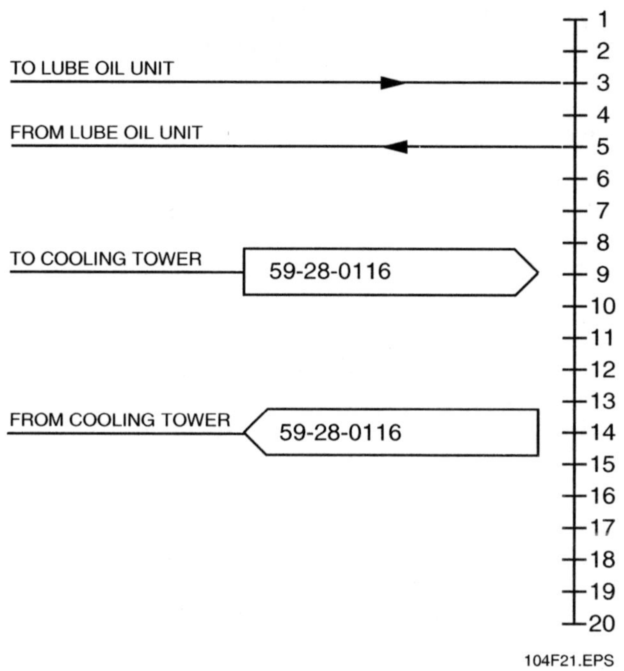

Figure 21 ◆ Match lines.

4.2.0 Piping Components

P&IDs show piping system components, such as valves, strainers, expansion joints, clean-outs, flexible connections, dampers, desuperheaters, removable **spools**, drains, and vents. P&IDs use symbols and abbreviations to describe these components. Always check the legend on the drawings to determine exactly what the symbols stand for. These symbols differ between engineering companies. *Figure 22* shows some commonly used piping component symbols.

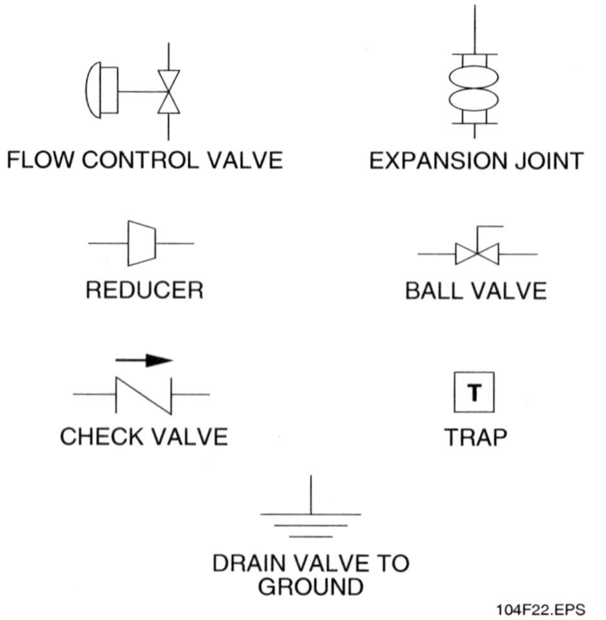

Figure 22 ◆ Commonly used piping component symbols.

4.3.0 Pipeline Facility

P&IDs show pipeline facility equipment, such as pumps, vessels, motors, turbines, condensers, and other major equipment in a system. P&IDs also give information about the size, capacity, horsepower, and operating parameters of the equipment. *Figure 23* shows commonly used equipment symbols.

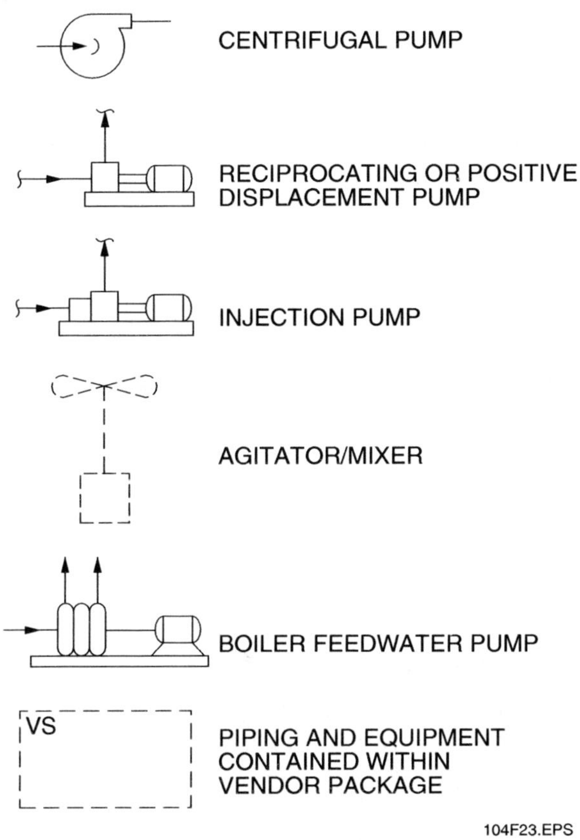

Figure 23 ◆ Commonly used equipment symbols.

4.4.0 Instrumentation

P&IDs show system instrumentation, such as control valve operators and analytical, flow, pressure, level, and temperature instruments. P&IDs use symbols and abbreviations to describe the instruments and the functions in a system. *Figure 24* shows typical instrumentation symbols and the meaning of each.

5.0.0 ◆ MACHINE DRAWINGS

Machine drawings contain all the information needed to manufacture a part exactly as it is designed, no matter what machine shop makes it and no matter where in the world it is made. The standardization of drawing practices and computer aided drawing (CAD) programs ensure that engineering drawings are a common language that provides consistent results. The basic parts of a machine drawing are the following:

- Lines
- Dimensions
- Notes and symbols
- Scale
- Revisions
- Vendor information
- Material specifications

Figure 25 shows a machine drawing of a direct drive pump.

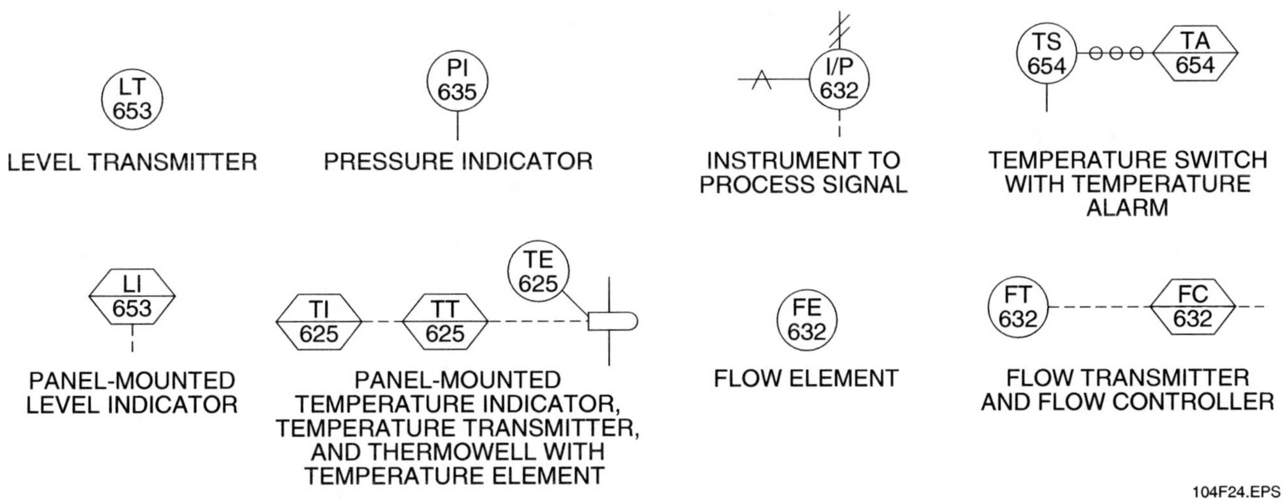

Figure 24 ◆ Typical instrumentation symbols.

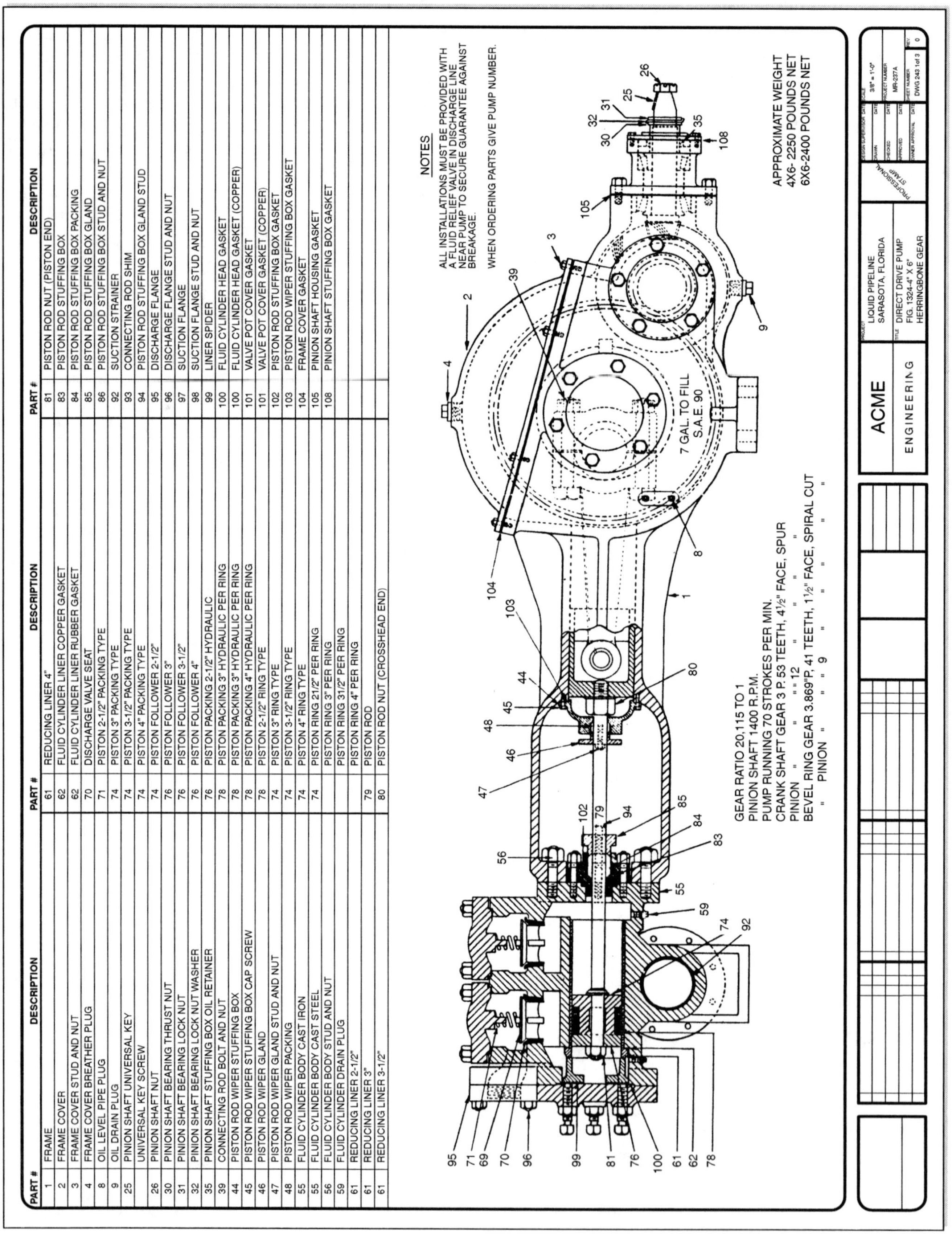

PART #	DESCRIPTION	PART #	DESCRIPTION	PART #	DESCRIPTION
1	FRAME	61	REDUCING LINER 4"	81	PISTON ROD NUT (PISTON END)
2	FRAME COVER	62	FLUID CYLINDER LINER COPPER GASKET	83	PISTON ROD STUFFING BOX
3	FRAME COVER STUD AND NUT	62	FLUID CYLINDER LINER RUBBER GASKET	84	PISTON ROD STUFFING BOX PACKING
4	FRAME COVER BREATHER PLUG	70	DISCHARGE VALVE SEAT	85	PISTON ROD STUFFING BOX GLAND
8	OIL LEVEL PIPE PLUG	71	PISTON 2-1/2" PACKING TYPE	86	PISTON ROD STUFFING BOX STUD AND NUT
9	OIL DRAIN PLUG	74	PISTON 3" PACKING TYPE	92	SUCTION STRAINER
25	PINION SHAFT UNIVERSAL KEY	74	PISTON 3-1/2" PACKING TYPE	93	CONNECTING ROD SHIM
	UNIVERSAL KEY SCREW	74	PISTON 4" PACKING TYPE	94	PISTON ROD STUFFING BOX GLAND STUD
26	PINION SHAFT NUT	74	PISTON FOLLOWER 2-1/2"	95	DISCHARGE FLANGE
30	PINION SHAFT BEARING THRUST NUT	76	PISTON FOLLOWER 3"	96	DISCHARGE FLANGE STUD AND NUT
31	PINION SHAFT BEARING LOCK NUT	76	PISTON FOLLOWER 3-1/2"	97	SUCTION FLANGE
32	PINION SHAFT BEARING LOCK NUT WASHER	76	PISTON FOLLOWER 4"	98	SUCTION FLANGE STUD AND NUT
35	PINION SHAFT STUFFING BOX OIL RETAINER	76	PISTON PACKING 2-1/2" HYDRAULIC	99	LINER SPIDER
39	CONNECTING ROD BOLT AND NUT	78	PISTON PACKING 3" HYDRAULIC PER RING	100	FLUID CYLINDER HEAD GASKET
44	PISTON ROD WIPER STUFFING BOX	78	PISTON PACKING 3" HYDRAULIC PER RING	100	FLUID CYLINDER HEAD GASKET (COPPER)
45	PISTON ROD WIPER STUFFING BOX CAP SCREW	78	PISTON PACKING 4" HYDRAULIC PER RING	101	VALVE POT COVER GASKET
46	PISTON ROD WIPER GLAND	78	PISTON 2-1/2" RING TYPE	101	VALVE POT COVER GASKET (COPPER)
47	PISTON ROD WIPER GLAND STUD AND NUT	74	PISTON 3" RING TYPE	102	PISTON ROD STUFFING BOX GASKET
48	PISTON ROD WIPER PACKING	74	PISTON 3-1/2" RING TYPE	103	PISTON ROD WIPER STUFFING BOX GASKET
55	FLUID CYLINDER BODY CAST IRON	74	PISTON 4" RING TYPE	104	FRAME COVER GASKET
55	FLUID CYLINDER BODY CAST STEEL	74	PISTON RING 21/2" PER RING	105	PINION SHAFT HOUSING GASKET
56	FLUID CYLINDER BODY STUD AND NUT		PISTON RING 3" PER RING	108	PINION SHAFT STUFFING BOX GASKET
59	FLUID CYLINDER DRAIN PLUG		PISTON RING 31/2" PER RING		
61	REDUCING LINER 2-1/2"		PISTON RING 4" PER RING		
61	REDUCING LINER 3"	79	PISTON ROD		
61	REDUCING LINER 3-1/2"	80	PISTON ROD NUT (CROSSHEAD END)		

Figure 25 ◆ Machine drawing of a direct drive pump.

5.1.0 Lines

The lines on a machine drawing describe the object to be made. The lines differ in their thickness, or weight, and in their construction so that different types of information can be detailed all around and on the part without becoming confusing. Each type of line has a specific use and furthers the goal of precise, universal communication. The lines may be continuous, dashed, or a combination of the two. The standard lines used on a machine drawing are as follows:

- *Object lines* – Heavy, continuous lines that show the outline or overall shape of the object. Object lines show every edge of the object that is visible from a particular view.
- *Hidden lines* – Dashed lines that show surfaces or edges of the object that are hidden or not visible in that view.
- *Center lines* – Thin lines that consist of alternating dashes and longer lines that show the center of a symmetrical or round object or part of the object for reference, clarification, or dimensioning.
- *Dimension lines* – Thin, continuous lines that terminate with an arrowhead at the extension lines and show the dimensions and geometric tolerances of the object.
- *Extension lines, or leaders* – Thin, continuous lines that extend from the object at the exact locations between which dimension lines are drawn.
- *Cutting plane lines* – Heavy, dashed, or continuous lines that show where the object is cut to make the cutaway section.
- *Break lines* – Dashed or continuous lines that may be freehand or machine-drawn to show cross sections or internal details.
- *Phantom lines* – Thin, dashed lines that denote alternate positions or details of an object that is not on that drawing.
- *Section lines* – Different patterns that denote various materials, such as aluminum, steel, or plastic. Section lines show a cutaway view of the inner structure of an object.

Figure 26 shows the lines commonly used on machine drawings.

5.2.0 Dimensions

The precise size of an object is shown on a detail drawing by using accurate dimensions. Dimensions are also used to show the size and locations of holes, slots, threads, and other detailed features of the object.

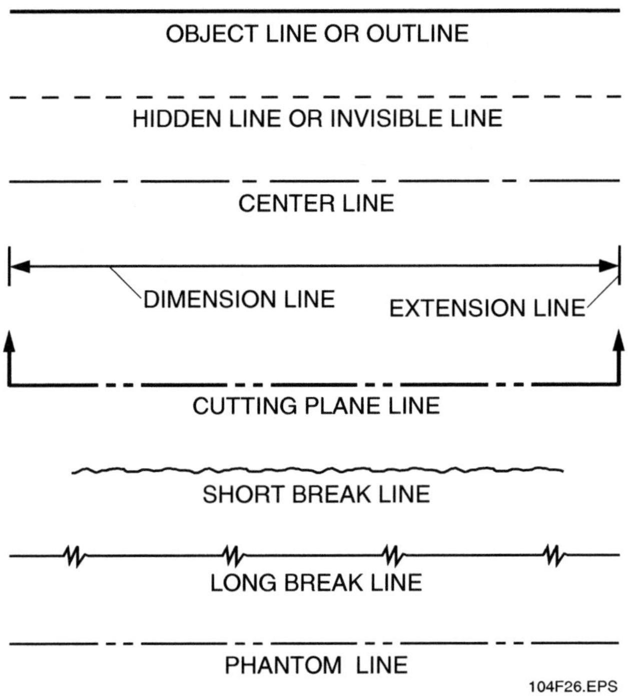

Figure 26 ◆ Lines commonly used on machine drawings.

There are several methods of dimensioning (*Figure 27*). The traditional method uses leaders coming off the object, with dimension lines connecting the leader to an arrowhead. A break in the dimension line allows the numerical value of the dimension to be inserted.

The standard dimensioning system used is ANSI Y14.5M-82, which is commonly known as *geometric tolerancing*. The purpose of this system is to standardize engineering drawings and provide a target dimensioning system that makes precise, consistent results easier to obtain from a drawing.

With the advent of computer-controlled machining comes datum or coordinate dimensioning, which uses two outside, 90-degree intersecting surfaces from which to reference all the dimensions. Tabular dimensioning uses a target designation on each feature that is listed in a chart that contains the X-Y location of the feature and, if required, a description of the feature.

The most common dimensions and tolerances found on a detail drawing are the following:

- *Fractional dimensions* – Given as fractional parts of an inch, fractional dimensions are used on objects that do not require very close tolerances, such as furniture or utility items.
- *Decimal dimensions* – Expressed in thousandths of an inch, decimal dimensions are used predominately and are more practical. The conversion to metric, the use of ANSI Y14.5M-82, and the almost complete dependency on computer

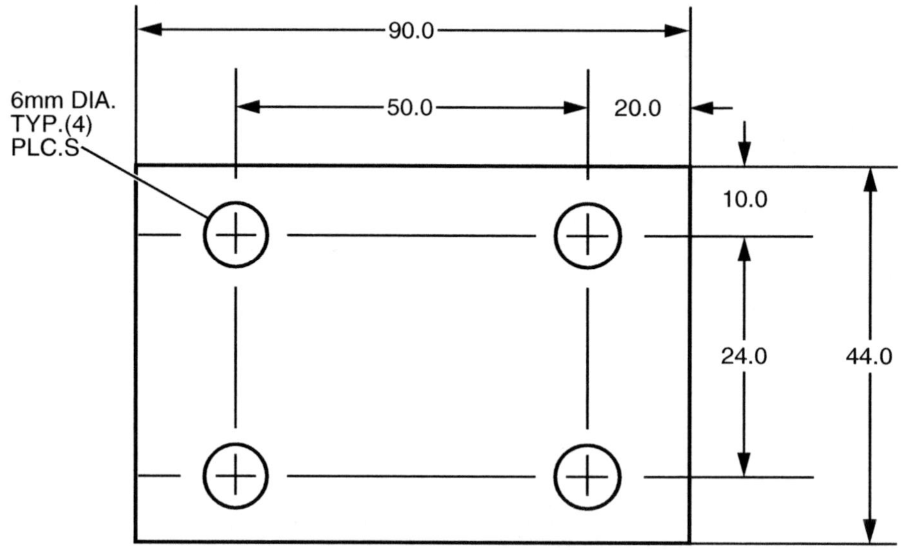

TRADITIONAL DIMENSIONING

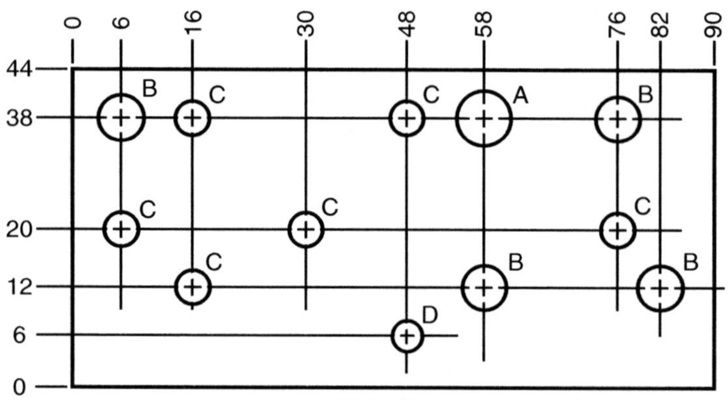

HOLE SYMBOL	HOLE ∅
A	6
B	5
C	4
D	3

COORDINATE DIMENSIONING

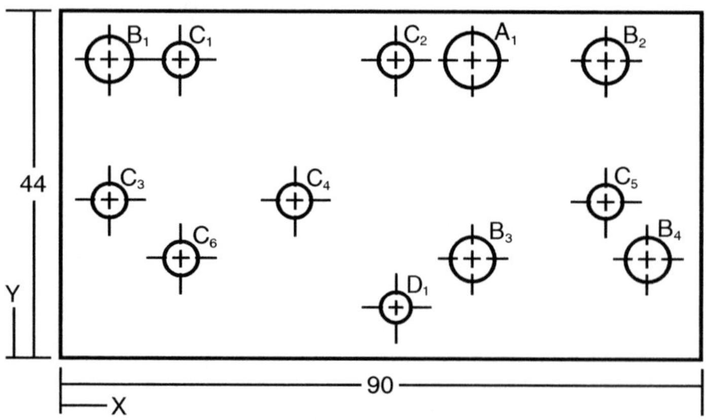

HOLE SYMBOL	HOLE ∅	LOCATION X	Y
A_1	6	58	38
B_1	5	6	38
B_2	5	76	38
B_3	5	58	12
B_4	5	82	12
C_1	4	16	38
C_2	4	48	38
C_3	4	6	20
C_4	4	30	20
C_5	4	76	20
C_6	4	16	12
D_1	3	48	6

TABULAR DIMENSIONING

104F27.EPS

Figure 27 ◆ Types of dimensioning.

interface has almost eliminated the use of fractions in machine design. Tolerances in thousandths or ten-thousandths of an inch are used on objects requiring greater precision.

- *Metric dimensions* – Both decimal dimensions and metric dimensions are used on many drawings so that they can be used outside the U.S. One of the dimensions is usually shown in parentheses. As more companies from other countries locate here and world manufacturing is becoming a reality, more design is generated with metric-dominant or metric-only dimensions.
- *Tolerances* – A tolerance is the total amount that a specific dimension may vary. These are standard limits specified in the title block or are specific high or low limits at individual dimensions. *Figure 28* shows different tolerances.

5.3.0 Notes and Symbols

Notes and symbols are used to provide information that is not conveyed by the views or dimensions on the drawing. Other than dimensions, this is the most important information on an engineering drawing. Three types of information that are conveyed by notes and symbols are:

- Finish marks
- Threads
- Geometric tolerancing

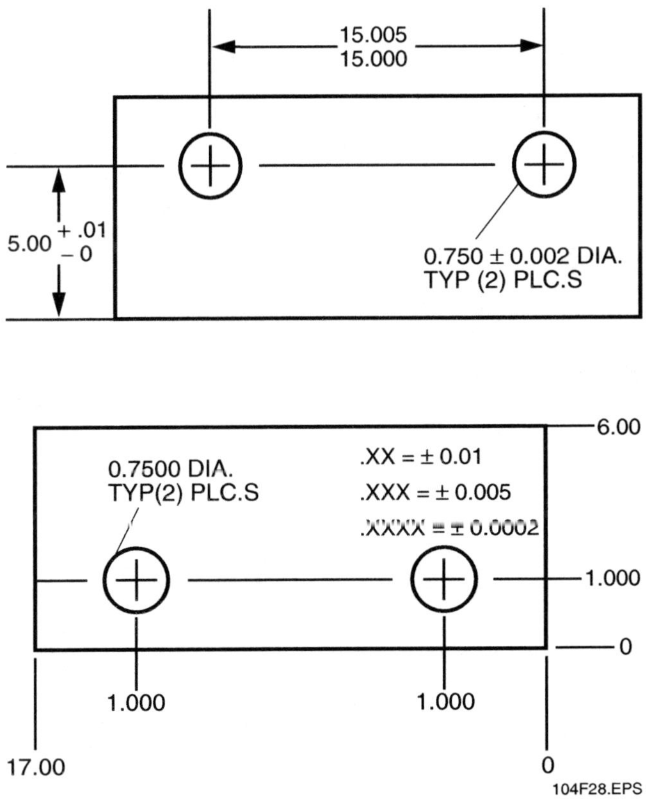

Figure 28 ◆ Different tolerances.

Finish marks designate an average grade of surface roughness in microinches or micrometers (see *Figure 29*). All machining processes cut the material, leaving grooves in the machined area. These grooves are measured and graded to provide a usable reference for the designer, machinist, and product-user to specify how parts fit or run together. The check symbol and surface grade are shown on the surface or on the dimension leader nearest the surface specified.

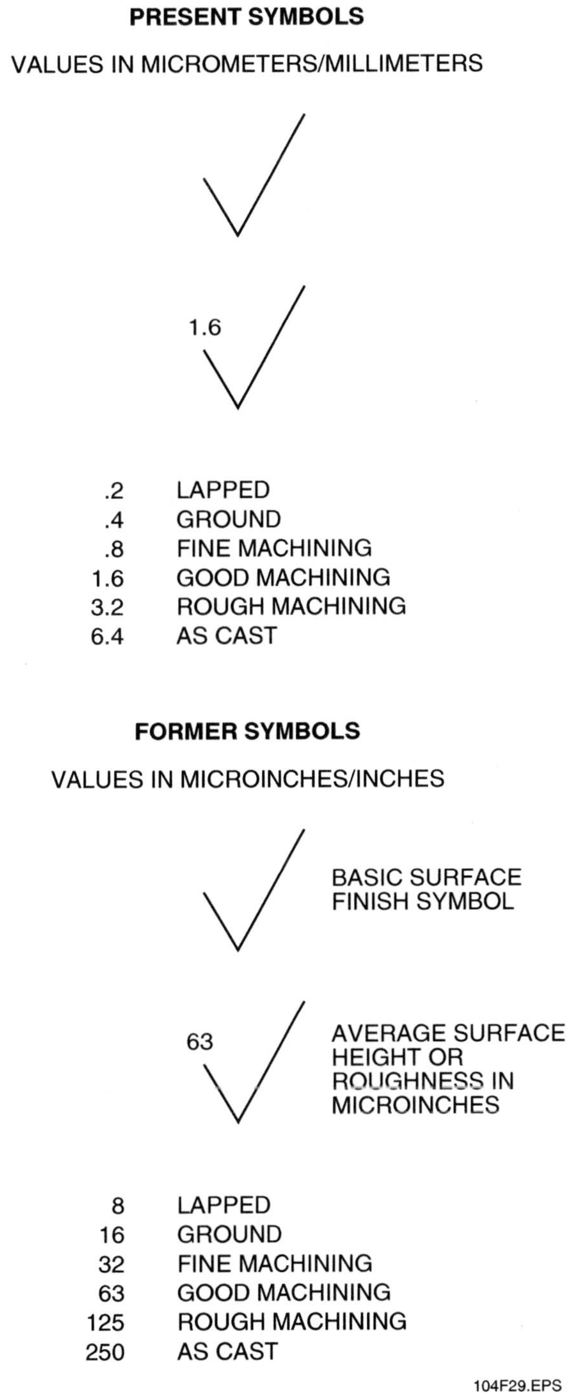

Figure 29 ◆ Finish marks.

Threads are used to hold components together and are tapped into parts or turned on cylindrical parts (see *Figure 30*). The method of specifying threads is with a leader or with a target. The callout shows major diameter, threads per inch (pitch), depth, and location. The thread pitch is coarse, fine, extra-fine, or pipe in both inch or metric.

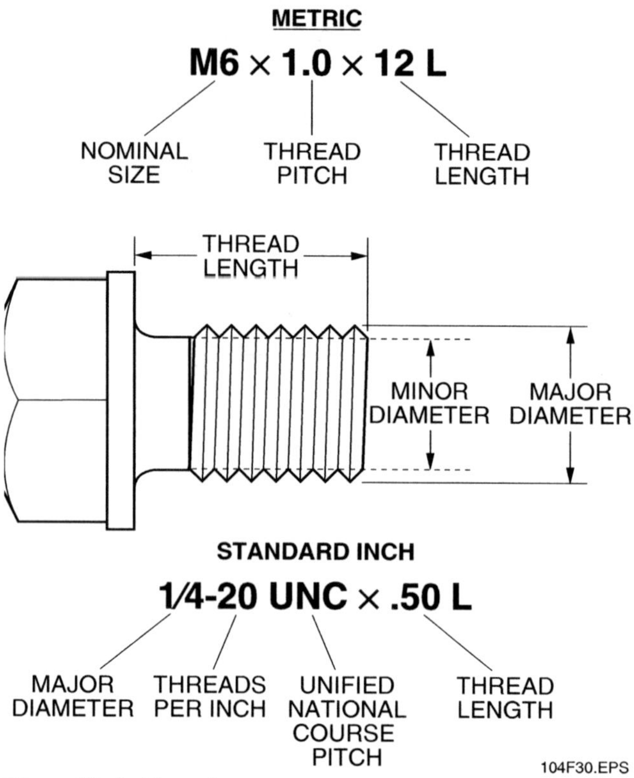

Figure 30 ◆ Threads.

Geometric tolerancing is also known as ANSI Y14.5M-82. This symbol and value standard standardizes the acceptable tolerance of features for all aspects of machining and assembly. These symbols are designed to provide the maximum allowable tolerance in feature dimensions and still maintain a good working part. Surface features, such as flatness, parallelism, and perpendicularity, define the tolerance without being overstrict. Target tolerances for diameters define a circular zone in which mating parts can vary and still fit accurately. It is important to understand what the symbols and tolerances mean because they are not meant to replace numerical, plus/minus tolerances but enhance the definition of the part. *Figure 31* shows geometric symbols and terms.

5.4.0 Scale

Whenever possible, objects are drawn to their actual size. However, some objects are so small or so large that it is meaningless to draw them to actual size. In either case, the object is drawn at some scale or exact proportion of its actual size. A drawing in which an object is represented as being enlarged, reduced, or actual size is known as a scale drawing. The ratio or proportion between the original and the scale drawing is known as the scale.

For example, if an object that is actually 12 inches measures 6 inches on the drawing, the drawing is half the size of the object. This drawing is drawn to a scale of ½. This scale is noted on the drawing in the following manner: Scale: 1" = 2" or 1:2. This means that 1 inch on the drawing equals 2 inches on the object.

If an object of 6 inches actual size measures 12 inches on the drawing, the drawing is said to be double scale, or 2 to 1. This scale is noted on the drawing in the following manner: Scale: 2" = 1" or 2:1. This means that 2 inches on the drawing is equal to 1 inch on the object. If the manufacturing floor is 300 feet by 175 feet but measures 37.5 inches by 21.875 inches on the drawing, the scale is 1 foot = ⅛ inch, or ⅛" = 1'.

No matter what the scale of the drawing, the dimensions are always the actual size of the object. The importance of accurate revision records cannot be overemphasized as it relates to keeping the dimensions accurate.

5.5.0 Revisions

If any person involved in the equipment or drawing package makes any change, revision, or modification to the equipment or drawing package without documenting it thoroughly, the whole system becomes unreliable and useless. If a part of a machine fails and someone redesigns or alters the part to correct the problem but fails to revise the drawings to detail what was done, it is very likely that the original mistake will be made again, which negates the effort and can be very costly. Working from sketches and marked-up prints is sometimes necessary to save time, but these changes have to be incorporated into the system as soon as possible.

Each drawing has a space in its title block for listing the revision history. This block should include the date, initials of who made the revision, a letter or number that appears beside the area where the correction was made, and a description of what was done. If the revision is too complicated or there is no room on the drawing, there are two alternatives. The drawing can be obsoleted, redrawn, and referenced, or it can be extended with the same number and title showing "Sheet 1 of 2."

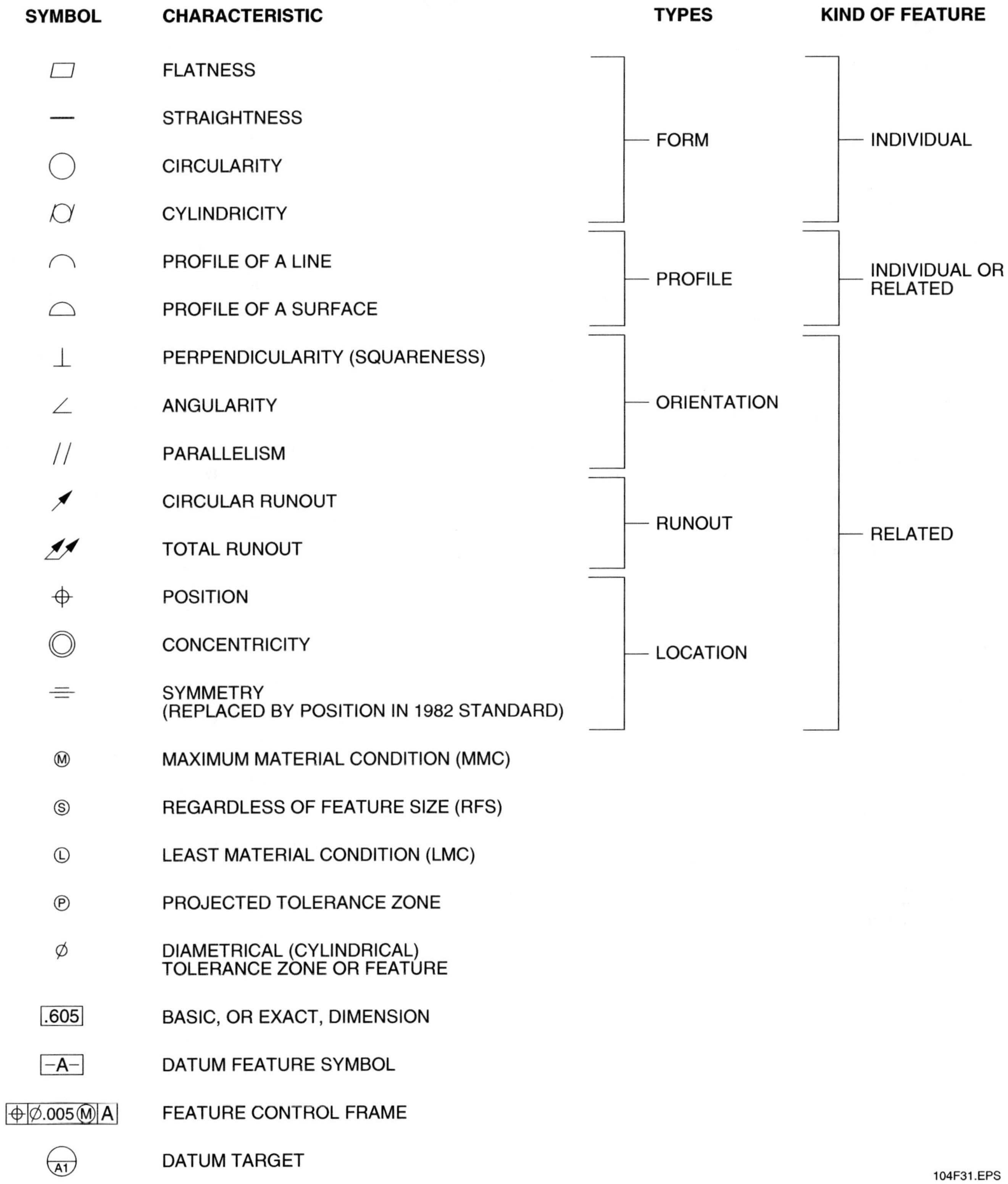

Figure 31 ◆ Geometric symbols and terms.

5.6.0 Vendor Information

Vendor information consists of specifications, part numbers, and sizes of components that are purchased to be used in the equipment. This information is listed in the bill of materials and appears on assembly or detail drawings as needed. The only rule for this type of information is that it should communicate what is necessary to facilitate the machining, assembly, and operation of the equipment. Vendor installation manuals, troubleshooting guides, or instructional courses may be listed as required resources. As manufacturing equipment develops with more complexity and is driven by computers, the need for accurate and concise vendor information is critical.

5.7.0 Material Specifications

Material specifications are found in the title block or in a note, and they spell out the details of the types of materials to be used. In all classes of materials, there are standards and specifications that correlate to performance or machine ability. These specified materials can have substitutes that perform the same, or the specification may require the use of one specific material only. For example, Delrin® and Celcon® are specific brand names for an acetal copolymer plastic. The products are very similar but differ in some ways. Therefore, the material specification on the blueprint may call for "Delrin® or equivalent" if the need is unspecific, or it may denote "Delrin®" if this is the only material acceptable.

Summary

A blueprint or drawing provides a detailed representation of a piping system or component within a piping system. Drawings provide information needed to build, install, and modify a piping system as well as information needed by pipeline mechanics to locate components, troubleshoot equipment failures, and make repairs to equipment and components within the system.

Although it is not necessary for pipeline mechanics to be able to design and generate blueprints and CAD drawings, it is critical that pipeline mechanics understand the information in a drawing system. The ability to find, read, revise, and help maintain the accuracy of the drawing system is important for many reasons. The future use of the drawings by mechanics, machinists, and engineers depends on how accurate and up to date the drawings for the piping system and related equipment are kept. Each company has its own system for incorporating modifications into the drawing system and departments involved in document control. Accurate revisions in the drawings ensure the useful functioning of the drawing system.

Review Questions

1. The information used to identify a project or any part of the project is found in the _____.
   a. notes
   b. coordinates
   c. title block
   d. drawing index

2. The distance from a given point to a base point is its _____.
   a. bench mark
   b. setback
   c. coordinate
   d. easement

3. A plan view normally includes a(n) _____, that references where the drawing fits into the overall system.
   a. equipment drawing
   b. P&ID
   c. key plan
   d. isometric drawing

4. _____ drawings show how the equipment is arranged and the location of all equipment used with the system.
   a. P&ID
   b. Isometric
   c. Piping arrangement
   d. Section

5. Drawings that show the process flow are known as _____.
   a. P&IDs
   b. section views
   c. plot plans
   d. isometric drawings

6. Drawings that include the fabrication, erection, and setting drawings necessary to install a piece of equipment are known as _____.
   a. isometric drawings
   b. equipment drawings
   c. detail sheets
   d. P&IDs

7. _____ are used on P&IDs to indicate the size of pipe, service of line, piping material specification group number, number of the pipe, and insulation required.
   a. Title blocks
   b. Notes
   c. Line numbers
   d. Section views

8. When two or more sheets are used to depict a large piping system, _____ are used to help trace the system from one sheet to another.
   a. line numbers
   b. match lines
   c. line weights
   d. detail sheets

9. On machine drawings, heavy, dashed, or continuous lines that show where the object is cut to make a cutaway section are known as _____ lines.
   a. dimension
   b. extension
   c. phantom
   d. cutting plane

10. Dimensions that are expressed in thousandths of an inch are known as _____ dimensions.
    a. fractional
    b. decimal
    c. metric
    d. US standard

GLOSSARY

Trade Terms Introduced in This Module

Dimension: A measurement between two points on a drawing.

Drawing package: A cohesive group of drawings comprised of assembly and detail drawings and a bill of material that contain all the information needed to build or assemble a piece of equipment.

Isometric drawing (ISO): A three-dimensional drawing of a piping system to provide clarity. It is not drawn to scale.

Line number: A group of abbreviations that specify size, service, material class/specification, insulation thickness, and tracing requirements of a given piping segment.

Piping and instrumentation diagrams (P&IDs): Diagrammatic layouts of the pipe and equipment to show the flow through the line, using standard piping and instrumentation symbols.

Revision: A change in a part on an engineering drawing that is noted on the drawing.

Spool: A piping segment of an ISO.

Title block: A section of an engineering drawing blocked off for pertinent information, such as the title, drawing number, date, scale, material, draftsperson, and tolerances.

CONTREN™ LEARNING SERIES — USER UPDATES

The NCCER makes every effort to keep these textbooks up-to-date and free of technical errors. We appreciate your help in this process. If you have an idea for improving this textbook, or if you find an error, a typographical mistake, or an inaccuracy in NCCER's Contren™ textbooks, please write us, using this form or a photocopy. Be sure to include the exact module number, page number, a detailed description, and the correction, if applicable. Your input will be brought to the attention of the Technical Review Committee. Thank you for your assistance.

Instructors – If you found that additional materials were necessary in order to teach this module effectively, please let us know so that we may include them in the Equipment/Materials list in the Instructor's Guide.

Write: Curriculum Revision and Development Department
National Center for Construction Education and Research
P.O. Box 141104, Gainesville, FL 32614-1104

Fax: 352-334-0932

E-mail: curriculum@nccer.org

Craft _______________ Module Name _______________

Copyright Date _______________ Module Number _______________ Page Number(s) _______________

Description _______________

____

____

____

(Optional) Correction _______________

____

____

(Optional) Your Name and Address _______________

____

____

Module 63105-02

Tubing, Threaded Pipe, and Hoses

Introduction to the Pipeline Operator Qualification Rule, API, and NCCER

It could be said that petroleum products keep America running. Petroleum products like oil and natural gas are the principal sources of energy in the United States and are used in many manufacturing and chemical processes. Petroleum products are primarily transported by pipeline. There are more than a million miles of liquid and natural gas pipeline in this country, ranging in diameter from two inches to four feet. These pipelines bring crude oil and natural gas from production fields to processing plants and refineries—and ultimately to the consumer.

Pipeline Operator Qualification

Petroleum products are considered hazardous, so pipeline personnel must take extra precautions and follow specific procedures when working with them. To help ensure the safe handling and transportation of petroleum products, the United States Department of Transportation (DOT) mandates safe practices as federal regulations.

One such regulation, the Operator Qualification (OQ) rule, requires pipeline operators to develop a written qualification plan to evaluate their personnel and their contract personnel (see *49 CFR Part 192 Subpart N—Gas*, and *49 CFR Part 195 Subpart G—Liquid*). The evaluation hinges on the operator's ability to perform certain covered tasks and to be able to recognize and respond to abnormal operating conditions that may be encountered while performing these covered tasks.

Covered Tasks

A pipeline performance task is deemed *covered* if the activity passes the following four-part test:

1. Is it performed on a pipeline facility?
2. Is it an operations or maintenance task?
3. Is it performed as a requirement of *49 CFR Part 195 (Liquid)* or *49 CFR Part 192 (Gas)*?
4. Does it affect the operation or integrity of the pipeline facility?

If the answer to all four questions is *yes*, then the activity is considered *covered* and must figure into the evaluation process of pipeline personnel.

American Petroleum Institute

The OQ rule and its related covered tasks provided the catalyst to develop a unified pipeline training and assessment program—so that pipeline operators and owners have ready access to the assessment and training tools they need. This effort was the direct result of the vision and leadership provided by The American Petroleum Institute (API) and API members.

Efforts to unify the industry are nothing new to API. Indeed, the impetus for forming the Institute in 1919 was the need to standardize engineering specifications for drilling and production equipment. Today, the most pressing issues revolve around public perceptions and government policies toward the industry. Speaking with one voice on these issues has become as essential as having interchangeable parts in the field. API's members determine the petroleum industry's positions on public policy as well as the standards that govern its day-to-day operations worldwide.

Consortium on Operator Qualification

API assembled a focus group whose primary purpose was to identify covered tasks and provide compliance recommendations to pipeline operators.

The Consortium on Operator Qualification (COOQ) became the guiding force for the identification and development of covered tasks for the liquid pipeline industry. Their work product, API publication 1161, *Guidance Document for the Qualification of Liquid Pipeline Personnel*, formed the basis for the resulting Pipeliner Training and Assessment Program.

The covered task list created in API publication 1161 has now evolved to include gas covered tasks and is maintained by NCCER. The most current covered task list can be downloaded at www.nccer.org.

National Center for Construction Education and Research

With the guidance and support of API, its members, and the Consortium on Operator Qualification, the National Center for Construction Education and Research (NCCER) was able to design and implement the Pipeliner Training and Assessment Program.

NCCER is a not-for-profit 501(c)(3) education foundation established in 1995 by the world's largest and most progressive construction and maintenance contractors and national contractor associations. NCCER was founded to address the

severe workforce shortage facing the construction and maintenance industries and to develop a standardized training process and curricula. Today, NCCER is supported by hundreds of leading construction and maintenance companies, manufacturers, and national associations and developed more than 35 construction and maintenance titles.

NCCER also maintains a National Registry, which contains the records of trainees and instructors who have received NCCER training. All completion information for every trainee is sent to the NCCER and filed within the Registry. The Registry can then confirm training and skills for workers as they move from company to company, state to state, or even within their own company.

Standardized Curriculum (NCCER's Contren™ Learning Series)

NCCER curricula are developed using the expertise of industry subject matter experts (SMEs). The use of SMEs as an integral part of every new development or revision ensures that the curriculum is current, industry-driven, and indicative of industry practices. The development and/or revision process is facilitated from start to finish by NCCER project managers, who work closely with the SMEs and a select team of technical writer(s) on the development and presentation of the curriculum.

Pipeline: Five Training Paths

The pipeline curriculum focuses on essential skills and competencies within five skill areas:

- Corrosion Control
- Electrical and Instrumentation
- Maintenance
- Mechanical
- Operations (Field and Control Center)

SMEs from the pipeline industry have determined the skills and competencies needed based on API's covered tasks and the consensus of the group/industry. Specific company information is generally not incorporated into the curriculum unless the SMEs determine that the information is the industry standard.

Training Modules and Covered Tasks

The smallest unit of instruction in the curriculum is a *module*. Each curriculum is composed of modules (the equivalent of stand-alone chapters) that are logically sequenced into complete *levels* of instruction. Every module is task-oriented and requires the successful completion of a written (knowledge-based) examination and a performance (hands-on) test. Modules may be individually selected and/or combined for efficient task training. Modules in the pipeline curricula, to the extent possible, will align with API-developed covered tasks. While not all modules will relate to covered tasks, all covered tasks will relate to at least one module.

Benefits of Training to the Pipeline Marketplace

Applying NCCER's standardized training process allows pipeline owners and contractors in the United States to establish the benchmark for acceptable skill and competency levels within the pipeline industry. This process will assist training organizations in developing and/or improving standardized pipeline training programs that effectively match their resources and operating conditions while responding to regulatory demands and to the needs of the industry. The process will also recognize training organizations that meet the minimum standards of competence, warranting regulatory, public, and professional confidence. This in turn will foster national unity within the pipeline industry for the mutual benefit of individual technicians, owners, contractors, and ultimately pipeline-users and society.

Additional Resources

To review the Pipeliner Training and Assessment Program Implementation and Resource Guide and the latest version of pipeline covered tasks, visit NCCER's Web site at www.nccer.org, or contact:

National Center for Construction Education and Research
PO Box 141104
Gainesville, FL 32614-1104
(888) 622-3720
Web site: www.nccer.org

For more information on API, please contact:

American Petroleum Institute
1220 L Street, Northwest
Washington, D.C. 20005
(202) 682-8000
Web site: www.api.org

To order *API Pub 1161*, contact Global Engineering Documents by calling 1-800-854-7149 or visiting their Web site at www.global.his.com.

Additional sources of information on the pipeline operator qualification regulation include:

DOT – Office of Pipeline Safety
www.ops.dot.gov

Transportation Safety Institute
www.tsi.dot.gov

COURSE MAP

This course map shows all of the modules in the first level of the Pipeline Mechanical curriculum. The suggested training order begins at the bottom and proceeds up. Skill levels increase as you advance on the course map. The local Training Program Sponsor may adjust the training order.

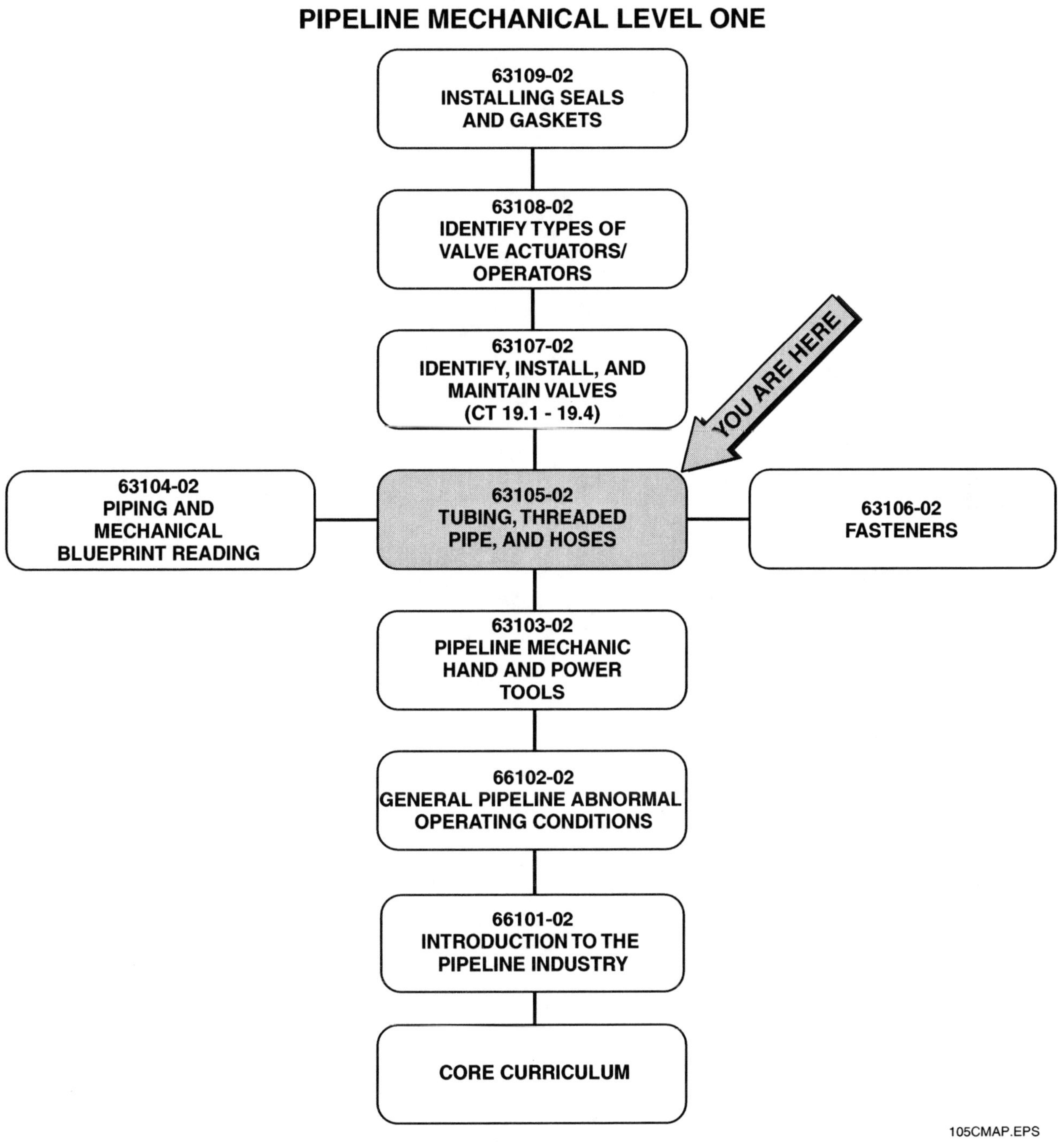

MODULE 63105-02 CONTENTS

Figures

Tables

Tubing, Threaded Pipe, and Hoses

Objectives

When you have completed this module, you will be able to do the following:

1. Identify and explain the materials used in tubing systems.
2. Identify, use, and care for tubing cutters, benders, and flaring tools.
3. Fabricate tubing systems.
4. Identify and explain the materials used in threaded piping systems.
5. Use and care for pipe threading tools.
6. Fabricate threaded piping systems.
7. Identify and select types and sizes of hoses.
8. Maintain hoses used in control systems.

Prerequisites

Before you begin this module, it is recommended that you successfully complete the following: Core Curriculum; Pipeline Core, Modules 66101 and 66102; Pipeline Mechanical Level One, Module 63103.

Required Trainee Materials

1. Pencil and paper
2. Appropriate personal protective equipment

1.0.0 ◆ INTRODUCTION

Tubing, piping, and hose systems are all used to transport a variety of fluids under various conditions. Tubing, piping, and hose systems are used in pipeline facilities to perform a number of functions such as carrying process fluids, steam, water, and hydraulic fluids; carrying pneumatic control signals, lubricating oil, and compressed air to control systems; and connecting instruments to process piping.

The terms *piping* and *tubing* are sometimes used to mean the same thing. Generally, though, tubing is considered to be more flexible than piping and has thinner walls. Because tubing is not as rigid as piping, it is easier to bend. Tubing installations usually require fewer fittings than a comparable piping **run** because bends can be made.

Threaded pipe and fittings are a common way to join pipe. Threaded piping systems require no specialized in-shop fabrication and can be easily fabricated on the job site. If a piping system is to be installed near flammable liquids or gases, threaded pipe is much safer and faster than welded systems because fire hazards can be avoided.

Hose is similar to tubing and pipe in its use in various control systems. The main difference between tubing or piping and hoses is the material from which it is made.

Hoses are made from flexible materials such as rubber, Teflon®, nylon, and other specially constructed materials. The flexibility of hose offers advantages over piping and tubing systems for certain applications. These advantages include its ability to be routed easily, to absorb vibration, to reduce sound, and to accommodate movement of connected components. However, hose lines must be used with caution in order to provide long service and to guard against potentially dangerous failure.

The ability to install tubing, threaded pipe, and hose systems in accordance with job requirements and specifications is a skill that every pipeline mechanic must develop. Throughout your career, you will encounter various tubing, threaded piping, and hose systems that must be maintained.

2.0.0 ◆ SIZES AND TYPES OF TUBING

It is important to use the proper size and type of tubing for specific applications. The engineering specifications, piping specifications, installation detail drawings, and other information for a project usually tell you what size and type of tubing to use. However, you must be able to identify and verify the size and type of the tubing once you know the requirements. Tubing is selected for specific applications according to a number of different factors, such as pressure and temperature conditions, corrosiveness of the environment, vibration in the system, system flow requirements, compatibility with the process fluids, and cost.

2.1.0 General Sizing Measurements for Tubing

Tubing is generally sized according to its **outside diameter** (OD) and **wall thickness**. The OD is normally expressed in fractions of an inch, such as ¼ inch, ⅜ inch, or ½ inch. The OD of most types of tubing ranges from 1⁄16 inch to 2 inches. The wall thickness of tubing is usually expressed in decimal units to the 1,000th of an inch, such as .035 inches or .068 inches. The normal range of wall thickness for tubing is .010 inches to .188 inches. *Figure 1* shows tubing size measurements.

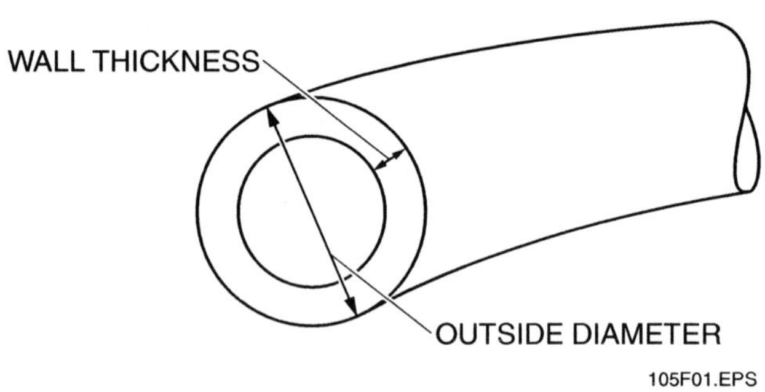

Figure 1 ◆ Tubing size measurements.

Normally, the manufacturer will print or stamp the OD and wall thickness on the tubing. However, if you need to measure the OD of tubing, you can use a rule, outside caliper, or caliper rule (see *Figure 2*).

> **NOTE**
> When using a rule to measure OD of tubing, remember to measure the diameter across the center of the tubing.

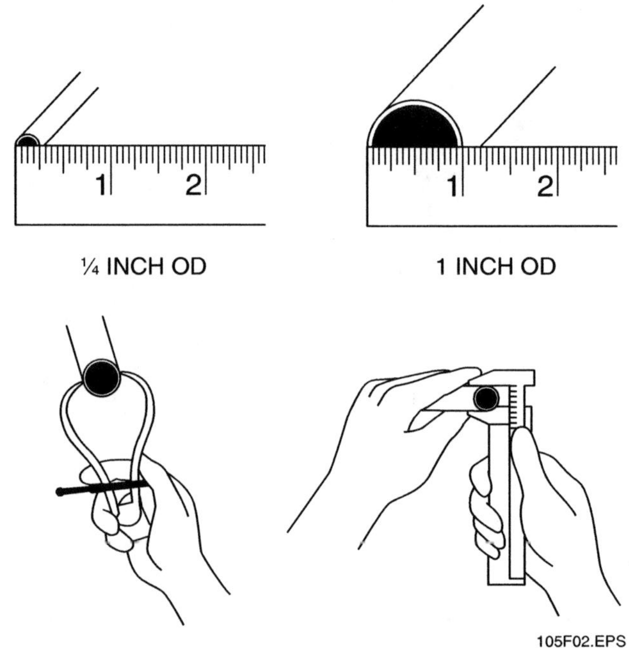

Figure 2 ◆ Measuring the outside diameter of tubing.

While most instrument tubing in the United States is measured in inches and fractions of an inch, sometimes tubing sizes will be marked in metric units, measured as millimeters. If you are working with metric-sized tubing, remember to use a metric rule or other metric measuring tools to get your measurements.

Another common measurement for instrument tubing is the **inside diameter** (ID) of the tubing. This is the distance across the middle of the hole in the tubing from inside wall to the opposite inside wall. The inside diameter of any tube is equal to the outside diameter minus twice the wall thickness. For example, tubing with a ½ inch OD and a wall thickness of .049 inches would have an ID of .402 inches, as calculated below.

TUBING ID = OD – (2 × wall thickness)

TUBING ID = .500 – (2 × .049)

TUBING ID = 402

For any given OD of tubing, the ID decreases as the wall thickness increases. In other words, as the wall of the tube gets thicker, the hole in the middle gets smaller. *Table 1* shows tubing sizes.

2.2.0 Tubing Materials

Tubing is made from a wide variety of materials, including different types of metals and poly. Each material has certain characteristics that make it desirable for particular applications.

Table 1 Tubing Sizes

Tube	Tube OD (in.)	Wall Thickness (in.)	Tube ID (in.)	Tube	Tube OD (in.)	Wall Thickness (in.)	Tube ID (in.)
1/16"	.062	.010	.042	3/4"	.750	.083	.584
		.015	.032			.095	.560
		.020	.022			.109	.532
1/8"	.125	.028	.069	7/8"	.875	.049	.777
		.035	.055			.065	.745
						.083	.709
3/16"	.187	.028	.131			.095	.685
		.035	.117			.109	.657
		.049	.089				
				1"	1.000	.065	.870
1/4"	.250	.028	.194			.083	.834
		.035	.180			.095	.810
		.049	.152			.109	.782
		.065	.120			.120	.760
5/16"	.312	.035	.242	1 1/4"	1.250	.083	1.084
		.049	.214			.095	1.060
		.065	.182			.109	1.032
						.120	1.010
3/8"	.375	.035	.305			.134	.982
		.049	.277			.156	.938
		.065	.245				
				1 1/2"	1.500	.095	1.310
1/2"	.500	.035	.430			.109	1.282
		.049	.402			.120	1.260
		.065	.370			.134	1.232
		.083	.334			.156	1.188
						.188	1.124
5/8"	.625	.049	.527				
		.065	.495	2"	2.000	.109	1.782
		.083	.459			.120	1.760
		.095	.435			.134	1.732
						.156	1.688
3/4"	.750	.049	.652			.188	1.624
		.065	.620				

105T01.EPS

Poly tubing is generally considered to be less expensive than metal tubing, easier to install, and easier to bend and reshape. However, it has limitations regarding its strength, compatibility to certain process fluids, and ability to withstand certain temperatures and pressures.

Metal tubing comes in two general types, welded and seamless. Welded tubing is made by rolling flat strips of metal and then forming the metal into a tube. The seam is then welded. While the welded seam may not be visible on the outside of the tubing, it can usually be seen on the inside surface of the tubing.

Seamless metal tubing is made by passing a cylinder of metal through a die which cuts out the inside of the tube. This process is called *extruding*. Usually seamless tubing is drawn through additional die sets to get the proper wall thickness and to meet other specifications for the tubing.

Metal tubing may also be annealed. Annealing is a process in which the metal is reheated, then slowly allowed to cool. Annealing makes the metal softer and less brittle, allowing for easier bending and flaring.

Metal tubing comes in straight lengths, normally 20 feet. Custom lengths can be obtained from 12 feet to 40 feet. Some types of tubing also come in rolls or coils, usually ranging from 50 feet to 100 feet.

The following describes materials used for tubing and some of the characteristics of the materials.

2.2.1 Copper Tubing

Copper is one of the most widely used metals for tubing. Copper tubing is identified by its color when new. The color of untarnished copper is reddish-brown, similar to the color of a new penny. As it ages, copper tarnishes and becomes dull and dark in color. Tarnished copper tubing is often mistaken for steel or polyvinyl chloride (PVC) tubing. When you are not sure if the tubing material is copper, polish a small area to bring back the original color.

Copper tubing is often used in pneumatic systems because it resists rust that could contaminate the air running through the tubing. It is lighter and easier to use than stainless steel, and it can be easily bent to form bends and angles. It holds up well in most environments, and it is strong enough to handle loads that would rupture plastic tubing. Copper tubing is usually used in systems when the temperature and pressure conditions are not extreme.

A disadvantage of copper tubing is that it is expensive in comparison to some other types of tubing. It also expands more than some other types, which can be a disadvantage in certain applications. Copper tubing also may not be suitable for systems that are exposed to extremely corrosive substances.

Copper tubing is sometimes covered with PVC to help resist attack by chemicals. PVC-covered copper tubing has all of the advantages of regular copper tubing with the added feature of being more tolerant of chemicals.

One way to identify copper tubing is by standard wall thicknesses designated by the American Society for Testing of Materials (ASTM). Each type, K, L, M, and DWV, represents a series of sizes with different wall thicknesses. The inside diameters depend on tubing size and wall thickness. The standard grades are as follows:

- K – Heavy grade (green)
- L – Medium grade (blue)
- M – Light grade (red)
- DWV – Used for drains, waste lines, and vents (yellow)

Straight lengths of hard copper tubing are marked by a color-coded band printed along the tubing. The color codes are indicated above for each specific grade of copper tubing. Soft copper tubing that comes in rolls is not color coded on the tubing. The color codes of soft copper tubing are usually marked on the cartons and shipping tags.

Copper tubing for air conditioning and refrigeration field service (ACR tubing) is designated by its actual outside diameter. ACR copper tubing is used with special fittings for connections, repairs, and alterations in air conditioning and refrigeration installations.

Copper tubing is also identified by its degree of hardness. Hard (drawn) copper tubing is more difficult to bend and cannot be **flared** without cracking. Hard copper tubing comes in straight sections only, not in rolls. Soft (annealed) copper tubing can be easily bent and flared. Soft copper tubing comes in either straight sections or rolls. Copper tubing types K, L, and M are available as either hard or soft. Type DWV is available only as hard tubing.

2.2.2 Steel Tubing

Steel is an alloy that contains iron plus a small amount of carbon. While other metals and nonmetals are also added to make steel, the presence of iron and carbon defines the metal as steel. The higher the carbon content, the harder and stronger the steel. Steel tubing is rolled to its finished size while it is red-hot. A black coating of iron oxide forms on the hot rolled steel after it is rolled. This black color helps identify steel tubing. Sometimes steel tubing is dipped into a solution of molten zinc to give it better corrosion-resisting qualities. This type of steel tubing is called *galvanized steel tubing*. Galvanized steel tubing has a shiny silver color.

Steel tubing can be cold- or hot-rolled. Where precise tubing dimensions are required (inside diameter or wall thickness), cold-rolled tubing is often used rather than the less precise hot-rolled steel tubing. Steel tubing is used in noncorrosive areas. It is very strong and is often used in high-pressure hydraulic systems. Steel tubing also works well in other high-pressure and high-temperature applications.

Steel tubing is less brittle than stainless steel tubing. Therefore, it is often used in systems that have a lot of vibration, because it will give and absorb the vibration. Steel tubing has great strength, yet it can be bent, flared, flattened, welded, and threaded to make turns and joints.

2.2.3 Stainless Steel Tubing

Stainless steel is a steel alloy that contains at least 12 percent chromium by weight. Stainless steel tubing is stiffer and stronger than copper tubing. It is used to handle many types of corrosive fluids. It performs well in high-pressure and high-temperature applications.

Stainless steel tubing is designated by a number that refers to the alloy content of the metal. Two common types of stainless steel tubing used in instrument work are identified by the numbers 304 and 316. These contain chromium and nickel but not manganese. The type of alloys that make up stainless steel tubing, its size, and its wall thickness determine the suitability of the tubing to varying temperature ranges and working pressures.

Even though stainless steel tubing is expensive in comparison to some other tubing materials, it has several important qualities that make it practical for specific applications. Since stainless steel is corrosion resistant and can easily be kept sterile to prevent contamination, it is often used in pharmaceutical, food processing, and dairy plants.

Stainless steel tubing is usually used in instrument work in sizes ranging from ⅛ inch to 2 inches. It is available in different wall thicknesses expressed in thousandths of an inch.

2.2.4 Aluminum Tubing

Aluminum is made from bauxite ore that contains aluminum oxide, iron, and silicon. Aluminum tubing has several advantages over other types of metal tubing for specific applications. It is softer, lighter in weight, and more flexible than most other metal tubing. It has a dull silver color and can be cut easily with a tubing shear or tube cutter. Aluminum tubing resists corrosion and is easy to cut and shape.

Aluminum tubing is often used where sanitary conditions are required, because it is easy to clean and sterilize. It is also used in many low-temperature applications to carry liquefied gases, because its strength increases as temperature decreases.

> **NOTE**
> Aluminum tubing breaks easily. It also tends to oxidize at the fittings, which may cause them to leak.

2.2.5 Monel® Tubing

Monel® is the trade name that International Nickel Corporation uses for its various nickel-copper alloys. Monel® tubing is highly resistant to corrosion from the atmosphere. It also shows good resistance to salt water, lye, sulfuric acid, most foods, and many chemicals.

Monel® tubing is similar to stainless steel in appearance, brittleness, and other properties. It is often used where high strength and ductility are required. Monel® tubing is considered expensive.

2.2.6 Inconel® Tubing

Inconel® is a trade name that International Nickel Corporation uses for its various nickel-iron-chromium alloys. Inconel® tubing is often used in high-temperature applications because it remains strong at very high temperatures in normal atmosphere.

Inconel® shows good resistance to acidic foods, such as fruit juices, and is often used in food processing plants. Inconel® also has applications in the pharmaceutical industry because it does not contaminate the contents of the tubing system. Additionally, Inconel® tubing is frequently used to convey dry chlorine gas, which is highly corrosive. Inconel® tubing is considered expensive.

2.2.7 Hastelloy® Tubing

Hastelloy® is a trade name that Union Carbide Corporation uses for its various nickel-based alloys. Hastelloy® tubing contains nickel plus molybdenum or silicon. It is highly resistant to corrosion, especially corrosion by hot concentrated acids or liquors.

Hastelloy® tubing is often used in the chemical processing industry to convey highly corrosive materials, such as boiling sulfuric acid. Hastelloy® tubing is considered expensive.

2.2.8 Poly Tubing

Poly tubing is a broad category that covers a variety of materials, each with its own specific properties. Tubing made from thermoplastics can be easily bent and reshaped. Poly tubing is used primarily in pneumatic control and junction boxes and in pneumatic control panels. Color-coded poly tubing is often used in pneumatic systems to show the difference between individual circuits. Black tubing is recommended when the tubing is exposed to sunlight or fluorescent light because it is less likely to deteriorate from the effects of the light.

> **NOTE**
> Black tubing can be color-coded with colored strips.

Poly tubing is generally considered easier to install than metal tubing because it does not require bending, flaring, or deburring. It is lightweight, pliable, and resistant to most chemicals and moisture. It is used in processes, such as chlorine and hydrochloric service, that would corrode metal tubing. Poly tubing does have limitations, including its inability to withstand high pressures and temperatures.

> **NOTE**
> Never use plastic tubing unless it is specifically called for in the project specifications.

A commonly used type of poly tubing is polyvinyl chloride (PVC) tubing. Like most other types of poly tubing, PVC tubing is flexible. It can be bent without using a bending tool. It is easily cut with snips, **tubing cutters**, or a pocketknife. It is often used inside pneumatic control panels where there is little chance of it being cut or broken. PVC tubing comes in a variety of colors, sizes, and wall thicknesses.

Polyethylene (PE) tubing is an inexpensive, flexible tubing that is widely used in laboratories, instrument air lines, and other applications where resistance to corrosion is important.

Polypropylene tubing is a flexible tubing that is much stronger than polyethylene and has good corrosion resistance. It can withstand higher pressures and temperatures than most other types of poly tubing.

TFE Teflon® tubing is made of a material that is much like the coating used on kitchen pots and pans to keep food from sticking. Because Teflon® is a very slick material, it creates little resistance to fluid flow. It is highly resistant to most chemical attacks and can be used in moderately high temperature applications. TFE Teflon® tubing is flexible, white, and soft.

Tygon tubing is a very soft, plasticized PVC tubing that is extremely flexible, corrosion resistant, and clear or milky white in color. It is used in many laboratory, medical, food, and pharmaceutical applications. Tygon tubing is often used when flexibility or transparency are important considerations. Tygon tubing is used in moderate temperature and pressure conditions.

> **NOTE**
> Poly tubing can use the same compression fittings as copper tubing. Be careful not to overtighten fittings used on poly tubing.

Nylon tubing is more rigid than other types of poly tubing. Nylon tubing can be used in higher temperature applications than most other types of plastic tubing. Nylon tubing doesn't require the use of special inserts for fittings.

2.3.0 Tubing Standards and Specifications

There are several organizations and professional societies that have published standards and specifications for tubing. These standards describe the various characteristics, applications, and capabilities of different types and sizes of tubing. Engineers and designers use these standards in selecting and identifying the type and size of tubing for particular applications. The American Society of Testing Materials (ASTM), the American National Standards Institute (ANSI), and the American Society of Mechanical Engineers (ASME) are three of the organizations that have published standards and specifications for tubing.

ASTM has a series of standards commonly used to identify and specify tubing. Some of the standards provide general requirements about certain types of tubing, while others contain specific requirements (such as finish, hardness, alloy, working pressures and temperatures, or method of manufacture) for particular types of tubing. In the ASTM system, each standard has an identification number consisting of a letter and a series of numbers. If the metal contains iron (ferrous metal), the identification number begins with an "A". If the metal does not contain iron (nonferrous metals), the identification number begins with a "B". After the letter, there is a number such as A213, which is a specification for seamless stainless steel tubing, or B68, which is a specification for high-quality, soft, annealed copper tubing.

An example of a general ASTM standard is ASTM A450, which covers general requirements such as OD tolerances for steel and stainless steel welded or seamless tubing. ASTM B165 is an example of a standard that refers to specific requirements for Monel® tubing. The ASTM identification number is often printed or stamped on the tubing.

When the project specifications list the required ASTM, ASME, or ANSI tubing identification number, you must use that type of tubing. You should review some of these specifications so that you know the type of information included in the specific tubing specifications. The following are typical ASTM identification codes used to specify particular types of tubing:

- Stainless Steel Tubing
  - ASTM - A213
  - ASTM - A249
  - ASTM - A269
  - ASTM - A450
  - ASTM - A632
- Carbon Steel Tubing
  - ASTM - A161
  - ASTM - A179
- Copper Tubing
  - ASTM - B68
  - ASTM - B75
  - ASTM - B88
  - ASTM - B251
- Aluminum Tubing
  - ASTM - B210
- Monel® Tubing
  - ASTM - B165
- Inconel®
  –ASTM - B167
- Hastelloy®
  –ASTM - B622

The ANSI Code for Pressure Piping B31.3 is often used by engineers and designers to determine the required wall thickness of tubing needed for different system working pressures. When the working temperature of the process increases, the allowable working pressure of the tubing decreases. Different pressure charts are used for different types of tubing such as aluminum, copper, stainless steel, and Monel®. *Table 2* shows typical working pressure and temperature charts. *Table 3* shows the factors used to determine tubing pressure ratings at elevated temperatures.

Table 2 Typical Working Pressure and Temperature Charts

ALUMINUM TUBING

Tube OD (in.)	Tube Wall Thickness (in.)				
	.035	.049	.065	.083	.095
⅛	8600				
3/16	5600	8000			
¼	4000	5900	**Working Pressure (psig)**		
5/16	3100	4600			
⅜	2600	3700			
½	1900	2700	3700		
⅝	1100	1600	2200		
¾		1300	1800	2200	
⅞		1100	1500		
1		900	1300	1700	2000

- Based on ultimate tensile strength 42,000 psi (289,400 kPa)
- For metal temperatures 20° to 100°F (29° to 37°C)
- Allowable working pressure loads calculated from S values 14,000 psi (96,500 kPa) as specified by ANSI B31.3. code.

105T02.EPS

3.0.0 ◆ PROPER AND SAFE METHODS FOR STORING TUBING

Proper and safe storage of tubing may sound simple, but you would be surprised at the amount of damage done to tubing, the number of personal injuries, and the time wasted due to improper storage techniques. There are correct ways to store tubing. Proper storage helps avoid costly damage, saves time in identifying stored tubing, and reduces the chance of personal injury to you and other workers.

Tubing racks are commonly used for storing tubing (see *Figure 3*). The racks provide support to prevent bending, sagging, distortion, scratching,

Table 3 Factors Used to Determine Tubing Pressure Ratings at Elevated Temperatures

Factors							
°F	°C	Aluminum	Copper	Steel	304SS	316SS	Monel®
200	93	1.00	.80	.95	1.00	1.00	.88
400	204	.40	.50	.86	.93	.96	.79
600	316	N/A	N/A	.77	.83	.85	.79
800	427	N/A	N/A	.58	.76	.79	.76
1000	538	N/A	N/A	N/A	.69	.76	N/A
1200	649	N/A	N/A	N/A	.30	.37	N/A

To determine allowable pressure for aluminum at elevated temperatures, multiply allowable working pressure from Aluminum Tubing Pressure Table (Table 2) by the appropriate factor from the table above. For example, the allowable working pressure for aluminum ½" OD × .035" wall is 1900 psi. At 400°F, 1900 psi × .40 = 760 psi. Therefore, the allowable working pressure for ½" OD × .035" wall tubing is 760 psi at 400°F.

105T03.EPS

or marring of tubing surfaces. Most racks have separate levels or compartments where different types and sizes of tubing can be stored for ease of identification and selection. Tubing racks are usually elevated to help avoid damage that might occur at floor level. Tubing that is stored at floor level is easily damaged or contaminated by people, materials, moisture, and equipment in the area.

Always inspect tubing before storing it to be sure it's clean and not damaged. The ends of stored tubing should be sealed to prevent damage. Tubing ends can be capped, taped, or **plugged** to seal them. In the case of soft copper tubing rolls, the ends are sometimes sealed by crimping during storage.

> **WARNING!**
> Be sure tubing is stored securely so when someone comes to get it, it does not fall or roll in any way that could cause injury.

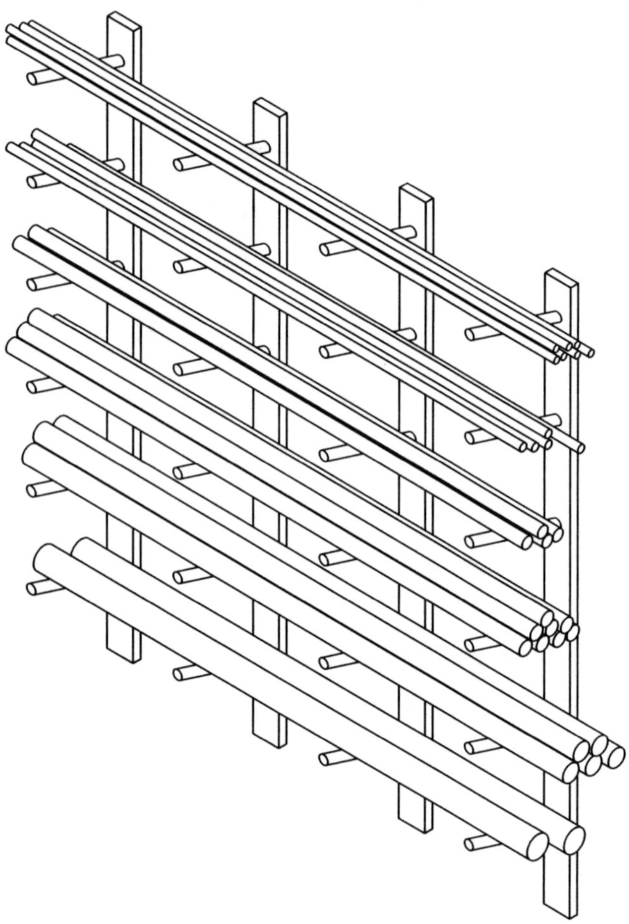

Figure 3 ◆ Typical tubing rack.

To prevent contamination and corrosion, tubing should be covered with a tarpaulin or other suitable covering. Some types of tubing should be separated from non-compatible materials. Covering stored tubing also helps prevent damage from sunlight and other environmental factors that especially affect some types of poly tubing.

4.0.0 ◆ PROPER AND SAFE METHODS FOR HANDLING TUBING

Tubing is made to strict specifications, but it can easily be degraded by careless handling. From the time tubing is delivered on a job site until installation is complete, proper and safe handling techniques will ensure quality tubing installations. The following are a few basic guidelines for handling tubing to avoid damaging or contaminating the tubing or injuring yourself or others:

- Never drag tubing off a delivery truck, off a storage rack, or on the ground or floor. Dragging tubing across any rough surface can scratch or dent the outside surface, damage the ends, and contaminate the tubing.

> **CAUTION**
> Scratches on tubing are paths for leaks.

- Keep tubing away from any material that might contaminate it during handling.
- Keep the end caps on during handling or transporting of tubing to avoid damage to ends and contamination.
- Wear gloves when handling tubing to avoid scraping or cutting yourself on rough or sharp edges of the tubing.
- Do not use a vise to hold tubing unless it is equipped with special jaws for holding and protecting round tubing. The jaws of conventional vises can damage tubing.
- Put flags on the ends of long lengths of tubing during handling and transporting so that you and others can easily see the tubing ends and avoid damage and/or accidents.
- Be aware of people, materials, and equipment around you when handling and transporting tubing so that you can avoid hitting anyone or anything with the tubing. This may cause injury to another person or damage to the tubing, other materials, or equipment.

- Place straight lengths of tubing inside PVC piping or an aluminum tray while transporting them to prevent bending or damage. This will also prevent you and others from being pinched by the tubing during moving.
- Handle soft tubing such as copper and aluminum with extreme care to avoid damage. It's very easy to dent or scratch soft tubing, and cause it to become out-of-round. This weakens the tubing and also makes it difficult to make good connections.
- Soft tubing that comes in rolls or coils must be carefully unrolled to prevent kinks from forming. The best way to unroll soft tubing is to first place the roll on a flat surface like a table top, bench, flat board, or the floor. If the flat surface is rough, put down some type of padding to protect the tubing from scratches, dents, or other damage. Hold the loose end of the tubing on the flat surface with one hand while holding the coil in your other hand. Unwind the tubing by rolling the coil about 2 feet at a time. Carefully slide your hand along the tubing as it unwinds onto the flat surface, straightening each section (see *Figure 4*).

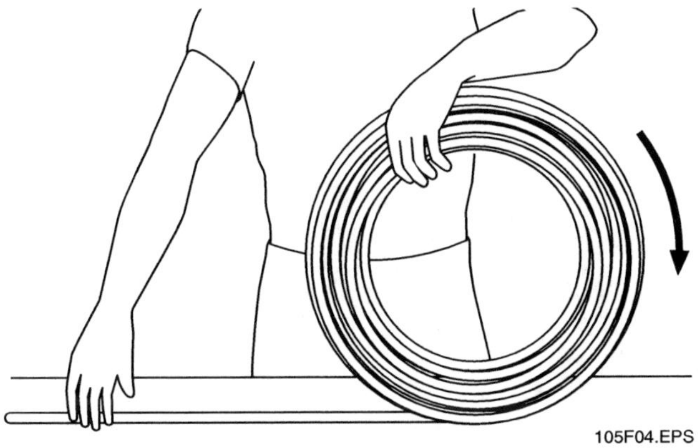

Figure 4 ◆ Unwinding rolled tubing.

Remember to unroll the coil, rather than pull it, to avoid twisting. Twisting might weaken the tubing or cause it to get out-of-round. Don't unroll more tubing than you need, since repeated uncoiling can distort, harden, or stiffen the tubing. A tape measure or folding rule lying next to the coil will help determine when you have enough tubing unrolled.

Sometimes tubing is stretched, or drawn, after it is unrolled to ensure that it is straight. Care must be taken, however, not to stretch the tubing so much that its OD is reduced. Tubing that is overstretched will not take fittings properly. This can cause leaks. Tubing can be stretched with a tube stretcher (see *Figure 5*).

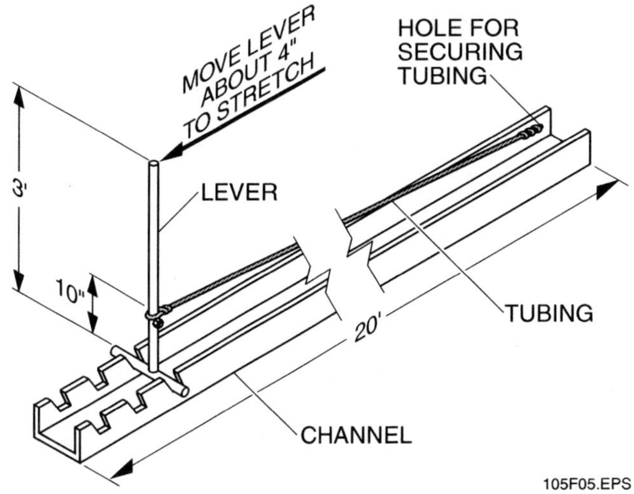

Figure 5 ◆ Typical tubing stretcher.

The tubing stretcher is made of a length of channel (usually 20 feet) notched on the end to accept a lever. The lever is made of a length of pipe (approximately 3 feet long and 1 inch in diameter) welded into a "T" shape. To operate the tubing stretcher, the tubing is tied to one end of the stretcher and to the lever at the other end. When the lever is pulled back about 4 inches (measured at the top of the lever), the tube is stretched enough.

If a tubing stretcher is not available, tubing can be stretched by tying one end to a fixed or anchored object and pulling against the other end with a lever. Remember to stretch the tubing close to the ground. If you do not, it could sag when you release the tension, causing it to bend.

NOTE

Be sure to support stretched copper tubing with a tray or PVC conduit when moving to the work area so it will not sag.

5.0.0 ◆ CUTTING TUBING

Before tubing can be installed, it must be cut squarely and cleanly to proper length. Tubing is often cut with a device called a tubing cutter. In some cases, a hacksaw or bandsaw may also be used to cut tubing. Some types of soft plastic tubing are cut with snips, a pocketknife, or a special "V" shaped cutting tool made especially for cutting plastic tubing.

5.1.0 Types and Sizes of Tubing Cutters

Tubing cutters are available in several different types and sizes. Generally, a tubing cutter consists of a cutting wheel, a handle, and a roller. On some types, the rollers are tightened to move the outside of the tubing against cutting wheel. Others have a sliding cutting wheel that is moved against the outside of the tubing. *Figure 6* shows different types of outside tubing cutters.

Some tubing cutters cut from the inside of the tubing. These internal tubing cutters are used where there is not enough room to use outside cutters.

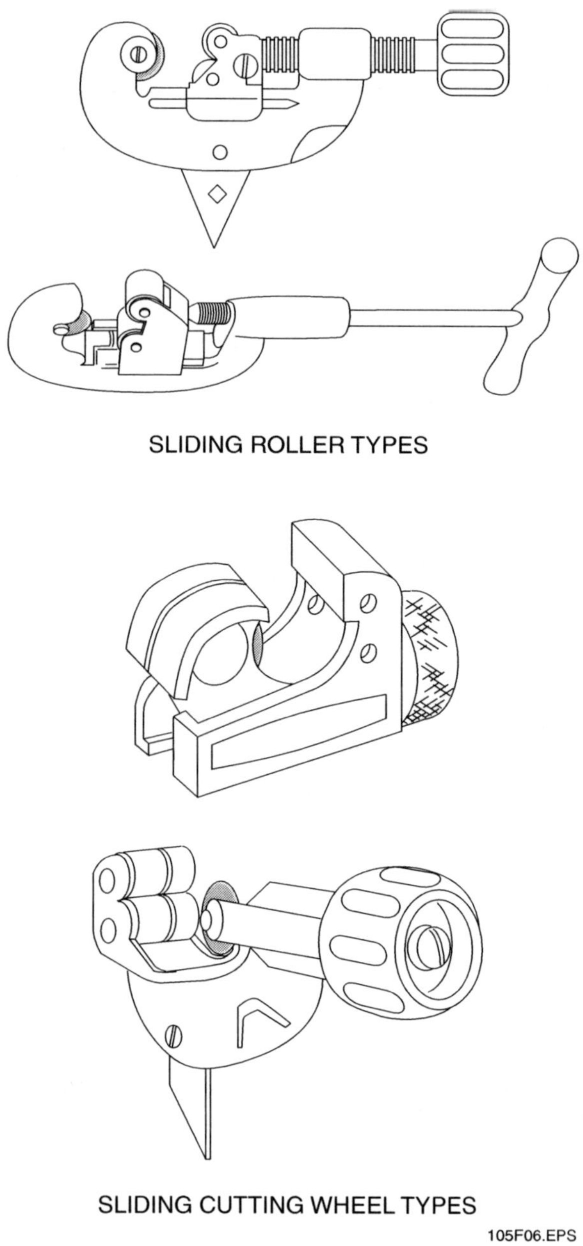

Figure 6 ◆ Different types of outside tubing cutters.

Tubing cutters are commonly used for cutting plastic, soft copper, aluminum, steel, and stainless steel tubing. Check the manufacturer's manual to make sure that the cutting wheel is designed to cut the type of tubing that you are cutting.

5.2.0 Cutting Tubing with a Tube Cutter

It's a good idea to check the manufacturer's instructions before using any tubing cutter to make sure that you know the proper and safe way to use the cutter. The manufacturer's instructions may also specify the type and size of cutter and wheel needed for the specific cutting job.

Before using a tubing cutter, the tubing should be marked at the exact place where the cut is to be made. A sharp pencil, colored felt tip pen, and silver marking pencil are good tools for marking tubing. They each make a nice thin line without scratching. It's best to make the mark all the way around the tubing so that it is always visible while you are cutting. A sleeve or ferrule can be used as a guide for making the mark around the tubing (see *Figure 7*).

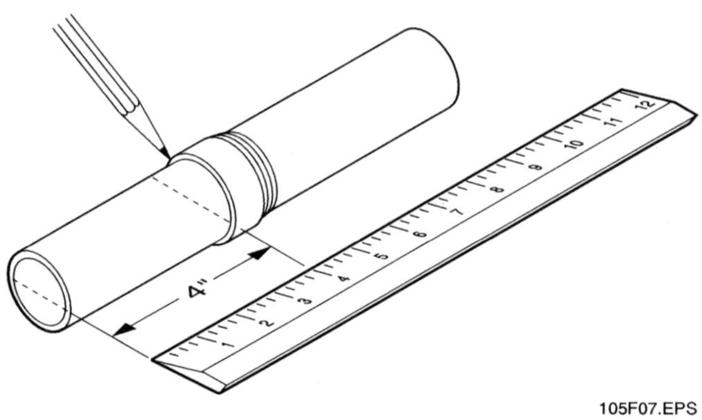

Figure 7 ◆ Marking tubing.

NOTE

Do not use grease pencils or crayons to mark metal tubing because the mark will be too wide. Wide marks may affect the accuracy of the cut. You should also not use sharp items like pens or scribes to mark metal tubing since the scratch marks are points where corrosion or stress can dangerously weaken the tubing.

Follow these steps to cut tubing with an outside tubing cutter:

WARNING!
Follow all safety precautions.

Step 1 Open the jaws of the cutter and place the tubing on the rollers of the cutter.

Step 2 Align the blade with the mark on the tubing.

Step 3 Turn the handle to tighten the blade (or rollers) until the blade is snug against the tubing (see *Figure 8*).

NOTE
The tubing should not be squeezed too tightly against the blade, which might dig into the tubing and flatten it.

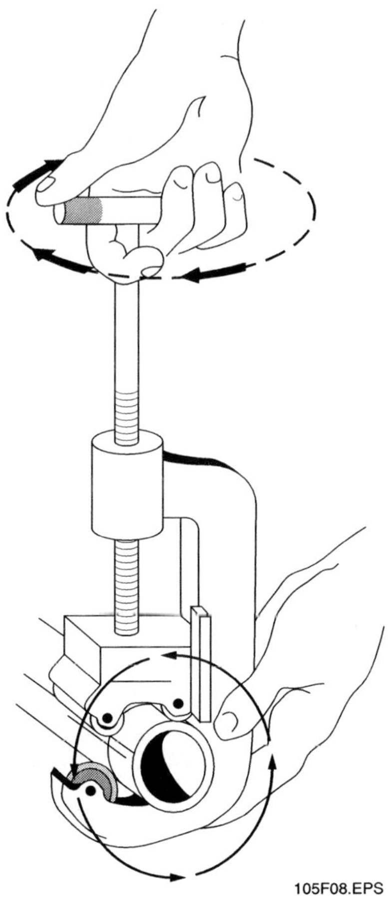

Figure 8 ◆ Using a tubing cutter.

Step 4 Rotate the cutter around the tubing, using a smooth motion and pushing the cutting wheel farther into the groove being cut.

Step 5 Gradually tighten the handle of the cutter every few turns, making sure the blade remains against the tubing at all times.

NOTE
Tighten the handle with the cutting wheel at different positions on the tubing each time.

Step 6 Support the end of the tubing so it will not fall when the cut is completed.

Step 7 Keep turning the cutter and tightening the handle until the tubing is cut entirely through.

NOTE
Do not try to rush the cutting procedure. Applying too much pressure on the tubing could result in a poor cut, cause excessive wear on the cutting wheel, or create flat spots on the tubing.

5.3.0 Cutting Tubing with a Hacksaw

Metal tubing is sometimes cut with a hacksaw. A hacksaw cut is preferred over a cut made with a tubing cutter for some types of tubing materials, for larger sizes of tubing, and when flared fittings are to be used. The hacksaw method is often used for cutting tubing made of hard materials. The heat and pressure that result when a tubing cutter is used can cause hard metals to become brittle near the cut, weakening the tubing. A tubing cutter can also bend in the edge of the tubing, slightly reducing the OD. This may make it difficult to get a good seal with some fittings.

When using a hacksaw to cut tubing, be sure to select the correct blade. Use an H.S.S. (high speed stainless) for stainless steel tubing. A general guideline is to choose a blade that allows two teeth of the blade to be in contact with the tubing. For thin-walled tubing, use a blade with smaller teeth. The distance between each tooth on a hacksaw blade is called the pitch. A pitch of 18 represents 18 teeth per inch (TPI). A hacksaw blade with at least 32 TPI is normally used to cut tubing.

Be sure to mark the tubing all the way around when cutting with a hacksaw. Keep the blade as square as possible to the tubing in order to produce an even, squared-off cut. Cut on the forward motion of the hacksaw. Keep light, even pressure on the tubing to avoid flattening.

Guideblocks are often used to ensure a square cut when cutting tubing with a hacksaw (see *Figure 9*). The guideblocks also help keep the tubing from flattening out. If the cut is not square when you have finished, use a flat file to square it off.

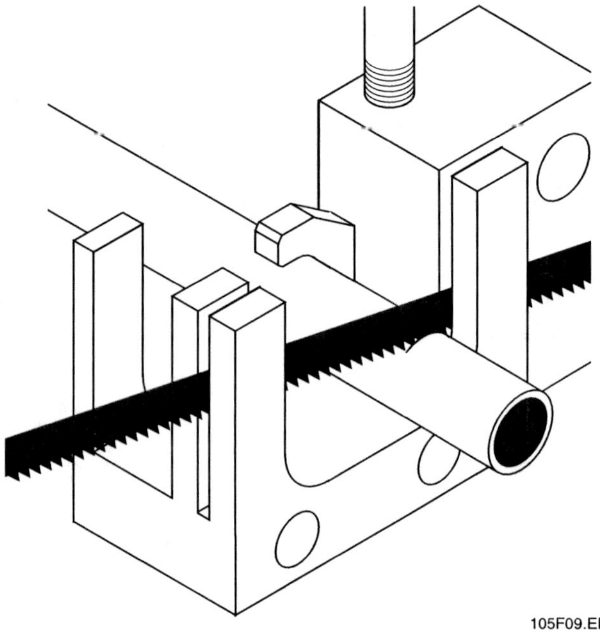

105F09.EPS

Figure 9 ◆ Cutting tubing with a hacksaw and guideblocks.

> **WARNING!**
> Always wear safety glasses when cutting tubing with a hacksaw to prevent metal chips from getting in your eyes.

5.4.0 Cutting Tubing with a Bandsaw

Bandsaws are sometimes used to cut tubing when numerous cuts need to be made. A bandsaw cuts much faster than a hacksaw or tubing cutter and makes good clean cuts. When cutting tubing with a portable bandsaw, be sure to place the brace, located below the saw blade, firmly against the tubing to be cut. The weight of the saw alone is enough to provide the leverage needed for cutting. Very little pressure is needed to make a good clean cut with a bandsaw.

Check the operating manual for the bandsaw you are using. The manual will help determine the correct saw speed and blade to use for the type of tubing being cut. As a rule of thumb when selecting the right blade for cutting tubing with a bandsaw, the softer the material the finer the tooth blade to use. For thin materials like tubing, at least two or three teeth should be in the cut. Where a smoother finish is important, select one of the finer tooth blades.

5.5.0 Cutting Poly Tubing with Snips

Soft, flexible poly tubing and PVC tubing are easy to cut using specially designed snips (see *Figure 10*). Before using snips to cut plastic tubing, mark the tubing at the exact place where the cut is to be made. A sharp pencil is a good tool for marking tubing because it makes a nice thin line. Colored pencils can be used if the color of a regular lead pencil is hard to see on the type or color of tubing being cut. It's best to make the marking all the way around the tubing so that the mark is always visible. Grease pencils, which make thicker lines than regular lead pencils, are sometimes used to mark plastic tubing because precise measurements are not as critical on plastic tubing runs. Grease pencil markings also show up well on the plastic tubing. When cutting tubing with snips, make a clean straight cut so the tubing will fit properly on the connector or fitting.

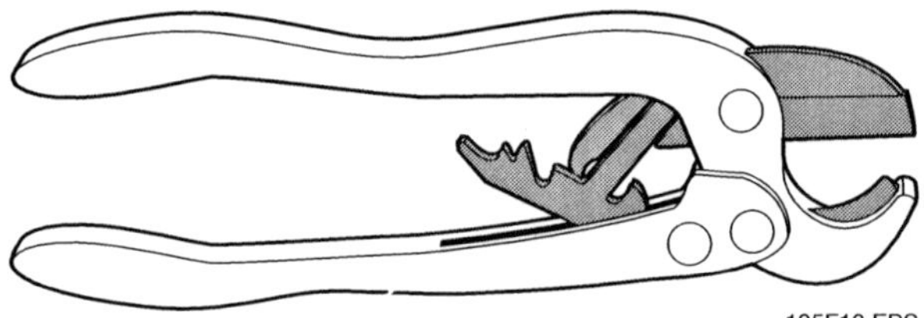

105F10.EPS

Figure 10 ◆ Tubing snips.

5.6.0 Deburring Tubing

When metal tubing is cut with a tubing cutter, hacksaw, or bandsaw, there may be burrs on the ID, OD, or both. Hacksaws are especially likely to leave burrs because the cutting action is not as smooth as that of other cutting tools. If not removed, burrs on the ID of tubing can break loose, causing metal chips and slivers to enter the tubing system. Chips and slivers can plug small pilot holes or vents, scratch valve seats and stem tips, or damage soft seals such as O-rings. Tubing ID can also be reduced because of the burrs. A smaller ID may affect the flow through the tubing.

Burrs on the OD of tubing make it difficult or impossible to properly install fittings to achieve a clean, leak-free tubing system. *Figure 11* shows tubing burrs.

Deburring the OD of tubing is usually done with a flat file. The file is normally held at a 45-degree angle, and the burr is removed with light forward strokes of the file on the tubing.

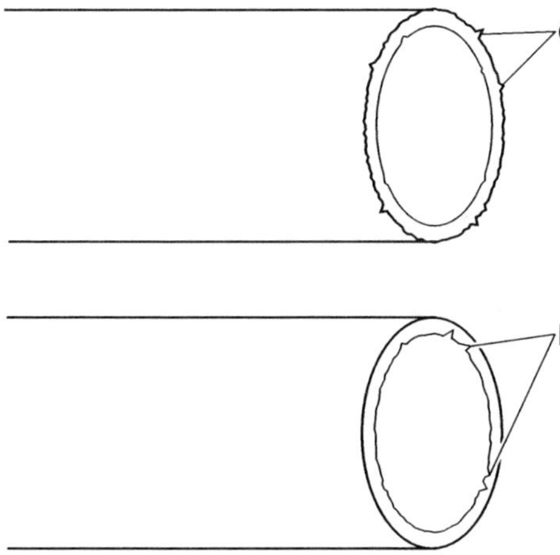

Figure 11 ◆ Tubing burrs.

NOTE
Be careful not to nick or scratch the tubing with the file, or you may have a problem installing the fitting.

A hand **reamer** is often used to remove the burr on the inside of tubing (see *Figure 12*). The reamer, which has blades on the outside of a cone-shaped end, is inserted into the cut end of the tubing. It is twisted clockwise to remove the ID burr. Reamers come in a range of types and sizes.

Sometimes round files and various types of scrapers with curved blades are used to remove ID burrs from tubing. While these tools can do a reasonable job, a reamer removes the burrs better and leaves a smoother ID.

NOTE
Tubing should be reamed only until the inside of the tubing is smooth and the burr is completely removed. Excessive reaming can weaken the tubing by thinning the tubing wall. You should not ream beyond the inside diameter of the tubing.

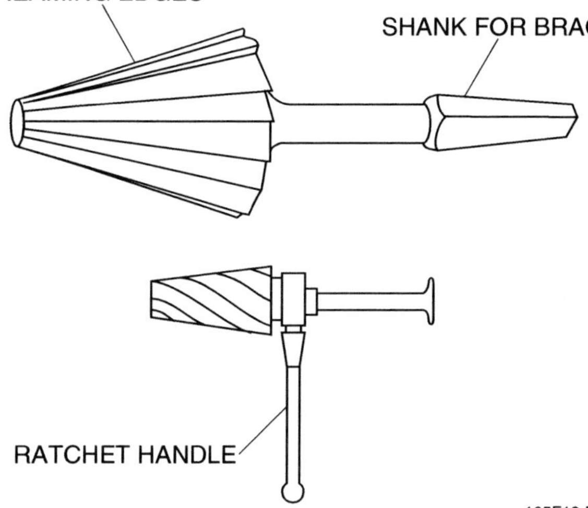

Figure 12 ◆ Tubing reamer.

A special device known as a *tube deburring tool* is used to deburr both the ID and OD of small, soft tubing (see *Figure 13*).

The tool is shaped like a barrel, with OD deburr on one end and ID deburr on the other end. The OD end has blades which contact the OD of the tubing. The ID end has protruding blades which contact the ID. The OD end of the deburring tool is placed over the end of the tubing and turned back and forth, removing any burrs on the OD. The tool is then reversed, using the ID end. The same twisting action is used to deburr the ID.

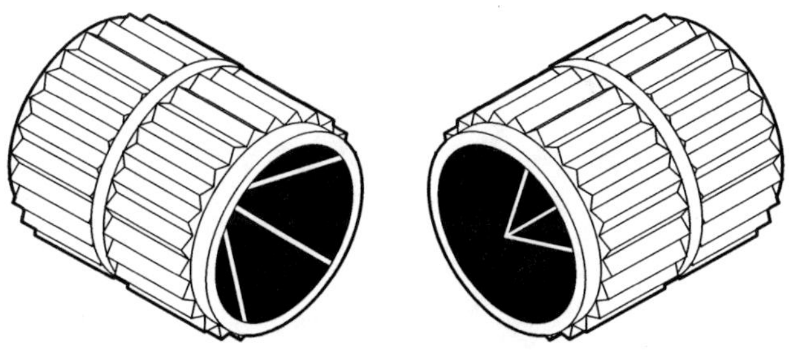

Figure 13 ◆ Tube deburring tool.

NOTE
After tubing is deburred, be sure to remove any filings or other particles from inside the tubing.

6.0.0 ◆ BENDING TUBING

One of the advantages of tubing is that it can be bent and shaped for various types of installations. Layout, cutting, bending, and attaching fittings to a tubing installation require a number of skills and plenty of practice to become expert.

6.1.0 Standard Tubing Bends

Tubing installations require a number of different types of bends. The following standard bends are used for specific purposes in tubing installations:

- *Angle bend* – Changes the direction of the tubing run
- *Offset bend* – Moves the tubing run to one side to avoid an interference
- *Crossover bend* – Allows room for other tubes to **cross**
- *U-bend* – Turns the tubing run around
- *Expansion bend* – Allows room for the tubing line to expand or absorb vibration

Figure 14 shows standard tubing bends.

6.2.0 Tubing Bending Methods

Tubing must be bent carefully and correctly. A good tubing bend is smooth all around the bend, providing even flow through the tubing. Good tubing bends are made by using the right equipment and proper methods. Whenever tubing is bent, the metal at the back of the bend is stretched and the metal in the throat is compressed. If the bend is too sharp, one of two things will happen:

- The metal at the back of the bend will be stretched too far and will break.
- The metal in the throat will collapse and close off the tubing.

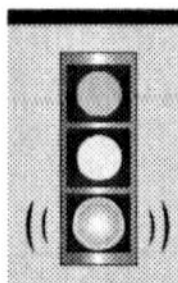

NOTE

A properly made bend will be smooth, stay in plane, and lie flat on a smooth surface.

Figure 15 shows damaging effects of overstretching and compressing in tubing bends.

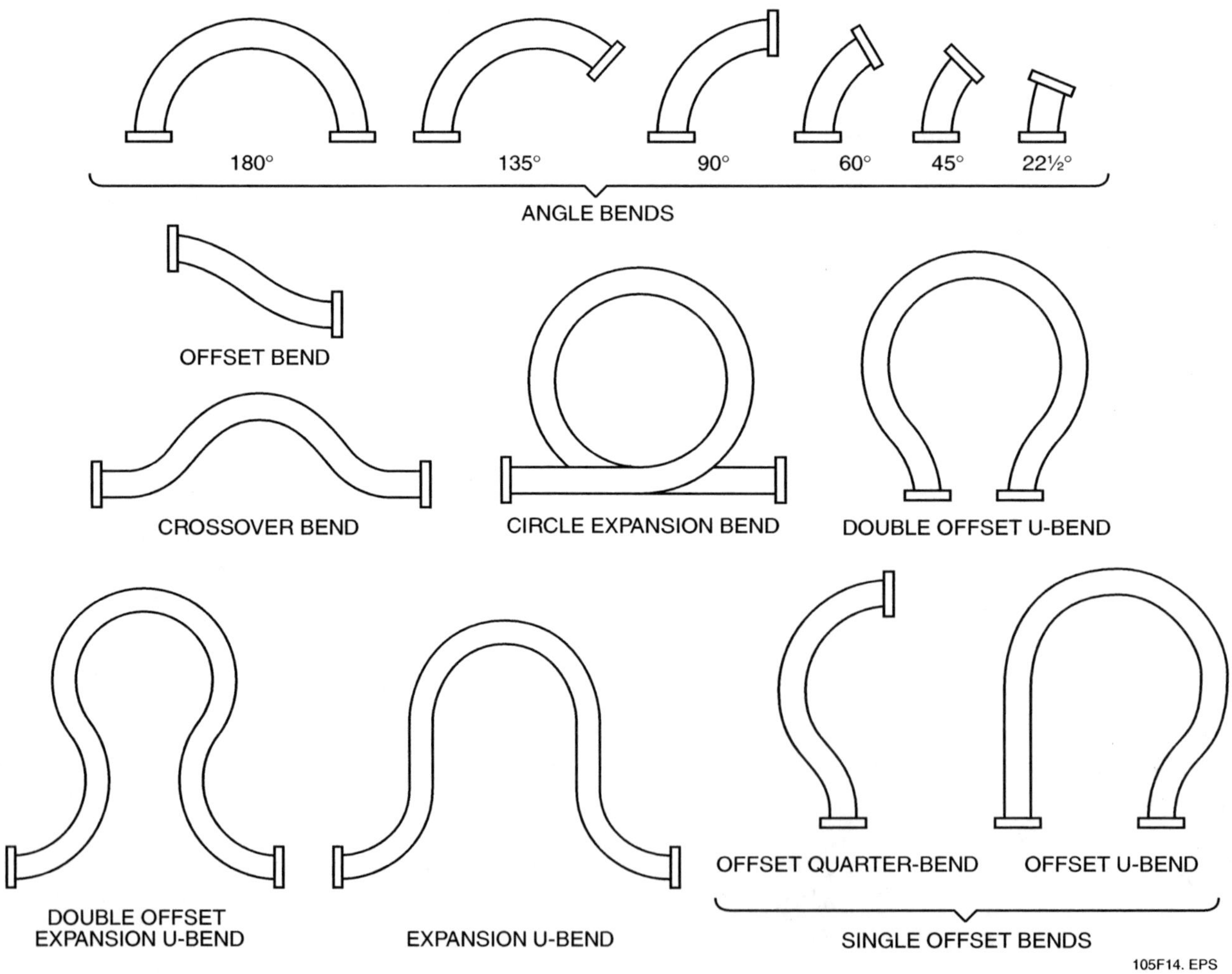

Figure 14 ◆ Standard tubing bends.

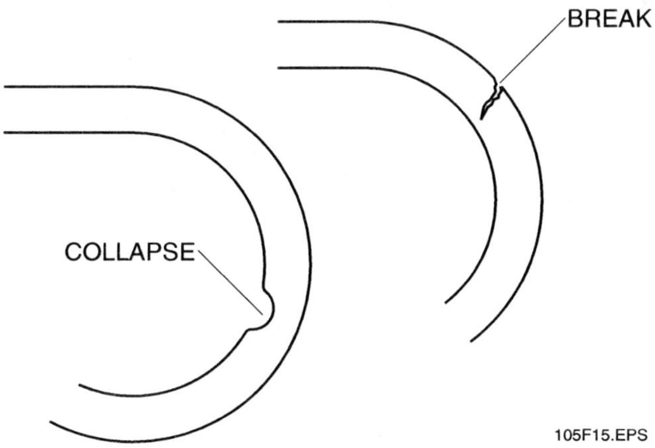

Figure 15 ◆ Damaging effects of overstretching and compressing in tubing bends.

A poor bend can weaken the tubing or restrict the flow through the tubing. A good tubing bend has no flattening of the tubing through the bend. *Figure 16* shows improper and proper tubing bends.

> **NOTE**
> If you make a poor bend in tubing, do not try to install the tubing. Do it again until you get it right.

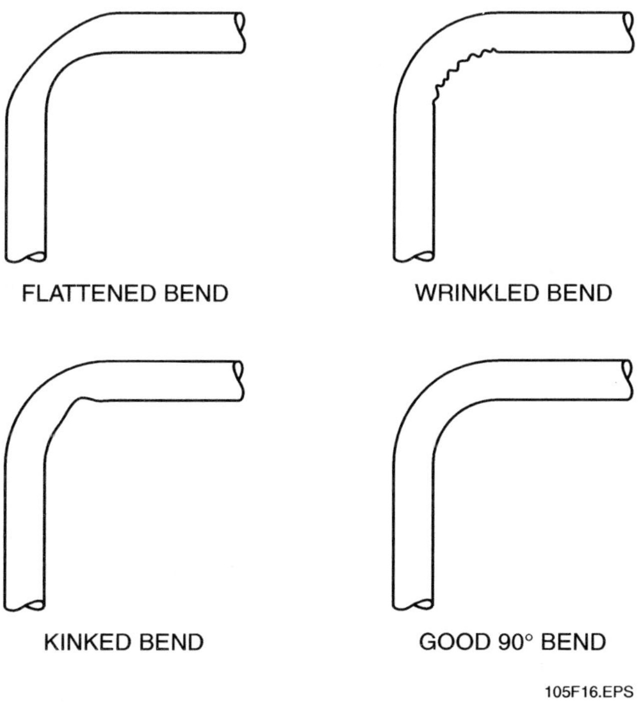

Figure 16 ◆ Improper and proper tubing bends.

6.3.0 Types of Tubing Benders

Tubing benders come in several different types and sizes. Hand benders are used to bend soft tubing such as soft copper, aluminum, and small diameter stainless steel. Most machine benders are powered by hydraulic pumps or hand-operated gear actions. Machine benders are used to bend steel and stainless steel tubing with larger ODs and heavier wall thicknesses, such as stainless steel or hard copper.

6.3.1 Using a Spring Tube Bender

The simplest hand bending tool is a spring tube bender (*Figure 17*).

Spring tube benders come in different sizes and are used to bend small diameter, soft metal tubing. *Figure 18* shows bending with a spring tube bender.

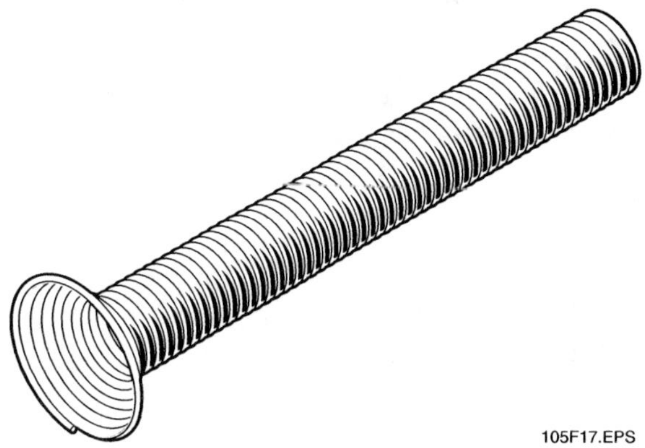

Figure 17 ◆ Spring tube bender.

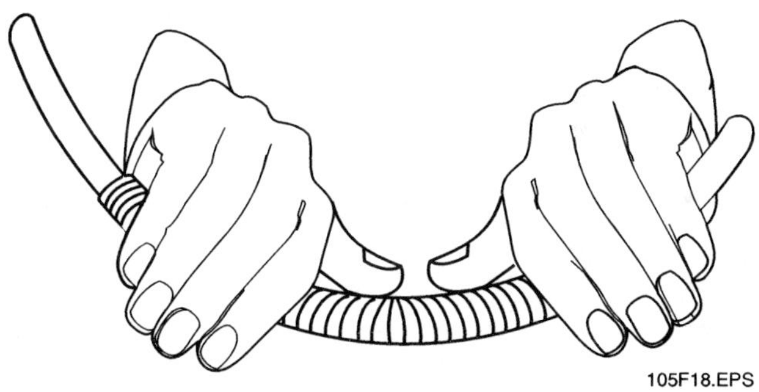

Figure 18 ◆ Bending with a spring tube bender.

Follow these steps when bending tubing with a spring tube bender:

> **WARNING!**
> Follow all safety precautions.

Step 1 Fit the spring bender tightly over the center of the area of the tubing to be bent.

Step 2 Bend the spring and tubing by hand until the proper angle and radius are formed.

Step 3 Slip the spring bender off the tubing.

Step 4 Check to see if you achieved the correct angle for the bend by using a protractor, angle finder, or tri-square.

> **NOTE**
> Spring tube benders have limited applications because it is difficult to get an accurate angle for bends.

6.3.2 Using a Compression-Type Hand Bender

A compression-type hand bender is a type of hand bender commonly used for bending soft tubing of ¾ inch or less (see *Figure 19*).

The compression-type hand bender, also called a *lever bender*, consists of a slide block, slide block handle, forming wheel, forming wheel handle, and tube clamp. Before using a tubing bender, it is important to check the manufacturer's manual to make sure you have a bender of the correct size for the tubing being bent.

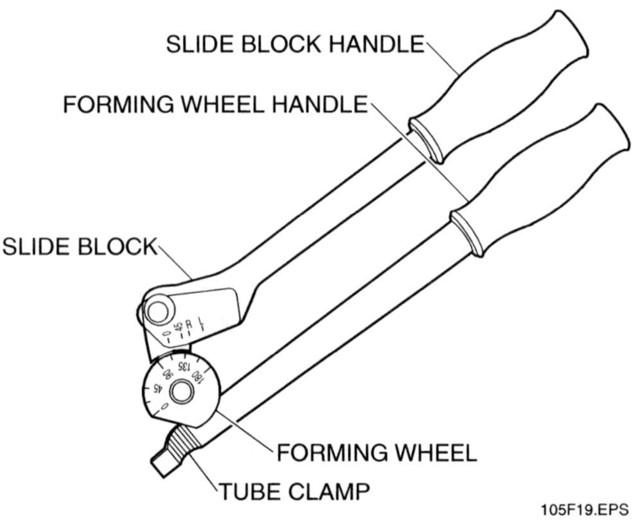

Figure 19 ◆ Compression-type hand bender.

The tubing to be bent is positioned in the groove in the forming wheel. It is held in place by the tube clamp. The numbers around the forming wheel refer to the angle of the bend that will be formed. The markings on the slide block are used for properly aligning the tubing in the bender. A groove along the outside of the forming wheel supports the tubing as it is bent. The bend is formed by the slide block moving around the outside of the tubing in the groove. As the slide block moves, it applies compression to bend the tubing. Follow these steps to form an initial bend with this type of bender:

> **WARNING!**
> Follow all safety precautions.

Step 1 Measure and mark the tubing exactly to get the desired bend.

Step 2 Insert the tubing in the bender, aligning the mark on the tubing with the "L" on the slide block.

Step 3 Clamp the tubing securely to the bender by dropping the clamp over the tubing to hold it in place.

Step 4 Align the "0" on the slide block with the "0" on the forming wheel. *Figure 20* shows the starting position for bending tubing.

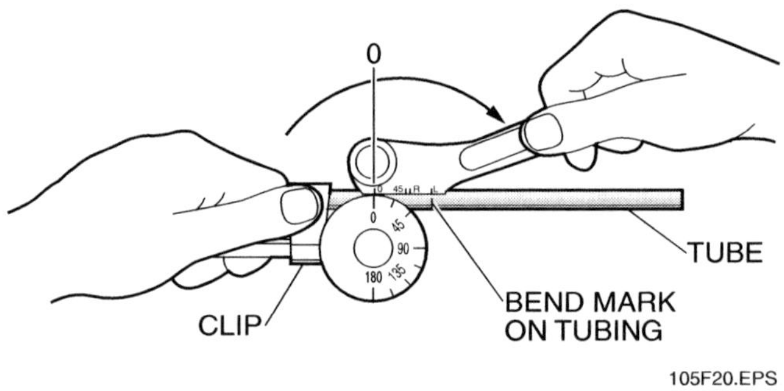

Figure 20 ◆ Starting position for bending tubing.

Step 5 Move the slide block handle around the tubing, forcing it to conform to the shape of the forming wheel until the "0" on the slide block aligns with the proper degree mark on the forming block. *Figure 21* shows making a 90-degree tubing bend.

> **NOTE**
> When bending harder metal tubing, bend it slightly past the mark to allow for springback.

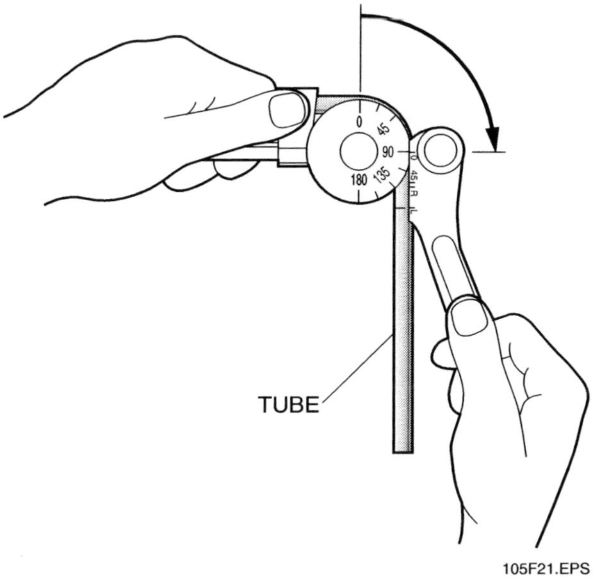

Figure 21 ◆ Making a 90-degree tubing bend.

Step 6 Lift the handle, raise the clamp, and remove the tubing.

Step 7 Check the tubing to be sure you achieved the correct angle for the bend. Also check that the length from the end of the tubing to the center of the bend is correct. The mark should be at the center of the bend.

CAUTION

Only bend the tubing in one direction so you do not overstretch or kink it.

6.3.3 Using Table- or Bench-Mounted Tubing Benders

There are several types and sizes of table- or bench-mounted tubing benders that are used to bend tubing of medium- and heavy-wall thicknesses (see *Figure 22*).

Table benders are often used for tubing that is ¾ inch or larger. They are also used when a large number of bends are being made. Most table benders work similar to a hand bender except that there is some type of gear assembly used to power the bender. The gear mechanism may be powered in a number of different ways including a ratchet or handcrank.

6.3.4 Using a Hydraulic Tubing Bender

Most hydraulic tubing benders have a cylinder and ram assembly which bends the tubing with power supplied by a hydraulic pump. Hydraulic tubing benders are used to bend harder metals and larger types of tubing. Hydraulic benders are usually used for ¾-inch to 2-inch hard tubing, such as steel or stainless steel with heavy wall thicknesses.

7.0.0 ◆ TUBING FITTINGS

After tubing has been measured, cut, bent, and deburred, it can be connected to other components with tubing fittings to make the tubing run. For example, a tubing fitting might be used to make the connection between tubing and tubing, tubing and piping, or tubing and a valve. The sizes and types of fittings to be used for a specific job are usually specified in the engineering specifications or on the instrument installation drawings for a project. Most tubing fittings are sized by the outside diameter of the tubing to which they are to be connected. For example, ½-inch tubing takes a ½-inch tubing fitting.

Fittings are specified to match the tubing material being used and the application. Factors such as pressure, temperature, and process characteristics are taken into consideration when selecting tubing fittings.

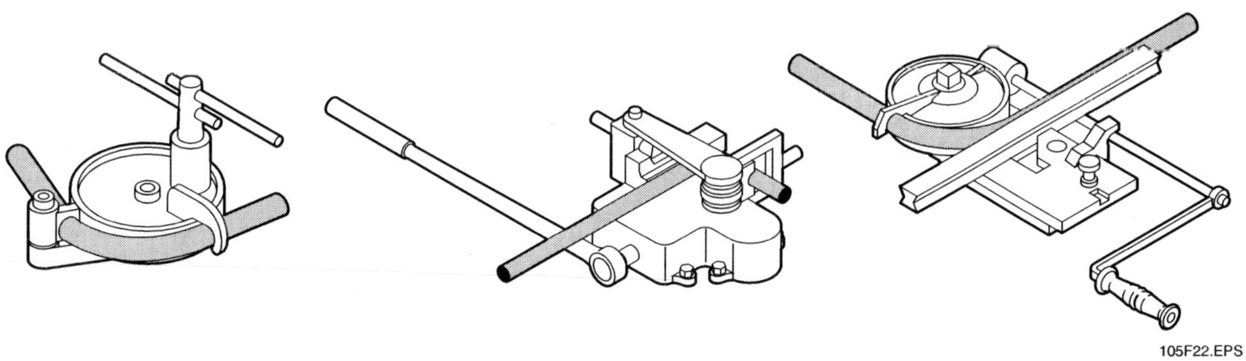

Figure 22 ◆ Table or bench benders.

Fitting manufacturers use codes to identify the fitting material. The codes are sometimes stamped on the fitting. These codes may also indicate the heat (batch) number of the traceable lots. Some of these identification codes and other specifications have been developed by ASTM, ASME, and ANSI, like the codes for tubing materials. Other codes are manufacturer-specific.

If you have any questions, you will need to check the fitting manufacturer's catalog to determine the material codes. The following is a partial list of material designator codes for fittings used by one manufacturer:

MATERIAL	MATERIAL CODE
Aluminum	A-
Brass	B-
Carbon Steel	S-
Monel®	M-
Nylon	NY-
316 Stainless Steel	SS-
347 Stainless Steel	347-
TFE	T-
Hastelloy® B2	HB-
Hastelloy® C276	HC-
Inconel® 600	INC-
Copper-Nickel 70-30	CN70-
Copper-Nickel 90-10	CN90-

A partial list of another manufacturer's material identification codes for fittings is presented below:

MATERIAL	MATERIAL CODE
Monel®	4
304 Stainless Steel	10
316 Stainless Steel	13
Brass	B
303/321 Stainless Steel	C
Steel	S

The four most common types of tubing fittings used in instrument work are flared, compression, socket-welded, and butt-welded.

7.1.0 Flared Fittings

A flared fitting gets its name from the fact that the end of the tubing is enlarged, or flared, slightly to give a rim against which the sleeve and body of the fitting can seal. Flared fittings can be assembled repeatedly and provide a good strong connection. They come in a variety of types, sizes, and materials for specific applications.

The fitting consists of three main parts: a body, a flare nut, and a flare sleeve (*Figure 23*). Not all flared fittings include a sleeve. Sleeves are used primarily in fittings for high-pressure applications.

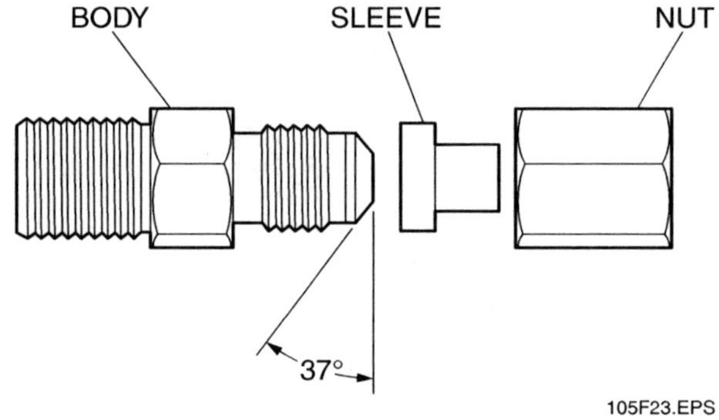

Figure 23 ◆ Parts of a flared fitting.

The fitting body of the flared fitting shown in *Figure 23* is part of a piping-to-tubing connector. The tip of the fitting body has a tapered angle of 37 degrees. Another commonly used fitting has a 45-degree angle taper. The tubing used with a particular flared fitting must be flared at the same angle as the tip of the fitting body.

> **NOTE**
> The 37-degree flared fittings are typically used for high-pressure applications such as hydraulic oil systems. The 45-degree flared fittings are usually used for low-pressure water and air systems.

The sleeve in a flared fitting acts as a buffer between the nut and the flared end of the tubing. The nut holds the tubing flare against the fitting body. When the nut is tightened, it applies pressure against the sleeve. The sleeve, in turn, applies pressure against the tubing, securing and sealing the tubing against the body. *Figure 24* shows an assembled flared fitting cutaway.

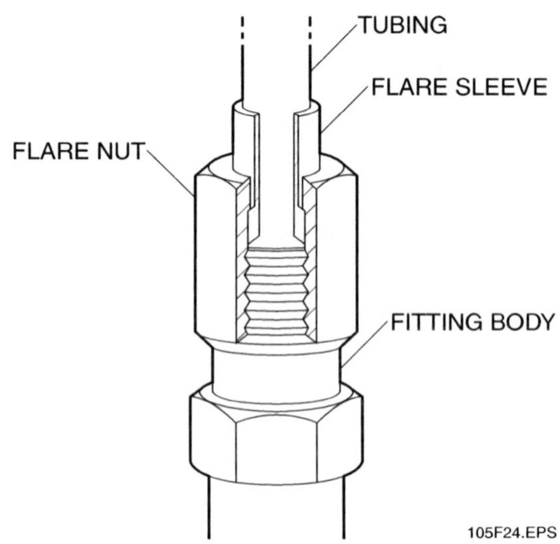

Figure 24 ◆ An assembled flared fitting cutaway.

> **WARNING!**
> To avoid possible rupture, always make sure that the flare angle matches the angle of the sleeve.

7.2.0 Compression Fittings

Compression fittings are often used in instrument work because they are considered reliable and relatively easy to install. Compression fittings can usually be disconnected and retightened without damaging the tubing or the fitting. They come in a variety of designs, sizes, and materials to fit different tubing applications. When using compression fittings, the end of the tubing is not flared. The tubing is fitted with a compression ring or ferrule that pinches, or compresses, the tubing as the locknut is tightened on the body of the fitting. Compression fittings have either one or two ferrules for each joint. The ferrule of the fitting must be harder than the tubing. *Figure 25* shows a one-ferrule compression fitting.

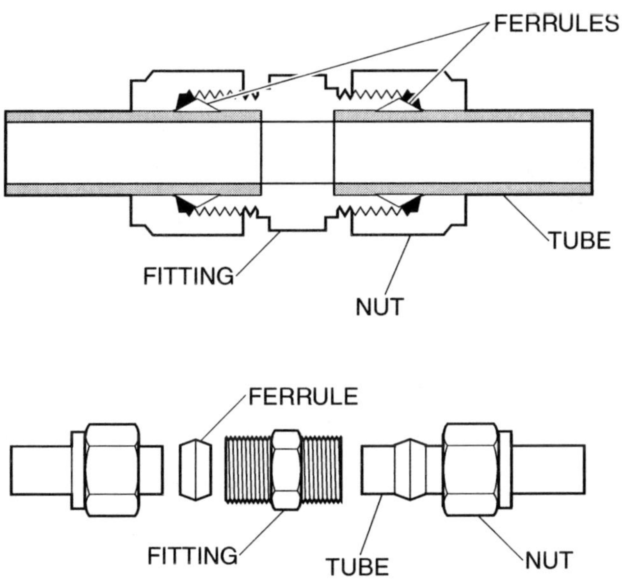

Figure 25 ◆ One-ferrule compression fitting.

Two-ferrule compression fittings have a back ferrule and a front ferrule for each joint (see *Figure 26*).

Compression fittings can also be used for connections on soft poly tubing such as PVC, Tygon, and Teflon®. Fittings used for poly tubing may differ slightly in design from those used on metal tubing. The fittings may have nylon ferrules rather than metal ferrules to prevent damage to the tubing as the ferrule is compressed. Most compression fittings used for poly tubing have a tube support insert (see *Figure 27*). The insert provides a rigid surface against which the tubing can be compressed so that it doesn't collapse.

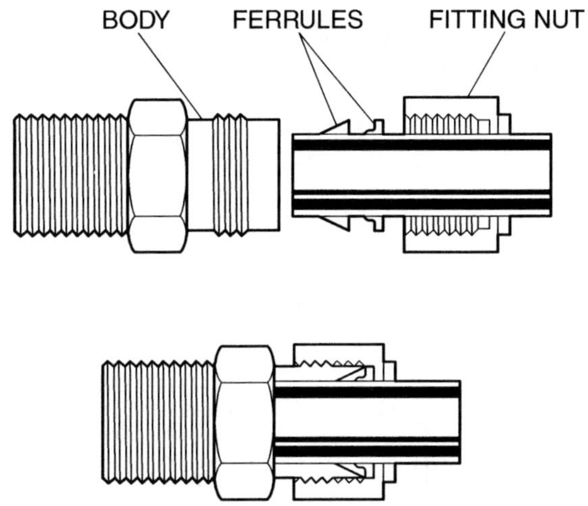

Figure 26 ◆ Two-ferrule compression fitting.

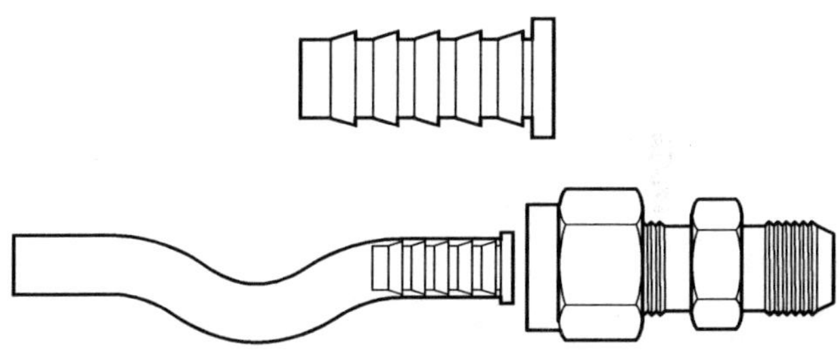

Figure 27 ◆ Insert for poly tubing compression fitting.

7.3.0 Types of Tubing Fittings

Numerous types of fittings are used to change direction of tubing runs, provide **branch** connections, close ends of tubing, and connect tubing to tubing, tubing to piping, and tubing to instruments. The following information describes several of the most commonly used types of tubing fittings.

7.3.1 Male/Female Connectors and Adapters

The general term *male fitting* refers to fittings with outside threads that connect tubing to a component with a female or internal thread. The general term *female fitting* refers to fittings with internal threads that connect tubing to a component with male or outside threads. While many types of fittings are referred to as male- or female-type fittings, two of the most common are male and female connectors and adapters. A male connector fitting has a male end that connects to a female threaded component. A female connector fitting has a female end that connects to a male threaded component. A male adapter fitting connects tubing to female piping. A female adapter connects tubing to male piping. *Figure 28* shows male/female connectors and adapters.

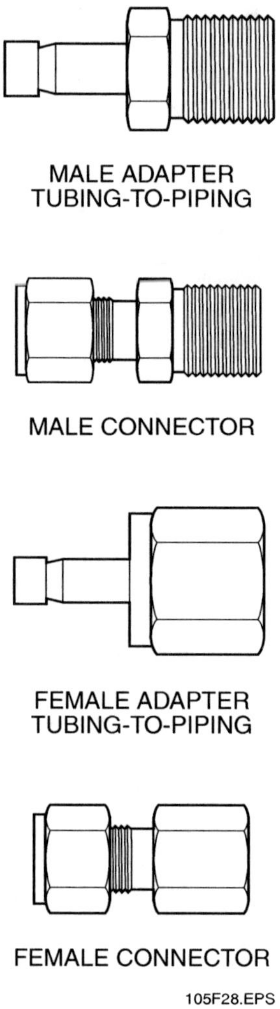

Figure 28 ◆ Male/female connectors and adapters.

7.3.2 Tee Fittings

Tee fittings are a commonly used type of branch fitting. They are available in a variety of designs and sizes (see *Figure 29*). All tee fittings are designed to branch tubing at a 90-degree angle. When all of the outlets of the tee are the same size, the tee is specified by that size, such as a 1-inch tee. When the main outlets and the branch outlets are different sizes, the tee is referred to by giving the main tubing run size first, and then the branch run size; for example, 1 inch × ¼-inch tee.

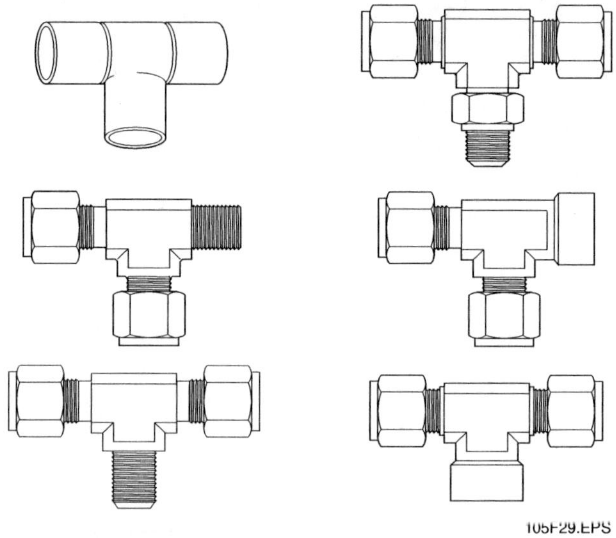

Figure 29 ◆ Tee fittings.

7.3.3 Elbow Fittings

Elbows are curved fittings that connect tubing to tubing or piping to tubing at various angles. An elbow fitting is commonly referred to as an ell (or L). The standard angles are 45 degrees and 90 degrees, although other angle sizes are available (see *Figure 30*).

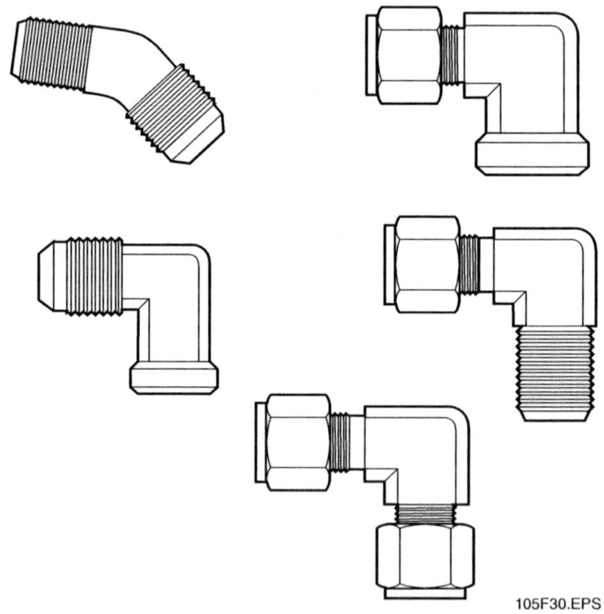

Figure 30 ◆ Elbow fittings.

7.3.4 Union Fittings

Union fittings connect two pieces of tubing or tubing and piping without branching or stopping the flow through the run (see *Figure 31*). Union fittings can be connected and disconnected without disturbing the position of the tubing or piping.

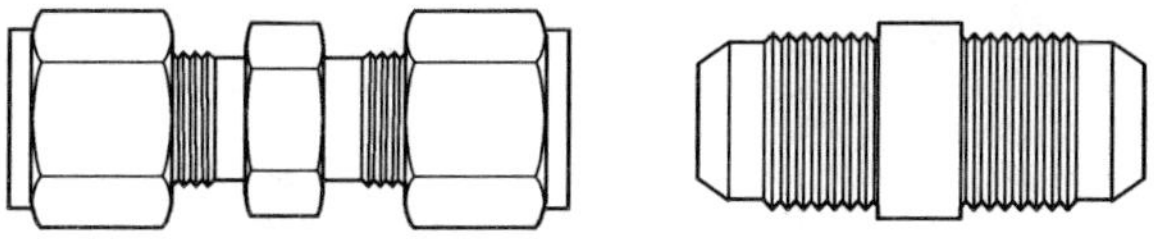

105F31.EPS

Figure 31 ◆ Union fittings.

7.3.5 Cross Fittings

The function of a cross fitting is to provide 90-degree outlets opposite each other. Cross fittings have two branch outlets that form the shape of a cross. Cross fittings come in a variety of sizes and designs. All four outlets in a straight cross fitting are equal in dimension. In a reducing cross fitting, the two branch outlets are smaller than the run outlets. This type of cross fitting is used to obtain the desired flow and pressure through the system. Cross fittings may be butt- or socket-welded or threaded. *Figure 32* shows a cross fitting.

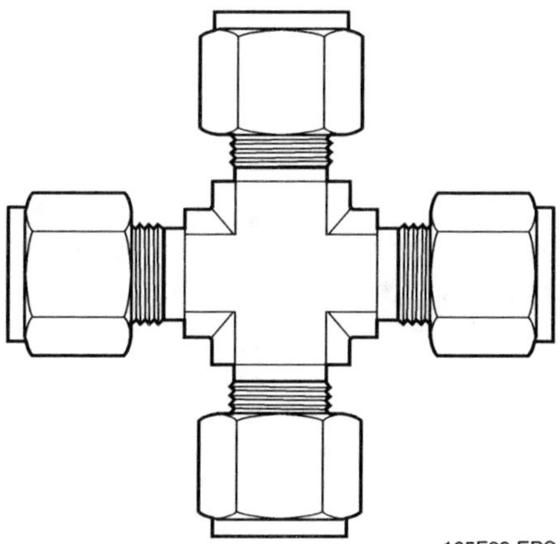

105F32.EPS

Figure 32 ◆ Cross fitting.

7.3.6 Caps and Plugs

Caps are used to close off the end of tubing. Plugs are used to close off an unused port of a tubing fitting. **Caps** and plugs come in various sizes and shapes. *Figure 33* shows cap and plug **fittings**.

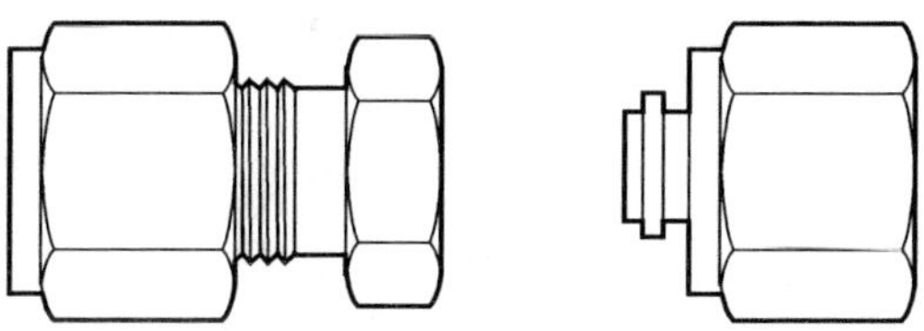

105F33.EPS

Figure 33 ◆ Cap and plug fittings.

7.3.7 Reducer Fittings

A **reducer** is designed to change the run or branch from a larger size to a smaller size of tubing or piping. This is usually done to increase flow pressure or to reduce the volume of flow in a tubing system. One face of the reducer matches the larger size of tubing or piping, and the other matches the smaller size of tubing or piping. A reducer may be a separate fitting or may be a built-in part of another fitting like a tee, cross, adapter, or connector (see *Figure 34*).

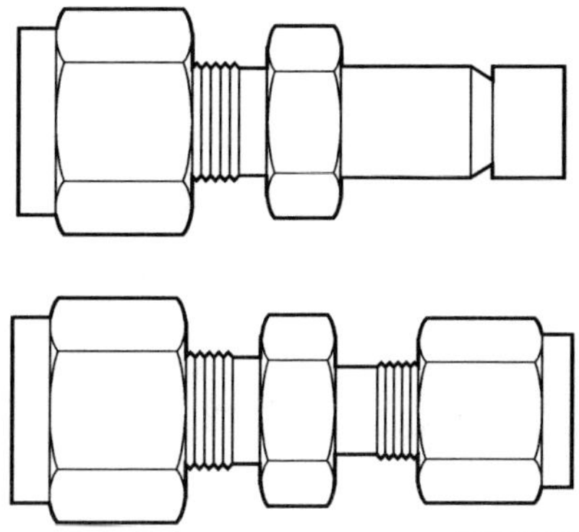

105F34.EPS

Figure 34 ◆ Reducer fittings.

7.3.8 Bulkhead Fittings

Bulkhead fittings are used to connect two pieces of tubing through a plate, junction box, or control panel by use of a lock-nut. Bulkhead fittings come in several types, sizes, and materials (see *Figure 35*).

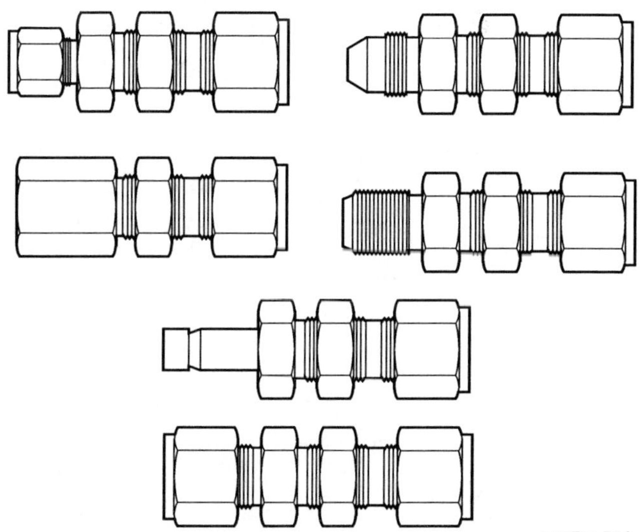

105F35.EPS

Figure 35 ◆ Bulkhead fittings.

7.3.9 Thermocouple Fittings

Thermocouples are widely used devices for measuring temperature in industrial processes. Most industrial thermocouples have a female connector at the end of the metal sheath that surrounds the thermocouple wires. To run the wires to a junction box or another instrument, the male connector of the thermocouple fitting is connected to the female connector of the thermocouple. The fitting then connects to tubing or a conduit that houses the wires until they terminate. *Figure 36* shows a thermocouple fitting.

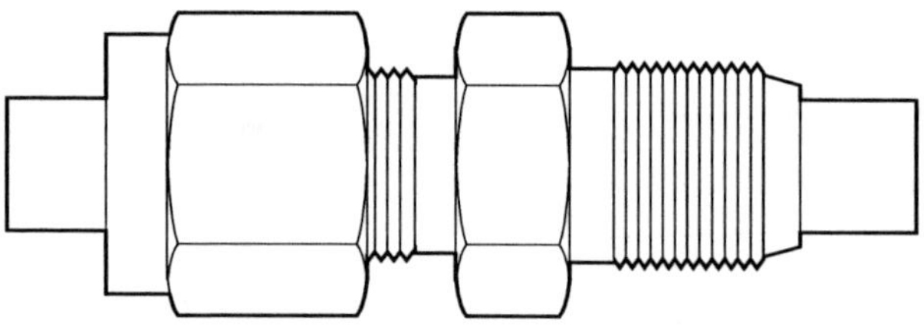

Figure 36 ◆ Thermocouple fitting.

8.0.0 ◆ FLARING TUBING

Flaring is a method used to connect flared fittings and tubing. The end of the tubing is spread, or flared, using a flaring tool. Flaring tools are available in several types including manually and hydraulically operated. Manual flaring tools are used for smaller, softer tubing. Hydraulic flaring tools are usually used with larger, harder tubing. All flaring tools consist of three basic parts: the die block, the cone, and the yoke (see *Figure 37*).

The holes in the die block correspond to the different sizes of tubing that the block can accommodate. Each hole has a flared side to create a flare at the desired angle. Flared connections used in high-pressure applications, like hydraulic systems, usually use a 37-degree flare angle. Low-pressure applications, like pneumatic systems, use a 45-degree angle for flared connections. Flaring cones come in different angles, with the degree of the angle usually stamped on the cone. If the wrong cone is used, the flare will have the wrong angle, and the tubing will not fit properly over the fitting body.

WARNING!
Always follow the manufacturer's instructions when using flaring tools.

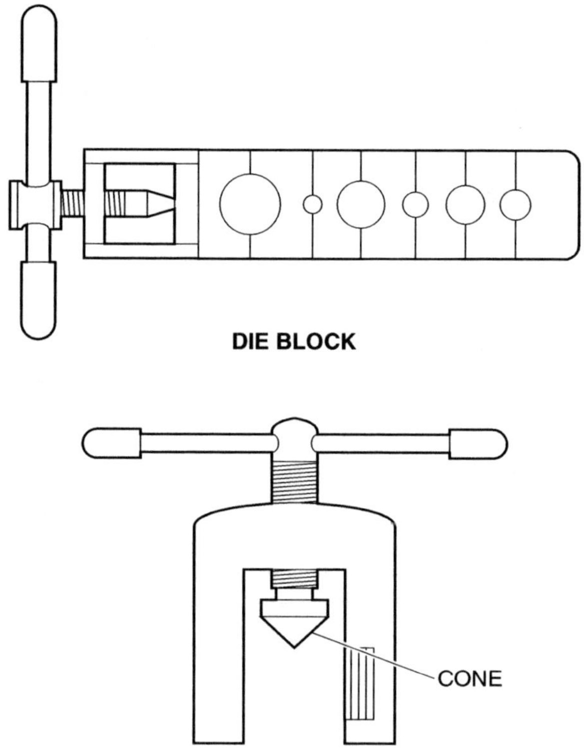

Figure 37 ◆ Flaring tool.

Tubing that is flared must be seamless tubing and should be cut with a hacksaw. Tubing made with seams tends to split along the seam during the flaring process. For a proper seal, the end of the tubing should be cut squarely. A square cut allows the tube to fit flush against the nose of the fitting body. The tubing should be free of burrs and ridges before starting to flare.

Follow these steps when flaring tubing with a manual flaring tool:

WARNING!
Follow all safety precautions.

Step 1 Cut the tubing to the desired length.

Step 2 Remove all burrs. This is very important to ensure metal-to-metal contact.

Step 3 Slide the fitting nut (and sleeve, if included with the fitting) over the tubing. This is important because the nut and the sleeve will not fit over the end of the tubing after it has been flared. The nut and the sleeve can be slipped out of the way while the tubing is being flared.

Step 4 Select a flaring tool that produces a flare of the correct angle for the fitting. The flaring tool will usually have the degree of its flare stamped on the cone. The angle of the tapered tip on the fitting body will have to be verified against the manufacturer's part number.

Step 5 Insert the end of the tubing into the appropriate hole in the die block so that the tubing extends about $\frac{1}{32}$ to $\frac{1}{16}$ inch.

Step 6 Tighten the wing nut on the die block to clamp it around the tubing.

Step 7 Slide the yoke over the die block so that the cone is aligned with the center of the tubing.

Step 8 Tighten the handle of the yoke, screwing the cone into the end of the tubing to form the flare. The handle is turned until the cone is snug inside the tubing and the handle is hard to turn.

Step 9 Loosen the yoke handle and remove the flaring tool and the tubing.

Step 10 Inspect the completed flare for cracks, debris, and smoothness.

Step 11 Check to see that the fitting nut and sleeve fit properly on the flare.

Step 12 Test the finished flare against the fitting body to make sure it fits properly.

Step 13 To complete the connection, place the tapered end of the fitting body into the flared end of the tubing. Slide the fitting nut and the sleeve down the tube. Screw the fitting nut into the fitting body by hand, ensuring that the pieces thread together properly. Use two wrenches to tighten the assembly. With one wrench holding the fitting body, tighten the nut with the other wrench.

CAUTION

It is important for the fitting to be tightened snugly, but do not overtighten, because this may damage the fitting or split the tubing at the flare.

9.0.0 ◆ MATERIALS USED IN THREADED PIPING SYSTEMS

Various types of materials can be used in threaded piping systems. These materials are classified according to the type of material, size of pipe, and schedule or wall thickness of pipe.

Carbon steel pipe is widely used throughout the pipefitting industry in various applications because it is ductile, durable, machinable, and less expensive than most other types of pipe.

Galvanized pipe is made by dipping carbon steel pipe into a mixture of hot zinc. The zinc coats the pipe both externally and internally, and protects the pipe from corrosion. Galvanized pipe is used to convey water, air, and other various fluids.

Stainless steel is an expensive material that can also be used in threaded systems. However, if at all possible, avoid using stainless steel pipe because it is difficult to correctly cut the threads and make up leak-proof joints. Stainless steel pipe is good for special applications that require the maximum resistance to corrosion, such as processing systems, instrument lines, and heat-transfer equipment. Stainless steel can also be kept sterile to prevent contamination of food, drugs, and dairy products. When cutting stainless steel threads, the best cutting lubricant is condensed milk.

9.1.0 Pipe Sizes

Pipe is listed in inches by its nominal size. For sizes of pipe up to and including 12 inches, nominal size is an approximation of the inside diameter. From 14 inches on, nominal size reflects the outside diameter of the pipe. *Figure 38* shows inside and outside diameters of pipe.

There are times when the nominal size of a pipe and its actual inside or outside diameter differ greatly, but nominal size is used to describe the pipe. The important thing to remember about size is that the sizes of different types of materials vary. A $\frac{3}{4}$-inch brass tube fitting will not fit a $\frac{3}{4}$-inch copper pipe. Brass is, however, interchangeable with steel pipes.

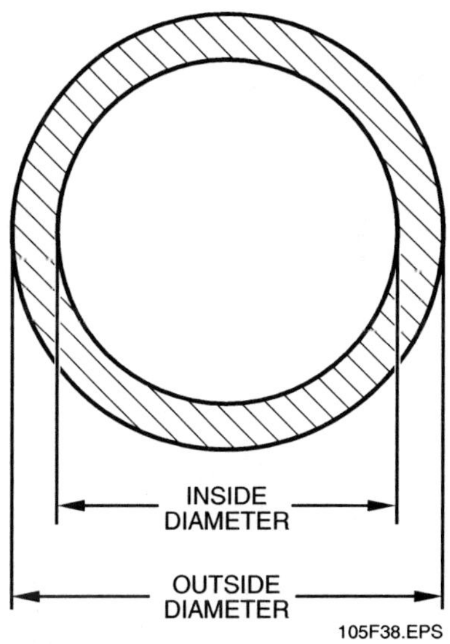

Figure 38 ◆ Inside and outside diameters of pipe.

9.2.0 Schedules and Wall Thicknesses

Wall thickness can be described in two ways. The first is by schedule. As the schedule numbers increase, the wall thickness gets larger; therefore, the pipe is stronger and can withstand more pressure. Schedule numbers for pipe range from 5 to 160, but no pipe smaller than schedule 40 should be threaded. It is important to remember that the schedule number refers only to the wall thickness of a pipe of a given nominal size. A ¾-inch schedule 40 pipe will not have the same wall thickness as a 1-inch schedule 40 pipe. Another way to describe pipe wall thickness is by manufacturer's weight. Three classifications are in common use today, which are from smallest to largest as follows:

- STD – Standard wall
- XS – Extra-strong wall
- XXS – Double extra-strong wall

Schedule numbers and wall thicknesses are somewhat interchangeable. Schedule 40 galvanized pipe and standard wall galvanized pipe have the same wall thickness for all sizes up to and including 10-inch nominal size. Schedule 80 and extra-strong wall galvanized pipe have the same wall thickness through 8-inch nominal size. Common wall thicknesses of carbon steel pipe are schedule 40, 80, and 160, and double extra-strong. Common wall thicknesses of stainless steel pipe range from schedule 5 to 160. *Table 4* shows commercial pipe dimensions.

10.0.0 ◆ PIPE FITTINGS

Piping systems consist of pipes, pumps, valves, and other parts, including fittings. Pipe fittings come in various sizes, materials, strengths, and designs to match the various piping systems and can be either plain fittings or banded fittings. The banded fittings are malleable iron and cast fittings, and the plain fittings are forged steel. The malleable iron fittings are banded to provide extra strength. Pipe fittings can be defined as fittings attached to pipe to change the direction of fluid flow, connect a branch line to a main line, close off the end of a line, or join together two pipes of the same or different sizes. Pipe fittings can be placed into different groups according to their use:

- Elbows, offsets, and return bends
- Branch connections
- Caps and plugs
- Line connections

10.1.0 Elbows, Offsets, and Return Bends

Elbows, offsets, and return bends are fittings used for changing the direction of fluid flow in a piping system. An elbow, also known as an *ell, L, 90,* or *45,* is a fitting that forms an angle between different connecting pipe (see *Figure 39*). This angle is always 90 degrees unless another angle is stated. Elbows come in several different angles. The 11¼-, 5⅝-, and 30-degree elbows are rarely used and hard to obtain. The 90- and 45-degree elbows are the most common.

Table 4 Commercial Pipe Dimensions

Nominal Size (inches)	Outside Diameter (inches)	Inside Diameter (inches)			
		Standard Schedule 40	Extra-Strong Schedule 80	Schedule 160	Double Extra-Strong
⅛	.405	.269	.215	—	—
¼	.540	.364	.302	—	—
⅜	.675	.493	.423	—	—
½	.840	.622	.546	.466	.252
¾	1.050	.824	.742	.614	.434
1	1.315	1.049	.957	.815	.599
1¼	1.660	1.380	1.278	1.160	.896
1½	1.900	1.610	1.500	1.338	1.100
2	2.375	2.067	1.939	1.689	1.503
2½	2.875	2.469	2.323	1.885	1.771
3	3.500	3.068	2.900	2.625	2.300
3½	4.000	3.548	3.364	—	—
4	4.500	4.026	3.826	3.438	3.152
5	5.562	5.047	4.813	4.313	4.063
6	6.625	6.065	5.761	5.187	4.897
8	8.625	7.981	7.625	6.813	6.875

105T04.EPS

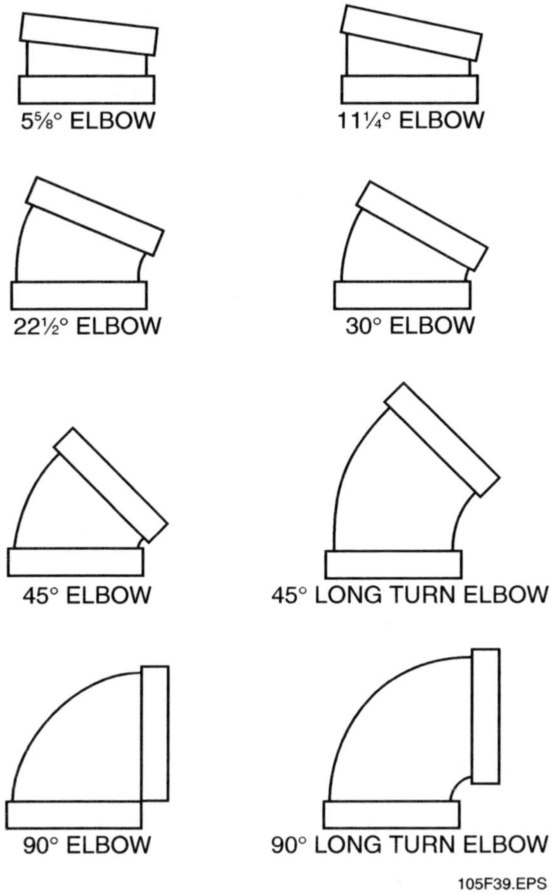

Figure 39 ◆ Elbows.

An offset changes the direction of flow for a short distance. The offset moves the pipe run to one side and then returns it to the same direction. An offset can be made by using two fittings, one at each end of the offset, or by using one special fitting (see *Figure 40*).

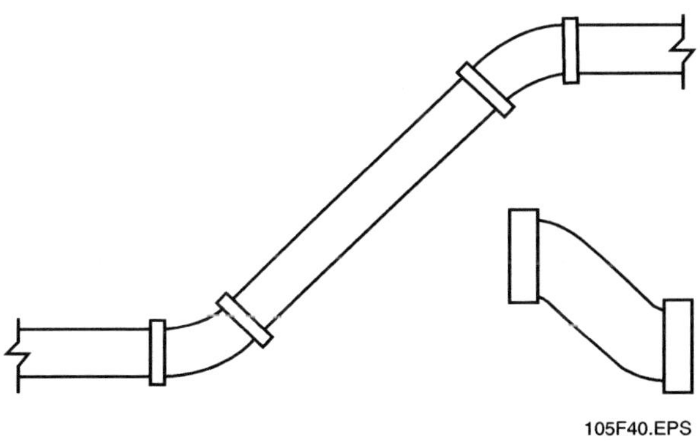

Figure 40 ◆ Offsets.

A return bend is a U-shaped fitting that sends the fluid back in the same direction from which it came. Return bends are often used in boilers, radiators, and systems in which the pipe must pass through the same area several times. Return bends are manufactured in various sizes and come in close, open, and wide patterns (see *Figure 41*). Return bends can be made with tapered threads to give a slant to the pipe.

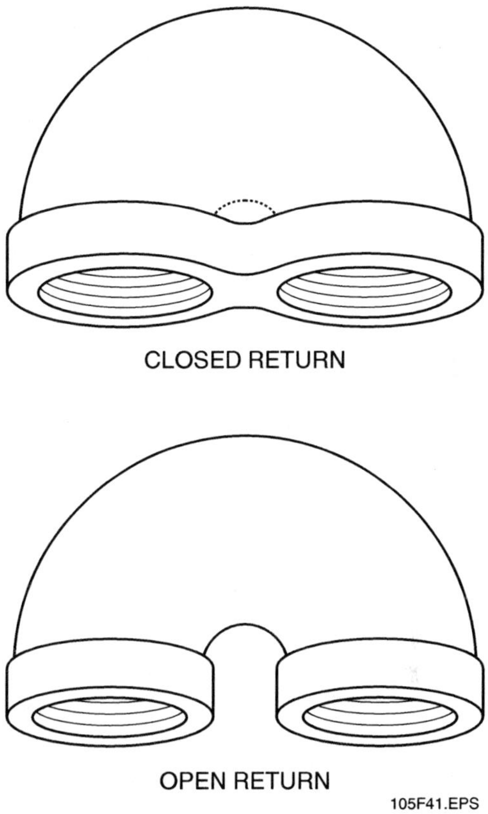

Figure 41 ◆ Return bends.

10.2.0 Branch Connections

Branch connections are fittings that divide the flow and send the fluid in two different directions. Tees are the most widely used of the branch connections. They are used for making a 90-degree branch in the main pipe. **Wyes** are another common branch connection that have a side opening set at a 45-degree angle. *Figure 42* shows some common branch connections.

10.3.0 Caps and Plugs

Caps and plugs are fittings used for closing off lines (see *Figure 43*). These have the same pressure ratings as other fittings. A cap fitting screws onto the pipe end, and a plug screws into the pipe or fitting end. Caps can either be plain caps with round heads or caps with a square projection on the body to allow for tightening with a wrench. Plugs can either be solid square-head plugs or countersink plugs, which have a square depression in which a square key can be inserted for tightening. A bull plug is a long shaft that is threaded on one end that may or may not have a hex head. Bull plugs are used to plug an insulated pipe line so that the head will extend out of the insulation for easy removal.

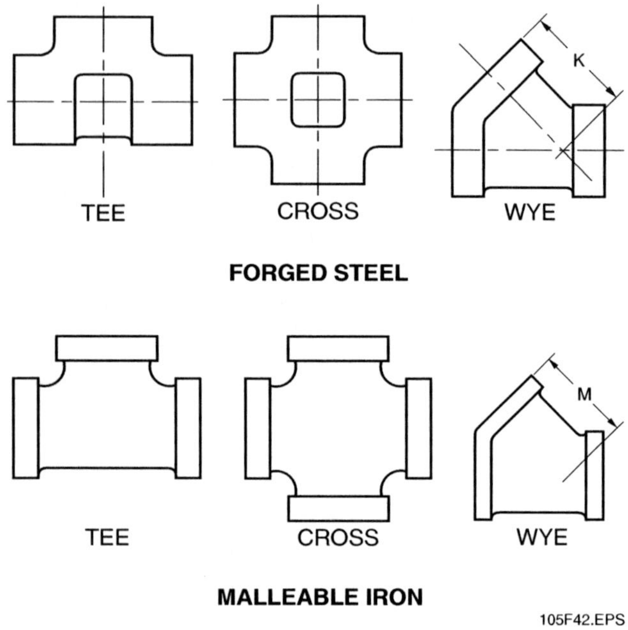

Figure 42 ◆ Common branch connections.

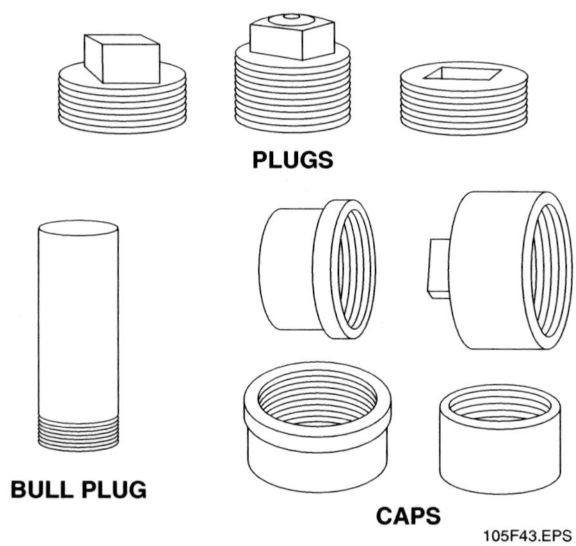

Figure 43 ◆ Caps and plugs.

10.4.0 Line Connections

Couplings and unions join pipes of the same size, and **bushings**, reducers, and reducing couplings join pipes of different sizes.

10.4.1 Couplings

Usually, all lengths of pipe less than 8 inches in diameter are provided by the manufacturer with one straight coupling attached. A coupling is a sleeve that is used to connect two straight pieces of pipe (see *Figure 44*). Both ends of a straight coupling are tapped with right-hand threads.

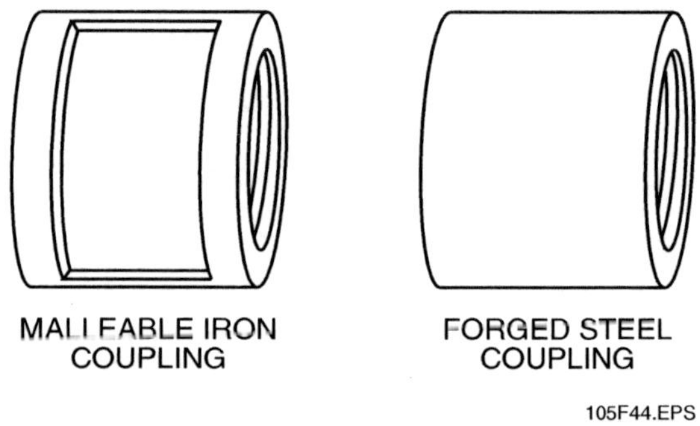

Figure 44 ◆ Couplings.

10.4.2 Unions

Unions are used either to join two ends of pipe which cannot be turned or to permit the disconnecting of pipes at some future time without cutting them. A union usually consists of three parts, including two sleeves to be threaded on the ends of the pipes and a threaded coupling ring to draw the sleeves together (see *Figure 45*). Some unions are insulated for cathodic protection.

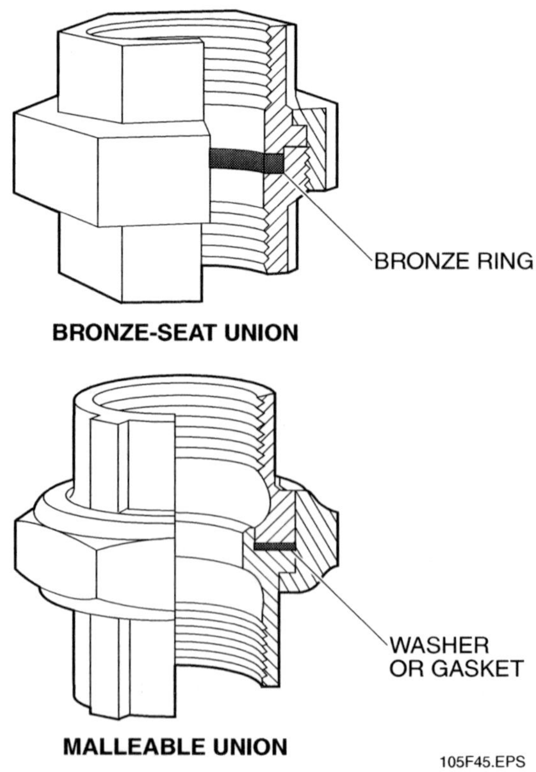

Figure 45 ◆ Unions.

10.4.3 Bushings

The function of a bushing is to connect the male end of a pipe to the female end of a fitting of a larger size. It consists of a hollow plug with male and female threads to suit the different diameters. Bushings are usually made with a hexagon top but can also be made without the head (see *Figure 46*). Those without a head are known as *face bushings* or *flush bushings* and can be screwed into the fitting to form a neat, flush finish. Bushings are used in low-pressure systems because they restrict the flow in the system.

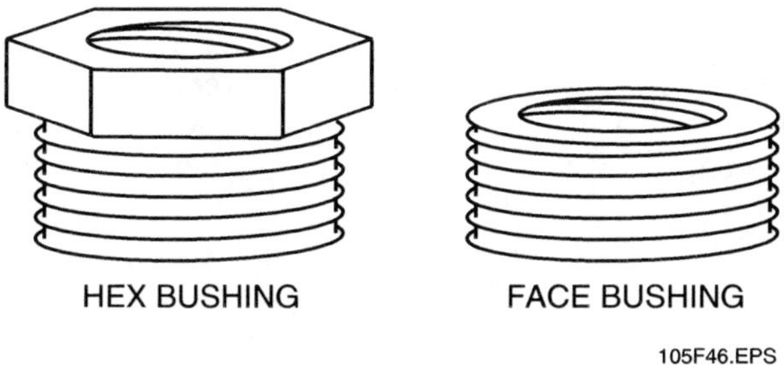

Figure 46 ◆ Bushings.

10.4.4 Reducers

A reducer is used to join the male ends of two different sizes of pipe. A reducer may be a separate fitting or may be a built-in part of another fitting. Most directional change and branch connection fittings are available with reduced outlet lines (see *Figure 47*). Malleable iron bell reducers are used in low-pressure systems, and swages are used in high-pressure systems. Both of these are made in concentric and eccentric types. A reducing coupling is a concentric high-pressure fitting that restricts flows. Reducing tees are also made in malleable iron or forged steel.

10.5.0 Nipples

Nipples are short pieces of pipe threaded on both ends that are used to make close connections between fittings. They are made in all sizes and types of pipe and are stocked in various lengths classified as close, shoulder, short, and long (see *Figure 48*). The longest nipples are 12 inches. Pipe longer than 12 inches is known as cut pipe

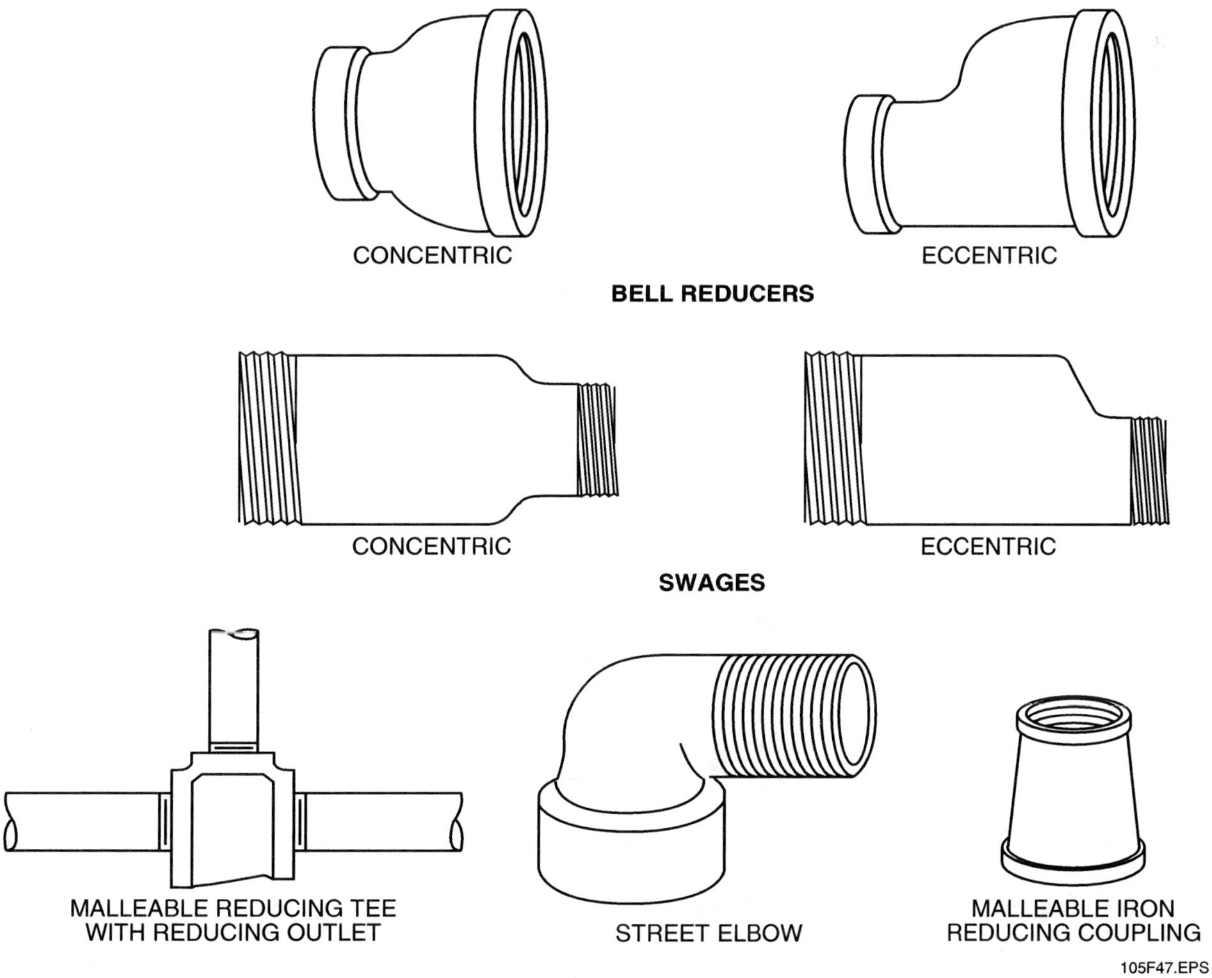

Figure 47 ◆ Reducers.

instead of as nipples. A **close nipple** is threaded end to end, and a shoulder nipple has a short section of unthreaded pipe in the center known as the shoulder.

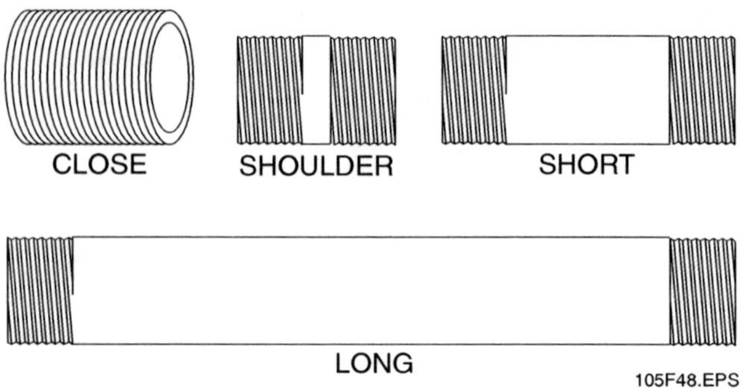

Figure 48 ◆ Types of nipples.

10.6.0 Flanges

The screw-on **flange** is the weakest type of flange because the threads cut into the pipe (see *Figure 49*). It is normally used only for low-pressure lines and pipes. Screw threads hold the flange to the pipe. As in all flanges, the bolt holes should be carefully aligned when making up the fitting.

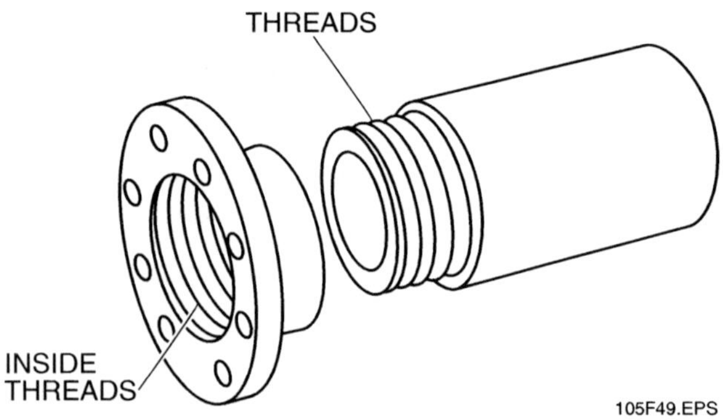

Figure 49 ◆ Screw-on flange.

Each type of flange is available in different face styles, and each flange face style uses a certain gasket. A set of two flanges of the same face style, called *companion flanges*, is used with the proper gasket. The two face styles of threaded flanges are the flat-face flange and the raised-face flange.

The flat-face flange is flat across the entire face. This flange uses a gasket that covers the entire flange face and has bolt holes cut in it (see *Figure 50*). Flat-face flanges are mainly used when joining two flanges of dissimilar metals.

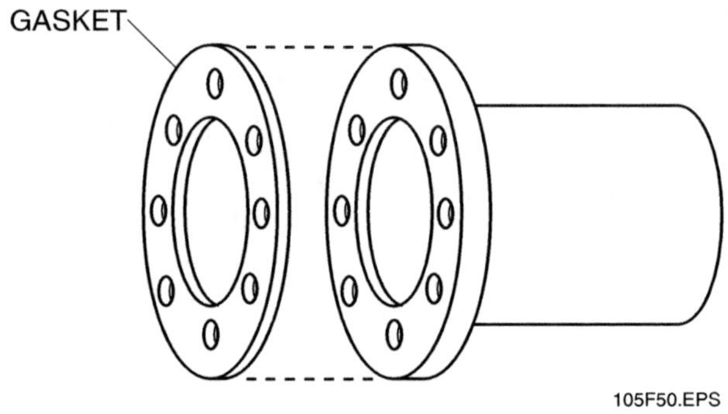

Figure 50 ◆ Flat-face flange and gasket.

The raised-face flange has a wide, raised rim around the center of the flange (see *Figure 51*). This flange uses a wide, flat ring gasket that is squeezed tightly between the flanges.

Typically, you should never use two different face styles at one joint, but this may be necessary when connecting to manufactured equipment. If you must connect a flat-face flange to a raised-face flange, you must confirm that the flat-face flange is not cast iron. The pressure of the bolts can crack or break the cast iron flange. If the flat-face flange is cast iron, change the raised-face flange to a flat-face flange with engineering approval.

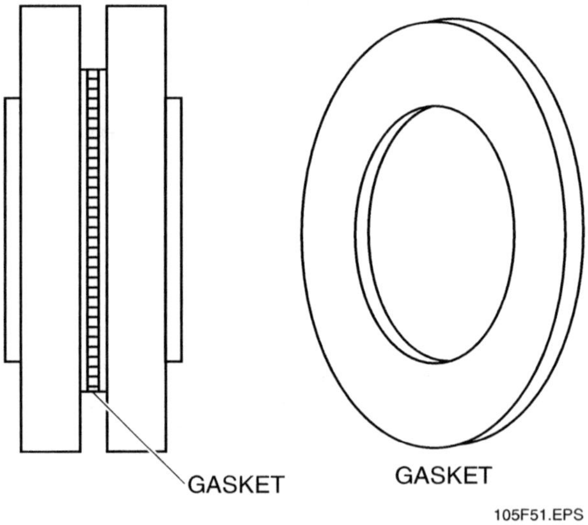

Figure 51 ◆ Raised-face flange.

11.0.0 ◆ THREADS

By definition, a pipe thread is an inclined plane wrapped around a cylinder. The thread angle is the angle formed by two inclined faces of the thread. The thread angle varies depending on the specific thread standard used. The pitch refers to the distance between adjacent screw threads when measured from center line to center line.

Pitch is also taken to mean inch when referring to pitch as threads per inch. The crest of the thread is the outermost part of the thread, or the top of the thread, when viewing it from the side. The root of the thread is the innermost part of the thread, or the bottom, if viewed from the side.

11.1.0 Types of Threads

The National Pipe Thread is the standard thread for pipe connections. The National Pipe Thread has a thread angle of 60 degrees and a slight flat at the crest and the root. While National Pipe Threads can be either straight or tapered, the type normally used has a tapered internal and external thread. The taper used is 1/16-inch per inch of threads. *Figure 52* shows the taper used in cutting pipe threads.

In addition to the widely used taper pipe thread, there are several other types of standard pipe threads for specific applications. All National Pipe Threads are designated by specifying in sequence the nominal size of the pipe, the number of threads per inch, and the symbols for thread series. For example, 3/8 - 18 - N.P.T., are threads for a 3/8-inch nominal size pipe having 18 threads per inch and are National Pipe Threads.

11.2.0 Taper Thread Connection

In order to have a strong, leakproof threaded joint, the threads must be clean and smoothly cut. They must have the correct pitch, lead, taper, and form, and the threads must be correctly sized. If these conditions are not met, the seal will leak.

The joining or makeup of a taper threaded pipe connection is performed in two distinct operations, known as *hand engagement* and *wrench makeup*. Hand engagement refers to how far the fitting can be tightened by hand. Wrench makeup is the additional turning of the fitting with a wrench to completely make up the joint. Hand engagement on properly cut threads is normally between three and four complete revolutions. Hand engagement is a good way to check the quality of threads after cutting. Wrench makeup is usually about three turns for a total of approximately seven revolutions for complete makeup. A greater number of turns are required for larger pipe. *Figure 53* shows hand engagement and wrench makeup.

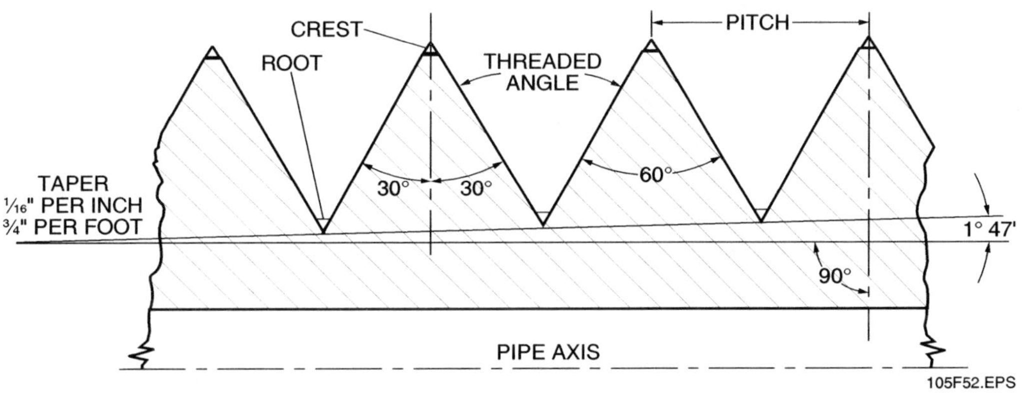

Figure 52 ◆ Taper used in cutting pipe threads.

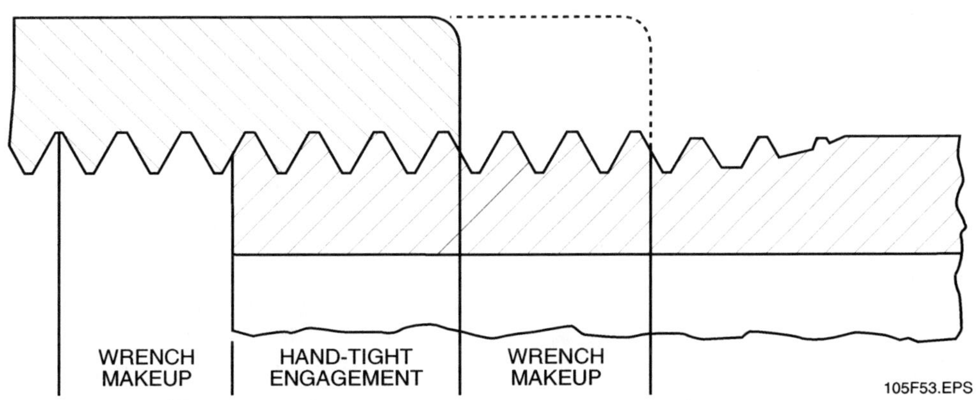

Figure 53 ◆ Hand engagement and wrench makeup.

The standard for taper thread pipe makeup gives basic thread dimensions and the number of rotations for hand engagement and wrench makeup. The taper thread makeup can vary plus or minus one turn from the standard thread makeup allowances. *Table 5* shows the standard for taper pipe thread makeup. Note that these thread makeups are for perfect threads cut under perfect conditions, which seldom occurs.

Table 5 Standard for Taper Pipe Thread Makeup

Nominal Pipe Size (inches)	Die Threads (per inch)	Thread Length (inches)	Number of Threads
1/8	27	3/8	10
1/4	18	5/8	11
3/8	18	5/8	11
1/2	14	3/4	10
3/4	14	3/4	10
1	11 1/2	7/8	10
1 1/4	11 1/2	1	11
1 1/2	11 1/2	1	11
2	11 1/2	1	11
2 1/2	8	1 1/2	12
3	8	1 1/2	12
3 1/2	8	1 5/8	13
4	8	1 5/8	13
5	8	1 3/4	14
6	8	1 3/4	14
8	8	1 7/8	15
10	8	2	16
12	8	2 1/8	17

105T05.EPS

12.0.0 ◆ HAND PIPE AND BOLT THREADERS

Pipefitters often have to use hand pipe and bolt threaders. The ratchet-type hand pipe threader with removable, interchangeable die heads is the most common. Die heads and dies are sized for specific diameters of pipe and bolts, and are specifically designed for the threading of a certain material. The size of the thread can be found on the face of the die heads. Special dies are available for threading brass, aluminum conduit, stainless steel, wrought iron, and cast iron and for left-handed threads. These dies should not be used on any material other than what they were designed for, because the dies could be damaged or produce imperfect threads. *Figure 54* shows hand pipe and bolt threaders.

Most pipe threads are **NPT** threads, which designate National Pipe Threads. In some cases you will find metric threads, though. The most common bolt threads are NC and NF, which designate National Coarse and National Fine. Always check the engineering specifications before selecting threading dies.

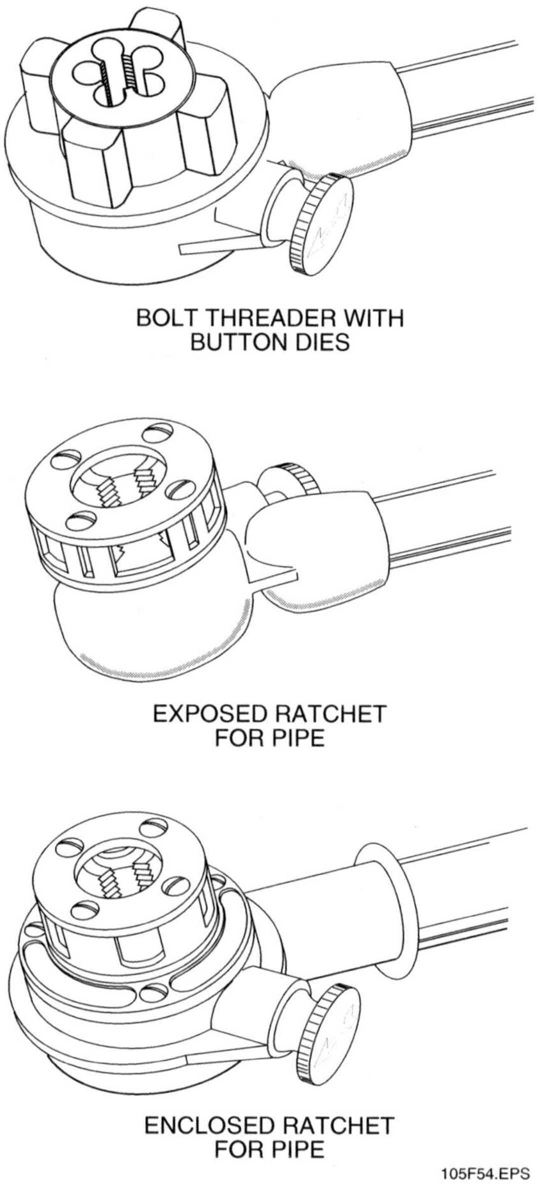

Figure 54 ◆ Hand pipe and bolt threaders.

12.1.0 Caring for Pipe and Bolt Threaders

To cut threads correctly, the pipe threader must be kept in good condition. Follow these guidelines when using hand pipe threaders:

- Inspect the tool regularly for broken, worn, or damaged parts.
- Inspect dies regularly to make sure they are sharp and not chipped.

- Use a good-quality cutting oil to lubricate the cutting edges and remove chips and curls.
- Use good techniques when cutting. Sloppy or careless work will not produce good threads.

12.2.0 Replacing Die Heads

Most handles will accept many different die heads for different jobs. Follow these steps to replace the die head on a pipe threader handle:

Step 1 Pull out the lock knob, and turn the knob so that the arrow points to the handle. *Figure 55* shows pipe threader components.

Step 2 Lift the die head straight out of the pipe threader handle.

Step 3 Insert a new die head into the pipe threader handle.

Step 4 Pull out and turn the lock knob so that the arrow is pointing away from the handle.

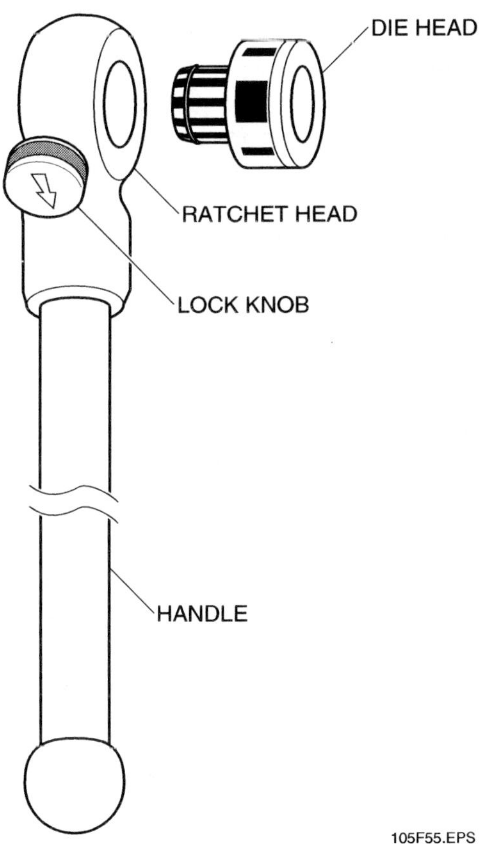

Figure 55 ◆ Pipe threader components.

12.3.0 Replacing Dies

The die head holds the dies in place. When the dies become worn or damaged, they must be replaced. Follow these steps to replace dies:

Step 1 Remove the die head from the handle.

Step 2 Remove the four screws in the top of the die head using a screwdriver.

Step 3 Remove the cover plate from the top of the die head. *Figure 56* shows die head components.

Step 4 Slide the dies out of the die head.

> **NOTE**
> Before replacing the dies, make sure the die head and the dies are clean, dry, and free of all metal chips. A small chip can raise a die and cause uneven pressure. This will tear the pipe thread and may break the die.

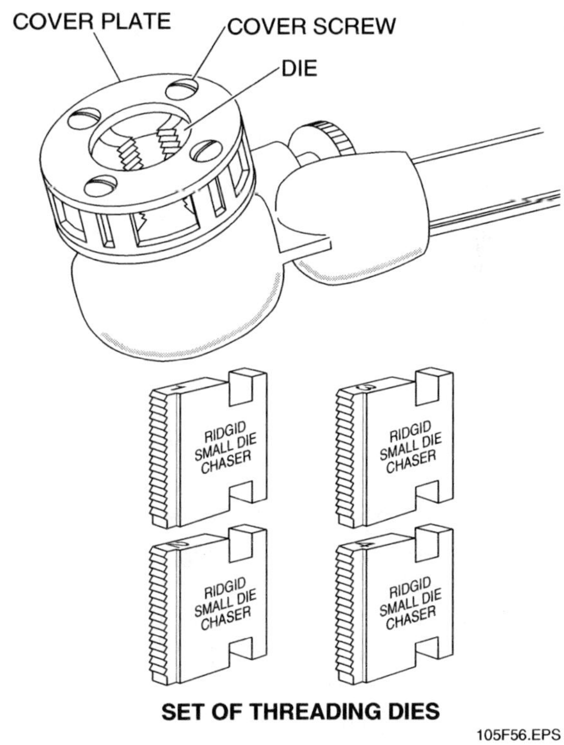

Figure 56 ◆ Die head components.

Step 5 Insert the new dies into the die head, making sure the number on each die matches the number on each die head slot.

> **NOTE**
> It is very important to keep the die sets together. Do not mix dies of one set with dies from another set. Usually, each die is marked with a serial number, and the four dies that are made together have the same number.

Step 6 Replace the cover plate onto the die head.

Step 7 Tighten the four screws into the die head.

Step 8 Replace the die head into the pipe threader handle.

12.4.0 Threading Pipe

Follow these steps to thread pipe:

Step 1 Identify the kind of material being threaded, such as pipe, tubing, or conduit and brass, copper, or steel.

Step 2 Identify the diameter of the material being threaded.

Step 3 Select the die head best suited for the threading process.

Step 4 Insert the die head into the pipe threader handle.

Step 5 Inspect the die and die handle to ensure that all parts are clean and in good condition.

Step 6 Secure the pipe in a vise.

Step 7 Make sure that the end of the pipe is square, clean, and deburred.

Step 8 Position your body so that you are balanced, facing the work area, and within easy reach of the work area.

Step 9 Position the die flush on the end of the pipe.

> **NOTE**
> Make sure that the arrow on the lock knob points in the direction in which you are threading the pipe.

Step 10 Press the die against the end of the pipe, and turn the die handle clockwise to start the threading process.

Step 11 Turn the die handle two full rotations around the pipe.

> **NOTE**
> Add cutting oil as necessary to keep a thin coat on the threads while cutting.

Step 12 Stop the work, pull out the lock knob, and turn the knob so that the arrow points in the opposite direction.

Step 13 Back the dies off the pipe.

> **NOTE**
> This allows you to make sure the threads are properly started and allows the threads to be cleaned from the cutting die.

Step 14 Clean the cutting threads from the die if necessary.

Step 15 Pull out the lock knob, and turn it so that the arrow points in the direction in which you are threading.

Step 16 Rethread the die onto the pipe, and continue the process until the threading is complete.

> **NOTE**
> The threading is complete when one full thread extends past the back edge of the dies.

> **CAUTION**
> Do not thread the pipe too far. This can cause an improper fit between the fitting and the pipe, which will result in leaks.

Step 17 Back the die off the pipe.

Step 18 Inspect the new threads and use a rag remove any debris from the threads.

> **WARNING!**
> Freshly cut threads are very sharp. Use care in handling to avoid cuts.

12.5.0 Checking Threads

To check the threads on a bolt, a thread gauge is used (see *Figure 57*). A thread gauge has many leaves. Each leaf measures 1 inch across and has a certain number of thread markers per inch as noted on the side of each leaf. This measures the number of threads per inch, also known as the pitch. The number of threads per inch depends on the diameter of the bolt. These are not commonly used on pipe threads.

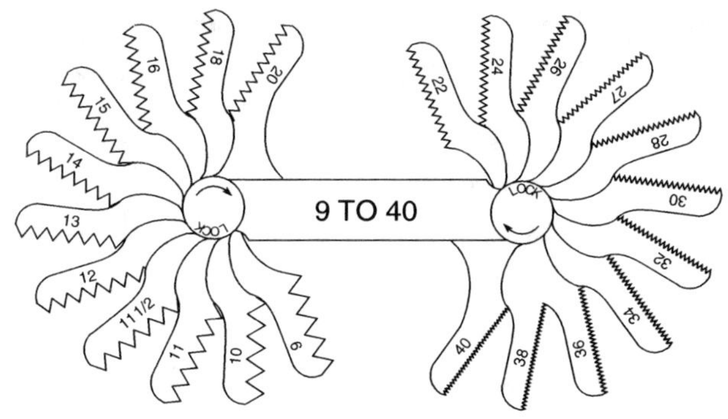

Figure 57 ◆ Thread gauge.

13.0.0 ◆ PIPE THREADING MACHINES

Pipe threading machines are multipurpose power tools that center, hold, and rotate pipe, conduit, or bolt stock while cutting, reaming, and threading operations are performed. All pipe threading machines are designed with a gear drive, chucks (which hold and turn the pipe), and a three-position power switch consisting of FORWARD, OFF, and REVERSE drive positions. Depending on the size and complexity of the machine, pipe threading machines can be mounted on tripods, tables, or special stands.

13.1.0 Types of Pipe Threading Machines

Pipe threading machines come in a variety of models, from the basic gear-drive type with front and rear pipe-holding chucks to the all-purpose model with built-in cutters, oilers, reamers, and a die for threading pipe. The Ridgid power drive 300 and the Ridgid threading machine 535 are covered in the following sections.

13.1.1 Ridgid Power Drive 300

The model 300 power drive unit consists of a ½-hp reversible motor, a locking speed chuck, a rear centering device, and two sliding heavy-duty support bars (see *Figure 58*).

A universal die head, pipe cutter, pipe reamer, moving carriage, and a tripod stand can be attached to the power drive to convert it into a portable pipe threading machine. The threading machine can thread pipe from ⅛ inch in diameter to 2 inches in diameter. A geared threader, used to thread pipe up to 6 inches in diameter, or a power-grooving machine, used to roll grooves into the pipe, can also be secured on the support bars. *Figure 59* shows a power drive converted into a threading machine.

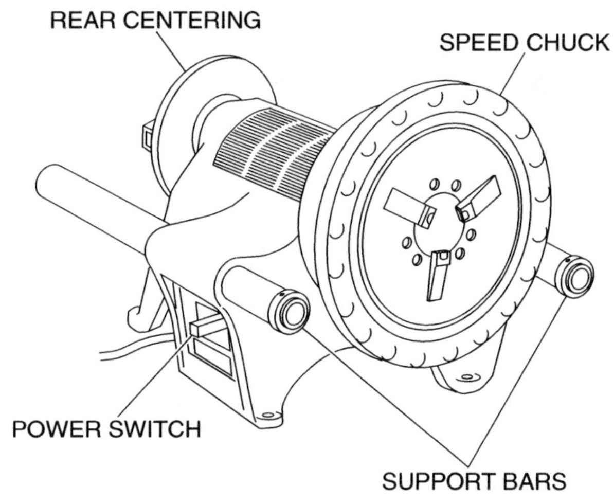

Figure 58 ◆ Ridgid power drive 300.

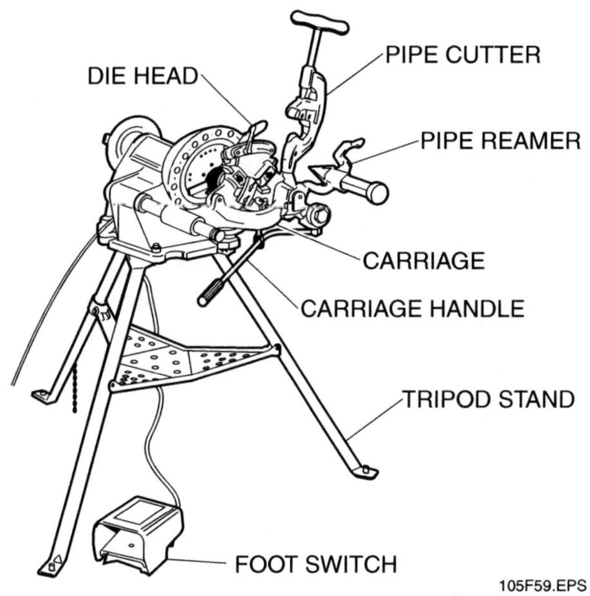

Figure 59 ◆ Power drive converted into a threading machine.

13.1.2 Ridgid Threading Machine 535

The Ridgid threading machine 535 is a complete pipe threading machine capable of threading pipe ranging from ⅛ inch to 2 inches in diameter. It can also be used to drive a geared threader to thread pipe up to 6 inches in diameter. It has a ½-hp reversible motor, a speed chuck with replaceable jaws, a rear centering device, a pipe cutter, a reamer, a universal die head, and a safety foot switch (see *Figure 60*). A handwheel located on the side of the machine allows the operator to move the tools into the rotating pipe and measure the distance that the tools travel. A built-in lubrication system automatically filters and delivers cutting oil to the pipe.

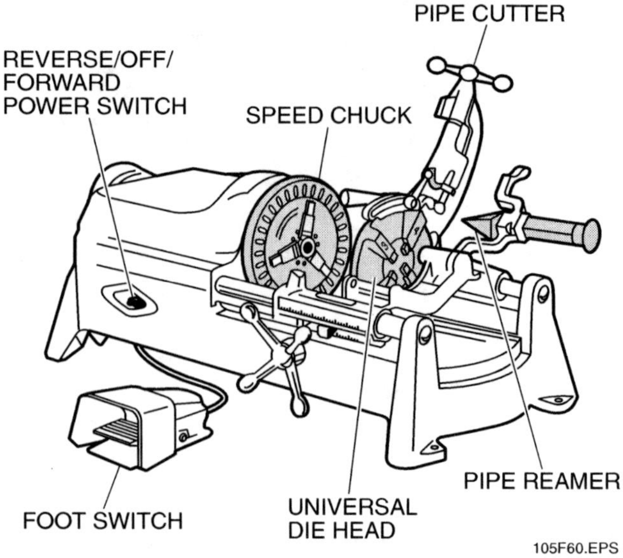

Figure 60 ◆ Ridgid threading machine 535.

13.2.0 Loading Pipe

On most threading machines, the pipe can be inserted from either end of the chuck. Some models, however, must be loaded from the rear. Either way, the operator must position the pipe so that the end to be threaded extends out away from the chuck. Most threading machines have a stop to help position the pipe properly. Follow these steps to load pipe into the threading machine:

Step 1 Measure and mark the length of pipe to be worked.

Step 2 Insert the pipe into the threading machine.

> **NOTE**
> If the pipe is long enough to be held by the centering device, insert the pipe through the front or rear of the machine. If the pipe is short, insert the pipe through the front of the machine. For long pipe, support the end of the pipe with an adjustable support.

Step 3 Make sure that the pipe is centered in the centering device.

Step 4 Tighten the centering device.

Step 5 Spin the handwheel counterclockwise to tighten the chuck jaws.

> **NOTE**
> This tightens the jaws on the pipe. A clockwise spin releases the jaws.

> **WARNING!**
> Do not wear loose clothing or jewelry. Remove safety gloves and keep hair pulled back while operating threading machine.

13.3.0 Cutting and Reaming Pipe

Follow these steps to cut and ream pipe:

Step 1 Determine the length of pipe needed.

Step 2 Measure the pipe to determine how much must be cut off.

Step 3 Mark the pipe at the point where it is to be cut.

Step 4 Swing the cutter, threader, and reamer up and out of the way.

Step 5 Load the pipe into the threading machine.

Step 6 Place pipe stands as needed.

Step 7 Turn the centering device to center the pipe in the machine.

Step 8 Spin the chuck handwheel counterclockwise to lock the pipe in place.

Step 9 Loosen the cutter until it fits over the pipe.

Step 10 Lower the cutter onto the pipe. *Figure 61* shows the cutter over the pipe.

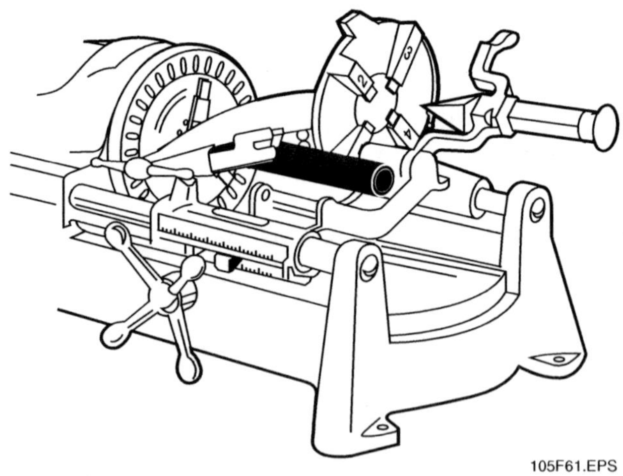

Figure 61 ◆ Cutter over the pipe.

Step 11 Rotate the handwheel to move the cutter wheel to the cutting mark on the pipe.

Step 12 Rotate the cutter handle to tighten the cutter on the pipe.

Step 13 Direct the cutting oil to the cutting area.

Step 14 Turn the pipe machine switch to the FORWARD position.

Step 15 Step on the foot switch to start the machine and turn the pipe.

Step 16 Rotate the cutter handle to apply pressure to the cutter until the cut is complete.

Step 17 Release the foot switch.

Step 18 Raise the cutter to an out-of-the-way position.

Step 19 Lower the reamer into place. *Figure 62* shows the reamer in place.

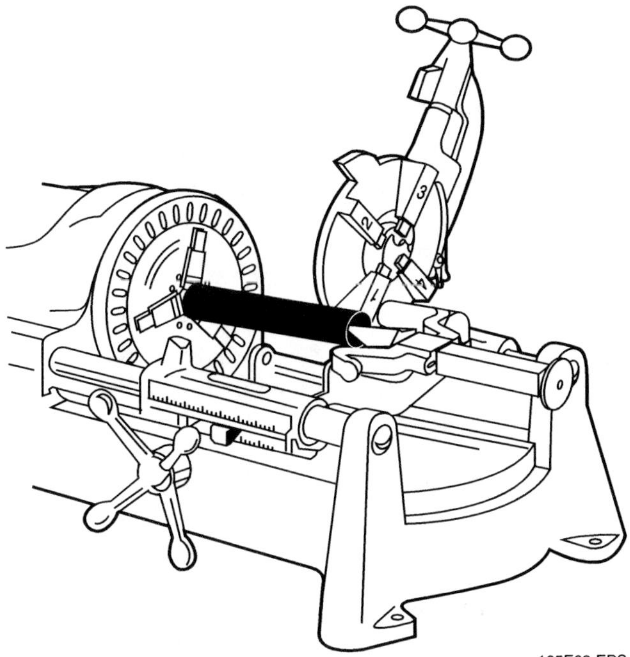
105F62.EPS

Figure 62 ◆ Reamer in place.

Step 20 Press the latch on the reamer, and slide the reamer bar toward the pipe until the latch catches.

Step 21 Step on the foot switch to start the machine.

Step 22 Rotate the handwheel to move the reamer into the end of the pipe.

Step 23 Ream the end of the pipe.

CAUTION

Do not overream the pipe. Only ream enough to remove the burrs.

Step 24 Rotate the handwheel to move the reamer out of the pipe.

Step 25 Release the foot switch.

Step 26 Slide the power switch to the OFF position.

Step 27 Raise the reamer to an out-of-the-way position.

Step 28 Loosen the chuck handwheel and the centering device.

Step 29 Remove the pipe from the machine.

13.4.0 Replacing Dies

Four sets of dies are used with pipe threading machines that thread pipe up to 2 inches in diameter, and another set is used for pipe 2½ inches and larger. Dies are classified according to the number of threads per inch. *Table 6* shows proper die selection.

Table 6 Proper Die Selection

Nominal Pipe Size (inches)	Threads Per Inch	Hand Tight		Wrench Makeup		Total Makeup	
		Dimension	Threads	Dimension	Threads	Dimension	Threads
1/8	27	3/16	4½	3/32	2½	9/32	7
1/4	18	7/32	4	3/16	3	13/32	7
3/8	18	1/4	4½	3/16	3	7/16	7½
1/2	14	5/16	4½	7/32	3	17/32	7½
3/4	14	5/16	4½	7/32	3	17/32	7½
1	11½	3/8	4½	1/4	3¼	5/8	7¾
1¼	11½	13/32	4¾	9/32	3¼	11/16	8
1½	11½	13/32	4¾	9/32	3¼	11/16	8
2	11½	7/16	5	1/4	3	11/16	8
2½	8	11/16	5½	3/8	3	1 1/16	8½
3	8	3/4	6	3/8	3	1 1/8	9

105T06.EPS

Once selected, the pipe thread dies must be installed and adjusted for different sizes of pipe. Pipe threading machines have universal die heads that adjust the dies for different sizes of pipe. *Figure 63* shows the die head details.

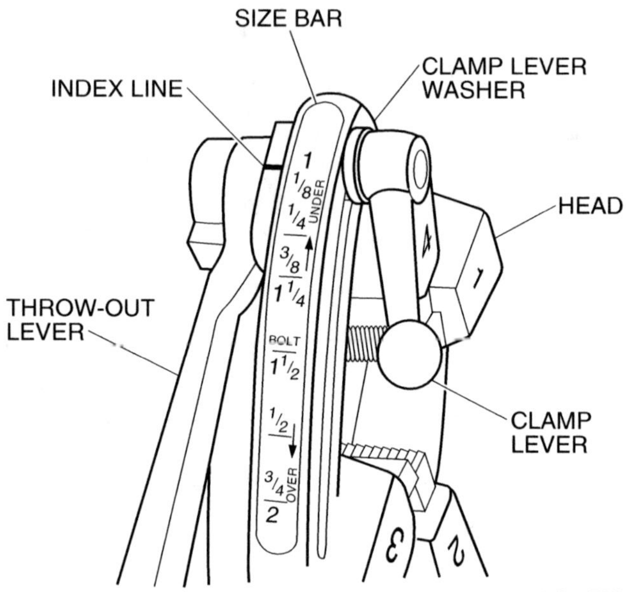

Figure 63 ◆ Die head details.

Each set of dies, except those used on ⅛-inch pipe, can be used for several sizes of pipe. To adjust the die head for different sizes of pipe, loosen the clamp lever and move the size bar until the line underneath the desired pipe size lines up with the index line. Tighten the clamp lever after the adjustment has been made. If oversized or undersized threads are required, set the index line in the direction of the over or under mark on the size bar.

The throw-out lever, which is marked OPEN and CLOSE, is a quick-release handle used to close the dies into the threading position and open the dies to retract the dies away from the pipe. After threading a pipe, the operator can lift this handle to the open position and quickly remove the dies from the pipe. Before cutting threads with the pipe threading machine, the proper dies must be installed in the die head. Follow these steps to replace the dies in a threading machine:

Step 1 Remove the die head from the machine and place it on a workbench with the numbers facing up.

Step 2 Open the throw-out lever.

Step 3 Loosen the clamp lever about three turns.

Step 4 Lift the tongue of the clamp lever washer up and out of the slot under the size bar.

Step 5 Slide the throw-out lever all the way to the end of the slot in the OVER direction of the slide bar.

Step 6 Remove the old dies from the die head.

Step 7 Select a matched set of replacement dies.

Step 8 Clean and inspect the new dies.

Step 9 Insert the new dies in the die head.

> **NOTE**
> The numbers 1 through 4 on the dies must match the numbers on the slots of the die head.

Step 10 Slide the throw-out lever back to where the tongue of the clamp lever washer drops in the slot under the size bar.

Step 11 Adjust the die head size bar until the index line is lined up with the proper size mark on the size bar.

Step 12 Tighten the clamp lever.

Step 13 Install the die head into the machine.

13.5.0 Performing Threading Operations

This section explains the threading operations that need to be performed to thread pipe using the Ridgid model 535 pipe threading machine. Before actually threading the pipe, the operator should check the thread cutting oil level and oil pump operation and prime the oil pump if necessary.

13.5.1 Checking Thread Cutting Oil Level

Before any threading operations are performed, it is extremely important to check the level of the special thread cutting oil. Follow these steps to check the thread cutting oil level:

Step 1 Slide the chip pan out from the base of the threading machine.

Step 2 Check to see if the oil is up to the fill-level line in the reservoir.

> **NOTE**
> Fill the reservoir with cutting oil as necessary.

Step 3 Lower the lubrication arm over the open reservoir.

> **NOTE**
> Oil cannot flow from the lubrication arm in its upright position.

Step 4 Turn the power selector switch to the FORWARD position.

Step 5 Step on the foot switch.

> **NOTE**
> Oil should flow from the lubrication arm. If it does not, it may be necessary to prime the pump.

Step 6 Turn the power selector switch to the OFF position to stop the threading machine.

Step 7 Raise the lubrication arm and slide the chip pan back into the base of the threading machine.

13.5.2 Priming Oil Pump

If the oil does not flow from the lubrication arm, it may be necessary to prime the pump. Follow these steps to prime the oil pump:

Step 1 Remove the button plug on the machine cover.

Step 2 Remove the primer screw through the opening.

Step 3 Fill the pump with cutting oil.

Step 4 Replace the primer screw and the button plug.

> **CAUTION**
> Failure to replace the primer screw and button plug causes the pump to drain itself immediately when the power is turned on.

Step 5 Turn the power on in the reverse direction, and check the flow of oil from the lubrication arm.

> **NOTE**
> Running the machine in reverse primes the pump more efficiently than running the machine forward.

13.5.3 Threading Pipe

After checking the cutting oil level and priming the pump, the operator is ready to thread the pipe. Follow these steps to thread pipe:

Step 1 Load the pipe into the threading machine.

Step 2 Place pipe stands under the pipe as needed.

> **WARNING!**
> Do not wear loose clothing or jewelry. Remove safety gloves and keep hair pulled back while operating threading machine.

Step 3 Cut and ream the pipe to the required length.

Step 4 Make sure that the proper dies are in the die head.

Step 5 Loosen the clamp lever.

Step 6 Move the size bar to select the proper die setting for the size of pipe being threaded.

Step 7 Lock the clamp lever.

Step 8 Swing the die head down to the working position.

Step 9 Close the throw-out lever.

Step 10 Lower the lubrication arm and direct the oil supply onto the die.

Step 11 Turn the machine switch to the FORWARD position.

Step 12 Step on the foot switch to start the machine.

Step 13 Turn the carriage to bring the die against the end of the pipe.

Step 14 Apply light pressure on the handwheel to start the die.

Step 15 Release the handwheel once the dies have started to thread the pipe.

> **CAUTION**
> Make sure that the die is flooded with oil at all times while the die is cutting to prevent overheating the die and the pipe. Overheating can cause damage to the die and to the pipe threads.

> **NOTE**
> The dies feed onto the pipe automatically as they follow the newly cut threads.

Step 16 Open the throw-out lever as soon as one full thread extends from the back of the dies.

Step 17 Release the foot switch to stop the machine.

Step 18 Turn the carriage handwheel to back the die off the pipe.

Step 19 Swing the die and the oil spout up and out of the way.

Step 20 Screw a fitting that is the same size as the pipe you are threading onto the end of the pipe to check the threads.

> **NOTE**
> If the threads are correct, you should be able to screw a fitting three and a half revolutions by hand onto the new threads. If the threads are not deep enough, adjust the dig using the clamp lever and repeat the threading operation.

Step 21 Turn the machine switch to the OFF position.

Step 22 Open the chuck and remove the pipe.

13.6.0 Threading Machine Maintenance

By following a few simple guidelines, you can expect a long life of service from a threading machine. This section describes the procedures to keep a threading machine in proper working condition.

- Never reverse the rotation of the machine while the machine is running. Always turn off the power and wait until the drive has stopped rotating before changing the direction of the drive mechanism.
- At the start of each day, clean the chuck jaws with a stiff brush to remove rust, scale, chips, pipe coating, or other foreign matter. Apply lubricating oil to the machine bedways, cutter rollers, and feed screw.
- Always use sharp dies, which produce smoother threads and require less motor power than dull dies.
- Refer to and follow the manufacturer's instructions for lubrication schedules for bearings and gears. The threading machine should be lubricated at least once for every 40 hours of running time.
- Ensure that there is always plenty of cutting oil in the machine. The cutting oil should be periodically cleaned to remove accumulated sludge, chips, and other foreign matter from the oil. Replace the cutting oil when it becomes dirty or contaminated.
- Remove the oil filter periodically and clean it using solvent. Blow the filter clean using compressed air. Do not operate the machine without the oil filter.
- Keep the chuck jaws in good condition, and replace the jaw inserts as necessary when they become worn.

14.0.0 ◆ SPECIAL THREADING APPLICATIONS

The special threading applications that can be performed with the threading machine include threading short nipples using a nipple chuck and threading pipe larger than 2 inches in diameter using a geared threader.

14.1.0 Cutting and Threading Nipples

A nipple is a piece of pipe less than 12 inches long that is threaded on each end. Threaded pipe longer than 12 inches is considered cut pipe. The nipple chuck is a useful tool for holding nipples in the threading machine when threading. A nipple chuck with inserts and adapters will thread nipples from ⅛ inch to 2 inches in diameter. *Figure 64* shows a nipple chuck kit.

Follow these steps to cut and thread nipples:

Step 1 Load the pipe into the threading machine.

Step 2 Ream the end of the pipe.

Step 3 Thread the pipe.

Step 4 Measure from the threaded end the desired length of the nipple and mark a line on the pipe at this point.

Step 5 Cut the pipe at the mark.

Step 6 Remove the pipe from the power drive chuck.

Step 7 Place the nipple chuck into the power drive chuck and tighten the jaws on the nipple chuck.

Step 8 Place the insert inside the end of the nipple chuck.

> **NOTE**
> Place the small end toward the outside if threading ⅛- to ¾-inch pipe. Place the large end toward the outside if threading 1-inch pipe. Use no insert with pipe that is 1¼ inch and larger.

Step 9 Select the proper size nipple chuck adapter depending on the size pipe you are threading.

> **NOTE**
> If you are threading 2-inch pipe, you do not need a nipple chuck adapter.

Step 10 Screw the nipple chuck adapter into the nipple chuck.

Step 11 Tighten the nipple chuck adapter into the nipple chuck using a chuck wrench.

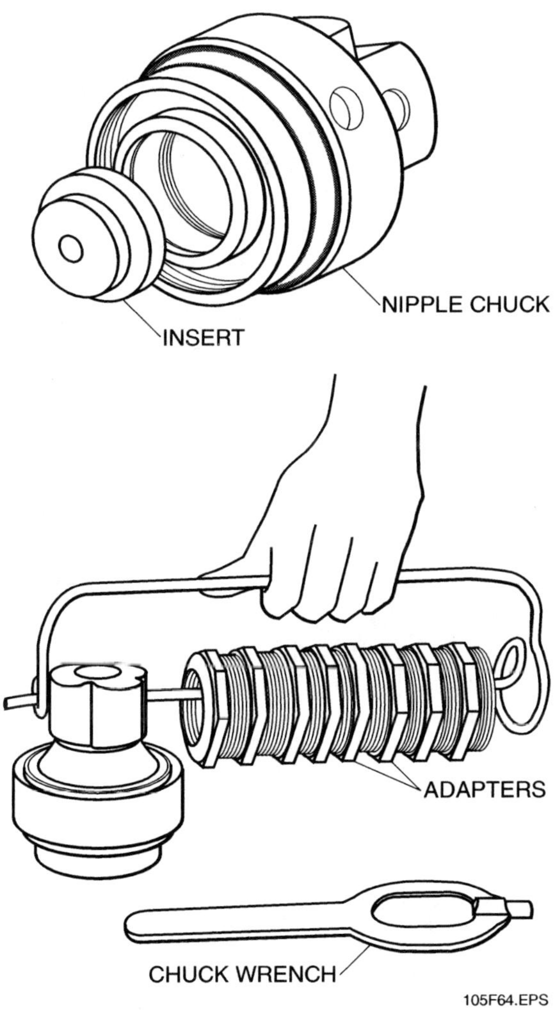

Figure 64 ◆ Nipple chuck kit.

> **WARNING!**
> Do not start the threading machine with the chuck wrench over the nipple chuck adapter.

Step 12 Screw the nipple into the end of the nipple chuck.

Step 13 Tighten the nipple chuck release collar using the chuck wrench. *Figure 65* shows tightening the nipple chuck.

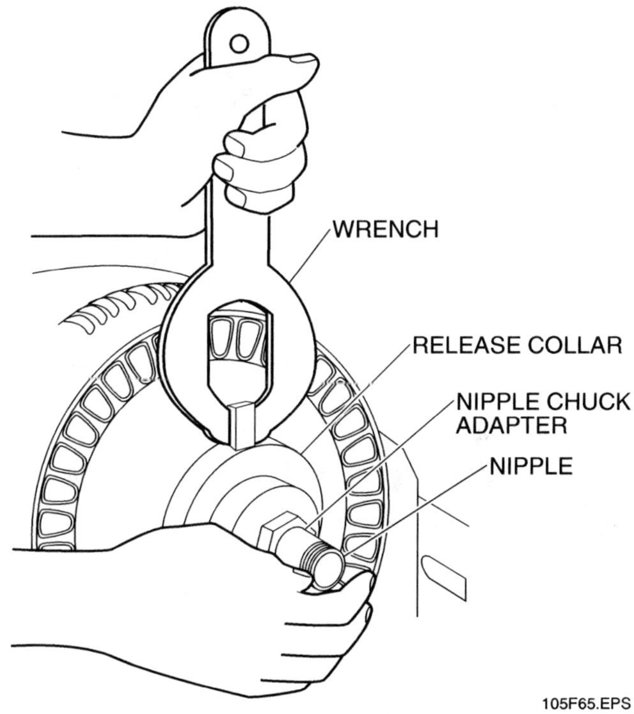

Figure 65 ◆ Tightening the nipple chuck.

Step 14 Ream the end of the nipple.

Step 15 Thread the nipple.

Step 16 Insert the chuck wrench into the release collar and turn the release collar to loosen the nipple from the nipple chuck.

Step 17 Unscrew the nipple from the nipple chuck.

> **WARNING!**
> Use a rag to grab the threads on the freshly cut nipple. These threads are extremely sharp and can cut you.

Step 18 Place the chuck wrench over the nipple chuck adapter and unscrew the adapter from the nipple chuck.

Step 19 Wipe off all excess cutting oil from the nipple chuck adapter and the insert, and store them with the nipple chuck kit.

Step 20 Release the power drive chuck to remove the nipple chuck from the threading machine.

Step 21 Wipe off the nipple chuck, and store it with the rest of the nipple chuck kit.

14.2.0 Using a Geared Threader

A geared threader, or mule, is used to thread pipe from 2½ to 6 inches in diameter. Two sets of dies are used with the geared threaders. One set is used for pipe that is 2½ to 4 inches in diameter and the other set is used for pipe that is 4 to 6 inches in diameter. Each die set contains five dies that are numbered just like the dies for the universal die head. The geared threader is connected to the pipe threading machine by a universal drive shaft (see *Figure 66*).

> **WARNING!**
> A geared threader produces an enormous amount of torque while threading. It can easily tip over the threading machine if the machine is not properly secured. If possible, bolt the threading machine to a steel table and secure the pipe in a vise that is bolted to a steel table. *Figure 67* shows the proper setup for a geared threader.

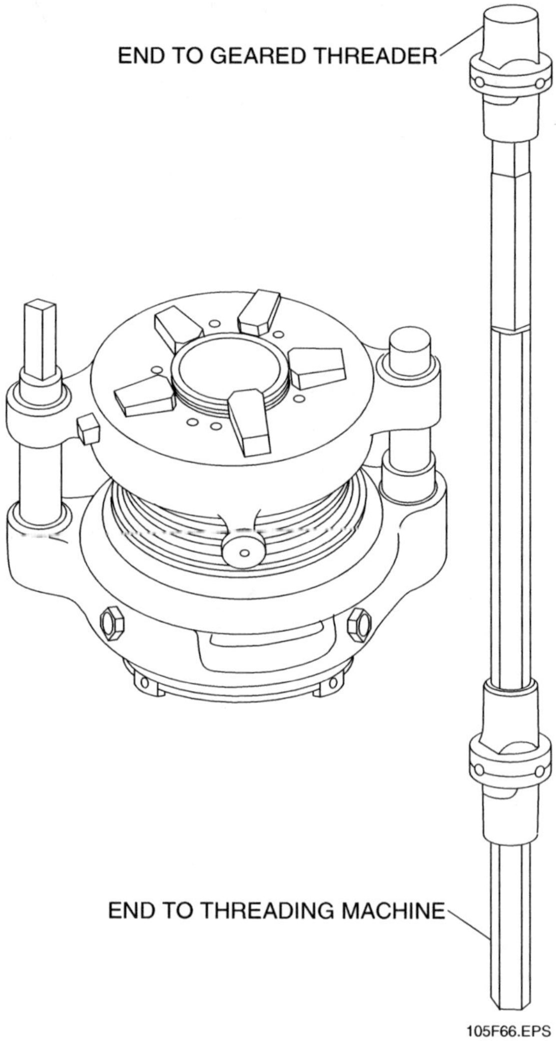

Figure 66 ◆ Geared threader and a universal drive shaft.

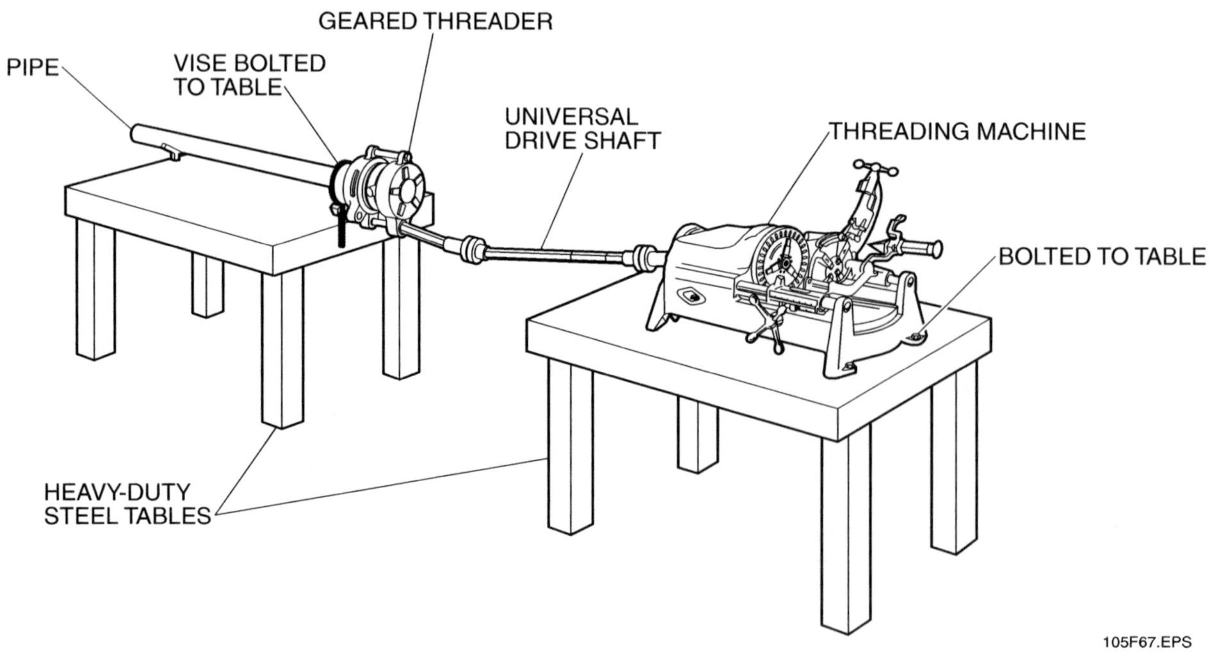

Figure 67 ◆ Proper setup for a geared threader.

15.0.0 ◆ PORTABLE POWER DRIVES

A portable power drive is a handheld power unit that can be used for various jobs, such as cutting threads, driving hoists and winches, and operating large valves (see *Figure 68*). An electric motor provides the power to rotate a ring in the end of the tool. Threading dies and other accessories can be inserted into the circular ring, which has grooved spines to lock these accessories into place.

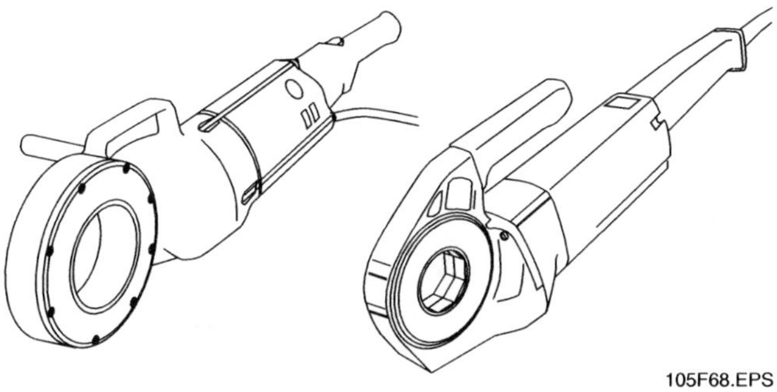

Figure 68 ◆ Portable power drives.

Threading die heads for pipe 2 inches and smaller either fit directly into the power drive or fit into a die head adapter and then into the power drive.

Usually, one hand can hold the power unit and control the power switch while oiling is done with the other hand. Follow these steps to thread pipe using a portable power drive:

Step 1 Secure the pipe to be threaded in a vise.

Step 2 Identify the size of the pipe to be threaded and insert the appropriate die into the portable power drive.

Step 3 Position the die head onto the end of the pipe with the face of the die stock facing away from the pipe.

> **NOTE**
> The drive unit should be positioned on the end of the pipe with the handle of the drive unit in a nearly vertical position above the pipe.

Step 4 Hold the handle of the drive unit with your right hand and center the die head onto the end of the pipe with your left hand.

Step 5 Pull the handle of the drive unit down sharply with your right hand while pushing the die stock against the end of the pipe with your left hand to make the dies bite into the end of the pipe.

Step 6 Continue to move the handle down to a position where you can straighten your elbow to provide a firm, strong grip on the drive unit.

> **NOTE**
> You should lower your right shoulder and balance your weight above the drive unit to resist the threading torque. If you are threading pipe larger than 1 inch, attach a torque arm to support the drive unit at this point. *Figure 69* shows the power drive support arm.

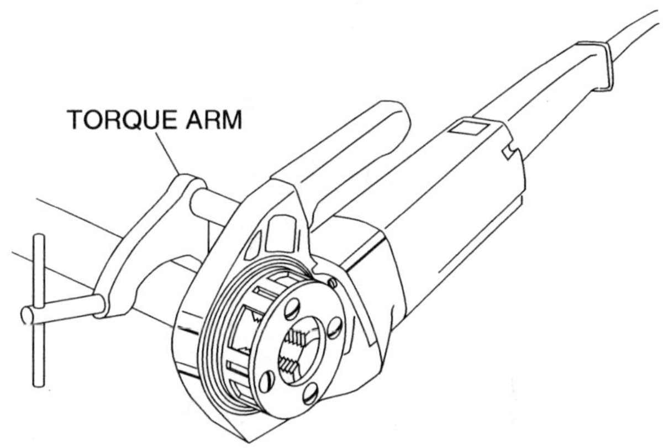

Figure 69 ◆ Power drive support arm.

Step 7 Start the power drive, and cut the thread.

> **CAUTION**
> Be sure to keep an adequate supply of cutting oil on the thread as it is being cut.

Step 8 Stop the power drive as soon as the thread has been cut to the required length.

Step 9 Turn the power switch to the reverse position, and back the dies off the thread.

Step 10 Inspect the threads to make sure they have been cut properly.

16.0.0 ◆ DETERMINING PIPE LENGTHS BETWEEN FITTINGS

In order to replace a piece of pipe between two fittings, the pipeline mechanic must be able to determine the take-out of the fittings to find the length of pipe needed. Take-out, commonly referred to as *makeup*, *take-up*, or *takeoff*, refers to the distance from the end of the pipe to the center of the fitting. It is essential to know the take-out for the fittings when figuring the lengths of straight pipe. Piping drawings usually show the dimensions of the overall run of pipe, but it is up to the pipeline mechanic to determine the lengths of the individual straight pipe within the run. Two of the most common methods for determining pipe lengths are the center-to-center method and the face-to-face method.

16.1.0 Center-to-Center Method

Most pipe drawings give the center-to-center dimensions for a run of pipe (see *Figure 70*). When using the center-to-center method, measurements are taken from the center of one fitting to the center of the next fitting. The actual length of pipe between the fittings is determined by subtracting the center-to-face dimension of each fitting from the overall length and then adding the fitting thread engagement to obtain the actual length that the pipe needs to be. *Figure 71* shows the dimensions of a common 90-degree elbow.

The thread engagement varies with the nominal size of the pipe. Thread engagement should always be measured for accuracy. Manufacturers of fittings do give dimensions of screwed fittings in makeup charts. *Figure 72* shows an example of a makeup chart.

Figure 73 shows the center-to-center method for determining the length of a pipe between two fittings.

Follow these steps to determine the length of a pipe between two fittings, using the center-to-center method:

Step 1 Find the center of each of the two fittings.

Step 2 Measure the distance between the centers of the two fittings.

Step 3 Measure the fitting to determine the center-to-face length of each of the fittings.

> **NOTE**
> This can be done by using the manufacturer's makeup chart or by actually measuring the threads using a rule. Larger fittings allow you to insert the rule inside the fitting to measure the thread engagement.

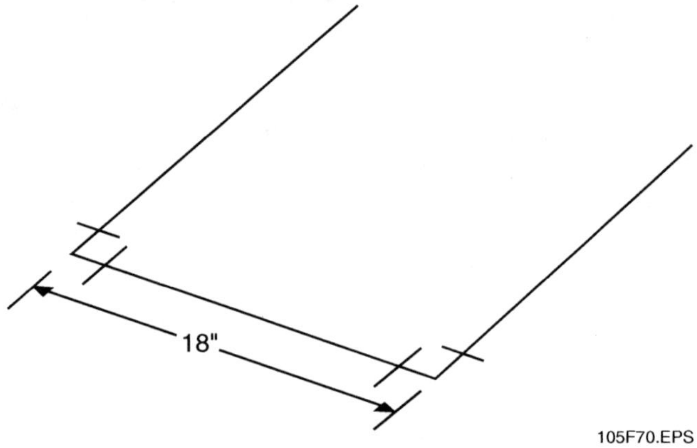

Figure 70 ◆ Pipe drawing showing the center-to-center dimension.

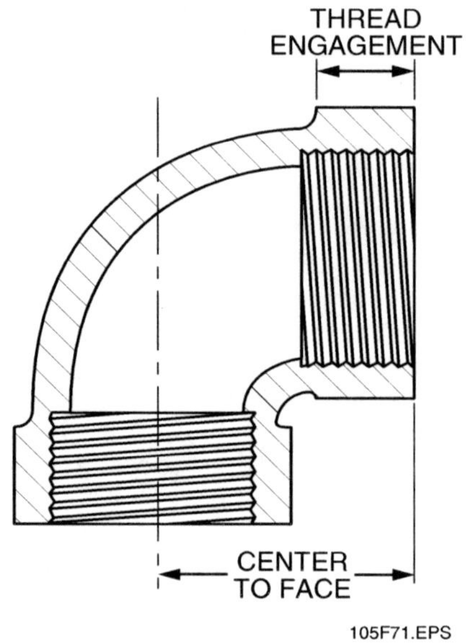

Figure 71 ◆ Dimensions of a common 90-degree elbow.

Step 4 Subtract the center-to-face length of both fittings from the center-to-center measurement taken in *Step 2*.

Step 5 Determine the thread engagement of the fittings.

Step 6 Add the thread engagement of both fittings to the length found in *Step 4* to find the length of pipe needed.

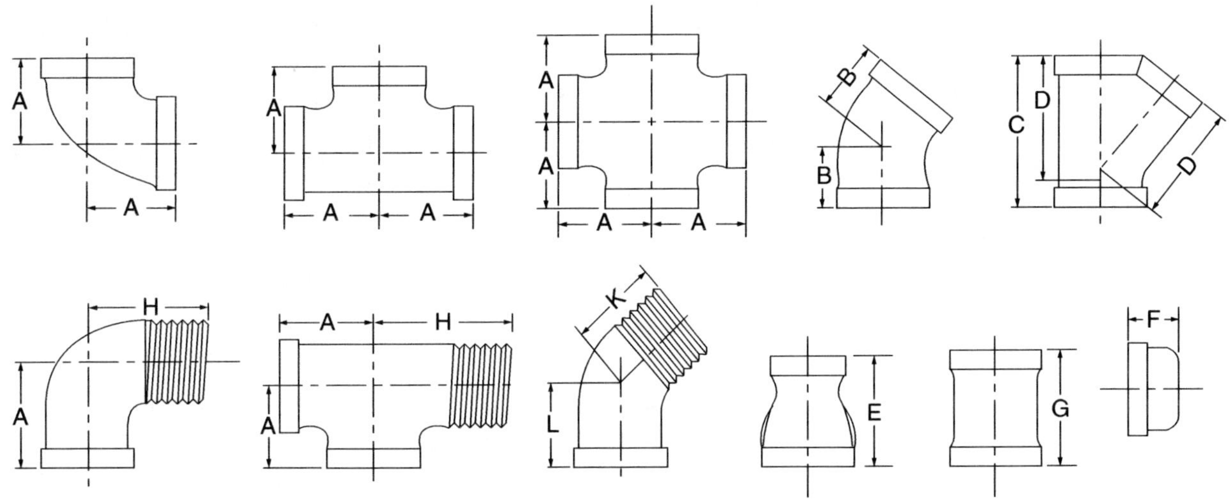

SIZE in.	1/8	1/4	3/8	1/2	3/4	1	1 1/4	1 1/2	2	2 1/2	3	3 1/2	4	4 1/2	5	6
A. in.	11/16	13/16	15/16	1 1/8	1 5/16	1 7/16	1 3/4	1 5/16	2 1/4	2 11/16	3 1/8	3 7/16	3 3/4	4 1/16	4 7/16	5 1/8
B. in.		3/4	13/16	7/8	1	1 1/8	1 5/16	1 7/16	1 11/16	1 15/16	2 3/16	2 3/8	2 5/8	2 13/16	3 1/16	3 7/16
C. in.			2 1/8	2 1/2	2 7/8	3 7/16	4 1/16	4 1/2	5 7/16	6 1/4	7 1/4		8 7/8			
D. in.			1 7/16	1 11/16	2	2 7/16	2 15/16	3 5/16	4 1/16	4 11/16	5 9/16		6 15/16			
E. in.		1	1 1/8	1 1/4	1 7/16	1 11/16	2 1/16	2 5/16	2 13/16	3 1/4	3 11/16	4	4 3/8			
F. in.	17/32	5/8	3/4	7/8	1 1/16	1 3/16	1 1/4	1 5/16	1 7/16	1 5/8	1 3/4	1 15/16	2		2 5/16	2 9/16
G. in.		1 1/16	1 3/16	1 5/16	1 1/2	1 11/16	1 15/16	2 1/8	2 1/2	2 7/8	3 3/16					
H. in.	1 1/8	1 5/16	1 7/16	1 5/8	1 7/8	2 1/8	2 1/2	2 11/16	3 3/16	3 13/16	4 1/2		5 11/16			
K. in.		15/16	1 1/16	1 3/16	1 5/16	1 1/2	1 11/16	1 7/8	2 1/4		3		3 3/4			
L. in.		5/8	11/16	13/16	15/16	1 1/16	1 1/4	1 3/8	1 11/16		2 1/8		2 1/2			

105F72.EPS

Figure 72 ◆ Makeup chart.

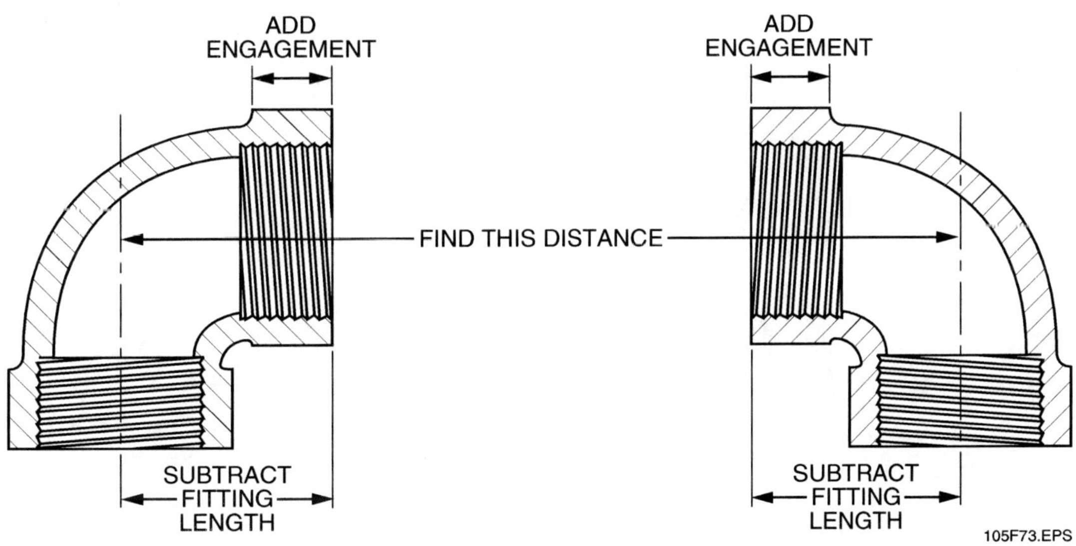

Figure 73 ◆ Center-to-center method.

16.2.0 Face-to-Face Method

Many times you will have to determine the actual pipe length using the face-to-face method. When the dimensions in a line drawing are shown between the end of one fitting to the end of another fitting, this method is known as the face-to-face method. *Figure 74* shows a line drawing of the face-to-face method.

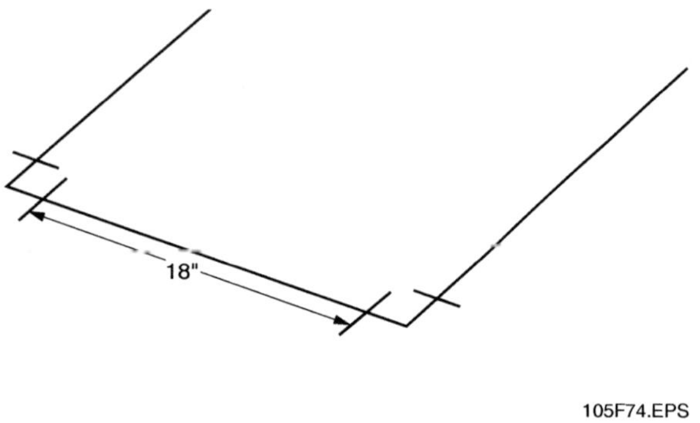

Figure 74 ◆ Line drawing of the face-to-face method.

The face-to-face measurement is taken between the faces of two fittings in a run. To obtain the actual cut pipe length, add the thread engagements for each fitting. *Figure 75* shows the face-to-face method for determining the length of a pipe between two fittings.

Follow these steps to determine the length of a pipe between two fittings, using the face-to-face method:

Step 1 Measure the distance between the faces of the two fittings.

Step 2 Determine the thread engagement of each of the fittings.

Step 3 Add the thread engagement of both fittings to the measurement found in *Step 1* to find the length of pipe needed.

17.0.0 ◆ ASSEMBLY TECHNIQUES

To assemble pipe, you must select the proper joint compound, select the proper tools, and fit the screwed fitting onto the pipe.

17.1.0 Pipe Joint Compounds

A good grade of pipe joint compound must be applied to the male threads joint to lubricate the joint to prevent **galling** of the threads and to ease the makeup of the joint. You must be careful about what compound you use to avoid product contamination or dangerous combinations of materials. Depending on the type of liquid being transferred through the pipes, sometimes no compound is permitted. Always check with the engineering specifications and requirements before applying any type of pipe joint compound. Some compounds that can be used include **Teflon® tape** and pipe dope.

17.1.1 Teflon® Tape

Teflon® tape is a special tape that is wrapped around threads to prevent metal-to-metal contact and to create a leakproof joint. Follow these steps to use Teflon® tape:

> **WARNING!**
> Never use Teflon tape on pipe that is to be welded or pipe that carries steam or other high-temperature service. When Teflon tape is heated, it emits a highly toxic gas that could be fatal.

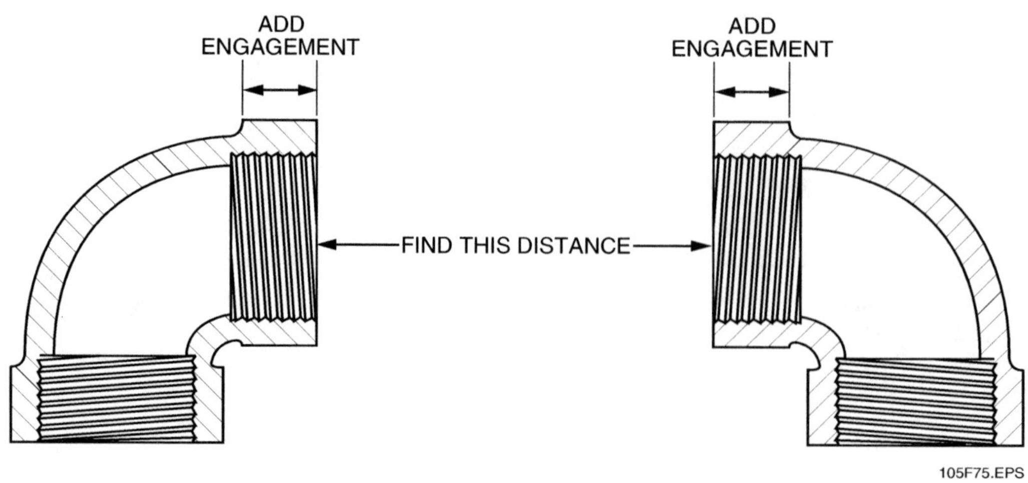

Figure 75 ◆ Face-to-face method.

Step 1 Remove all excess cutting oil from the threads to improve the grip of the tape on the threads.

> **NOTE**
> Use ½-inch wide Teflon tape on pipe that is ¾-inch nominal size and smaller. Use ¾-inch wide tape for pipe that is 1-inch nominal size or larger.

Step 2 Start the tape from the end of the thread, leaving the first full thread bare to prevent the tape from bunching up at the beginning of the thread.

Step 3 Wrap the tape around the pipe in the direction that the joint is to be assembled.

> **NOTE**
> Tape should be wrapped around right-hand threads in a clockwise direction and around left-hand threads in a counterclockwise direction. *Figure 76* shows wrapping a normal right-hand thread with Teflon® tape.

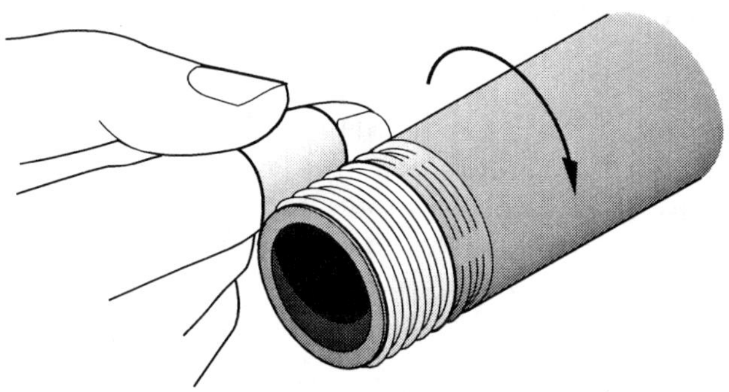

Figure 76 ◆ Wrapping a normal right-hand thread with Teflon® tape.

Step 4 Continue to wrap the tape around the joint, overlapping the edges of each wrap until all remaining threads have been covered.

Step 5 Press the tape against the threads to seal it to the threads and prevent the tape from slipping off the threads once you start to make up the joint.

17.1.2 Liquid Teflon®

Liquid Teflon® is a Teflon®-based pipe joint compound is painted onto the male threads using a small brush. It can be used on all types of materials and can withstand temperatures up to 400°F.

17.1.3 Pipe Dope

Pipe dope is manufactured for sealing threaded pipe joints. If manufactured pipe dope is not available, a mixture of graphite paste and water, petroleum jelly and zinc dust, or white or red lead will work as a joint compound. Make sure that the pipe dope is suitable for the type of pipe you are using. Dopes are made specifically for stainless steel pipe and for carbon steel threads. Try to avoid dopes that claim to be suitable for both stainless steel and carbon steel because they tend to fail more often than others. Pipe dope is relatively easy to apply. It is simply brushed onto the male threads to be fitted. Do not apply pipe dope to the female threads, and do not apply dope to the first thread of the pipe end. This is to keep the pipe dope out of the pipeline and avoid contaminating the liquid inside the pipeline.

17.2.0 Selecting Wrenches

When you assemble threaded pipe joints, first tighten the fitting by hand, and then make up the joint completely using a pipe wrench. The pipe should either be secured in a vise, or you should use two pipe wrenches: one to hold the pipe and the other to turn the fitting. In selecting and using the pipe wrench, use the following suggestions as a guide:

- Pipe wrenches should be in good working condition and adjusted properly to avoid damaging the pipe.
- Use a strap wrench when working with brass, chrome-plated, or other specially finished pipe to avoid scratching or marring the surface.
- Select a wrench that is the proper size for the pipe with which you are working. Wrenches that are too short make your job much harder. Wrenches that are too long strip the threads, crack the fitting, or crush the pipeline.
- Threads should never bottom out of the fitting or adjacent pipe.
- Never use a power drive or threading machine to tighten fittings onto pipes.
- Be sure to install unions in accordance with flow requirements.
- Never use a cheater bar on the end of a wrench or strike the wrench handle with a hammer.

17.3.0 Fitting Screwed Pipe and Fittings

Properly fitting pipe and fittings is a skill that must be learned through experience. It is crucial to fabricating a quality piping system. Poor fitting practices can result in leaks and other defects in the system. When tightening fittings onto a piece of pipe, the technician must be aware of the end position that the fitting must face. If the fitting is turned past the desired end position and then turned back, the fitting will leak. Make sure that the union is placed in the direction of flow. Follow these steps to fit screwed pipe and fittings:

Step 1 Determine the type, size, and schedule of pipe being used in the system.

Step 2 Determine the length of pipe needed.

Step 3 Cut the pipe to the desired length.

Step 4 Ream the pipe to remove all internal burrs.

Step 5 Thread the pipe, taking caution not to cut the threads too deep.

Step 6 Select the proper size, shape, and type of fittings needed.

Step 7 Clean the pipe and fittings thoroughly inside and out.

> **NOTE**
> All sand, dirt, and oil must be removed from the inside of the pipe and fittings to avoid contaminating the system or clogging the line. The fittings and pipe ends may be cleaned with a clean rag soaked in nonflammable solvent.

Step 8 Check the threads on the pipe and the fitting to ensure that they are properly cut and not damaged.

Step 9 Apply joint compound or Teflon® tape to the pipe threads.

Step 10 Start the fitting on the pipe by hand.

Step 11 Tighten the fitting slowly, using a pipe wrench.

> **CAUTION**
> Tightening the fitting quickly causes excessive heat due to friction, which could cause the threads or the fitting to expand. Do not allow the threads to bottom out into the fitting. This can damage the threads and the fitting.

18.0.0 ◆ HOSE STANDARDS AND SPECIFICATIONS

While the flexibility of hose offers some advantages over pipe and tube, there are limitations to the use of hose. Several factors must be considered when hose and hose fittings are being selected for a particular application. In determining the suitability of hose and hose fittings for a particular application, designers and engineers must carefully consider many properties of the system, including:

- *Temperature* – Process temperature and ambient temperature have a definite effect on hose and fitting selection. Care must be taken to ensure that process fluid temperatures and ambient temperatures, both static and transient, do not exceed the limitations of the hose and fitting. Also, special care must be taken when routing hoses near hot equipment or high-temperature areas.
- *Pressure* – Hose and fitting selection must be made so that the recommended maximum operating pressure is equal to, or greater than, the system pressure. Surge pressures higher than the maximum operating pressure must be taken into account when selecting the proper hose and fittings. Also, hoses and fittings that are used for suction applications must be able to withstand the negative pressure of the system.
- *Compatibility* – Hose and fitting selection must also ensure compatibility of the hose tube, cover, and the material being transported. Care must be taken to ensure that the hose and fittings are either compatible with, or protected from, the environment to which they are exposed. Environmental conditions such as ultraviolet light, ozone, salt water, chemicals, and air pollutants can cause degradation and premature failure of hose.

Hoses of different types and sizes are made to handle a variety of applications. Lightweight hose is commonly used in low pressure (usually below 250 psi) air systems for ventilation and air supplies. Medium-grade hose is used in moderate pressure systems (approximate range from 250 to 3,000 psi) to carry water and other fluids that normally are not subject to high pressure. For use in industrial plants, heavy-grade hoses are made to withstand high pressures (approximate range 3,000 to 6,000+ psi), high temperatures, adverse process fluids, and extreme environmental conditions. Heavy grade hoses are used to carry fluids in industrial hydraulic, chemical, and steam systems.

Several organizations and professional societies have published standards and codes for hoses. Engineers and designers use these standards in selecting and identifying the type and size of hose for particular applications. The Society of Automotive Engineers (SAE), U.S. Military Specification (**MIL**), Department of Defense, and the Food and Drug Administration (FDA) are three of the organizations that have published standards, codes, and specifications for hoses.

When the project specifications list the required SAE, MIL, or FDA hose identification number, you must use that type of hose. You should review some of these codes so that you know the type of information included in the hose specifications.

The recommendations for selection, installation, and maintenance of hose and assemblies established by SAE J1273 are intended as a guide to assist system designers and users in the selection, installation, and maintenance of hose. Designers and users make a systematic review of each application and then select, install, and maintain the hose to fulfill the requirements of the application.

All SAE hoses are branded (marked) with a line parallel to the longitudinal axis as a visual aid to prevent installing hose in a twisted position. As part of this line, the hose is marked with the applicable SAE specification number and the size.

18.1.0 Sizing Measurements for Hoses

It is important to use the proper size and type of hose for specific applications. Hose size is normally identified by the inside diameter (ID) of the hose. The ID size of hose is usually expressed in inches and fractions of an inch, such as ½-inch hose, 1-inch hose, or 1½-inch hose.

The dash numbering system is used by most hose and fitting manufacturers to identify sizes of hose and hose fittings. To determine a dash size, the hose ID size is simply converted into 16ths of an inch. The numerator (top number) of the fraction is the dash size.

For example, a ¼-inch hose size is converted into the fraction $\frac{4}{16}$. Since the numerator is 4, this hose size is expressed as -4 (dash 4). A 1¼-inch hose size is converted to $\frac{20}{16}$ and is expressed as -20 (dash 20).

Hose can be connected to pipe or tube by matching the ID of the hose with the ID of the pipe or tube to which it is to be connected. By using same ID of hose and tube or pipe in a run, the fluid flow through the system will be smooth and uninterrupted.

18.2.0 Pressure Ratings for Hoses

Generally, the minimum burst pressure of a hose should be four times the rated working pressure of the hose. For example, if a hose has a maximum rated working pressure of 400 psi, it should not burst until at least 1,600 psi has been reached. This safety margin is designed into hoses to help prevent ruptures, which could cause injury to personnel or damage to equipment.

19.0.0 ◆ TYPES OF HOSES

Hoses are made from a wide variety of materials including different types of metal, rubber, Teflon®, and poly. Each hose material has certain characteristics that make it desirable for particular applications. Hoses can be classified by material, by type of service (such as hydraulic or acid resistant), by pressure rating, and by type of construction.

Elastomers mixed with various chemicals provide a wide range of hose materials with physical properties required for specific service needs. For example, hose materials like Teflon®, polyethylene, nylon, neoprene, and vinyl nitrile are used to carry different types of process fluids at differing pressure and temperature conditions.

19.1.0 Metallic Hoses

Flexible metal hoses are used as general all-purpose hose for conveying liquids and gases. Metallic hose is considered pressure-tight. It is often used for applications involving vibration or flexure. Other common applications for metallic hose include saturated and superheated steam lines, gas and oil lines, and lubricating lines.

Metallic hoses are generally available in stainless steel, carbon steel, bronze, aluminum, Monel®, and other corrosion-resistant metals. Some metal hoses have a liner of flexible, corrosion-resistant material, such as Teflon®.

Series SS (standard gauge) hoses are constructed of close pitch, annular corrugations in different types, such as hose only, hose with single **braid**, and hose with double braid. Standard sizes range from 1 inch to 12 inches.

Series B (bronze) hoses are constructed with annular corrugations from seamless tin bronze tube usually covered with a bronze wire braid. Standard sizes range from ¼ inch to 3 inches.

As the working temperature of metal hose increases, the pressure rating of the hose decreases. Always consult the vendor's manual for the hose specifications, temperature conversion factors, and maximum service temperatures of different metal hoses.

Figure 77 shows the construction of a braided, flexible metal hose.

Flexible metal hoses are typically constructed with annular corrugations from stainless steel, carbon steel, or seamless tin bronze tubing. The tube is usually covered with a woven wire braid for protection against abrasion and to provide increased resistance to pressure.

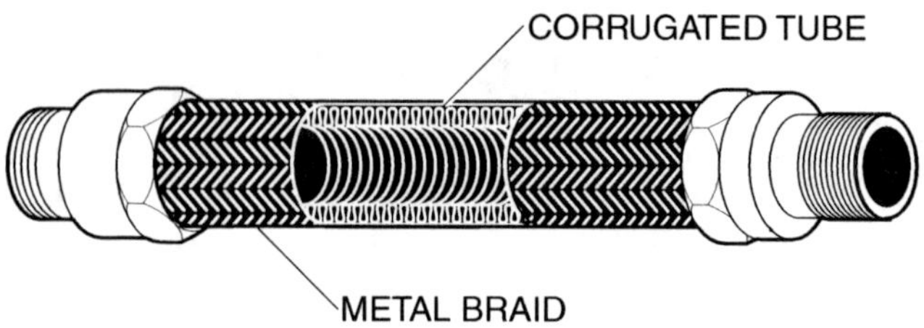

Figure 77 ◆ Construction of a braided, flexible metal hose.

Two general categories of metal hose are commonly used:

- An all-metal hose is used when no plastics or synthetics are permitted in the system. A corrugated metal bellows-type hose is covered with a metal braid. They are welded to each other at each end at a weld ring. The tube fittings, pipe fittings, weld fittings, or tube adapter ends are then welded to the hose ends. This type of metal hose is ideally suited to applications where vibration, thermal expansion, misalignment, or intermittent flexing are present.

> **NOTE**
> Do not exceed the minimum bend radius because this type of metal hose is easily **kinked**.

  Single braided and double braided stainless steel hoses are available for use at pressures up to approximately 3,825 psi and temperatures to +1,200°F.

- Metal braided hose lined with Teflon® is used where an extremely smooth, corrosion-resistant interior is required. Instead of annular corrugations, the metal hose is lined with Teflon® tubing. An outer braid of metal encloses the TFE tube. If made out of 304 stainless steel, this type of metal hose has a temperature range of approximately –100° to +450°F and a pressure rating to 4,000 psi.

19.2.0 Nonmetallic Hoses

Many hoses are made from nonmetallic materials such as Teflon®, nylon, rubber, neoprene, vinyl nitrile, and vinyl.

19.2.1 Teflon® Hoses

Teflon® is the DuPont Company's registered trademark for its TFE (tetrafluoroethylene) fluorocarbon resins. There are only two classes of chemicals known to attack Teflon®: molten alkali metals, such as sodium and potassium, and fluorochemicals, such as chlorine trifluoride, oxygen difluoride, and fluorine gas.

Teflon® is a very good hose material for a variety of applications. Usually, a wire overbraid is added to the Teflon® core to increase pressure ratings. Depending upon the type of wire braid, Teflon® hose can be used for pressures up to 3,500 psi.

Teflon® hose has a long life because its inner core is resistant to damage from steam, chemicals, solvents, heat, pressure impulses, flexing, vibration, and aging. It stands up well to continuous flexing and vibration. Teflon® hose is capable of handling a wide range of temperatures as well as rapid temperature changes. It is easily cleaned, allowing one hose to be used for several types of services. Teflon® is non-contaminating and FDA-approved.

19.2.2 Nylon Hoses

Nylon hose is constructed using nylon braided tubes covered with neoprene. Nylon hoses are resistant to abrasion, oil, and the weather. They can be used for transferring chemicals, solvents, paints, glues, and petroleum products. Nylon hoses are also commonly used to convey air, water, and oils.

Nylon hoses come in various colors. They generally have temperature ratings ranging from –30° to +160°F. Some types of nylon hose can withstand pressures up to 750 psi.

19.2.3 Natural Rubber Hoses

Hoses made from natural rubber are used to carry acids, chemicals, and gases. Rubber hose is often used in laboratories. Natural rubber hoses should not be used for petroleum products because natural rubber is not oil-resistant.

19.2.4 Natural Latex Hoses

Natural latex hoses can be sterilized. They are often used in hospitals, food handling operations, and laboratories.

19.2.5 Silicon Rubber Hoses

Hoses made from silicone rubber can be used in higher temperature applications and with certain types of chemicals. Silicone rubber hose has a temperature rating usually ranging from approximately –150° to +600°F.

19.2.6 Neoprene Hoses

Neoprene hoses are used in applications similar to natural rubber, but where increased resistance to ozone, weathering, gasoline, or oil is needed.

19.2.7 Vinyl Nitrile Hoses

Hoses made from vinyl nitrile are resistant to abrasion, animal fat, oil, and the weather. They can be used to transport non-dairy food products and bulk liquids such as sugar, syrups, and vegetable oils. Vinyl nitrile hose meets FDA requirements.

When reinforced with wire, vinyl nitrile hoses have pressure ratings up to 150 psi. They have a temperature rating usually ranging from –40° to +180°F.

19.2.8 Vinyl Hoses

Flexible vinyl hose (tubing) meets the criteria in 3-A plastic standard and complies with FDA requirements. Vinyl hose is not affected by dairy products. It is often used in dairies, food and beverage plants, laboratories, hospitals, and chemical and industrial installations.

20.0.0 ◆ HOSE CONSTRUCTION

Most hoses are made from more than one material. The hose materials are constructed in layers that reinforce the hose to increase strength and pressure capabilities. The type of **reinforcement** differentiates between low-pressure, medium-pressure, and high-pressure hoses.

Low-pressure hose is typically reinforced with a fabric braid. Medium- and high-pressure hoses have a single or multiple wire reinforcing braid. Special wire wrap hoses are also used for extra high pressure applications.

The inner tube or lining of hoses is usually made from some type of rubber, synthetic rubber, or plastic. The tube or lining is surrounded by a braided material (carcass), which is covered by rubber or cotton. The type of braiding used is determined by the application of the hose.

20.1.0 Vertical Braiding

Hoses reinforced with vertical braiding are strengthened against pressure applied at right angles to the tube or lining.

Vertical braided hose is constructed of a seamless rubber tube with one or more layers of braided yarn wrapped around the tube (see *Figure 78*).

Vertical braided hoses are typically used in low-pressure systems. Depending upon the type of tube and braiding, vertical braided hoses can be used to carry water, air, fuel, lubricating oils, antifreeze solutions, and diesel fuels.

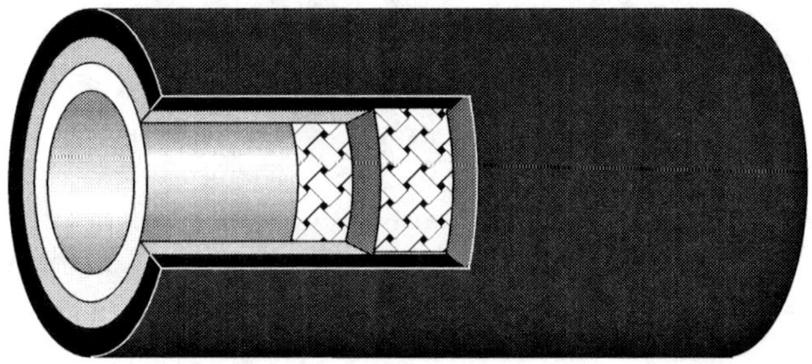

Figure 78 ◆ Vertical braided hose.

20.2.0 Horizontal Braiding

Hoses reinforced with horizontal braiding are strengthened to resist expansion and contraction. Horizontal braided hoses are used for high-pressure or high-suction (vacuum) systems.

Horizontal braided hoses are made by wrapping a seamless tube with one or more layers of braided fibers or wire. Depending upon the type of tube and braiding, horizontal braided hose can be used in higher pressure systems for hydraulic, gasoline, fuel oil, lubricating oil, air, and water applications.

Horizontal braided wire reinforced hoses may include one or more layers of wire braid between two or more layers of fiber braid. This type of construction makes the hose mechanically strong for use in high pressure or high vacuum systems. Depending upon the type of tube and braiding, wire reinforced hoses can be used for water-based liquid fertilizers and pesticides transfer; pumping, suction, and discharge of water and slurries; and petroleum applications.

20.3.0 Wrapped Hose

Wrapped hose is made by wrapping the tube with several layers (plies) of fabric (see *Figure 79*).

The tube may be made from Teflon® or synthetic rubbers. The type of tube selected is dependent on the application.

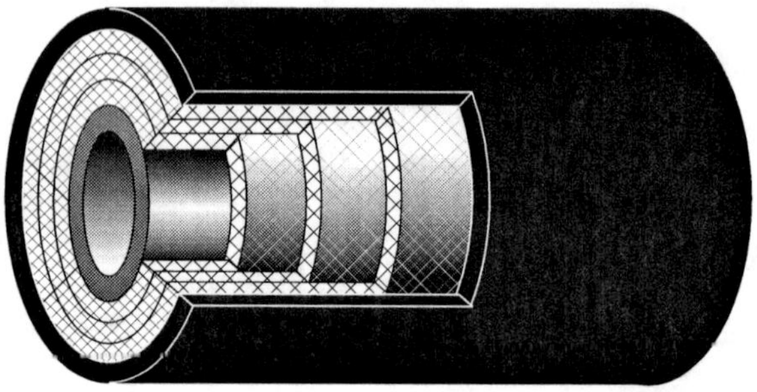

105F79.EPS

Figure 79 ◆ Wrapped hose.

20.4.0 Wire Reinforced Hose

Wire reinforced hose is made by winding braided wire around the tube (see *Figure 80*).

Wire reinforced hose may include layers of fabric. When fabric is used, the wire braid reinforcement layers are inserted between the fabric layers. Wire reinforced hose is typically used in high-pressure systems because of its strength.

Wire woven hose is made by spirally wrapping the tube with woven wire. *Figure 81* shows a wire woven hose.

Layers of fabric may be inserted between the layers of woven wire reinforcement. Woven wire hose stands up well under severe operating conditions, including high-pressure hydraulic systems and high vacuum applications.

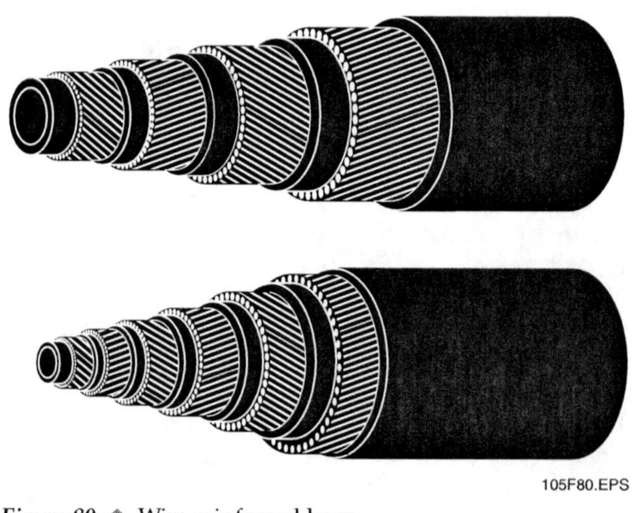

105F80.EPS

Figure 80 ◆ Wire reinforced hose.

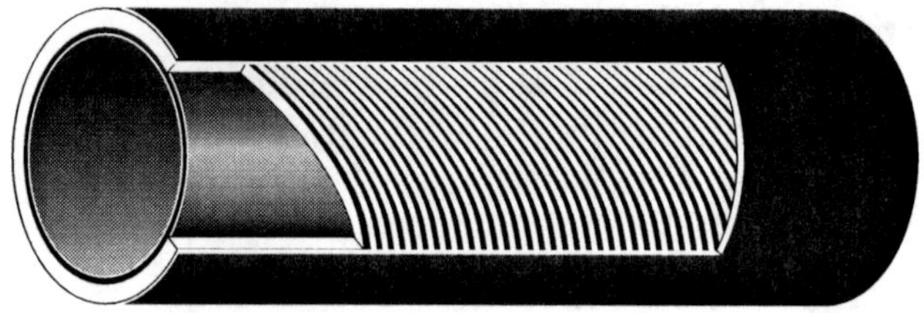

105F81.EPS

Figure 81 ◆ Wire woven hose.

21.0.0 ◆ HOSE FITTINGS

There are several types of hose fittings. Each type is designed to perform a specific job. For example, a hose fitting might be used to make the connection between hose and hose, hose and pipe, hose and tubing, or between hose and a valve. Hose fittings can be either permanent or reusable. Some types of fittings are made to connect and disconnect quickly, an advantage and requirement in many applications.

Fittings are matched with the hose material being used and the application. As with hoses, factors such as pressure, temperature, and process characteristics are taken into consideration when selecting hose fittings.

Fitting manufacturers use codes that are usually stamped on the fitting to identify the fitting material and size. High-pressure fittings are similar to medium-pressure fittings. Both have a socket with a size designation to fit the hose. Most manufacturers identify high-pressure fittings from medium-pressure fittings by stamping "HP" on high-pressure fittings and "MP" on medium-pressure fittings. In addition, the sockets of high-pressure fittings are notched. Medium-pressure fitting sockets are not notched.

Some fitting identification codes and other specifications have been developed by the SAE, MIL, and FDA. The fitting codes are similar to the codes used for identifying hose materials. Other codes for fittings are manufacturer specific. Check the manufacturer's catalog if you have any question about the material, properties, or proper use of a particular fitting.

21.1.0 Permanently Crimped (Skived) Fittings

Permanent **crimped** (skived) fittings are one-piece fittings consisting of a crimp nipple and crimp socket that are fastened together (see *Figure 82*).

This type of fitting is used for high-pressure applications such as hydraulic fluid; crude, fuel, and lubricating oils; and water. Permanently crimped (skived) fittings are generally considered good for pressures up to 5,500 psi depending on the dash size.

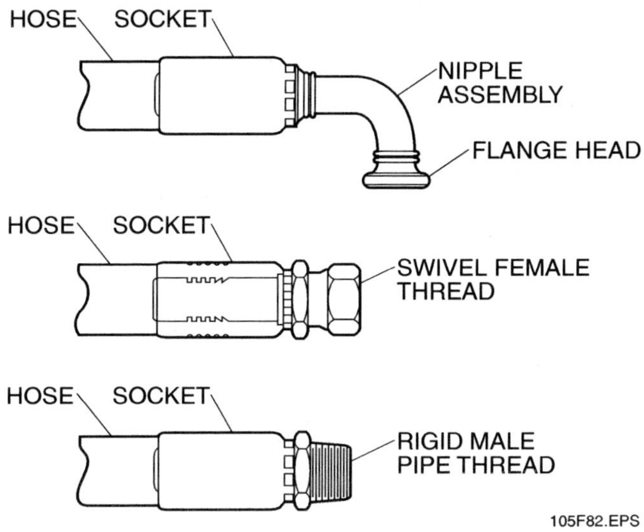

Figure 82 ◆ Crimped (skived) fittings.

NOTE

Some manufacturers do not recommend reusing this type of socket.

21.2.0 Reusable (No-Skived) Fittings

Reusable (no-skived) fittings are two-piece fittings that consist of a nipple and socket (see *Figure 83).*

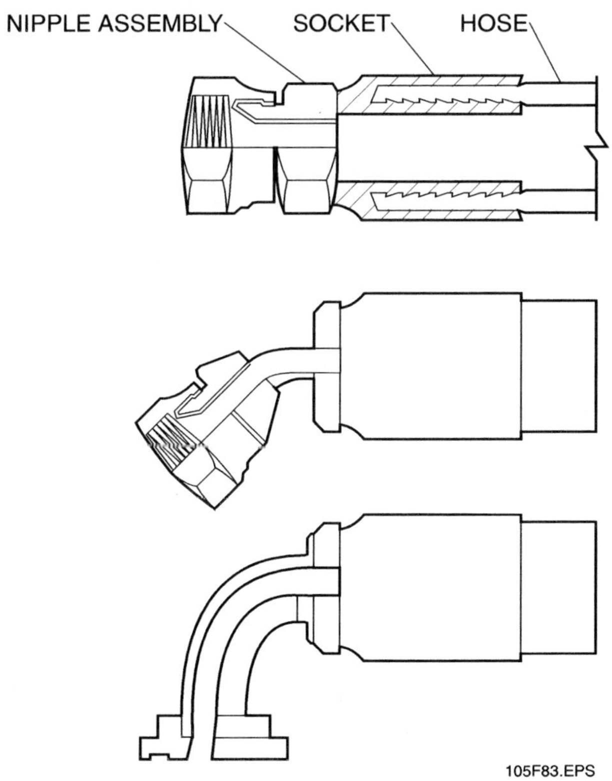

Figure 83 ◆ Reusable (no-skived) fittings.

The nipple can be reused if the fitting needs to be replaced. This type of fitting is often used for high-pressure hydraulics, fuel and lubricating oil, gasoline, and water glycol base fluids. These fittings are generally considered good for pressures up to 5,500 psi depending on the dash size.

21.3.0 Push-On Fittings

Push-on fittings are primarily used for general, low-pressure, low-temperature air and water service (see *Figure 84*).

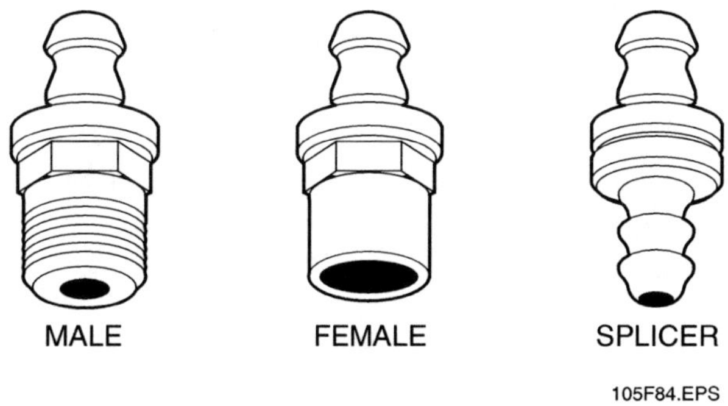

Figure 84 ◆ Push-on fittings.

WARNING!

Push-on fittings are NEVER used for high pressure or chemical process applications.

21.4.0 Swivel Joint Fittings

Swivel joint fittings are not rigid (see *Figure 85*). Therefore, they are capable of absorbing some shock incurred when a system pulses.

The swivel action of the joint prevents the hose from twisting and kinking when used on moving equipment. Swivel joints are typically available in brass/steel, steel, and stainless steel. Depending upon the process and the size of the fitting, swivel joint fittings are generally capable of handling a range of 25 in/Hg (inches of mercury) vacuum to 10,000 psi operating pressure.

WARNING!

To be safe, you should check the vendor's selection chart before using any swivel joint.

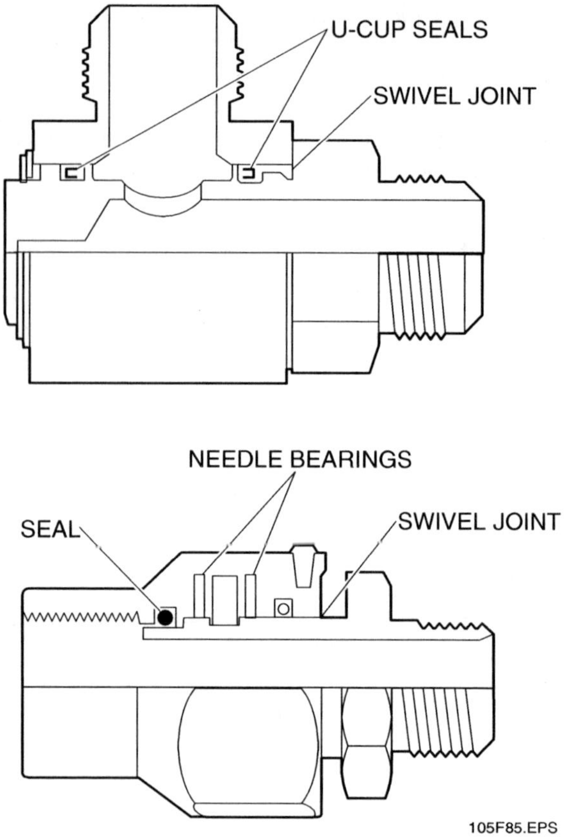

Figure 85 ◆ Swivel joint fittings.

21.5.0 Quick-Disconnect Fittings

Quick-disconnect fittings (couplings) are connecting devices that permit easy, immediate connection and separation of fluid lines. When installed in a fluid system, quick-disconnect couplings save time by eliminating the necessity to close valves, bleed the system, recharge with fluid, and purge entrapped air whenever an accessory is being replaced. Dependability is assured because the coupling valves of a quick-disconnect fitting automatically open and close when connected or disconnected. The automatic valve actions also minimize the possibility of air, dirt, and moisture being trapped in the system.

Quick-disconnect couplings may be used in systems to help align components. The swivel feature of quick-disconnect fittings helps prevent twisting of hose assemblies. However, quick-disconnects are not intended to be used as swivel joints in applications subjected to constant rotation.

There are two basic valve options for quick-disconnect fittings. The valve type used depends on the application requirements.

- Double valve quick-disconnects have spring-loaded poppet valves in each half to provide an immediate positive seal (see *Figure 86*). The double valve design prevents escape of liquids and gases upon disconnection. Double valve couplings are generally used for fuel, oil, pneumatic, and hydraulic systems or where the fluid provides little or no lubrication.

  A small cavity between the valve halves permits some air inclusion and loss of some fluid upon connection and disconnection.

> **NOTE**
> Often a pusher is required in the non-valved half to open the poppet in the valved half.

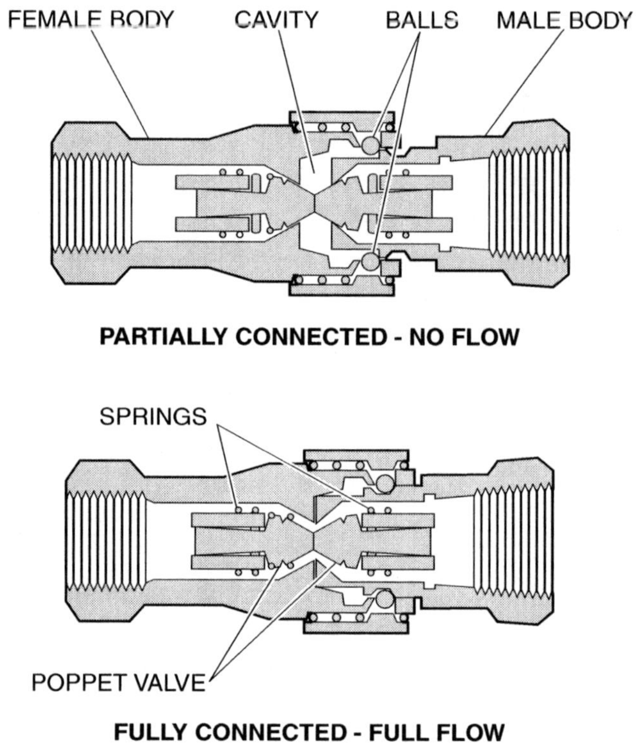

Figure 86 ◆ Double-valve coupling.

- Single valve quick-disconnects provide positive shutoff in only one-half of the fitting. Either the male half or the female half of the quick-disconnect can be the valved half. A spring-loaded tubular valve or poppet valve closes upon disconnection, eliminating escape of fluids from the valved half. The non-valved half is open, allowing for fluid drainage upon disconnection.

  In a spring-loaded valve, there is no cavity ahead of the valve to cause air inclusion or fluid spillage. A single valve coupling incorporates a spring-operated sleeve to close the ports of the valve (see *Figure 87*).

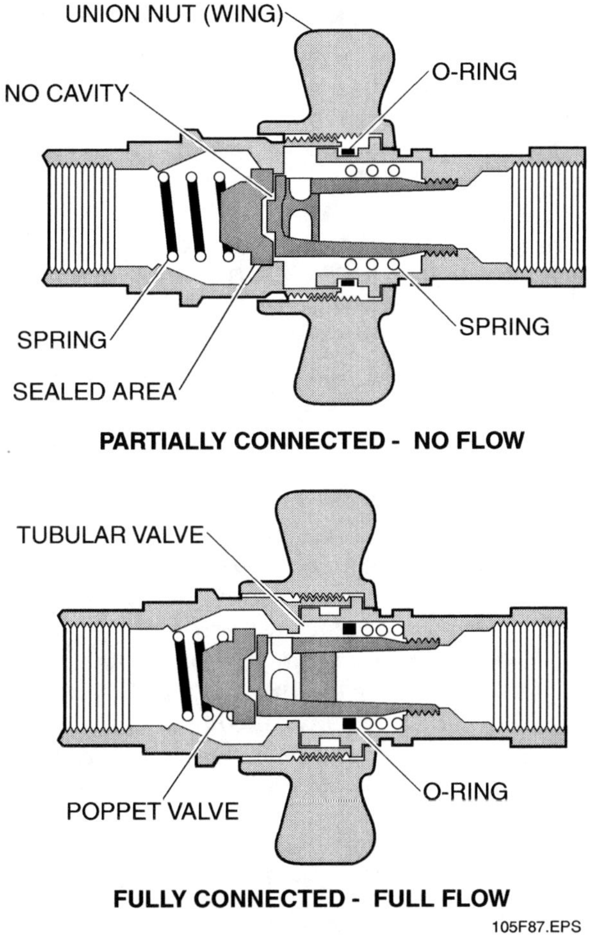

Figure 87 ◆ Single valve coupling.

> **WARNING!**
> Both halves are open to fluid drainage when disconnected.

Typical applications of single valve quick-disconnects include fluid transfer, hydraulic, fuel, oil, or any application where the agent provides lubrication.

Straight-through-quick-disconnect couplings, cam lever couplings, quick acting couplings, and interlocking clamp couplings do not incorporate valves. Straight-through-quick-disconnect couplings can be used for alignment of components to prevent twisting of hose assemblies. Also, the large inside diameters through the coupling allow maximum flow with minimum pressure drop.

> **NOTE**
> Quick acting couplings can be wired together as an added safety precaution.

Straight-through-quick-disconnect cam lever couplings provide for fast, simple connections (see *Figure 88*).

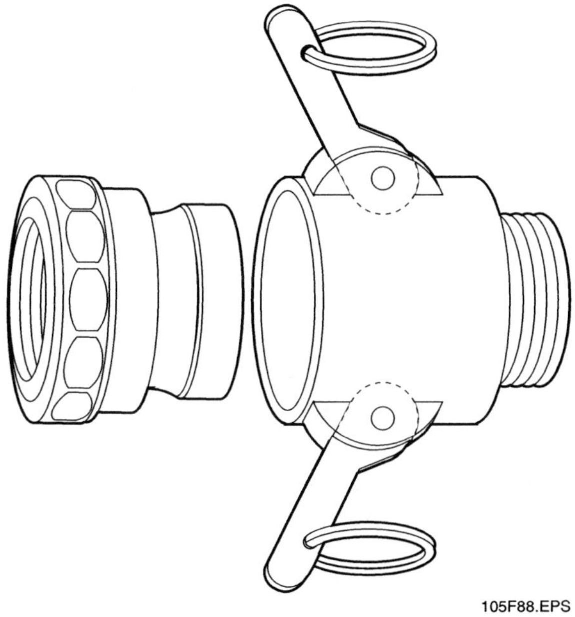

Figure 88 ◆ Cam lever coupling.

Cam lever couplings can be used in a number of different applications, such as transporting and blending raw materials used in processing, on trucks used in vacuum cleaning, to transport flour from a railroad car to the plant for processing, and for delivering gasoline from a tank truck to a service station. Cam lever couplings are often on low- and medium-pressure water, petroleum, and chemical transfer.

Quick acting couplings are often used for low- to medium-pressure air, water, or oil service (see *Figure 89*).

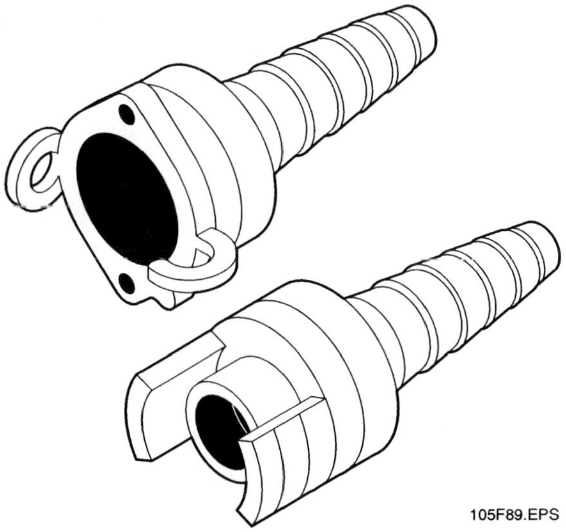

Figure 89 ◆ Quick acting couplings.

Interlocking clamp couplings are used for heavy-duty high pressure applications, such as air, steam, water, spray, LPG, and anhydrous ammonia (see *Figure 90*).

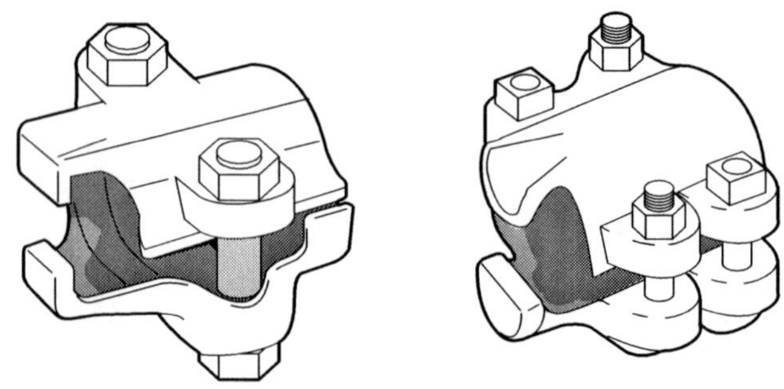

Figure 90 ◆ Interlocking clamp couplings.

22.0.0 ◆ PROPER AND SAFE METHODS FOR STORING HOSES

Proper and safe storing methods for hoses may sound like a simple task, but you would be surprised at the amount of damage done to hoses, personal injuries, and time wasted due to improper storage techniques. There are correct ways to store hose that will help avoid costly damage, save time in identifying stored hoses, and reduce the chance of personal injury to you and other workers.

The appropriate method for storing hose depends to a great extent on the hose size (diameter and length), hose construction, the quantity to be stored, and the way in which it is packaged.

Hose should not be piled or stacked to such an extent that the weight of the stack creates distortions on the lengths stored at the bottom. Because hose products vary considerably in size, weight, and length, it is not practical to establish definite recommendations for storage. Hose with a very light wall will not support as much load as hose with a heavier wall or hose with wire reinforcement. Hose shipped in coils or bales should be stored so that the coils are in a horizontal plane.

Whenever feasible, rubber hose products should be stored in their original shipping containers. This is especially true when the containers are wooden crates or cardboard cartons that provide some protection against the deteriorating effects of oils, solvents, and corrosive liquids. Shipping containers also afford some protection against deterioration caused by ozone and sunlight.

To prevent contamination and damage of stored hoses, the hose should be wrapped with burlap or other suitable material. Covering hoses in storage also helps prevent damage from sunlight and other environmental factors that are known to be harmful to rubber products.

Certain rodents and insects will damage rubber hose products. Adequate protection should be provided.

The ideal temperature for the storage of rubber products ranges from 50° to 70°F, with a maximum limit of 100°F. If stored below 32°F, some rubber products become stiff and require warming before being placed in service. Rubber products should not be stored near sources of heat, such as radiators or base heaters, nor should they be stored under conditions of high or low humidity.

To avoid the adverse effects of high ozone concentration, rubber hose products should not be stored near electrical equipment that may generate ozone. They also should not be stored for any lengthy period in geographical areas of known high ozone concentration. Exposure to direct or reflected sunlight, even through windows, should be avoided. In addition, uncovered hose should not be stored under fluorescent or mercury lamps, which generate light waves harmful to rubber.

Cotton-jacketed hose should be protected against fungal growths if the hose is to be stored for prolonged periods under conditions in which humidity exceeds 70%.

NOTE

Extreme care must be taken when flushing out a chemical hose with water. Some chemicals, such as concentrated acids, may react with the water and cause spattering. When flushing a hose, disposal of the fluid must be made in such a manner that environmental problems are not created.

Storage areas should be relatively cool and dark and free of dampness and mildew. Hose should be stored and used on a first-in, first-out basis. Even under the best of conditions, an unusually long shelf life can deteriorate certain rubber products.

Hose racks provide support to prevent bending, sagging, distortion, scratching, or marring of hose surfaces. Most racks have separate levels or compartments where different types and sizes of hose can be stored for ease of identification and selection. Hose racks are usually elevated to help avoid damage that might occur at floor level. Hose stored at floor level is easily damaged or contaminated by people, other materials, moisture, or equipment in the area.

Before placing a used chemical hose in storage, the hose must be completely drained and any potentially explosive vapors or corrosive residues flushed out.

Chemical hose should be stored so that air can circulate through it. It should be laid straight on a solid support. The hose should be stored in a cool, dark, dry place at a temperature less than 100°F.

Additional information on proper and safe methods for storing hose can be found in the Rubber Manufacturer's Association (RMA) Handbook - Hose.

23.0.0 ◆ PROPER AND SAFE METHODS FOR HANDLING HOSES

Hoses are made to strict specifications that can easily be degraded by careless handling. From the time the hose is delivered on a job site until installation is complete, proper and safe handling techniques will help ensure quality hose installation. There are a few basic guidelines for handling hose that will help you avoid damaging or contaminating the hose or injuring yourself or others.

- Do not subject hose to any form of abuse. A normal, reasonable amount of care should be exercised in handling hose.
- Never drag hose over sharp or abrasive surfaces unless the hose was specifically designed for such service.
- Care should be taken not to inflict severe end pull for which the hose and coupling assembly was not designed.
- Hose should neither be kinked nor run over by equipment.
- Never drag hose on the ground or floor. Dragging hose can cause damage to the ends or contaminate the hose.
- Keep hose away from any material that might contaminate it during handling.
- Keep the end caps on during handling or transporting of hose to avoid damage to ends and contamination.
- Wear gloves when handling hose to avoid scrapes and cuts from hose assemblies.
- If possible, coil long lengths of hose and transport it on a pallet using a fork truck.
- Be aware of people, materials, and equipment around you when handling or transporting hose so that you can avoid hitting anyone or anything with the hose. Hitting someone or an object with the hose may cause injury to another person, yourself, or damage to the hose, fittings, other material, or equipment.
- Dollies or fork trucks should be used whenever possible when handling large hose. Slings or handling rigs should be used to support heavy hose.

Additional information on proper and safe methods for handling hose can be found in the RMA Handbook - Hose.

24.0.0 ◆ REUSABLE FITTING INSTALLATION

There are several different types of reusable fittings. Each is installed in a particular way. In the following section, the procedure for installing a standard reusable fitting is described.

24.1.0 Standard Reusable Fitting Installation

The following procedure shows the general steps for installing a standard reusable hose fitting:

WARNING!
Follow all safety precautions.

Step 1 Select the proper hose and fitting for the application.

Step 2 Cut the hose off square with a fine tooth blade hacksaw. *Figure 91* shows cutting hose with hacksaw.

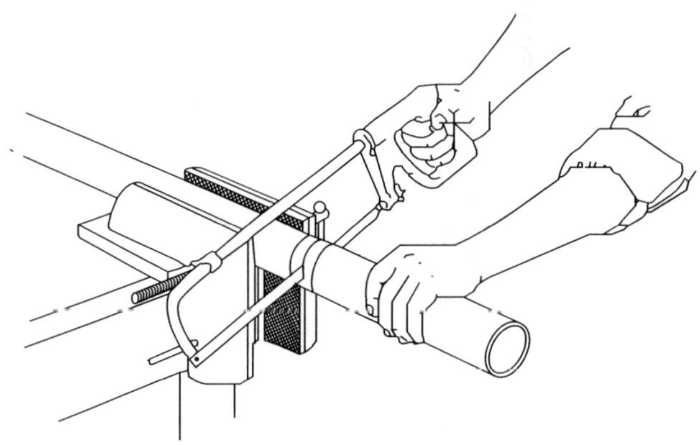

105F91.EPS

Figure 91 ◆ Cutting hose with hacksaw.

Step 3 Clean out the inside of the hose using compressed air.

WARNING!
Use safety goggles and point the hose down toward the floor when blowing it out.

Step 4 Put the socket in a vise.

Step 5 Liberally lubricate the hose cover with hose assembly lube or heavy weight oil.

NOTE
Read the vendor's manual for the proper lubricant to use. Heavy-weight oil cannot be used on some types of hose.

Step 6 Screw the hose counterclockwise into the socket until the hose bottoms on the shoulder of the socket (see *Figure 92*).

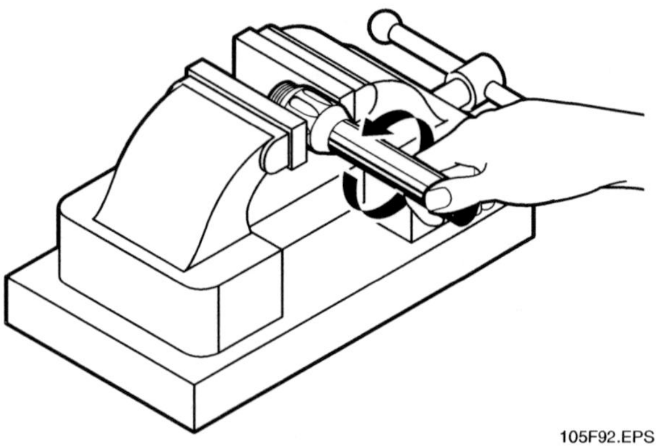

105F92.EPS

Figure 92 ◆ Inserting hose into socket.

NOTE
When assembling long lengths of hose, it may be preferable to put the hose in a vise just tight enough to prevent it from turning. The socket is then screwed onto the hose counterclockwise until the shoulder of the socket bottoms out on the hose.

Step 7 Liberally lubricate the nipple threads and the inside of the hose with a heavy weight oil or hose assembly lube (see *Figure 93*).

Step 8 Screw the nipple clockwise into socket and hose (see *Figure 94*). Leave about 1/32 inch to 1/16 inch clearance between the nipple hex and the socket.

Step 9 Clean the hose assembly by blowing it out with clean compressed air.

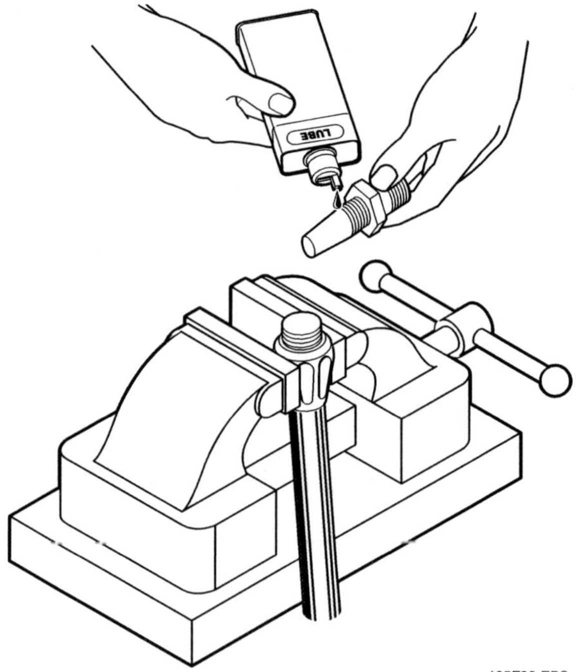

105F93.EPS

Figure 93 ◆ Lubricating nipple threads.

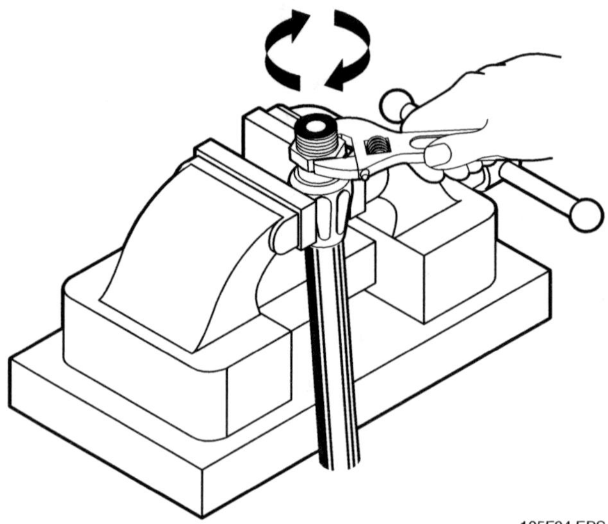

105F94.EPS

Figure 94 ◆ Inserting a nipple into a socket and hose.

NOTE
Some assemblies may be rinsed out with R66 mineral spirits if the tube stock (Teflon®, nitrile, neoprene) is compatible with oil. If not, use hot water at a maximum temperature of +180°F.

WARNING!
Use safety goggles and point the hose down toward the floor when blowing it out.

Summary

This module introduced the various types and applications of tubing and tubing fittings, how to properly handle and store tubing, and procedures for cutting, deburring, bending, flaring, and joining tubing.

This module also introduced threaded pipe systems. Threaded pipe is a versatile, widely used method of joining pipe that is relatively simple to install. All pipe is threaded according to national standards. The tapered pipe thread is the most common type of thread used on pipe and fittings. The threaded pipe system requires a sealant and proper assembly techniques to ensure a tight, leakproof joint.

You also learned about the various types and applications of hose and hose fittings, how to properly handle and store hoses, and procedures for cutting hose and installing a reusable hose fitting.

Tubing installation, threaded pipe fabrication, and hose system installation are important parts of a pipeline mechanical technician's work. The knowledge and skills presented in this module, along with practice, study, and on-the-job experience will help you to install and maintain all of these systems efficiently.

Review Questions

1. Tubing is generally sized according to its _____.

a. outside diameter and wall thickness
b. ASTM identification number
c. ASME identification number
d. outside and inside diameters

2. The wall thickness of tubing is normally expressed _____.

a. in fractions of an inch such as ½, ¼, or ¾ inch
b. in decimal units such as .034, .047, or .109 inches
c. by ASTM identification code such as ASTM A249, ASTM A632, or ASTM B68
d. in nominal pipe sizes

3. To make tubing easier to bend, the metal is _____.

a. annealed
b. flared
c. extruded
d. welded

4. Stainless steel tubing is often used in _____ applications.

a. high-pressure and high-temperature
b. noncorrosive environment
c. air conditioning and refrigeration
d. pneumatic signal line

5. When storing tubing, it is good work practice to _____.

a. crimp the ends of all types of tubing
b. place the tubing in a sunny area so it does not get damp and corrode
c. lay the tubing on the floor so it does not fall on anyone causing personal injury
d. keep the same size and types of tubing together

6. When unrolling coils of soft tubing, it is good work practice to _____.

a. twist the tubing as you unroll it to help keep it straight
b. set a heavy object on one end of the tubing while unrolling the coil in a smooth motion
c. unroll only as much tubing as you need
d. unroll the entire coil, cut the amount needed, and then roll up the remaining tubing for future use

7. Tubing snips are often used to cut _____ tubing.

a. soft copper
b. aluminum
c. stainless steel
d. poly or plastic

8. A _____ is often used to bend heavy-walled tubing.

a. spring tube bender
b. compression-type hand tubing bender
c. hydraulic tubing bender
d. tubing stretcher-type bender

9. A _____ fitting may have one or two ferrules that pinch the tubing as the locknut is tightened on the body of the fitting.
   a. flared
   b. compression
   c. butt-welded
   d. socket-welded

10. A _____ fitting is designed to change the run from a larger size to a smaller size of tubing or piping.
   a. tee
   b. cap
   c. reducer
   d. drain

11. For special applications that require the maximum resistance to corrosion, _____ pipe is used.
   a. copper
   b. carbon steel
   c. galvanized
   d. stainless steel

12. The strongest and best pipe fittings are _____.
   a. galvanized
   b. carbon steel
   c. plain
   d. banded

13. To join pipes of different sizes, _____ and reducers are used.
   a. couplings
   b. unions
   c. bushings
   d. wyes

14. The National Pipe Thread has an internal and external thread taper of _____ inch.
   a. $\frac{1}{16}$
   b. $\frac{1}{10}$
   c. $\frac{1}{12}$
   d. $\frac{1}{8}$

15. The distance from the end of the pipe to the center of the fitting is called the _____.
   a. run
   b. specifications
   c. pitch
   d. takeout

16. Never use _____ on pipe that is to be welded or on pipe that carries steam because it emits a toxic gas.
   a. pipe joint compound
   b. Teflon® tape
   c. pipe dope
   d. graphite paste

17. Surge pressures _____ must be taken into account when selecting the proper hose and fittings.
   a. lower than the maximum operating pressure
   b. higher than the minimum operating pressure
   c. lower than the minimum operating pressure
   d. higher than the maximum operating pressure

18. Dash size is calculated by converting the ID size of the hose into _____ of an inch.
   a. 8ths
   b. 16ths
   c. 4ths
   d. 32nds

19. If a hose has a maximum rated working pressure of 400 psi, it should not rupture until at least _____ psi has been reached.
   a. 2,000
   b. 1,600
   c. 1,000
   d. 800

20. Hoses made from _____ are used in higher temperature applications ranging up to 600° F.
   a. natural rubber
   b. silicone rubber
   c. vinyl nitrile
   d. nylon

21. The inner section or core of a hose is called the _____.
   a. tube
   b. carcass
   c. braid
   d. reinforcement

22. _____ strengthens the hose along its length, giving it resistance to expansion and contraction.
   a. Vertical braiding
   b. Circumference braiding
   c. Parallel braiding
   d. Horizontal braiding

23. _____ fittings are used only for general, low-pressure air and water service.

a. Push-on
b. Swivel joint
c. Crimped
d. Flanged

24. When transporting hose, it is good work practice to _____.

a. transport only one hose at a time so it is easy to identify
b. coil long lengths of hose and transport it on a pallet using a fork truck
c. fill it with water to keep it from kinking
d. cover the hose with a tarpaulin and drag it along the floor

25. Quick acting couplings are often used for _____ service.

a. low-pressure
b. high-pressure
c. low-to-medium pressure
d. medium-to-high pressure

GLOSSARY

Trade Terms Introduced in This Module

Braid: The reinforcement material put over the hose.

Branch: The outlet or inlet of a fitting not in line with the run but which may make any angle.

Bushing: A pipe fitting with external threading large than its internal threading. It allows two different size pipes to be joined.

Cap: A tubing fitting used with a compression nut to close off the end of tubing.

Close nipple: A very short piece of pipe that is threaded along its entire length with no shoulder.

Coupling: A metal collar with inside threads used to join threaded pipes.

Crimping: A method used for attaching fittings to a hose. It is done by pressing a set of dies against the outside of the fitting and compressing the shell around the hose.

Cross: A pipe fitting with four branches all at right angles to each other.

Elbow: A fitting that makes an angle between adjacent pipes. An elbow is always 90 degrees unless another angle is stated. Also known as an *ell*.

Fitting: A device, such as an elbow, tee, or union, used with tubing to connect tubing to tubing, tubing to instruments, tubing to piping; to close off tubing; and/or to branch tubing.

Flanges: Projecting rims or edges with drilled holes for connecting to other flanged fittings.

Flare: A tubing end that has been spread open to properly fit a flared fitting.

Galling: Deformity of the threads in which the threads are stripped.

Inside diameter (ID): The distance across the middle of the hole in the tubing from inside wall to the opposite inside wall.

Kinking: A temporary or permanent distortion of a hose induced by winding or doubling upon itself.

MIL: An abbreviation used to indicate a standard developed by the military. The military issues standards relating to the use of hoses.

Nipple: A pipe section inserted in a hose.

NPT: Designates National Pipe Thread, which is the United States' standard for pipe threads. This thread has a 1⁄16-inch taper per inch from back to front.

Outside diameter (OD): The distance from one outside wall of tubing across the middle of the opening to the other outside wall.

Plug: A tubing fitting used to close off a part of another fitting.

Reamer: A device used to remove burrs and ridges from the inside of tubing.

Reducer: A pipe fitting with female threads on each end, with one end being one or more sizes smaller than the other end.

Reinforcement: The material put over the tube of a hose to provide resistance to pressure, either from the inside or outside.

Run: A length of pipe made up of more than one length of pipe.

Tee: A fitting that has one side outlet 90 degrees to the run.

Teflon® tape: Tape made of Teflon® that is wrapped around the male threads of a pipe before it is screwed into a fitting to serve as a lubricant and sealant.

TFE Teflon®: Teflon® is the DuPont Company's registered trademark for its TFE (tetrafluoroethylene) fluorocarbon resins. Teflon® is used in highly corrosive environments to protect component parts.

Tubing bender: A device used to bend tubing without *kinking* or *crimping*. Tubing benders may be hand-operated, bench- or table-mounted, or hydraulically powered.

Tubing cutter: A device with a cutting wheel(s) and rollers that is rotated around tubing to cut it.

Union: A coupling device that allows pipes to be connected without being rotated. The mating surfaces are pulled together by a flanged, threaded collar on the union.

Wall thickness: Half of the difference between the outside diameter and the inside diameter of tubing.

Wye: A fitting that has a side outlet of 45 degrees to the run; also referred to as a lateral.

CONTREN™ LEARNING SERIES — USER UPDATES

The NCCER makes every effort to keep these textbooks up-to-date and free of technical errors. We appreciate your help in this process. If you have an idea for improving this textbook, or if you find an error, a typographical mistake, or an inaccuracy in NCCER's Contren™ textbooks, please write us, using this form or a photocopy. Be sure to include the exact module number, page number, a detailed description, and the correction, if applicable. Your input will be brought to the attention of the Technical Review Committee. Thank you for your assistance.

Instructors – If you found that additional materials were necessary in order to teach this module effectively, please let us know so that we may include them in the Equipment/Materials list in the Instructor's Guide.

Write: Curriculum Revision and Development Department
National Center for Construction Education and Research
P.O. Box 141104, Gainesville, FL 32614-1104

Fax: 352-334-0932

E-mail: curriculum@nccer.org

Craft ______________________ Module Name ______________________

Copyright Date ______________ Module Number ______________ Page Number(s) ______________

Description

____

____

____

____

(Optional) Correction

____

____

____

(Optional) Your Name and Address

____

____

____

Module 63106-02

Fasteners

Introduction to the Pipeline Operator Qualification Rule, API, and NCCER

It could be said that petroleum products keep America running. Petroleum products like oil and natural gas are the principal sources of energy in the United States and are used in many manufacturing and chemical processes. Petroleum products are primarily transported by pipeline. There are more than a million miles of liquid and natural gas pipeline in this country, ranging in diameter from two inches to four feet. These pipelines bring crude oil and natural gas from production fields to processing plants and refineries—and ultimately to the consumer.

Pipeline Operator Qualification

Petroleum products are considered hazardous, so pipeline personnel must take extra precautions and follow specific procedures when working with them. To help ensure the safe handling and transportation of petroleum products, the United States Department of Transportation (DOT) mandates safe practices as federal regulations.

One such regulation, the Operator Qualification (OQ) rule, requires pipeline operators to develop a written qualification plan to evaluate their personnel and their contract personnel (see *49 CFR Part 192 Subpart N—Gas,* and *49 CFR Part 195 Subpart G—Liquid*). The evaluation hinges on the operator's ability to perform certain covered tasks and to be able to recognize and respond to abnormal operating conditions that may be encountered while performing these covered tasks.

Covered Tasks

A pipeline performance task is deemed *covered* if the activity passes the following four-part test:

1. Is it performed on a pipeline facility?
2. Is it an operations or maintenance task?
3. Is it performed as a requirement of *49 CFR Part 195 (Liquid)* or *49 CFR Part 192 (Gas)*?
4. Does it affect the operation or integrity of the pipeline facility?

If the answer to all four questions is *yes,* then the activity is considered *covered* and must figure into the evaluation process of pipeline personnel.

American Petroleum Institute

The OQ rule and its related covered tasks provided the catalyst to develop a unified pipeline training and assessment program—so that pipeline operators and owners have ready access to the assessment and training tools they need. This effort was the direct result of the vision and leadership provided by The American Petroleum Institute (API) and API members.

Efforts to unify the industry are nothing new to API. Indeed, the impetus for forming the Institute in 1919 was the need to standardize engineering specifications for drilling and production equipment. Today, the most pressing issues revolve around public perceptions and government policies toward the industry. Speaking with one voice on these issues has become as essential as having interchangeable parts in the field. API's members determine the petroleum industry's positions on public policy as well as the standards that govern its day-to-day operations worldwide.

Consortium on Operator Qualification

API assembled a focus group whose primary purpose was to identify covered tasks and provide compliance recommendations to pipeline operators.

The Consortium on Operator Qualification (COOQ) became the guiding force for the identification and development of covered tasks for the liquid pipeline industry. Their work product, API publication 1161, *Guidance Document for the Qualification of Liquid Pipeline Personnel,* formed the basis for the resulting Pipeliner Training and Assessment Program.

The covered task list created in API publication 1161 has now evolved to include gas covered tasks and is maintained by NCCER. The most current covered task list can be downloaded at www.nccer.org.

National Center for Construction Education and Research

With the guidance and support of API, its members, and the Consortium on Operator Qualification, the National Center for Construction Education and Research (NCCER) was able to design and implement the Pipeliner Training and Assessment Program.

NCCER is a not-for-profit 501(c)(3) education foundation established in 1995 by the world's largest and most progressive construction and maintenance contractors and national contractor associations. NCCER was founded to address the

severe workforce shortage facing the construction and maintenance industries and to develop a standardized training process and curricula. Today, NCCER is supported by hundreds of leading construction and maintenance companies, manufacturers, and national associations and developed more than 35 construction and maintenance titles.

NCCER also maintains a National Registry, which contains the records of trainees and instructors who have received NCCER training. All completion information for every trainee is sent to the NCCER and filed within the Registry. The Registry can then confirm training and skills for workers as they move from company to company, state to state, or even within their own company.

Standardized Curriculum (NCCER's Contren™ Learning Series)

NCCER curricula are developed using the expertise of industry subject matter experts (SMEs). The use of SMEs as an integral part of every new development or revision ensures that the curriculum is current, industry-driven, and indicative of industry practices. The development and/or revision process is facilitated from start to finish by NCCER project managers, who work closely with the SMEs and a select team of technical writer(s) on the development and presentation of the curriculum.

Pipeline: Five Training Paths

The pipeline curriculum focuses on essential skills and competencies within five skill areas:

- Corrosion Control
- Electrical and Instrumentation
- Maintenance
- Mechanical
- Operations (Field and Control Center)

SMEs from the pipeline industry have determined the skills and competencies needed based on API's covered tasks and the consensus of the group/industry. Specific company information is generally not incorporated into the curriculum unless the SMEs determine that the information is the industry standard.

Training Modules and Covered Tasks

The smallest unit of instruction in the curriculum is a *module*. Each curriculum is composed of modules (the equivalent of stand-alone chapters) that are logically sequenced into complete *levels* of instruction. Every module is task-oriented and requires the successful completion of a written (knowledge-based) examination and a performance (hands-on) test. Modules may be individually selected and/or combined for efficient task training. Modules in the pipeline curricula, to the extent possible, will align with API-developed covered tasks. While not all modules will relate to covered tasks, all covered tasks will relate to at least one module.

Benefits of Training to the Pipeline Marketplace

Applying NCCER's standardized training process allows pipeline owners and contractors in the United States to establish the benchmark for acceptable skill and competency levels within the pipeline industry. This process will assist training organizations in developing and/or improving standardized pipeline training programs that effectively match their resources and operating conditions while responding to regulatory demands and to the needs of the industry. The process will also recognize training organizations that meet the minimum standards of competence, warranting regulatory, public, and professional confidence. This in turn will foster national unity within the pipeline industry for the mutual benefit of individual technicians, owners, contractors, and ultimately pipeline-users and society.

Additional Resources

To review the Pipeliner Training and Assessment Program Implementation and Resource Guide and the latest version of pipeline covered tasks, visit NCCER's Web site at www.nccer.org, or contact:

National Center for Construction Education and Research
PO Box 141104
Gainesville, FL 32614-1104
(888) 622-3720
Web site: www.nccer.org

For more information on API, please contact:

American Petroleum Institute
1220 L Street, Northwest
Washington, D.C. 20005
(202) 682-8000
Web site: www.api.org

To order *API Pub 1161*, contact Global Engineering Documents by calling 1-800-854-7149 or visiting their Web site at www.global.his.com.

Additional sources of information on the pipeline operator qualification regulation include:

DOT – Office of Pipeline Safety
www.ops.dot.gov

Transportation Safety Institute
www.tsi.dot.gov

COURSE MAP

This course map shows all of the modules in the first level of the Pipeline Mechanical curriculum. The suggested training order begins at the bottom and proceeds up. Skill levels increase as you advance on the course map. The local Training Program Sponsor may adjust the training order.

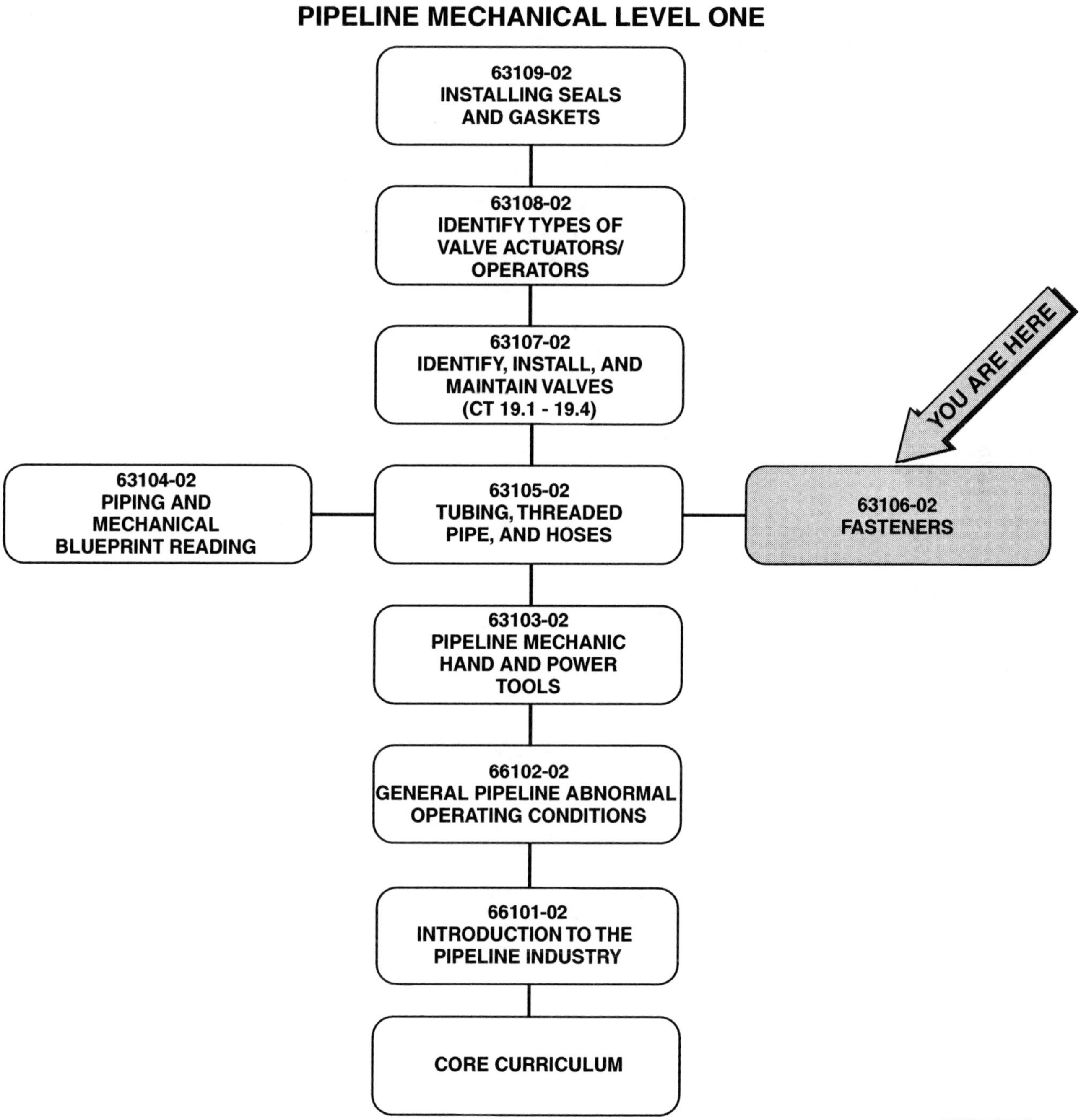

106CMAP.EPS

MODULE 63106-02 CONTENTS

Figures

Tables

MODULE 63106-02

Fasteners

Objectives

When you have completed this module, you will be able to do the following:

1. Identify and explain threaded and nonthreaded fasteners.
2. Identify and explain insulation fasteners.
3. Install fasteners.

Prerequisites

Before you begin this module, it is recommended that you successfully complete the following: Core Curriculum; Pipeline Core, Modules 66101 and 66102; Pipeline Mechanical Level One, Module 63103.

Required Trainee Materials

1. Paper and pencil
2. Appropriate personal protective equipment

1.0.0 ◆ INTRODUCTION

Fasteners are used by pipeline mechanical technicians to assemble and install many kinds of parts and equipment. Fasteners are also used to attach insulation to pipes, vessels, and equipment. Fasteners may include screws, bolts, pins, and retainers. A skilled pipeline mechanical technician should be familiar with many different types of fasteners and be able to select and use the proper type for each job.

2.0.0 ◆ THREADED FASTENERS

Threaded fasteners are the most commonly used fasteners in maintenance work. Many of these are assembled with nuts and washers, while others are installed into threaded holes. There are many types of threaded fasteners (see *Figure 1*).

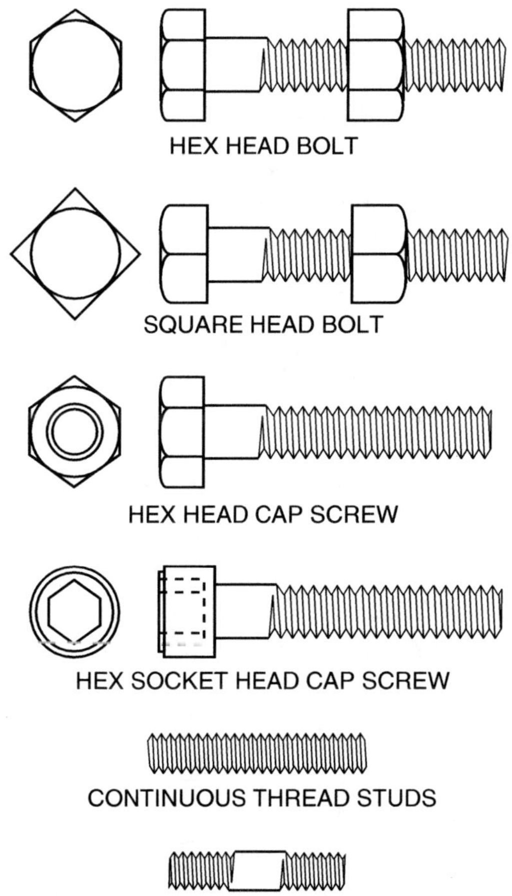

Figure 1 ◆ Common threaded fasteners.

The following three qualities should be considered when selecting threaded fasteners to do a job:

- Thread type
- Fastener grade
- Fastener type

2.1.0 Thread Type

Choosing the proper thread type is important when selecting a fastener because the fastener must be suited to the job requirements. The following five qualities should be considered when selecting thread type:

- Thread standards
- Thread series
- Thread class
- Thread designation
- Thread design

2.1.1 Thread Standards

Fastener threads are made to established standards. The most common standard is the unified, or American, standard. Unified standards are used to establish thread series and classes.

2.1.2 Thread Series

Unified standards are established for three series of threads, depending on the number of threads per inch for a certain diameter fastener. These three series and their abbreviations are as follows:

- *Unified National Coarse Thread* (*UNC*) – Used for bolts, screws, nuts, and other general purposes. Fasteners with UNC threads are used for rapid assembly or disassembly of parts where corrosion or slight damage may occur.
- *Unified National Fine Thread* (*UNF*) – Used for bolts, screws, nuts, and other purposes where a finer thread than UNC is required.
- *Unified National Extra Fine Thread* (*UNEF*) – Used for thin-walled tubes, nuts, ferrules, and couplings.

2.1.3 Thread Class

Thread classes are also established by unified standards. Classes 1A, 2A, and 3A apply to external threads (bolts) only. Classes 1B, 2B, and 3B apply to internal threads (nuts) only. Thread classes are distinguished from each other by the amount of **tolerance** they provide. Classes 3A and 3B provide a minimum clearance; classes 1A and 1B provide a maximum clearance.

The most commonly used classes are 2A and 2B. Classes 3A and 3B are used when close tolerances are required. Classes 1A and 1B are used when a large tolerance for assembly over dirty or damaged threads and quick assembly are needed.

2.1.4 Thread Designation

Bolt and screw threads are designated by a standard method. The following are the standard designations for bolt and screw threads:

- *Nominal size* – Diameter
- *Number of threads per inch* – Standard for different diameters
- *Thread series symbol* – The unified standard thread used (UNC, UNF, or UNEF)
- *Thread class symbol* – The closeness of fit between the bolt threads and nut threads
- *Left-hand thread symbol* – Specified by the symbol LH. (Unless specified LH, all other threads are right-hand.)

Figure 2 shows the thread designation for a common fastener.

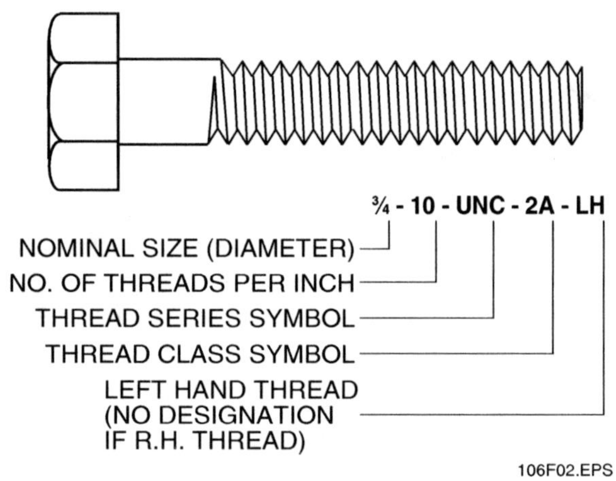

Figure 2 ◆ Thread designation for a common fastener.

2.1.5 Thread Design

There are several types of thread design. The thread design selected depends on the fastener purpose.

Power transmission threads are special threads that are used to move machine parts for adjusting, setting, and transmitting power. Three common transmission threads are:

- *Square thread* – The strongest and most useful of all thread forms. It is also the most difficult to make because its parallel sides must be accurately machined (see *Figure 3*).

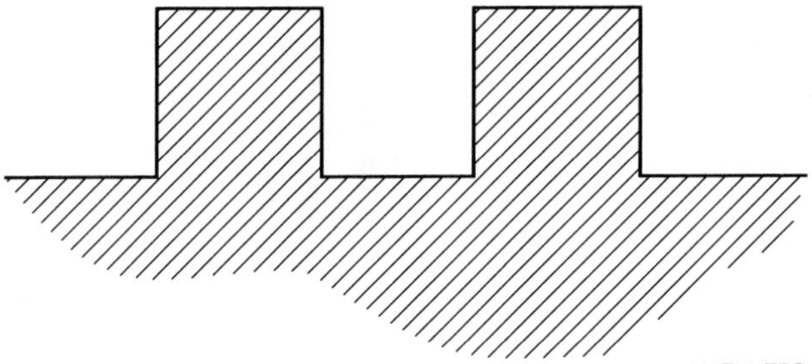

106F03.EPS

Figure 3 ◆ Enlargement of a square thread.

- *Acme thread* – Has replaced the square thread in most cases because it is easier to machine (see *Figure 4*).

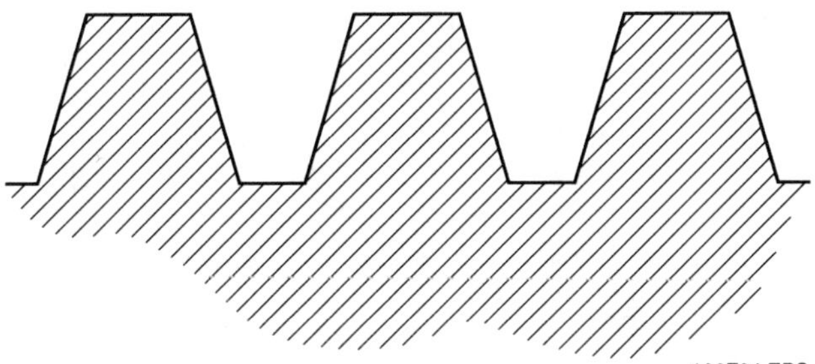

106F04.EPS

Figure 4 ◆ Enlargement of an acme thread.

- *Buttress thread* – Has one side of the thread cut square and the other side cut at a slant. It has great strength along the thread axis in one direction only. This thread form is used to screw thin, tubular parts together (see *Figure 5*).

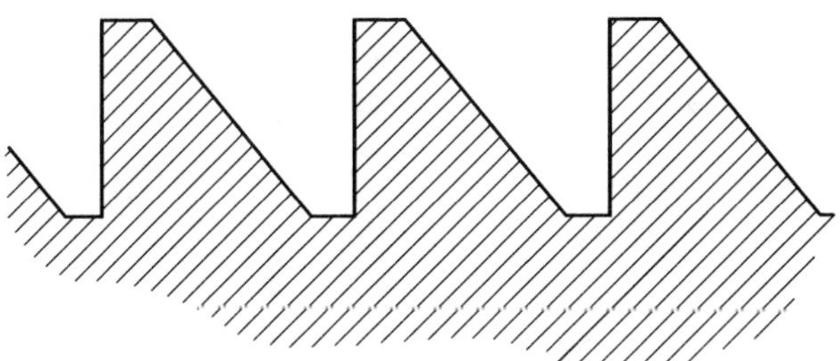

106F05.EPS

Figure 5 ◆ Enlargement of a buttress thread.

2.2.0 Fastener Grade

The strength and quality of a fastener can be determined by special grade markings on the head of the fastener. These markings are standardized by the Society of Automotive Engineers (SAE) and the American Society for Testing of Materials (ASTM). Grade markings are sometimes called *line markings*. *Table 1* shows the ASTM and SAE markings for steel bolts and screws.

> **NOTE**
> There are many counterfeit bolts on the market that do not have the ASTM grade markings and may be inferior bolts. Do not use bolts that do not have the grade markings.

2.3.0 Fastener Type

Threaded fasteners are made in different shapes and sizes. They are identified by the type of head used or other special characteristics. The following sections explain these types of fasteners:

- Machine screws
- Machine bolts
- Cap screws
- Setscrews
- Stud bolts
- Thread-forming screws
- Thread-cutting screws
- Drive screws
- Eyebolts
- Toggle bolts
- Anchor bolts
- Self-tapping bolts
- Nuts
- Jam nuts
- Castellated, slotted, and self-locking nuts
- Acorn nuts
- Wing nuts
- Inserts
- Flat washers
- Lock washers

Table 1 ASTM and SAE Markings for Steel Bolts and Screws

Grade Marking	Specification	Material
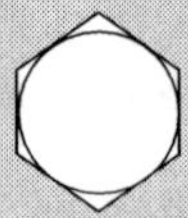	SAE-Grade 0	Steel
	SAE-Grade 1 ASTM-A 307	Low-Carbon Steel
	SAE-Grade 2	Low-Carbon Steel
	SAE-Grade 3	Medium-Carbon Steel, Cold Worked
	SAE-Grade 5 ASTM-A 449	Medium-Carbon Steel, Quenched and Tempered
	ASTM-A 325	Medium-Carbon Steel, Quenched and Tempered
	ASTM-A 354 Grade BB	Low-Alloy Steel, Quenched and Tempered
	ASTM-A 354 Grade BC	Low-Alloy Steel, Quenched and Tempered
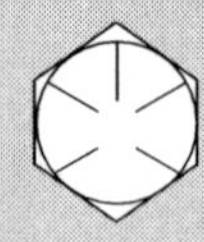	SAE-Grade 7	Medium-Carbon Alloy Steel, Quenched and Tempered, Roll Threaded After Heat Treatment
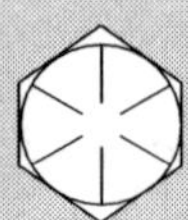	SAE-Grade 8	Medium-Carbon Alloy Steel, Quenched and Tempered
	ASTM-A 354 Grade BD	Alloy Steel, Quenched and Tempered
	ASTM-A 490	Alloy Steel, Quenched and Tempered

ASTM Specifications
- A 307 - Low-carbon steel, externally and internally threaded standard fasteners
- A 325 - High-strength steel bolts for structural steel joints, including suitable nuts and plain hardened washers
- A 449 - Quenched and tempered steel bolts and studs
- A 354 - Quenched and tempered alloy steel bolts and studs with suitable nuts
- A 490 - High-strength alloy steel bolts for structural steel joints, including suitable nuts and plain hardened washers

SAE Specifications
- J 429 - Mechanical and quality requirements for threaded fasteners

106T01.EPS

2.3.1 Machine Screws

Machine screws are used in general assembly. They may have slotted or recessed heads. Machine screws are available in diameters from #0 (0.060 inch) to ½ or 0.500 inch and in lengths from ⅛ inch to 3 inches. Metric sizes are also made. Nuts for machine screws (square or hexagonal) are usually sold separately. *Figure 6* shows different types of machine screws.

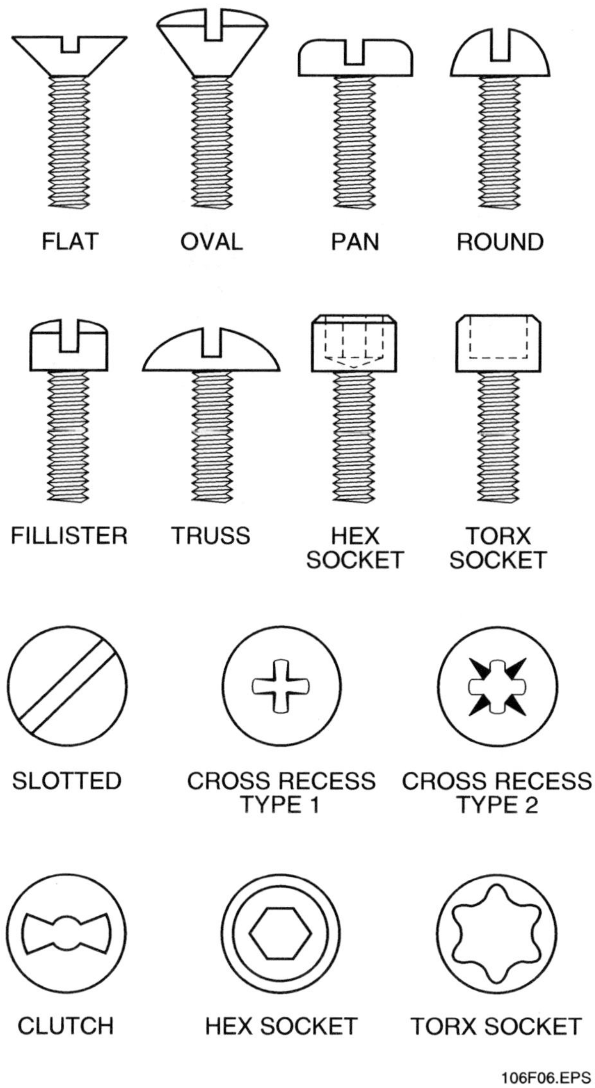

Figure 6 ◆ Machine screws.

2.3.2 Machine Bolts

Machine bolts are used to assemble parts that do not require close tolerances. They are made with square and hexagonal heads. The diameters range from ¼ inch to 3 inches and the lengths from ½ inch to 30 inches. The nuts are similar in shape to the bolt heads. They are usually furnished along with the machine bolts (see *Figure 7*).

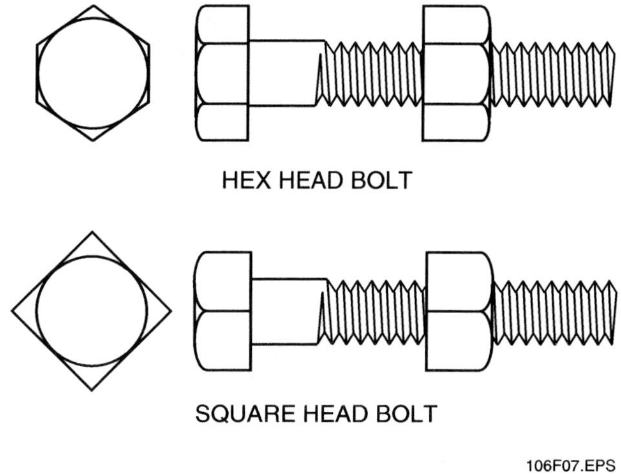

Figure 7 ◆ Machine bolts.

2.3.3 Cap Screws

Cap screws are used on high-quality assemblies requiring a finished appearance. The cap screw passes through a clearance hole in one of the assembly parts and screws into a threaded hole in the other part. The clamping action occurs when the cap screw is tightened.

Cap screws are made to close tolerances and are provided with a machined or semifinished bearing surface under the head. They are stocked in coarse and fine thread series and in diameters from ¼ inch to 2 inches. Lengths from ⅜ inch to 10 inches are available. Cap screws are also made in metric sizes (see *Figure 8*).

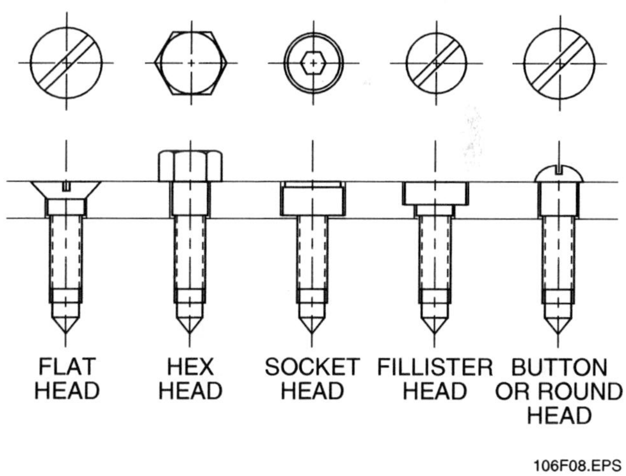

Figure 8 ◆ Cap screws.

2.3.4 Setscrews

Setscrews are usually made of heat-treated steel. They are used to prevent pulleys from slipping on shafts, to hold collars in place on shafts, and to hold shafts in place.

Setscrews are classified by head styles and point styles (see *Figure 9*).

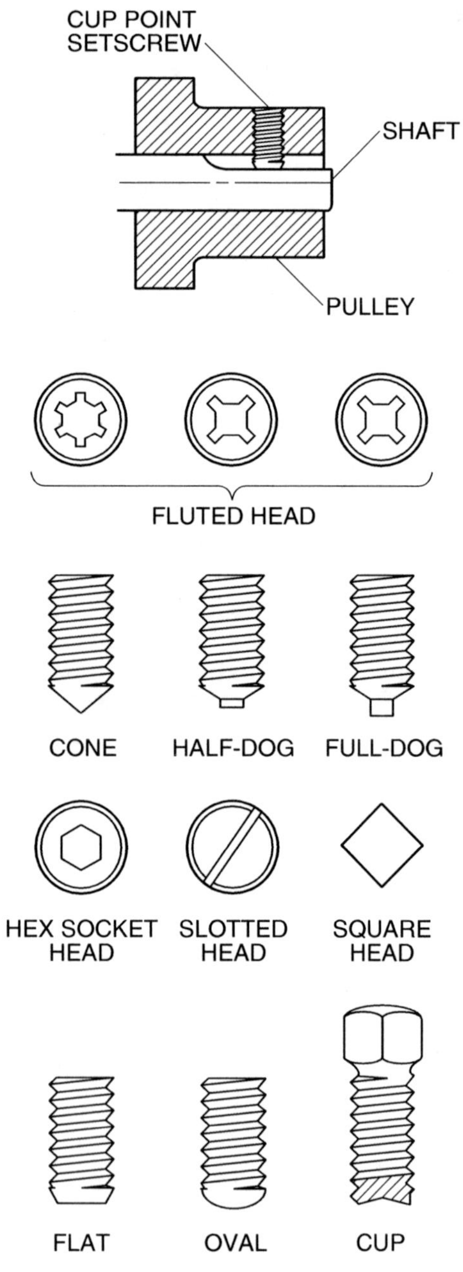

Figure 9 ◆ Several setscrew heads and point styles.

2.3.5 Stud Bolts

Stud bolts are headless bolts threaded either the entire length or on both ends (see *Figure 10*). One end is screwed into a tapped hole. The part to be clamped is fitted over the stud, and a nut is screwed on to clamp the two parts together. Some inspection covers and bearing caps are held together by stud bolts.

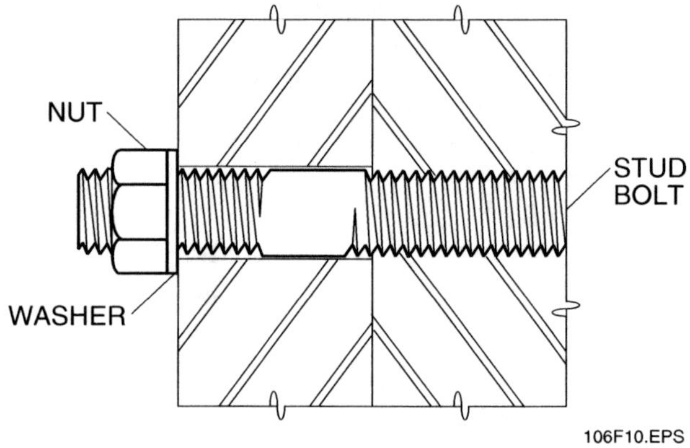

Figure 10 ◆ Stud bolt.

2.3.6 Thread-Forming Screws

Thread-forming screws form a thread as they are driven (see *Figure 11*). This eliminates tapping. Some thread-forming screws drill their own holes, which eliminates drilling, punching, and aligning of parts. These screws are used mainly to fasten light-gauge metal parts.

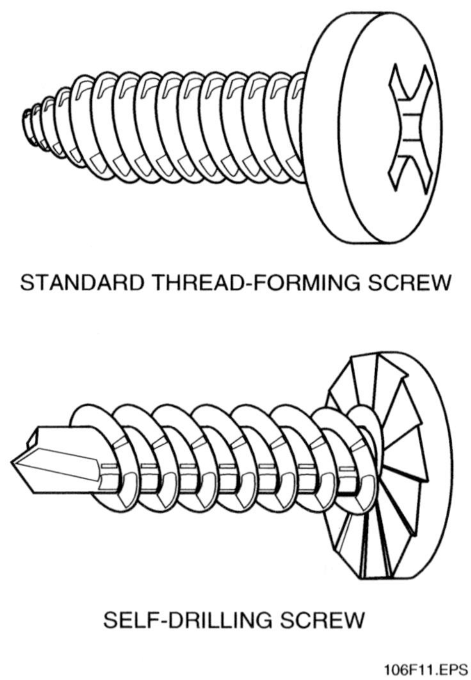

Figure 11 ◆ Thread-forming screws.

2.3.7 Thread-Cutting Screws

Thread-cutting screws actually cut threads into the metal as they are driven into a pilot hole (see *Figure 12*). They are made of hardened steel and used to join heavy-gauge sheet metal and **nonferrous** metal parts.

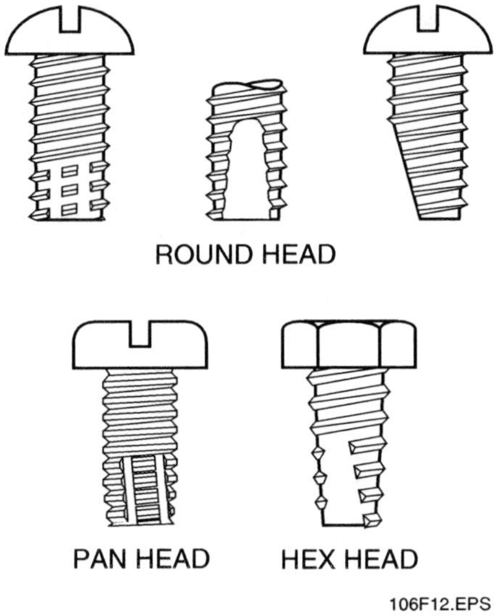

Figure 12 ◆ Thread-cutting screws.

2.3.8 Drive Screws

Drive screws (see *Figure 13*) are hammered into a drilled or punched hole of the proper size. They are mostly used to attach permanent name plates to electric motors and other types of equipment.

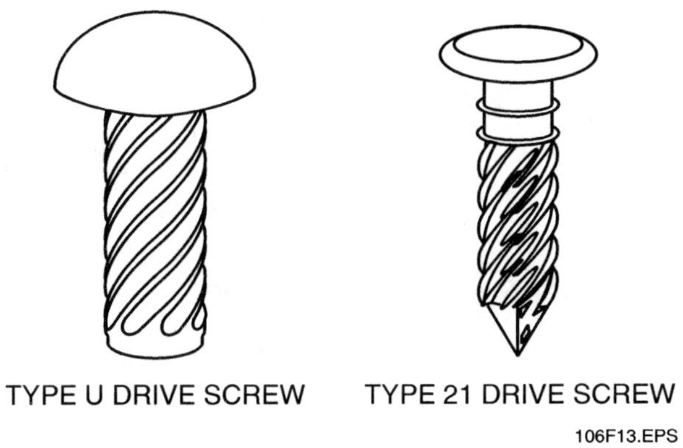

Figure 13 ◆ Drive screws.

2.3.9 Eyebolts

Eyebolts are made with an eye, or loop, at one end and threads on the other end. There are many types of eyebolts (see *Figure 14*). Some have a formed and welded eye, and others are forged. They are commonly used as lifting devices and guides for wires, cables, and cords.

2.3.10 Toggle Bolts

Toggle bolts are used to fasten a part to a hollow wall or panel. *Figure 15* shows three types of toggle bolts.

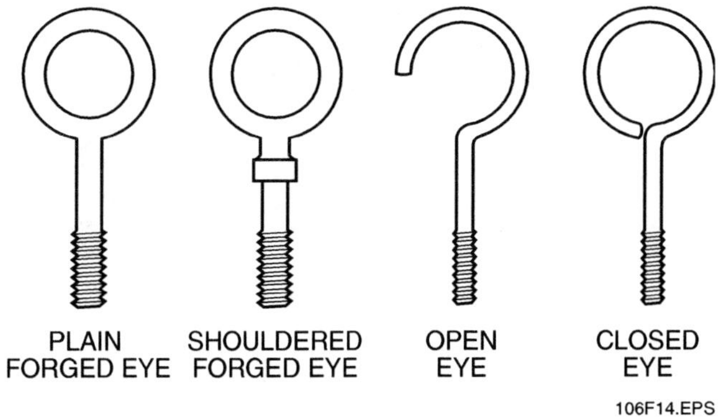

Figure 14 ◆ Eyebolts.

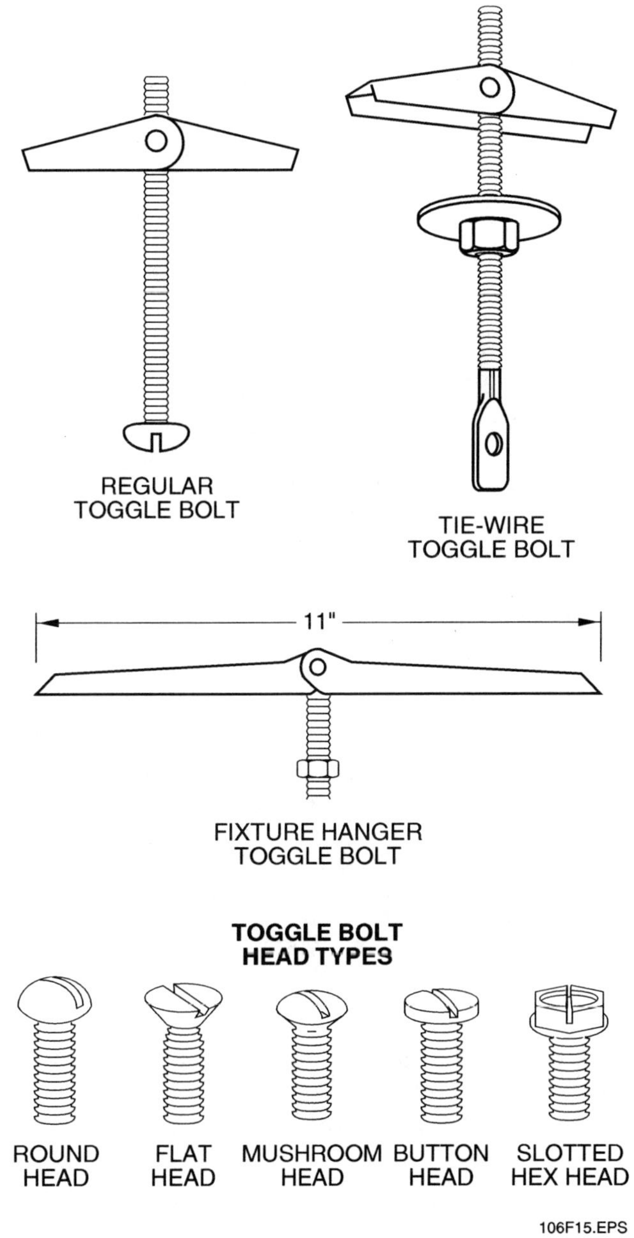

Figure 15 ◆ Three types of toggle bolts.

2.3.11 Anchor Bolts

Anchor bolts are used to fasten parts, machines, and equipment to concrete or masonry foundations, floors, and walls. There are several types for different uses (see *Figure 16*).

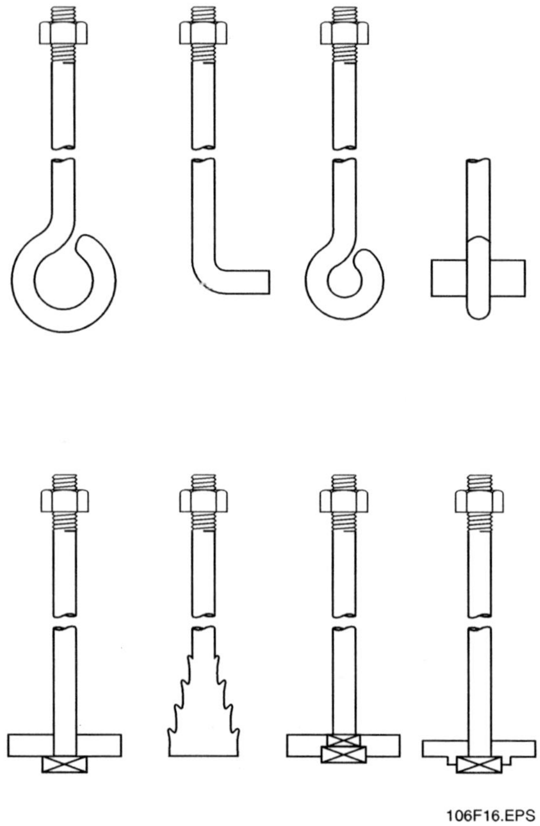

Figure 16 ◆ Anchor bolts for installation.

Expansion anchor bolts are installed in concrete after it has hardened. These bolts are made in different sizes with different heads. *Figure 17* shows an expansion anchor bolt. When the nut is tightened, the bolt base expands to provide the holding force.

2.3.12 Self-Tapping Bolts

Self-tapping bolts are used to fasten parts to concrete, block, and brick. A special drill bit is used to drill the hole for the bolt. The bolt is then screwed into the hole, forming its own locking threads. *Figure 18* shows two types of self-tapping bolts.

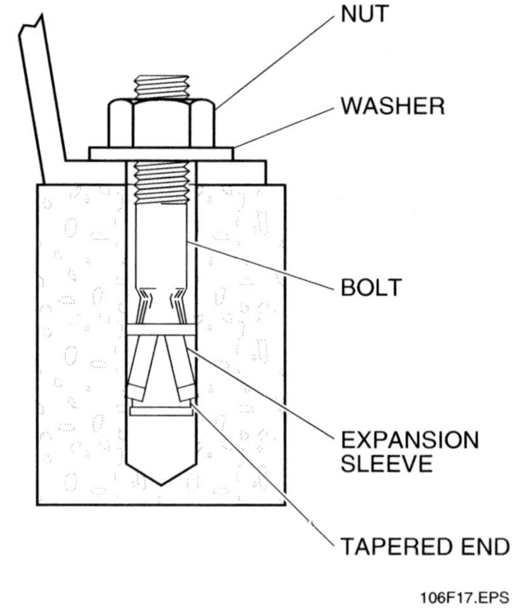

Figure 17 ◆ Expansion anchor bolt.

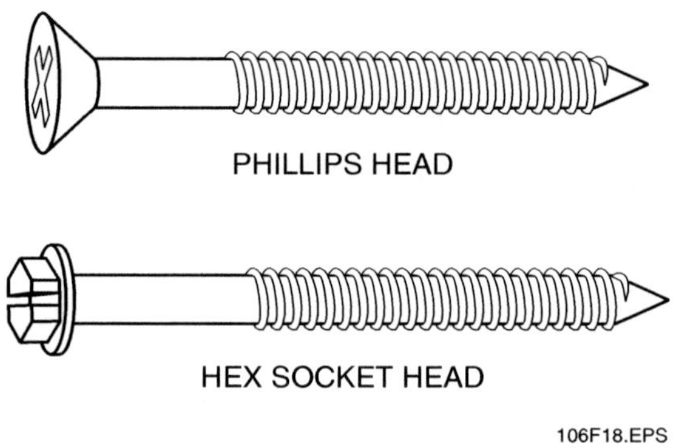

Figure 18 ◆ Two types of self-tapping bolts.

2.3.13 Nuts

For most threaded fasteners, nuts have hex (hexagonal) or square shapes and are used with bolts having the same shaped head. *Figure 19* shows different types of nuts for threaded fasteners.

The standard nut may be regular, semifinished, or finished (see *Figure 20*). Regular nuts are machined only on the threads. Semifinished nuts are machined on the bearing face to make a truer surface for the washer. Finished nuts are the same as semifinished nuts, but they are made to closer tolerances. The standard machine screw nut is regular finished.

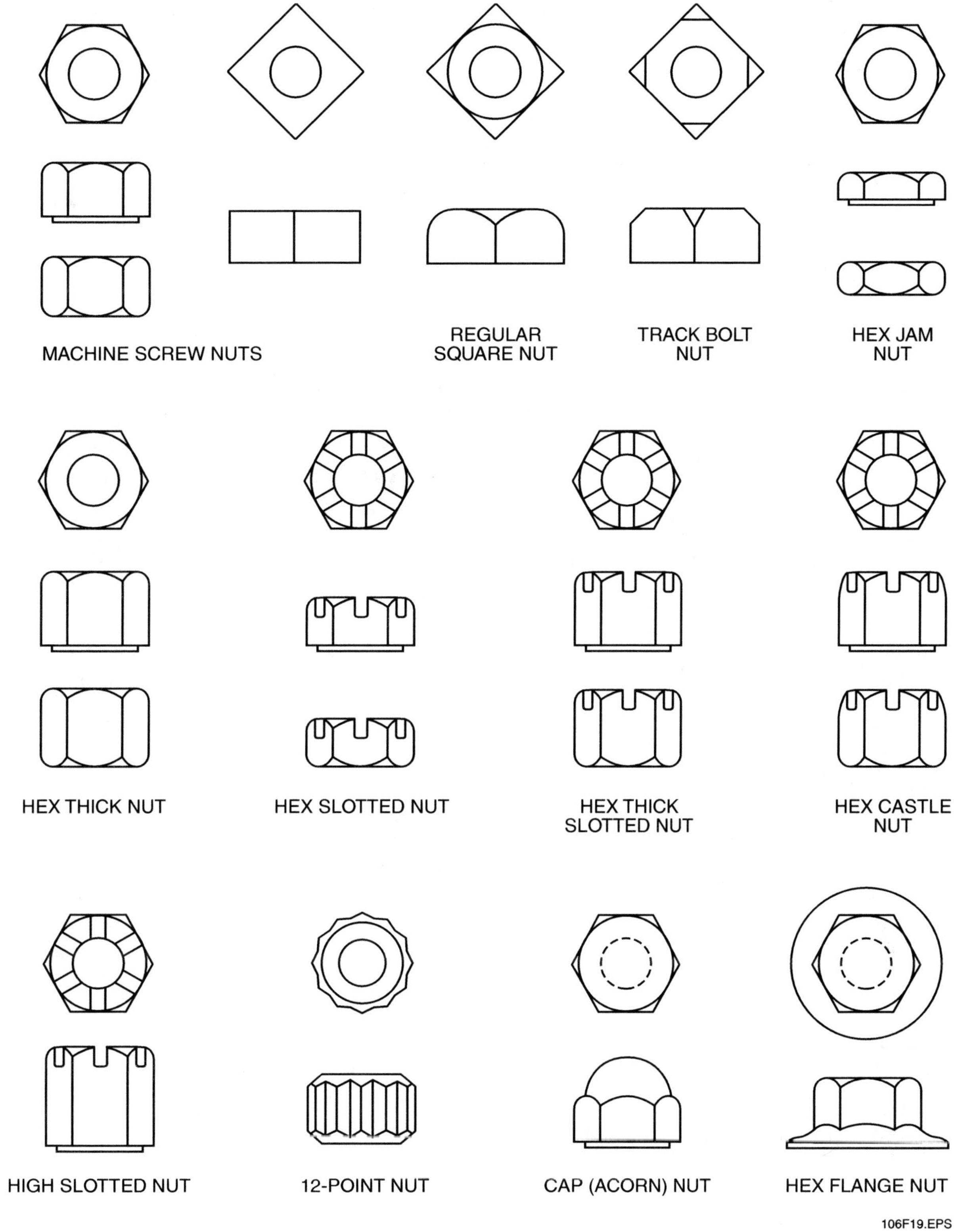

Figure 19 ◆ Different types of nuts for threaded fasteners.

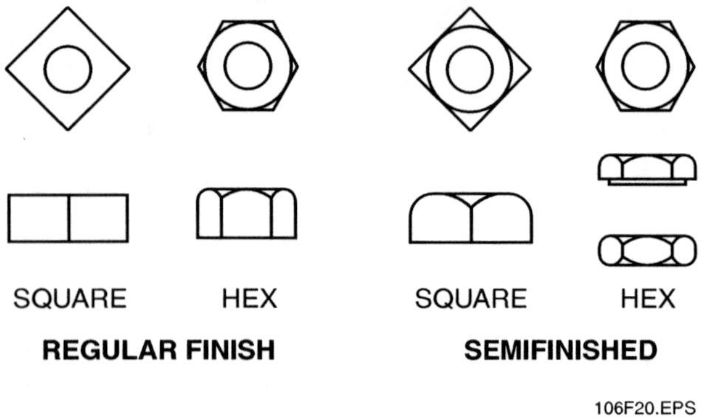

Figure 20 ◆ Regular and semifinished nuts.

2.3.14 Jam Nuts

The jam nut is a thin nut used to lock a standard nut in place. *Figure 21* shows a jam nut.

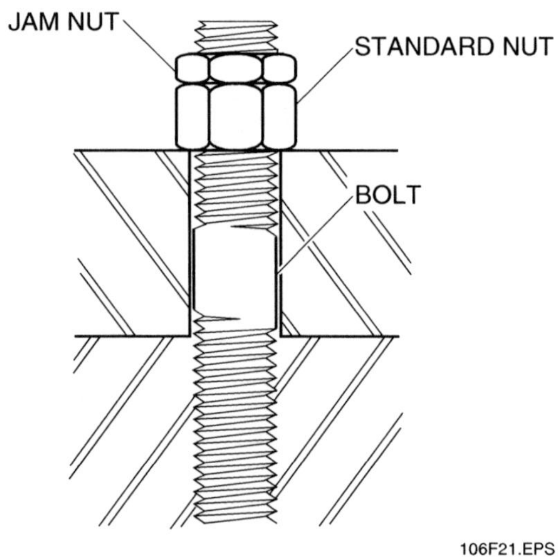

Figure 21 ◆ Jam nut.

2.3.15 Castellated, Slotted, and Self-Locking Nuts

Castellated, or castle, and slotted nuts are slotted across the flats. After the nut has been tightened, a cotter pin is fitted in one set of slots and through a hole in the bolt. The cotter pin keeps the nut from loosening under working conditions. These nuts are being replaced on many parts by self-locking nuts. Self-locking nuts have nylon inserts or are slightly deformed so they cannot work loose. No hole through the bolt is needed for self-locking nuts. *Figure 22* shows castellated, slotted, and self-locking nuts.

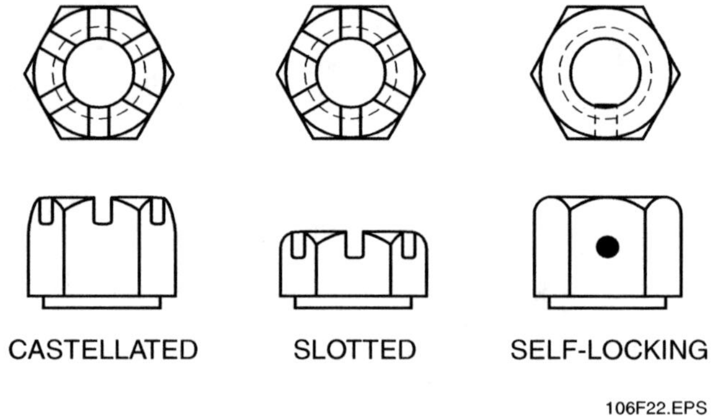

Figure 22 ◆ Castellated, slotted, and self-locking nuts.

2.3.16 Acorn Nuts

Acorn, or cap, nuts are used when appearance is important or when exposed, sharp thread edges on the fastener must be avoided (see *Figure 23*).

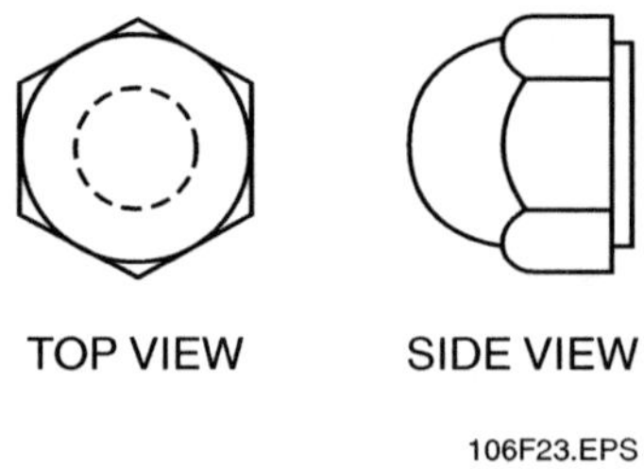

Figure 23 ◆ Acorn nut.

2.3.17 Wing Nuts

Wing nuts allow rapid loosening and tightening of a fastener without the need for a wrench (see *Figure 24*). Wing nuts are used where frequent adjustments and service are necessary.

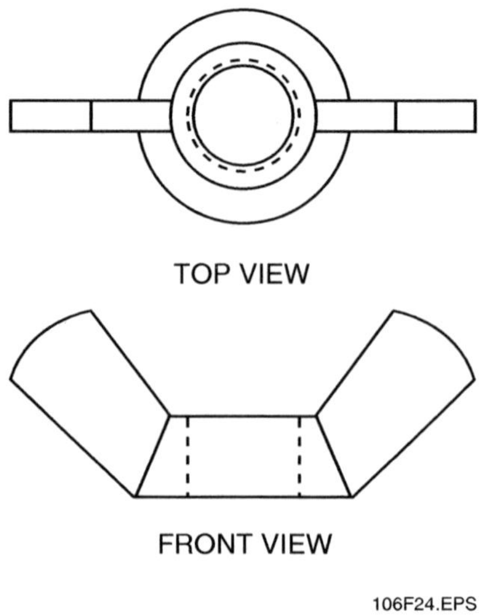

Figure 24 ◆ Wing nut.

2.3.18 Inserts

Inserts, or heli-coils, are a special kind of nut used to provide high-strength threads in soft metals and plastics (see *Figure 25*). They are also used to replace damaged or stripped threads in a tapped hole. Internal threads are made in standard sizes and forms.

Figure 25 ◆ Thread insert.

2.3.19 Flat Washers

Washers provide an enlarged surface for bolt heads and nuts. They also distribute the load over a larger area and prevent **marring**. The standard washer is made in light, medium, heavy-duty, and extra heavy-duty series. *Figure 26* shows a standard flat washer.

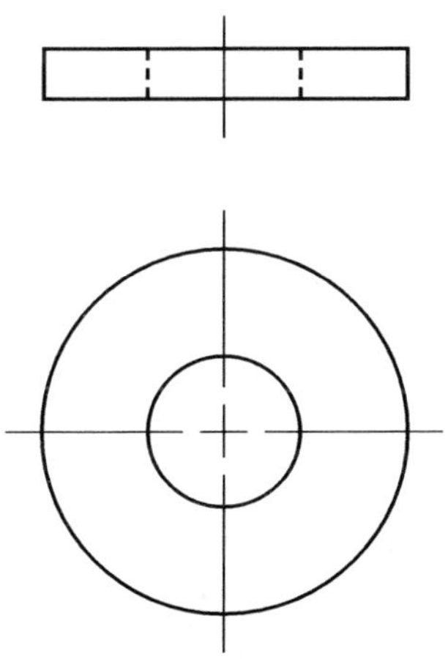

Figure 26 ◆ Standard flat washer.

2.3.20 Lock Washers

Lock washers are used to keep bolts or nuts from working loose (see *Figure 27*).

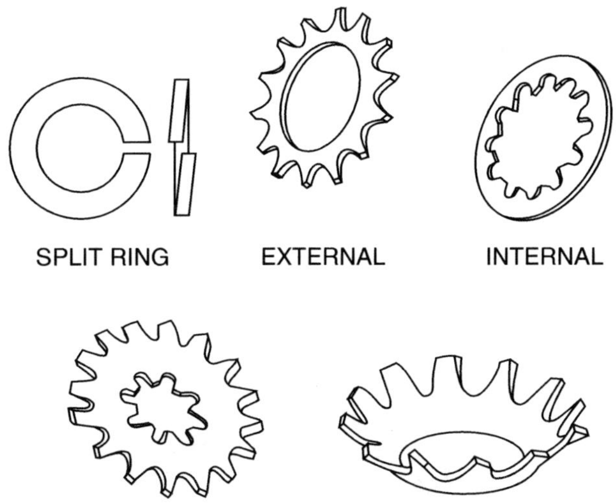

Figure 27 ◆ Several types of lock washers.

Each type of lock washer has a specific purpose:

- *Split ring* – Common type for bolts and cap screws
- *External* – For the greatest resistance
- *Internal* – For small screws
- *Internal-external* – For oversized mounting holes
- *Countersunk* – For flat or oval-head screws

3.0.0 ◆ NONTHREADED FASTENERS

Nonthreaded fasteners have many uses in maintenance and construction. The following sections explain these types of nonthreaded fasteners:

- Retainer rings
- Keys
- Pin fasteners
- Rivets

CAUTION

When using a nonthreaded fastener, be sure to follow all manufacturer's recommendations for proper use.

3.1.0 Retainer Rings

Retainer rings are designed for both internal and external applications. Some retainer rings are seated in grooves. Others are self-locking and do not require a groove. Special pliers are used to move internal and external rings. *Figure 28* shows typical retainer rings.

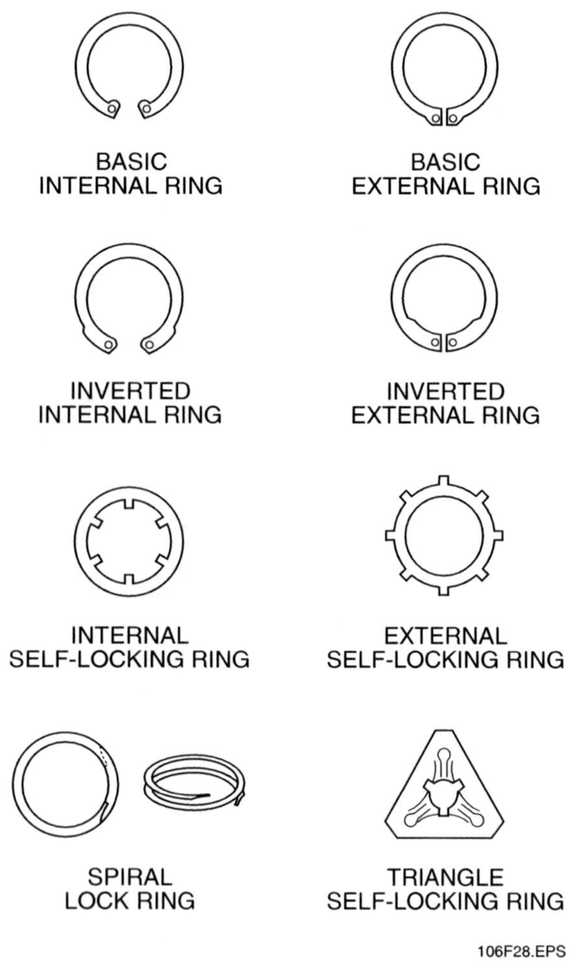

Figure 28 ◆ Typical retainer rings.

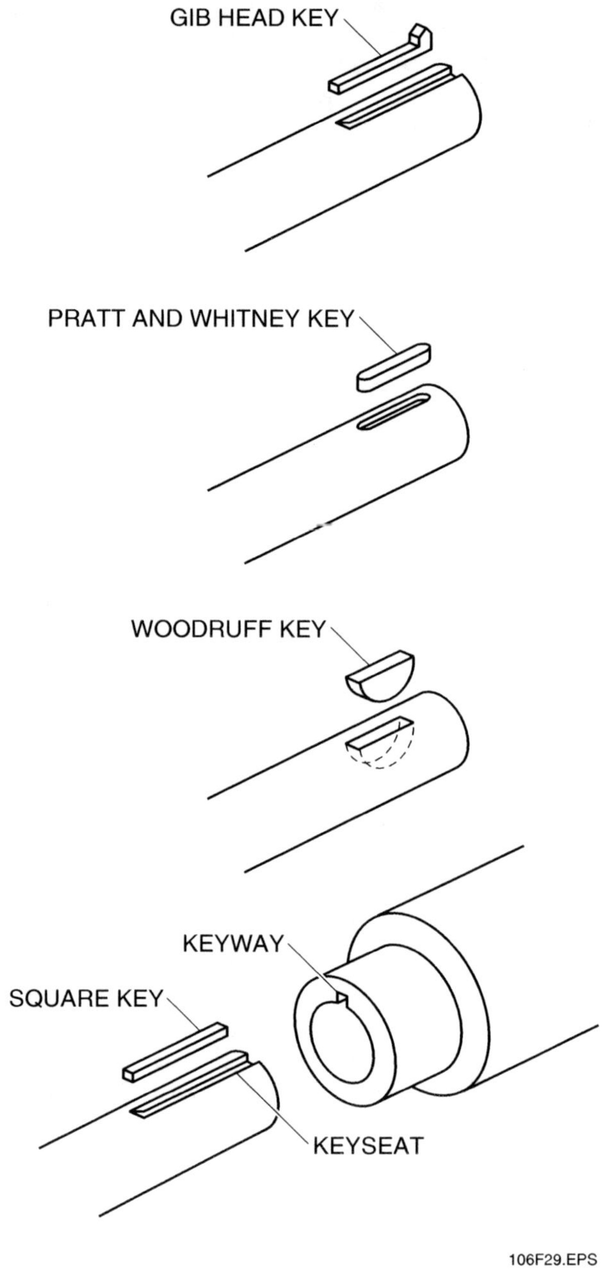

Figure 29 ◆ Several types of keys.

3.2.0 Keys

Keys are metal parts used to prevent a gear or pulley from rotating on a shaft (see *Figure 29*). One half of the key fits into a keyseat on the shaft. The other half fits into a **keyway** in the hub of the gear or pulley.

Types of keys include:

- *Square key* – Usually one-fourth the shaft diameter. It may be slightly tapered on the top for easier fitting.
- *Gib head key* – Interchangeable with the square key. The head design allows easy removal from the assembly.
- *Pratt and Whitney key* – Similar to the square key but rounded at both ends. It fits into a keyseat of the same shape.
- *Woodruff key* – Semicircular shape that fits into a keyseat of the same shape. The top of the key fits into the keyway of the mating parts.

3.3.0 Pin Fasteners

Pin fasteners are used to align mating parts, to hold gears and pulleys on shafts, and to secure slotted nuts (see *Figure 30*). They are also used for hinges.

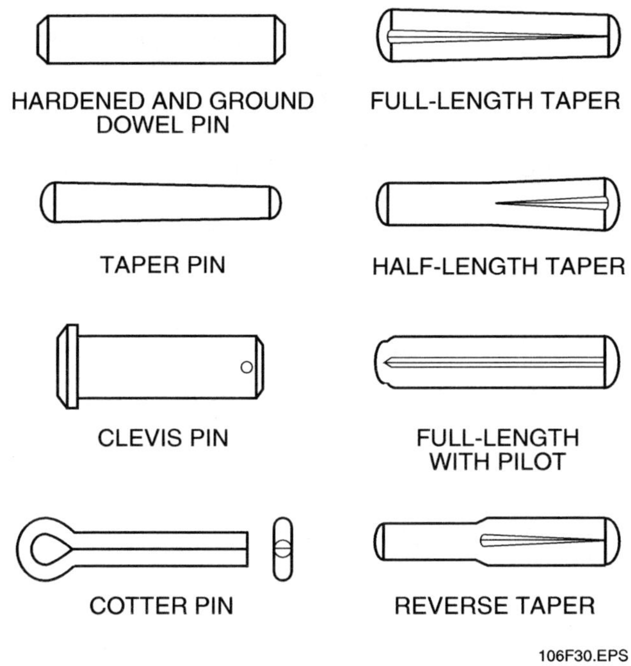

Figure 30 ◆ Examples of pin fasteners.

Several different types of pins are explained in the following sections:

- Dowel pins
- Taper and spring pins
- Cotter pins

3.3.1 Dowel Pins

Dowel pins are fitted into reamed holes to position mating parts (see *Figure 31*). They also support a portion of the load placed on the parts.

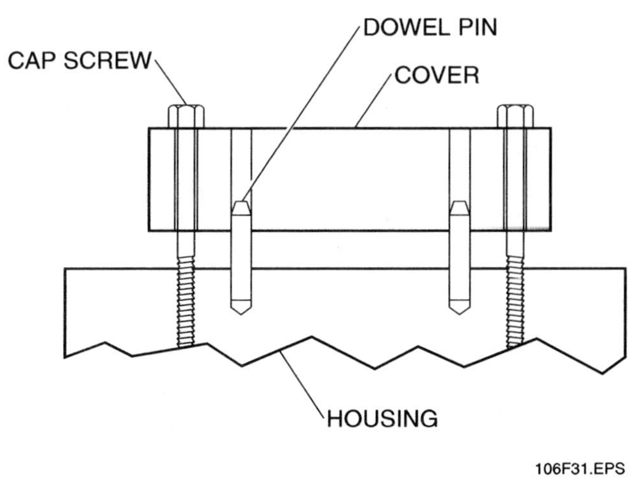

Figure 31 ◆ Dowel pins used to position mating parts.

3.3.2 Taper and Spring Pins

Taper and spring pins are used to fasten gears, pulleys, and collars to a shaft (see *Figure 32*).

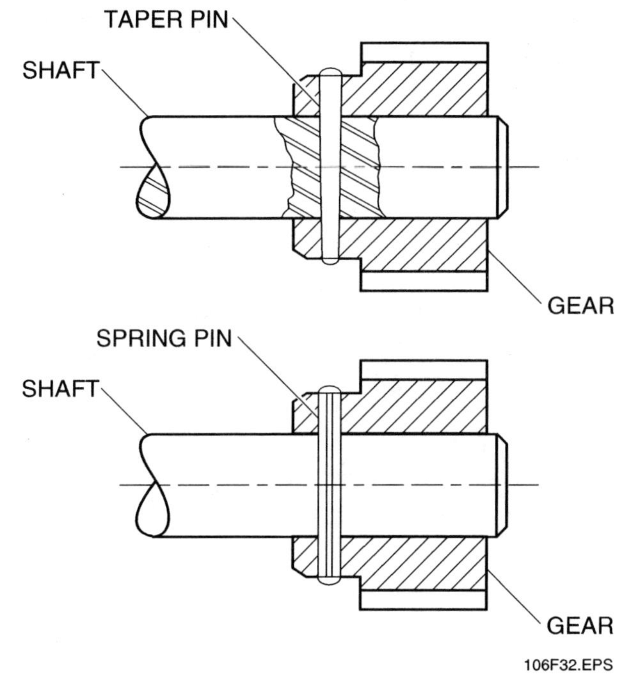

Figure 32 ◆ Taper and spring pin uses.

3.3.3 Cotter Pins

A cotter pin is fitted into a hole drilled crosswise through a shaft (see *Figure 33*). It is used to prevent parts from slipping on or off the shaft. Cotter pins are also used to keep slotted nuts from working loose.

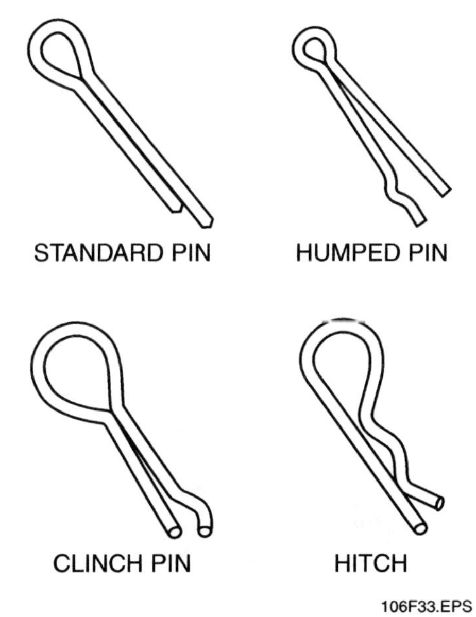

Figure 33 ◆ Cotter pins.

3.4.0 Rivets

Rivets are metal fasteners that look like threadless bolts. They are made of soft iron, steel, aluminum, copper, or brass and are available in many shapes and sizes. They may be solid, tubular, or of some special form. The following sections explain Tinner's rivets and blind rivets.

3.4.1 Tinner's Rivets

Tinner's rivets are designed to join two pieces of material together permanently. Tinner's rivets are sized in ounces or pounds per 1,000. For example, a 6-ounce rivet means that 1,000 of them weigh 6 ounces. As the weight increases, so does the diameter and length of the rivet. *Figure 34* shows some common tinner's rivets.

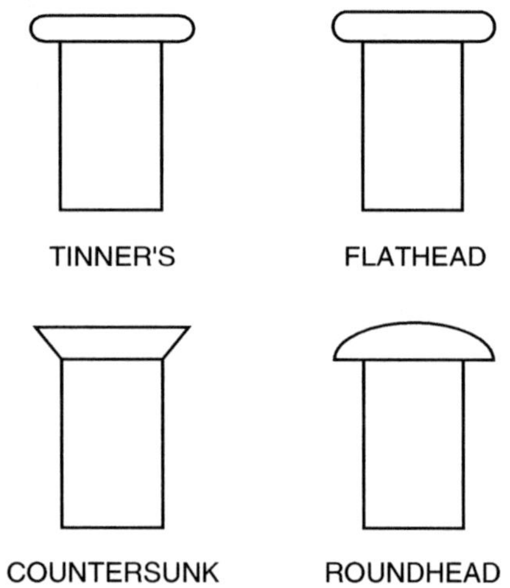

Figure 34 ◆ Common tinner's rivets.

To install a tinner's rivet, the rivet is inserted into a hole that has been drilled or punched. The small end of a solid tinner's rivet is hammered, or set, in the form of a head. Hollow tinner's rivets are clinched at the small end with a special tool. As a rule, rivets should not be placed closer together than 3 times the rivet diameter, nor farther apart than 24 times the rivet diameter.

3.4.2 Blind Rivets

Blind rivets are used to fasten sheet metal, fiberglass, plastics, and belting. Blind rivets are made in various lengths and diameters.

Blind rivets are used when the joint can only be reached from one side. To install a blind rivet, insert the rivet into a hole that has already been drilled or punched; then set, or pop, the rivet using a special tool (see *Figure 35*).

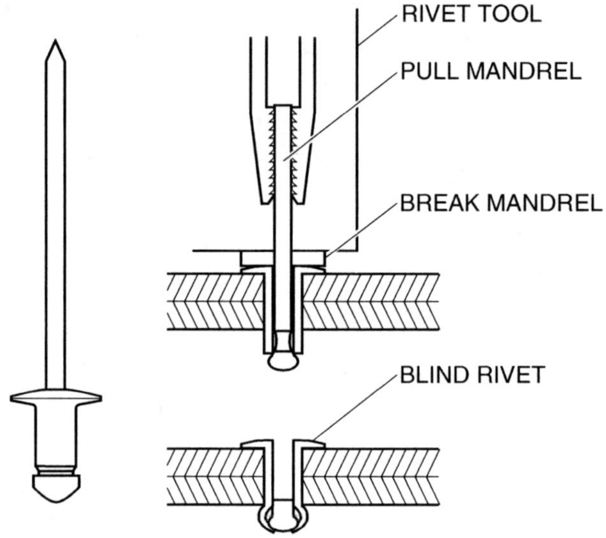

Figure 35 ◆ Blind rivet installation.

4.0.0 ◆ INSULATION FASTENERS

Many types of fasteners are used to secure insulation and jacketing. These fasteners may be welded, twisted, screwed, or snapped on. Some common fasteners described in the following sections are:

- Annealed wire
- Bands and clips
- Staples
- Stick clips
- Studs and pins
- Bonding adhesives
- Insulation tapes

4.1.0 Annealed Wire

Annealed wire, usually 16-gauge stainless steel, is used to fasten pipe insulation. Copper or galvanized steel wire may also be used. Stainless steel has the best working properties, however, and is used more often than the others.

4.2.0 Bands and Clips

Galvanized and stainless steel bands are also used to hold insulation in place. Bands are fastened with metal clips. *Figure 36* shows bands and clips.

To use bands and clips, place the band roll on a dispenser reel, if available. This lets you unroll the correct length without tangling or kinking. Cut the bands 4 inches longer than the required length. Place the bands on pipe insulations at 9- or 12-inch centers.

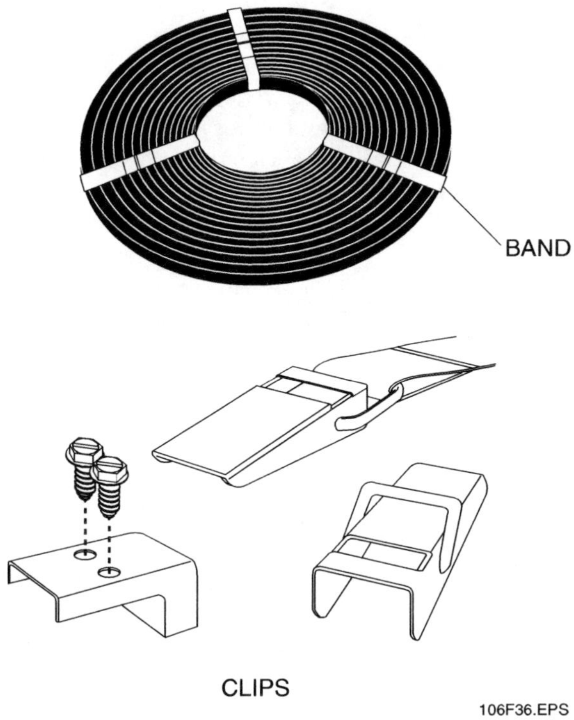

Figure 36 ◆ Bands and clips.

Clips on bands should be placed on the top of the pipe or, if possible, out of sight. All clips should be straight and even. Clips are placed on the ends of bands, and the bands are folded back ½ inch over the clip. The other end of the band is placed through the clip, and a band puller is used to tighten the band. The loose end is then folded back over the seal.

In certain cases, when bands are placed on vertical pipe, a screw can be placed under the clip to prevent the band from sliding down.

4.3.0 Staples

Staples come in different sizes, with ½ inch the most common size. Staples are generally used only for tacking purposes. *Figure 37* shows staples and a staple gun.

Keep the staple gun in good working condition, and replace the head if it does not flare staples properly.

To use the staple gun, press the head loosely against the surface being stapled. Do not jam or bump the head into the surface, because this causes breaks in the insulation material.

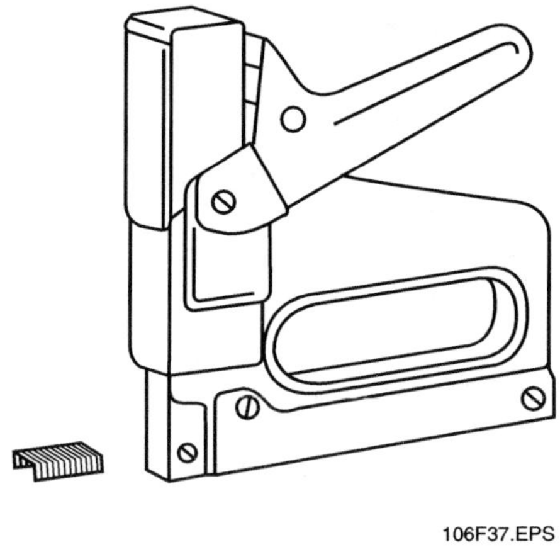

Figure 37 ◆ Staples and a staple gun.

4.4.0 Stick Clips

A stick clip, also called a *speed clip* or *insulation anchor*, is a fastener with a metal pin and a special glue on the back. *Figure 38* shows stick clips and washers.

Most stick clips come with the adhesive already applied and covered by a peel-off tape. Other stick clips come without the adhesive, which must be applied before use. If you apply it to the stick clip, the adhesive must be allowed to dry for 24 hours before use.

Stick clips must only be applied to clean and dry surfaces. Heavy coats of paint should be removed. Stick clips should be applied in a straight line, on center, to fit the size of the insulation being used.

The insulation is pushed onto the stick clip, and a washer is used to hold the material in place. The stick clip is then cut off, using end-cutting nippers. At times, the stick clip may be bent down flush with the surface of the insulation.

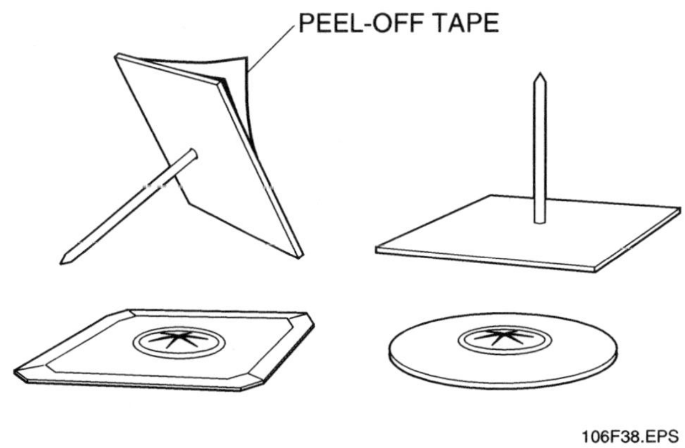

Figure 38 ◆ Stick clips and washers.

4.5.0 Studs and Pins

Studs and pins are similar to stick clips except that they are welded directly to a metal surface (see *Figure 39*). A stud and pin weld gun is used to weld these holders to the surface. Studs usually have threads, but pins do not.

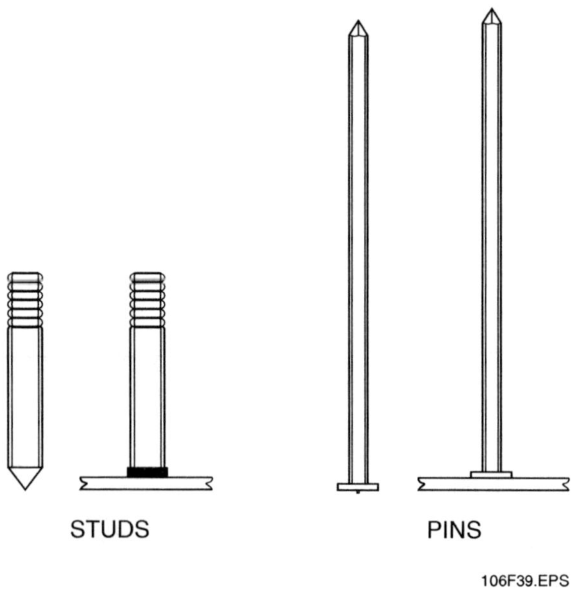

Figure 39 ◆ Studs and pins.

The following three types of weld pins and studs are used:

- Copper-clad
- Stainless steel
- Aluminum

Copper-clad pins and studs are used on galvanized or carbon steel surfaces. Stainless steel pins and studs are used on stainless steel surfaces. Aluminum pins and studs are used on aluminum surfaces. Make sure you are using the right pin and stud for the surface.

4.6.0 Bonding Adhesives

Adhesives are used to hold insulation materials together. Contact and noncontact adhesives are commonly used in insulation work.

4.6.1 Contact Adhesives

Contact adhesives make a tight bond upon contact. Most contact adhesives are highly flammable when applied. They should be used only in well-ventilated areas. Contact adhesive containers should be kept tightly closed when not in use.

WARNING!

Use contact adhesives in well-ventilated areas and be careful not to breathe the vapors, which can be harmful. If adhesives must be used in closed areas, have a second person stand outside the work area and observe the person applying the adhesive.

Before a contact adhesive is applied, the surface must be clean and dry. Apply the adhesive to both surfaces to be joined, and allow it to dry to the touch before bonding.

4.6.2 Noncontact Adhesives

With noncontact adhesives, the surfaces being joined must be held in place until the adhesive sets. Noncontact adhesives are used to seal laps on jacketing materials.

WARNING!

Be careful not to breathe the adhesive vapors or spill the adhesive on the floor or on equipment.

5.0.0 ◆ INSTALLING FASTENERS

To install fasteners, you must know the proper type of fastener and the correct method to use. You must also understand the tightening sequence and torque specifications needed. The following sections explain how to install two common types of fasteners.

5.1.0 Installing Threaded Fasteners

Commonly used threaded fasteners are bolts and nuts, screws, and specialized fasteners. To correctly install fasteners, the installer must understand the strength, design intent, torque (tightening) limits, and other specifications of the fasteners being used. Follow these steps to install a threaded fastener.

Step 1 Locate the object to be fastened.

Step 2 Select the proper type and grade of fastener for the job.

Step 3 Select the number of fasteners needed for the job.

Step 4 Inspect each fastener, looking for damaged threads both on the bolt or screw and inside the matching nut, if available, and checking for dirt or rust on the threads or grease on the head.

Step 5 Position the items to be fastened, using a tool if necessary, and eliminating the largest gap first.

NOTE

If the threads are damaged, replace the damaged item. If the threads are dirty or rusty, clean and lightly lubricate them. If the fastener is being installed in a threaded hole, inspect the hole threads for damage. If the threads are damaged the hole should be retapped. Be sure to use the correct size tap for the fastener being used. Tap drill size charts are used to select the proper size drill and tap. *Table 2* shows a sample tap drill size chart.

Table 2 Tap Drill Size Chart

Screw Thread		Commercial Tap Drills		Screw Thread		Commercial Tap Drills	
Outside Diameter-Pitch	Root Diameter	Size or Number	Decimal Equivalent	Outside Diameter-Pitch	Root Diameter	Size or Number	Decimal Equivalent
1/16-64	0.0422	3/64	0.0469	27	0.4519	15/32	0.4687
72	0.0445	3/64	0.0469	9/16-12	0.4542	31/64	0.4844
5/64-60	0.0563	1/16	0.0625	18	0.4903	33/64	0.5156
72	0.0601	52	0.0635	27	0.5144	17/32	0.5312
3/32-48	0.0667	49	0.0730	5/8-11	0.5069	17/32	0.5312
50	0.0678	49	0.0730	12	0.5168	35/64	0.5469
7/64-48	0.0823	43	0.0890	18	0.5528	37/64	0.5781
1/8-32	0.0844	3/32	0.0937	27	0.5769	19/32	0.5937
40	0.0925	38	0.1015	11/16-11	0.5694	19/32	0.5937
9/64-40	0.1081	32	0.1160	16	0.6063	5/8	0.6250
5/32-32	0.1157	1/8	0.1250	3/4-10	0.6201	21/32	0.6562
36	0.1202	30	0.1285	12	0.6418	43/64	0.6719
11/64-32	0.1313	9/64	0.1406	16	0.6688	11/16	0.6875
3/16-24	0.1334	26	0.1470	27	0.7019	23/32	0.7187
32	0.1469	22	0.1570	13/16-10	0.6826	23/32	0.7187
13/64-24	0.1490	20	0.1610	7/8-9	0.7307	49/64	0.7656
7/32-24	0.1646	16	0.1770	12	0.7668	51/64	0.7969
32	0.1782	12	0.1890	14	0.7822	13/16	0.8125
15/64-24	0.1806	10	0.1935	18	0.8028	53/64	0.8281
1/4-20	0.1850	7	0.2010	27	0.8269	27/32	0.8437
24	0.1959	4	0.2090	15/16-9	0.7932	53/64	0.8281
27	0.2019	3	0.2130	1-8	0.8376	7/8	0.8750
28	0.2036	3	0.2130	12	0.8918	59/64	0.9219
32	0.2094	7/32	0.2187	14	0.9072	15/16	0.9375
5/16-18	0.2403	F	0.2570	27	0.9519	31/32	0.9687
20	0.2476	17/64	0.2656	1 1/8-7	0.9394	63/64	0.9844
24	0.2584	I	0.2720	12	1.0168	1 3/64	1.0469
27	0.2644	J	0.2770	1 1/4-7	1.0644	1 7/64	1.1094
32	0.2719	9/32	0.2812	12	1.1418	1 11/64	1.1719
3/8-16	0.2938	5/16	0.3125	1 3/8-6	1.1585	1 7/32	1.2187
20	0.3100	21/64	0.3281	12	1.2668	1 19/64	1.2969
24	0.3209	Q	0.3320	1 1/2-6	1.2835	1 11/32	1.3437
27	0.3269	R	0.3390	12	1.3918	1 27/64	1.4219
7/16-14	0.3447	U	0.3680	1 5/8-5 1/2	1.3888	1 29/64	1.4531
20	0.3726	25/64	0.3906	1 3/4-5	1.4902	1 9/16	1.5625
24	0.3834	X	0.3970	1 7/8-5	1.6152	1 11/16	1.6875
27	0.3894	Y	0.4040	2-4 1/2	1.7113	1 25/32	1.7812
1/2-12	0.3918	27/64	0.4219	2 1/8-4 1/2	1.8363	1 29/32	1.9062
13	0.4001	27/64	0.4219	2 1/4-4 1/2	1.9613	2 1/32	2.0312
20	0.4351	29/64	0.4531	2 3/8-4	2.0502	2 1/8	2.1250
24	0.4459	29/64	0.4531	2 1/2-4	2.1752	2 1/4	2.2500

106T02.EPS

> **WARNING!**
> Do not use your finger to check bolt-hole alignment. This can cause severe injury or amputation.

> **CAUTION**
> Check to see that the materials to be fastened are properly aligned, or damage may occur to the materials.

Step 6 Install one fastener loosely by hand.

> **NOTE**
> Install fasteners so that they each point in a direction that allows easy access for later repairs.

Step 7 Check for cross-threading.

Step 8 Check the alignment again after the first fastener is installed.

Step 9 Install the remaining fasteners, and tighten them by hand.

Step 10 Tighten each fastener snugly, following the proper tightening sequence for that bolt pattern and fastener type. *Figure 40* shows the proper tightening sequence for flanges, which are often used for fastening pipes.

Step 11 Use a suitable torque wrench to tighten the fasteners to their torque specifications. *Table 3* shows the proper torque specifications.

> **CAUTION**
> Tighten the fasteners only small amounts at a time, following the proper tightening sequence for that bolt pattern and fastener type, before reaching the final torque. Failure to follow this sequence or overtightening can result in damage to either the fasteners or the object being fastened.

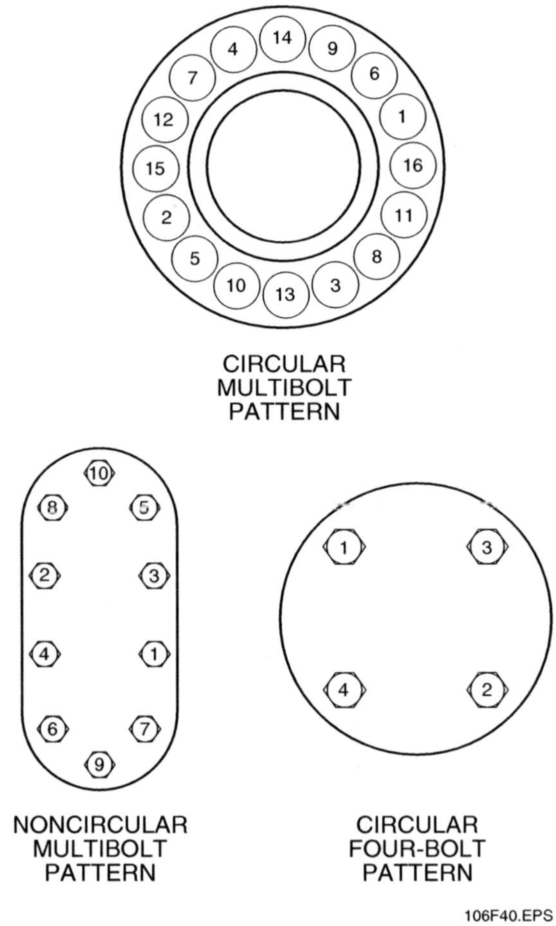

Figure 40 ◆ Proper tightening sequence for flanges.

Step 12 Secure the fasteners, using jam nuts, cotter pins, or safety wire as needed.

Step 13 Clean the tools that were used, and inspect them for damage before storing them.

Step 14 Clean the work area, and dispose of any debris.

5.2.0 Installing Fasteners (Anchor Bolts) in Hardened Concrete

Anchor bolts come in either expandable or nonexpandable form. The following steps describe installing both kinds of fastener. Follow these steps to install an anchor bolt in hardened concrete.

Table 3 Proper Torque Specifications

Bolt Diameter	Threads per Inch	Foot-Pounds Torque		
		Mild Steel	**Stainless Steel 18-8**	**Alloy Steel**
¼	20	4	6	8
5/16	18	8	11	16
⅜	16	12	18	24
7/16	14	20	32	40
½	13	30	43	60
9/16	12	40	68	90
⅝	11	60	92	120
¾	10	100	128	200
⅞	9	160	180	320
1	8	245	285	490

	Grade	SAE 1 or 2	SAE 5	SAE 6	SAE 8
	Tensile Strength	64,000 psi	105,000 psi	130,000 psi	150,000 psi
	Grade Marking		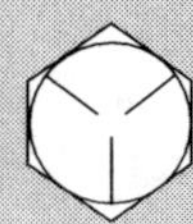	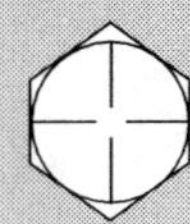	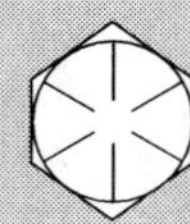
Bolt Diameter	**Threads per Inch**	**Foot-Pounds Torque**			
¼	20	5	7	10	10
5/16	18	9	14	19	22
⅜	16	15	25	34	37
7/16	14	24	40	55	60
½	13	37	60	85	92
9/16	12	53	88	120	132
⅝	11	74	120	169	180
¾	10	120	200	280	296
⅞	9	190	302	440	473
1	8	282	466	660	714

106T03.EPS

Step 1 Identify the item to be fastened.

Step 2 Locate the area where the item to be fastened must sit or be attached.

CAUTION

The area where the item is to be fastened should be smooth so that the item will have solid footing. Uneven footing might cause the item being fastened to twist, warp, or not tighten properly, and vibrations may result.

Step 3 Make a layout **template** of the bolt holes to be drilled.

Step 4 Use the template to lay out and mark where each hole must be located.

Step 5 Set up a core drill and bit for drilling the hole in the concrete.

WARNING!

Carefully inspect the core drill and bit to make sure they are in good operating condition. If they are not, you could be injured by parts that chip and fly off during operation.

Step 6 Identify the type (expandable or nonexpandable) and size of fastener needed.

Step 7 Select a drill bit to match the fastener chosen.

NOTE

The drill bit selected for a nonexpandable fastener must be slightly larger in diameter than the head of the fastener and the flat washer used with the fastener. The washer must be able to fit down inside the hole, just below surface level. The hole-cutting device for the expandable fastener is the fastener shaft itself (called the cutting sleeve).

Step 8 Put on safety glasses, ear protectors, and gloves.

WARNING!

Drilling in concrete generates noise, dust, and possible flying objects. Make sure others in the area also wear their protective equipment.

Step 9 Assemble the drill and drill bit.

Step 10 Drill the holes for the nonexpandable fasteners. The drill bit may be lubricated with water during the drilling process.

Step 11 Clean out the hole when finished drilling.

Step 12 Assemble and lay out the fastener and the parts for each hole.

5.2.1 Installing Nonexpandable Fasteners

The following describes how to install nonexpandable fasteners.

Step 1 Position a nonexpandable anchor bolt into the drilled hole. The nonexpandable fastener anchor bolt should extend out of the hole far enough for the threads and a little of the unthreaded bolt to be above the surface level.

Step 2 Prepare a portion of filler mixture (**grout**).

Step 3 Fill the hole around bolt until it is about three-fourths filled. The grout will later harden and lock the bolt in the hole.

Step 4 Work the bolt up and down in the hole to settle the filler material and to eliminate air pockets.

Step 5 Finish filling the hole with grout and work it in well, but leave enough room at the top for the fastener washer to be installed.

Step 6 Slip the washer down over the bolt, and settle it down into the hole on top of the filler material. *Figure 41* shows a properly installed anchor bolt.

Step 7 Clean any excess filler from around the hole and off the bolt threads and shaft.

Step 8 Run a matching nut down the shaft a few turns to make sure the threads are good.

Step 9 Remove the nut.

Step 10 Allow the filler mixture to dry fully before mounting anything to the fastener.

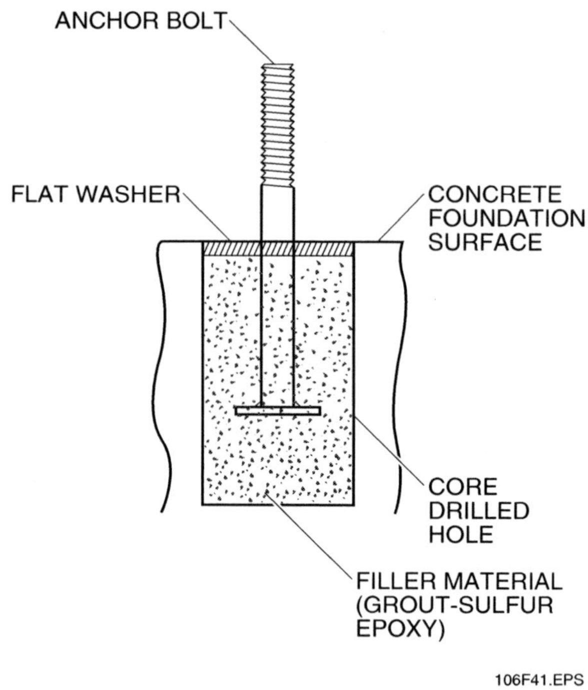

Figure 41 ◆ Properly installed anchor bolt.

CAUTION

Make sure the bolt is left straight after working it around in the hole and installing the washer. The washer centers the bolt and holds it until the filler material hardens. If the grout sets and the bolt is not straight, the bolt will be unusable, and the job will have to be repeated after the bolt has been removed.

5.2.2 Installing Expandable Fasteners

The next steps are for installing the expandable fastener. The expandable fasteners, called *red-heads*, are widely used. The redhead fastener uses a cutting sleeve that first acts as a drill bit and later becomes the expandable fastener itself. The size of an expandable fastener is based on the size of the bolt that will be used in it. A redhead drill is used with the cutting sleeve to cut the hole and later to install the fastener. This tool hammers the cutting sleeve into the hole as it rotates and cuts. Later, the process is repeated, except that the sleeve will have the expander in the tip. As the fastener is hammered down, it hits the bottom where the tapered expander (redhead) causes the fastener to expand and lock into the hole. *Figure 42* shows the parts of an expandable fastener and how they connect with the redhead drill used to cut and install the fastener.

Step 1 Locate the marked area for the hole that is to receive an expandable fastener.

Step 2 Acquire a redhead drill, and connect an expandable fastener cutting sleeve to it.

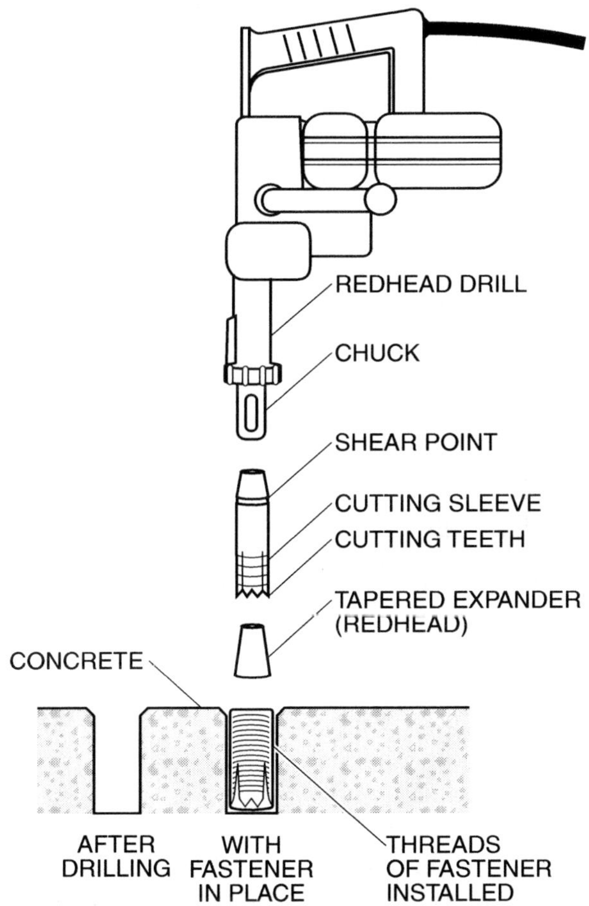

Figure 42 ◆ Expandable (redhead) fastener process.

> **NOTE**
> The tapered expander (redhead) is not inserted or used at this point.

Step 3 Put on safety glasses, ear protectors, and gloves.

Step 4 Center the cutting sleeve over the marked spot, place the drill selector switch to the drill and hammer position, and drill the hole.

> **NOTE**
> Drill the hole deep enough so that the fastener cutting sleeve shear point is just below the surface of the concrete. (The portion above the shear point will be broken off later.)

Step 5 Keep the cutting sleeve rotating, and withdraw it from the hole.

Step 6 Stop the drill, and lay it aside with the cutting sleeve still attached.

Step 7 Clean out the hole, the area around the hole, and the cutting sleeve.

Step 8 Insert the tapered expander, small end first, into the cutting end of the cutting sleeve.

> **NOTE**
> Slightly jam the redhead into the cutting sleeve so that it will not fall out when the cutting sleeve is pointed downward.

Step 9 Pick up the drill with the cutting sleeve and the tapered expander attached, and align it in or over the hole opening.

Step 10 Set of position the drill selector switch to hammer only, and operate the hammer.

> **CAUTION**
> Keep the hammer and fastener going straight into the hole, or the fastener, the hammer, or the concrete may be damaged.

Step 11 Hammer the fastener down into the hole until the shear point is at or below the concrete surface.

Step 12 Stop hammering, but do not remove the drill.

Step 13 Lean the drill to the left or the right, and snap the top of the fastener off at the shear point.

NOTE

The top should snap off and leave the top of the fastener exposed and level with the surface of the concrete.

Step 14 Release the drill chuck, and eject the part of the fastener that was broken off.

Step 15 Inspect the hammer for damage, and clean it if necessary.

Step 16 Inspect, deburr, and clean the installed expandable fastener. Check to see that it is flush with the concrete surface.

Step 17 Screw a matching bolt into the fastener a few turns to make sure the fastener threads are usable.

Step 18 Remove the bolt.

Step 19 Store the excess fasteners, tools, and other equipment, and clean the area.

Summary

Fasteners are used in most jobs that the pipeline mechanical technician performs. Fasteners come in many shapes and sizes for different uses. It is essential to use the proper fastener for the job to ensure safety and efficiency. The strength of fasteners can vary to a great extent. The item fastened is no stronger than the fastener used to fasten it. It is also essential that fasteners be installed properly for the same reasons.

Review Questions

1. The most common thread standard is the _____ standard.
   a. national
   b. acme
   c. unified
   d. U.S.

2. The strongest and most useful thread design is the _____ thread.
   a. acme
   b. square
   c. buttress
   d. unified

3. To prevent pulleys from slipping on shafts, to hold collars in place on shafts, and to hold shafts in place, _____ are used.
   a. stud bolts
   b. setscrews
   c. drive screws
   d. cap screws

4. A thin nut used to lock a standard nut into place is a(n) _____.
   a. acorn nut
   b. castellated nut
   c. jam nut
   d. insert

5. Metal parts used to prevent a gear or pulley from rotating on a shaft are called _____.
   a. inserts
   b. retainer rings
   c. rivets
   d. keys

6. Rivets should not be placed closer together than _____ times the rivet diameter.
   a. 2
   b. 3
   c. 10
   d. 24

7. To fasten pipe insulation, _____ is/are used.
   a. rivets
   b. retainer rings
   c. annealed wire
   d. setscrews

8. Copper-clad pins and studs are used on _____ surfaces.
   a. stainless steel
   b. galvanized or carbon steel
   c. aluminum
   d. copper or copper-clad

9. If _____ adhesives are being used in a closed area, have a second person stand outside the work area and observe the person applying the adhesive.
   a. contact
   b. lagging
   c. noncontact
   d. insulation

10. The expandable fastener that uses a cutting sleeve that first acts as a drill bit and then becomes the actual fastener is known as a(n) _____.
    a. anchor bolt
    b. redhead
    c. rawl plug
    d. retainer ring

GLOSSARY

Trade Terms Introduced in This Module

Grout: A thin mortar used to secure leveling shims and to provide a support surface for a bedplate.

Keyway: A slot in a shaft for a key.

Marring: Damaging or defacing something.

Nonferrous: A metal that contains no iron.

Template: A pattern.

Tolerance: The amount of variation allowed from a standard.

CONTREN™ LEARNING SERIES — USER UPDATES

The NCCER makes every effort to keep these textbooks up-to-date and free of technical errors. We appreciate your help in this process. If you have an idea for improving this textbook, or if you find an error, a typographical mistake, or an inaccuracy in NCCER's Contren™ textbooks, please write us, using this form or a photocopy. Be sure to include the exact module number, page number, a detailed description, and the correction, if applicable. Your input will be brought to the attention of the Technical Review Committee. Thank you for your assistance.

Instructors – If you found that additional materials were necessary in order to teach this module effectively, please let us know so that we may include them in the Equipment/Materials list in the Instructor's Guide.

Write: Curriculum Revision and Development Department
National Center for Construction Education and Research
P.O. Box 141104, Gainesville, FL 32614-1104

Fax: 352-334-0932

E-mail: curriculum@nccer.org

Craft ______________________ Module Name ______________________

Copyright Date ______________ Module Number ______________ Page Number(s) ______________

Description

____

____

____

____

(Optional) Correction

____

____

____

(Optional) Your Name and Address

____

____

____

Module 63107-02

Identify, Install, and Maintain Valves

(*CT 19.1-19.4*)

Introduction to the Pipeline Operator Qualification Rule, API, and NCCER

It could be said that petroleum products keep America running. Petroleum products like oil and natural gas are the principal sources of energy in the United States and are used in many manufacturing and chemical processes. Petroleum products are primarily transported by pipeline. There are more than a million miles of liquid and natural gas pipeline in this country, ranging in diameter from two inches to four feet. These pipelines bring crude oil and natural gas from production fields to processing plants and refineries—and ultimately to the consumer.

Pipeline Operator Qualification

Petroleum products are considered hazardous, so pipeline personnel must take extra precautions and follow specific procedures when working with them. To help ensure the safe handling and transportation of petroleum products, the United States Department of Transportation (DOT) mandates safe practices as federal regulations.

One such regulation, the Operator Qualification (OQ) rule, requires pipeline operators to develop a written qualification plan to evaluate their personnel and their contract personnel (see *49 CFR Part 192 Subpart N—Gas*, and *49 CFR Part 195 Subpart G—Liquid*). The evaluation hinges on the operator's ability to perform certain covered tasks and to be able to recognize and respond to abnormal operating conditions that may be encountered while performing these covered tasks.

Covered Tasks

A pipeline performance task is deemed *covered* if the activity passes the following four-part test:

1. Is it performed on a pipeline facility?
2. Is it an operations or maintenance task?
3. Is it performed as a requirement of *49 CFR Part 195 (Liquid)* or *49 CFR Part 192 (Gas)*?
4. Does it affect the operation or integrity of the pipeline facility?

If the answer to all four questions is *yes*, then the activity is considered *covered* and must figure into the evaluation process of pipeline personnel.

American Petroleum Institute

The OQ rule and its related covered tasks provided the catalyst to develop a unified pipeline training and assessment program—so that pipeline operators and owners have ready access to the assessment and training tools they need. This effort was the direct result of the vision and leadership provided by The American Petroleum Institute (API) and API members.

Efforts to unify the industry are nothing new to API. Indeed, the impetus for forming the Institute in 1919 was the need to standardize engineering specifications for drilling and production equipment. Today, the most pressing issues revolve around public perceptions and government policies toward the industry. Speaking with one voice on these issues has become as essential as having interchangeable parts in the field. API's members determine the petroleum industry's positions on public policy as well as the standards that govern its day-to-day operations worldwide.

Consortium on Operator Qualification

API assembled a focus group whose primary purpose was to identify covered tasks and provide compliance recommendations to pipeline operators.

The Consortium on Operator Qualification (COOQ) became the guiding force for the identification and development of covered tasks for the liquid pipeline industry. Their work product, API publication 1161, *Guidance Document for the Qualification of Liquid Pipeline Personnel*, formed the basis for the resulting Pipeliner Training and Assessment Program.

The covered task list created in API publication 1161 has now evolved to include gas covered tasks and is maintained by NCCER. The most current covered task list can be downloaded at www.nccer.org.

National Center for Construction Education and Research

With the guidance and support of API, its members, and the Consortium on Operator Qualification, the National Center for Construction Education and Research (NCCER) was able to design and implement the Pipeliner Training and Assessment Program.

NCCER is a not-for-profit 501(c)(3) education foundation established in 1995 by the world's largest and most progressive construction and maintenance contractors and national contractor associations. NCCER was founded to address the

severe workforce shortage facing the construction and maintenance industries and to develop a standardized training process and curricula. Today, NCCER is supported by hundreds of leading construction and maintenance companies, manufacturers, and national associations and developed more than 35 construction and maintenance titles.

NCCER also maintains a National Registry, which contains the records of trainees and instructors who have received NCCER training. All completion information for every trainee is sent to the NCCER and filed within the Registry. The Registry can then confirm training and skills for workers as they move from company to company, state to state, or even within their own company.

Standardized Curriculum (NCCER's Contren™ Learning Series)

NCCER curricula are developed using the expertise of industry subject matter experts (SMEs). The use of SMEs as an integral part of every new development or revision ensures that the curriculum is current, industry-driven, and indicative of industry practices. The development and/or revision process is facilitated from start to finish by NCCER project managers, who work closely with the SMEs and a select team of technical writer(s) on the development and presentation of the curriculum.

Pipeline: Five Training Paths

The pipeline curriculum focuses on essential skills and competencies within five skill areas:

- Corrosion Control
- Electrical and Instrumentation
- Maintenance
- Mechanical
- Operations (Field and Control Center)

SMEs from the pipeline industry have determined the skills and competencies needed based on API's covered tasks and the consensus of the group/industry. Specific company information is generally not incorporated into the curriculum unless the SMEs determine that the information is the industry standard.

Training Modules and Covered Tasks

The smallest unit of instruction in the curriculum is a *module*. Each curriculum is composed of modules (the equivalent of stand-alone chapters) that are logically sequenced into complete *levels* of instruction. Every module is task-oriented and requires the successful completion of a written (knowledge-based) examination and a performance (hands-on) test. Modules may be individually selected and/or combined for efficient task training. Modules in the pipeline curricula, to the extent possible, will align with API-developed covered tasks. While not all modules will relate to covered tasks, all covered tasks will relate to at least one module.

Benefits of Training to the Pipeline Marketplace

Applying NCCER's standardized training process allows pipeline owners and contractors in the United States to establish the benchmark for acceptable skill and competency levels within the pipeline industry. This process will assist training organizations in developing and/or improving standardized pipeline training programs that effectively match their resources and operating conditions while responding to regulatory demands and to the needs of the industry. The process will also recognize training organizations that meet the minimum standards of competence, warranting regulatory, public, and professional confidence. This in turn will foster national unity within the pipeline industry for the mutual benefit of individual technicians, owners, contractors, and ultimately pipeline-users and society.

Additional Resources

To review the Pipeliner Training and Assessment Program Implementation and Resource Guide and the latest version of pipeline covered tasks, visit NCCER's Web site at www.nccer.org, or contact:

National Center for Construction Education and Research
PO Box 141104
Gainesville, FL 32614-1104
(888) 622-3720
Web site: www.nccer.org

For more information on API, please contact:

American Petroleum Institute
1220 L Street, Northwest
Washington, D.C. 20005
(202) 682-8000
Web site: www.api.org

To order *API Pub 1161*, contact Global Engineering Documents by calling 1-800-854-7149 or visiting their Web site at www.global.his.com.

Additional sources of information on the pipeline operator qualification regulation include:

DOT – Office of Pipeline Safety
www.ops.dot.gov

Transportation Safety Institute
www.tsi.dot.gov

COURSE MAP

This course map shows all of the modules in the first level of the Pipeline Mechanical curriculum. The suggested training order begins at the bottom and proceeds up. Skill levels increase as you advance on the course map. The local Training Program Sponsor may adjust the training order.

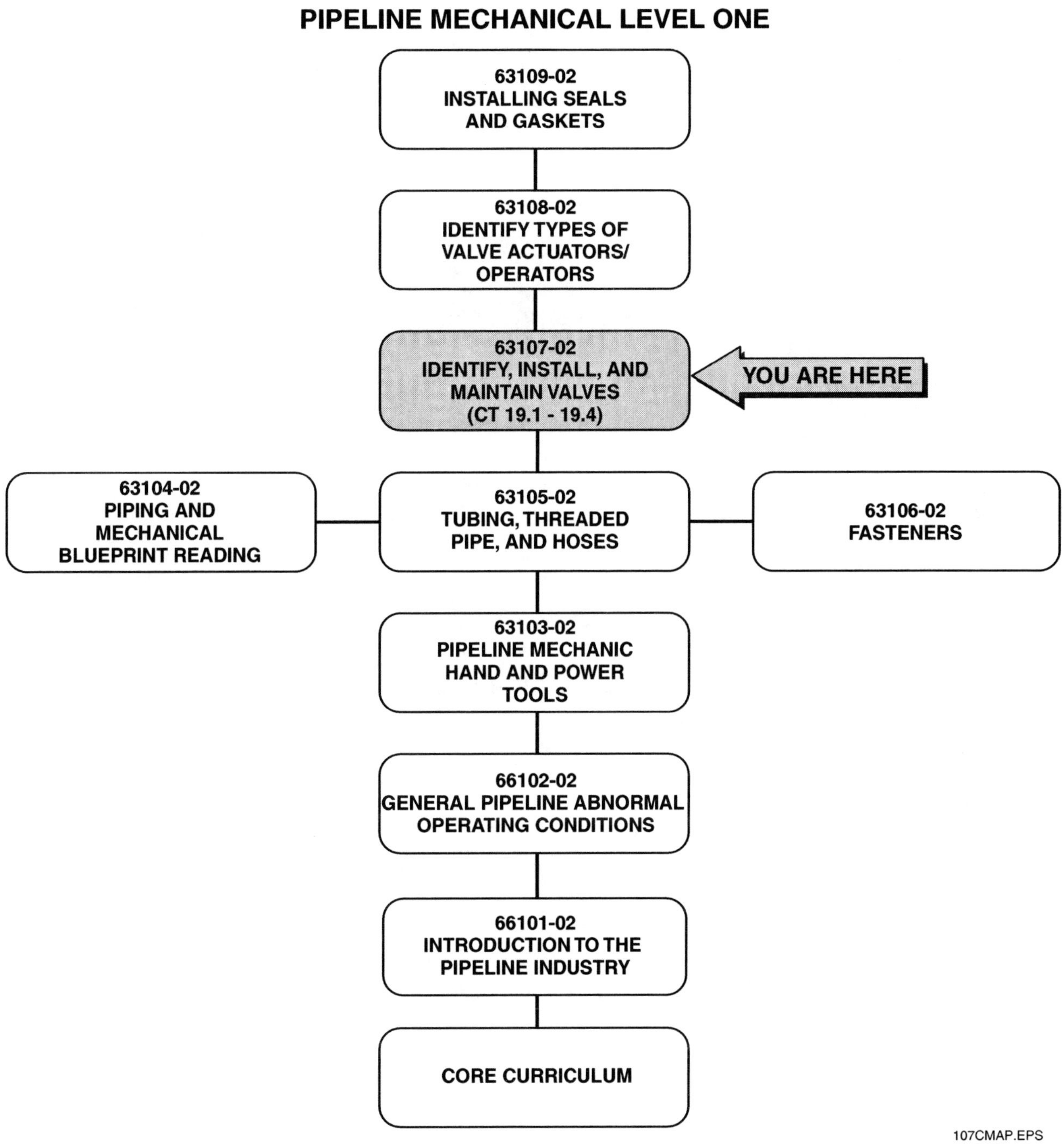

MODULE 63107-02 CONTENTS

Figures

Tables

MODULE 63107-02

Identify, Install, and Maintain Valves (CT 19.1-19.4)

Objectives

When you have completed this module, you will be able to do the following:

1. Identify the types of valves used to start and stop flow on a pipeline and regulate flow on a pipeline.
2. Identify and explain the functions of the various types of relief valves.
3. Properly isolate and purge a valve.
4. Understand and perform preventive maintenance and lubrication on pipeline valves (CTs 19.2, 19.3, 19.4).
5. Winterize valves (CT 19.1).
6. Install threaded and flanged valves.

Prerequisites

Before you begin this module, it is recommended that you successfully complete the following: Core Curriculum; Pipeline Core, Modules 66101 and 66102; Pipeline Mechanical Level One, Modules 63103 through 63106.

Required Trainee Materials

1. Pencil and paper
2. Appropriate personal protective equipment

1.0.0 ◆ INTRODUCTION

Valves are the basic control devices of the pipeline system. They are mechanical devices that start and stop product flow, control the rate of flow, and act as safety devices on the pipeline. Some are manually operated while others are automatic. The entire system is controlled by the method, sequence, and timing of the opening and closing of respective valves.

Although there are many variations for particular applications, basic valve design is relatively uniform. All valves have two common characteristics. Every valve has a passageway in the **body** through which fluid passes and a moveable part that moves into the passageway to open, close, or change the size of the passageway.

A valve can be used as a shutoff device for on-off service, a **throttling** device for controlling flow, a pressure relief device for protecting the pipeline, or as a safety device to prevent reversal of flow. The pipeline achieves efficient operation according to the way the various valves and pipes are installed and operated.

Valves are rated for maximum operating pressure (MOP) of the valve. Valve ratings are provided by the American National Standards Institute (ANSI) Code B31.5 and API 60. Each valve must have the ANSI rating stamped on the body of the valve. Valve selection is made by using a valve-rating chart (see *Table 1*).

Table 1 ANSI Valve-Rating Chart

ANSI	MOP (psi)
150	275
300	720
400	960
600	1,440
900	2,160
1,500	3,600

Pipeline designers select valves carefully to ensure that the integrity of the pipeline is maintained. A pipeline is like a chain in that it is only as strong as its weakest link. A poor selection of valves could greatly reduce the efficiency of the entire pipeline system. For example, imagine a pipeline was designed to use ANSI 900 valves and be rated at MOP of 2,160 pounds per square inch (psi). By mistake, an ANSI 400 valve was installed at some point. The MOP of the entire pipeline section would be reduced to 960 psi or 44% of the planned MOP, because the ANSI 400 valve would fail long before normal operating pressure was reached.

This module provides guidance in the selection of valves to be used on the pipeline. An explanation of proper installation, maintenance, service, and winterizing techniques is also provided. Guidelines for removing a valve from service, servicing, and returning a valve to service and the proper types of documentation and record keeping are also presented.

2.0.0 ◆ VALVES THAT START AND STOP FLOW

Pipelines use six basic types of valves to start and stop product flow. These types of valves close completely and seal off all flow. These valves are not suitable for throttling flow as they are designed to be completely open or closed.

2.1.0 Gate Valves

Gate valves (see *Figure 1*) are one of the most common types of valves used on the pipeline. It is the best valve for services that do not require frequent valve operation and for services that require the valve to be either fully open or fully closed. Such valve action is required for pigging operations.

Gate valves should not be used in applications that require throttling of pipeline flow. Close control of flow through a gate valve is almost impossible. Gate valves have very little resistance to flow and exert minimal pressure drop across the valve.

In a gate valve, a gate-like **disc**, actuated by a **stem** and turned by a handwheel, moves up and down across the path of flow through the valve and **seats** against two seat faces to shut off flow. Gate valves have four major components: the body, **bonnet**, gate, and stem.

2.1.1 Body

The body of the valve is connected directly to the pipeline and is the main section of the valve. It is the area through which fluid flows. The body of the valve is sized by casting or machining to nominal size, corresponding to the size of the pipeline in which it is to be installed. The size of the pipe and the size of the valve body must be the same.

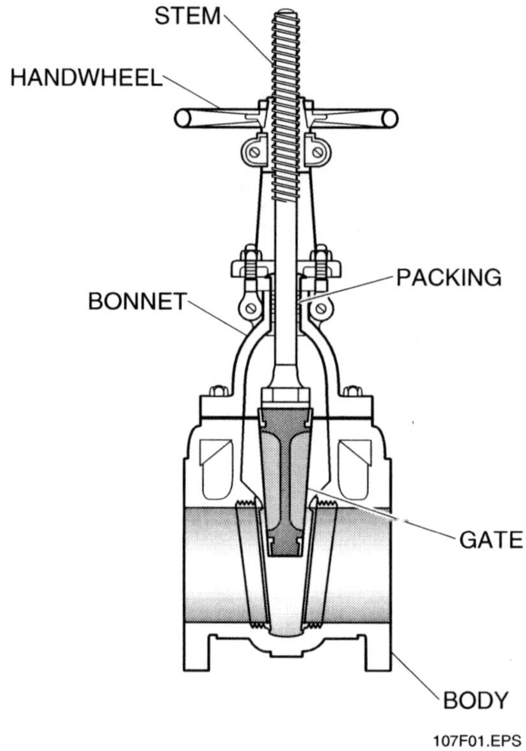

Figure 1 ◆ Typical gate valve.

2.1.2 Bonnet

The bonnet is located above the body and encloses the **trim**, including the stem, disc, seat ring, disc holder or guide, wedge, and bushings. Various types of bonnets are used with gate valves. In high-pressure applications like a pipeline, most gate valves have a flanged, pressure-seal bonnet that is bolted to the body using a ring-type gasket (see *Figure 2*).

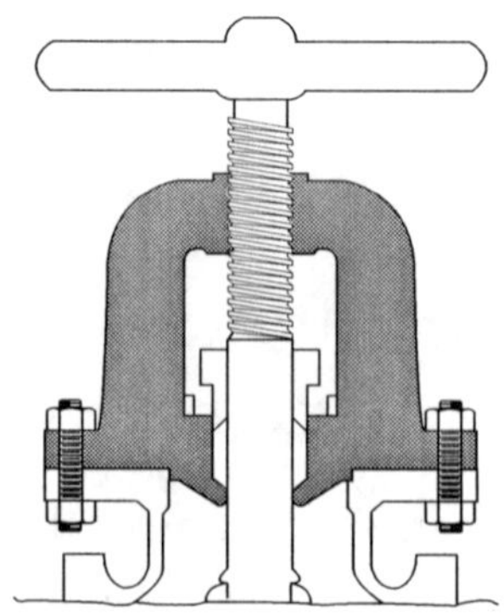

Figure 2 ◆ Flanged, bolted bonnet.

Valves with bolted bonnets are commonly used for line sizes larger that 1¼ inch. Such equipment can withstand temperatures up to 925°F. The bolted bonnet gate valve is also the preferred choice in severe applications where dismantling is frequent. This type of valve must be adjusted after the pipeline is placed in service because application of pressure to the valve relaxes the bolting. This type of valve should be inspected regularly for seal leakage. *Figure 3* shows gate valves with flanged bonnet.

107F03.EPS

Figure 3 ◆ Gate valves with flanged bonnet.

2.1.3 Gate

The gate in a gate valve is the component that actually stops the flow through the valve when the valve is closed. **Full port** gate valves provide the maximum internal opening through the valve. This opening will match the internal diameter of the joining pipes.

The types of gates used in gate valves include the solid-wedge gate, the double-disc gate, and the simple slab gate (see *Figure 4*). The solid-wedge gate is the simplest and most common type of gate. It has good stress resistance and does not trap product in the bonnet. It is made of one piece that is accurately machined on each side to form a leak-proof seal with the seat faces.

The double-disc gate consists of two discs that are forced against the seat faces by a spreader located between the discs. Double-disc gates are used for moderate temperatures and are only found in bronze and iron valves. Specially designed gates are also available from various valve manufacturers.

The simple slab gate valve is an efficient, full-port gate valve that uses two floating seats and hardened steel to provide a complete seal with the gate. In corrosive environments, simple slab gate valves with Teflon® inserts and elastomer O-rings are recommended to prolong the life of the valve. As pressure is applied across the port opening, the floating action of the gate and seats ensure a tight seal on both the **upstream** and downstream sides of the gate.

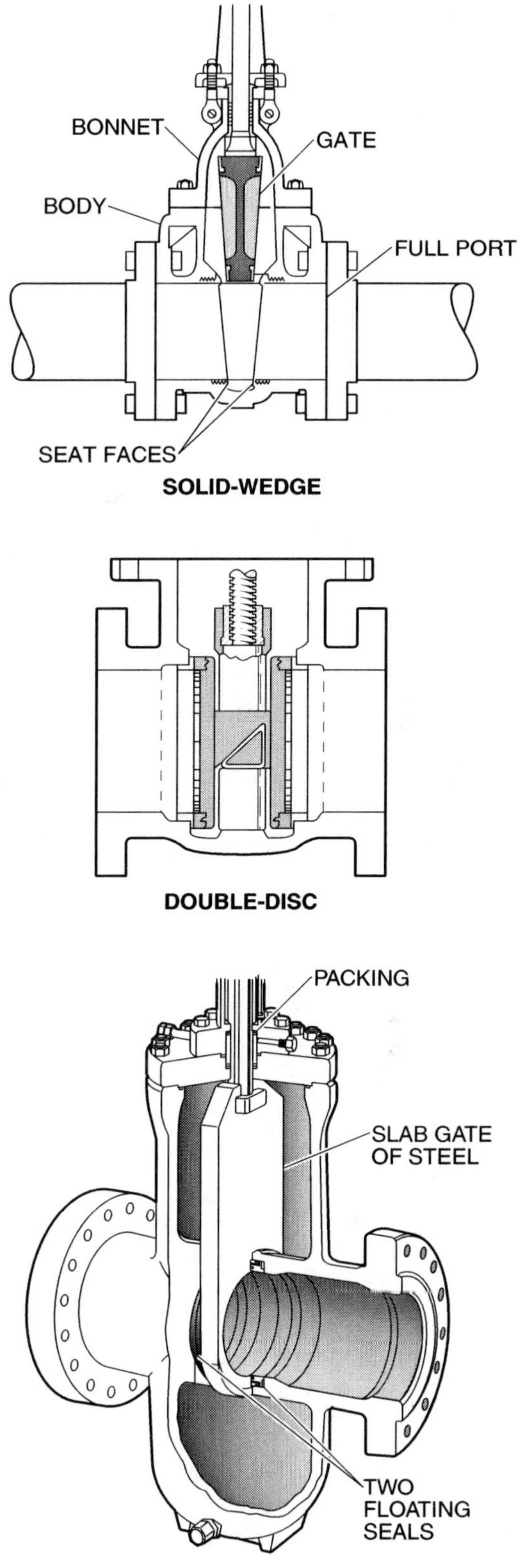

107F04.EPS

Figure 4 ◆ Types of gate valve gates.

2.1.4 Stem

The valve stem provides a link between the handwheel and the gate. Many times, the installation crew has to attach the handwheel to the stem after the valve is installed. Some handwheels bolt onto the stem. Other handwheels slip over a keyway cut into the stems and are secured to the stems by a nut or machine screw. Handwheels vary according to the type of stem they fit. Three general types of stems include the rising stem, the nonrising stem, and the outside screw-and-yoke (OS&Y) stem (see *Figure 5*).

When a rising stem valve is opened, both the handwheel and the stem rise. The height of the stem gives an approximation of how far the valve is open. These stems can only be installed where there is sufficient clearance for the handwheel to rise. Rising stems are exposed to the product flowing through the valve. Rising stem valves are only used with products that will not harm the threads, such as hydrocarbons, water, and steam.

Nonrising stems are suitable where space is limited. As the name implies, neither the stem nor the handwheel rises when the valve is opened. A spindle inside the valve body is turned when the handwheel is turned and the gate is raised. Stem wear is held to a minimum. This type of stem is exposed to the product flowing through the valve.

The OS&Y stem is suitable for corrosive products because the stem is not exposed to the product in the pipeline. As the handwheel is turned, the stem moves up through the handwheel. The height of the stem exposed indicates how far the valve is open.

2.2.0 Ball Valves

A **ball valve** has a ball with a hole through its center that acts as the control element (see *Figure 6*). This ball is located in the center of the valve body and is attached to a valve stem. A handle is attached to the valve stem. Ball valves turn one-quarter turn, or 90 degrees, to move the hole in line with the flow through the valve. Turning the handle one-quarter turn in the opposite direction moves the solid part of the ball in line with the valve inlet and outlet, stopping the flow through the valve.

Ball valves are quick-opening and easy and inexpensive to repair. The position of the handle indicates whether the valve is opened or closed. If the handle is in line with the flow line of the pipe, the valve is open. If the handle is at a right angle to the flow line of the pipe, the valve is closed.

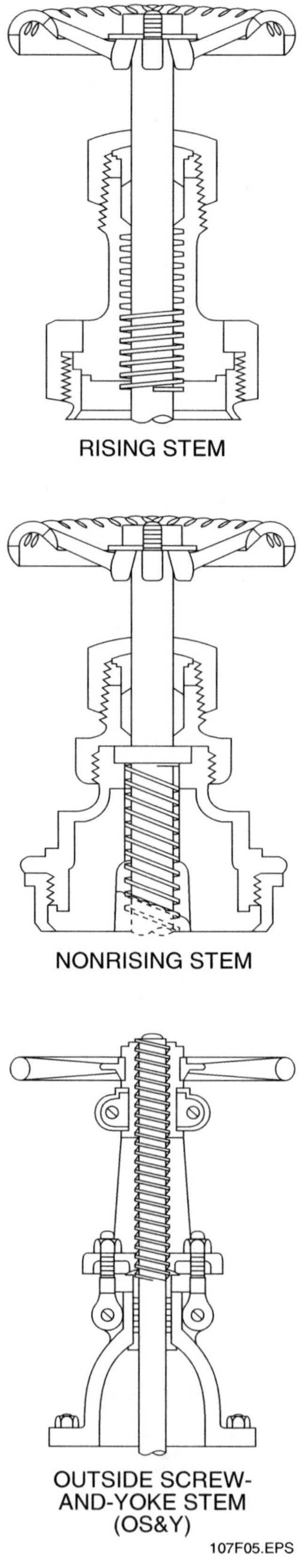

Figure 5 ◆ Types of stems.

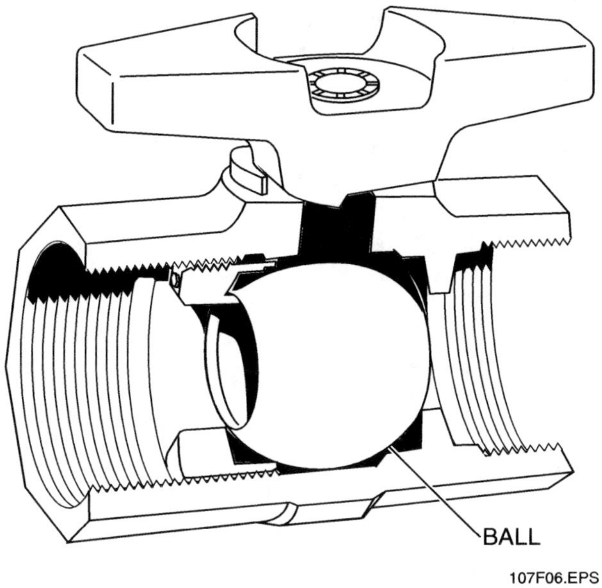

Figure 6 ◆ Ball valve.

2.3.0 Plug Valves

The **plug valve**, also known as a *cock* or *petcock*, consists of a cone-shaped plug that fits into a machined hole. This plug has a hole through it to allow product to flow through the valve when it is in the open position. Plug valves are not used to regulate flow but only to turn the flow completely on or off.

The plug valve can be operated using a handwheel or a wrench attached to the top of the stem. The top of the stem is often a square plug to which a wrench is attached for opening and closing only. The wrench is removed until it is needed again. There are three types of plug valves (see *Figure 7*). The regular plug valve pattern provides the largest **port** opening available in plug valves. The port is the same diameter as the inside of the pipeline. Short-pattern plug valves offer the largest port valve opening available in a size that is consistent with the face-to-face dimensions of a gate valve. The venturi pattern has a smaller port than the other two patterns and provides smooth flow angles into and out of the valve to minimize pressure drop through the valve.

Plug valves are easy to operate and require a minimum amount of installation space. The advantages of a plug valve include the following:

- A fully opened valve exposes no part of the machined surfaces of the plug to erosion or corrosion by the product flowing.
- A fully opened plug valve provides a smooth passageway that does not allow sludge or particles to accumulate.

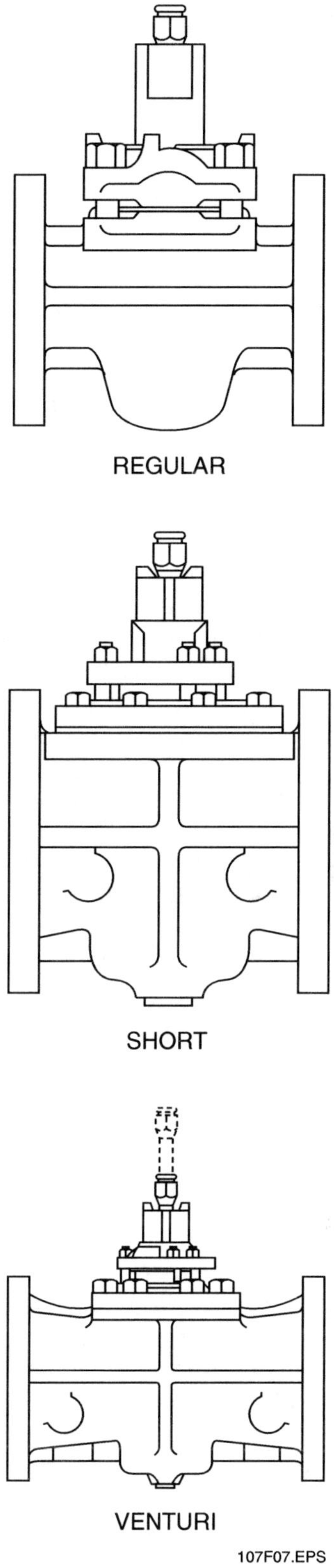

Figure 7 ◆ Three types of plug valves.

- A quarter turn of the wrench or handwheel will open or close the valve.
- A closed valve exposes only a very small area to the product in the pipeline.

2.3.1 Lubricated Plug Valves

Lubricated plug valves have a lubricating screw at the top of the valve that forces lubricant into grooves in the plug and to the bottom lubricant chamber. These plugs are used in many critical situations that ordinarily require the plug to be repacked with grease under pressure. There are several disadvantages to using lubricated plugs:

- The lubricant may contaminate the product in the pipeline.
- Lubrication requires extra maintenance.
- The type of lubricant determines the maximum service temperature of the valve.

2.3.2 Nonlubricated Plug Valves

Nonlubricated plug valves have a cam-crank mechanism that lifts the plug and turns it without creating friction between the plug and the seat. These valves are used for service where lubrication is inconvenient or temperature exceeds lubricated plug limits. They are excellent for installations where corrosion requires special alloys and linings. The disadvantages of nonlubricated plug valves are that they cannot be repacked while under pressure and that the seal they provide is not as positive as that provided by a lubricated plug valve.

2.3.3 Three-Way Valves

Three-way valves are multi-port plug valves that are installed at the intersection of three lines. They are used to direct the flow between two of the lines only and to block off the third line. Situations requiring three-way valves include alternating the connections of two supply lines to a common delivery line or diverting a line into one or the other of two possible directions. Three-way valves are designed so that when the plug is turned from one position to another, the channels previously connected are completely closed off before the new channels begin to flow, preventing mixture of fluids or loss of pressure. Three-way valves come in several different arrangements. They can be two- or three-port valves with stops that limit the turning of the plug to two, three, or four positions (see *Figure 8*).

2.3.4 Four-Way Valves

A four-way diverter valve is another variation of the plug valve designed for specialized use and is widely used in the pipeline industry. They are sometimes called **double block-and-bleed** valves. The four-way valve is central to operation of bi-directional pipe prover systems. *Figure 9* shows a four-way valve installed in a pipe prover system.

These valves are constructed to very strict requirements. A diverter valve used in a prover system is required to cycle as many as 300 times each day, as compared to some valves that are only cycled weekly or yearly. Valves used in proving operations must cycle extremely quickly—in less than 10 seconds, compared to as much as a minute cycle time for ordinary valves. API Liquid Measurement Manual 2531 requires that such valves must be able to demonstrate zero leakage during each proving run. Four-way valves must demonstrate proof of sealing at very low differential pressures, 2 to 5 psi. Other valves require high differential pressures, 40 to 400 psi, to activate their seating mechanisms, not available in bi-directional prover systems. *Figure 10* shows a four-way valve operating a prover system.

2.4.0 Expanding Gate Valves

Expanding gate valves were developed to accomplish **block-and-bleed** operations without the necessity of two gate valves separated by a spool (see *Figure 11*). Originally, the spool was fitted with a bleed valve to guarantee product separation. Expanding gate valves are designated double block-and-bleed valves because they accommodate the same requirements with a single valve.

The design of expanding gate valves eliminates seal abrasion. Design allows the valve to be turned more easily, and when fully closed, a mechanically created, metal-to-metal seal is completed, reinforcing the primary seal. An airtight seal is created both upstream and **downstream**. These valves have no reliance on line pressure for complete seal.

Sealing action is accomplished by two independent seals mounted on a tapered plug. During closing, the mechanical wedge in the plug forces the sealing areas against the machined surfaces of the valve body. The plug and seals do not contact the surfaces until the plug is completely lowered and in position for sealing. After the plug is completely in position, the valve action forces the seals against the valve body, forming complete and separate seals on both sides of the valve. No scrubbing or wear on the seal is possible during to this action. The valve may then be bled, and no contamination is possible.

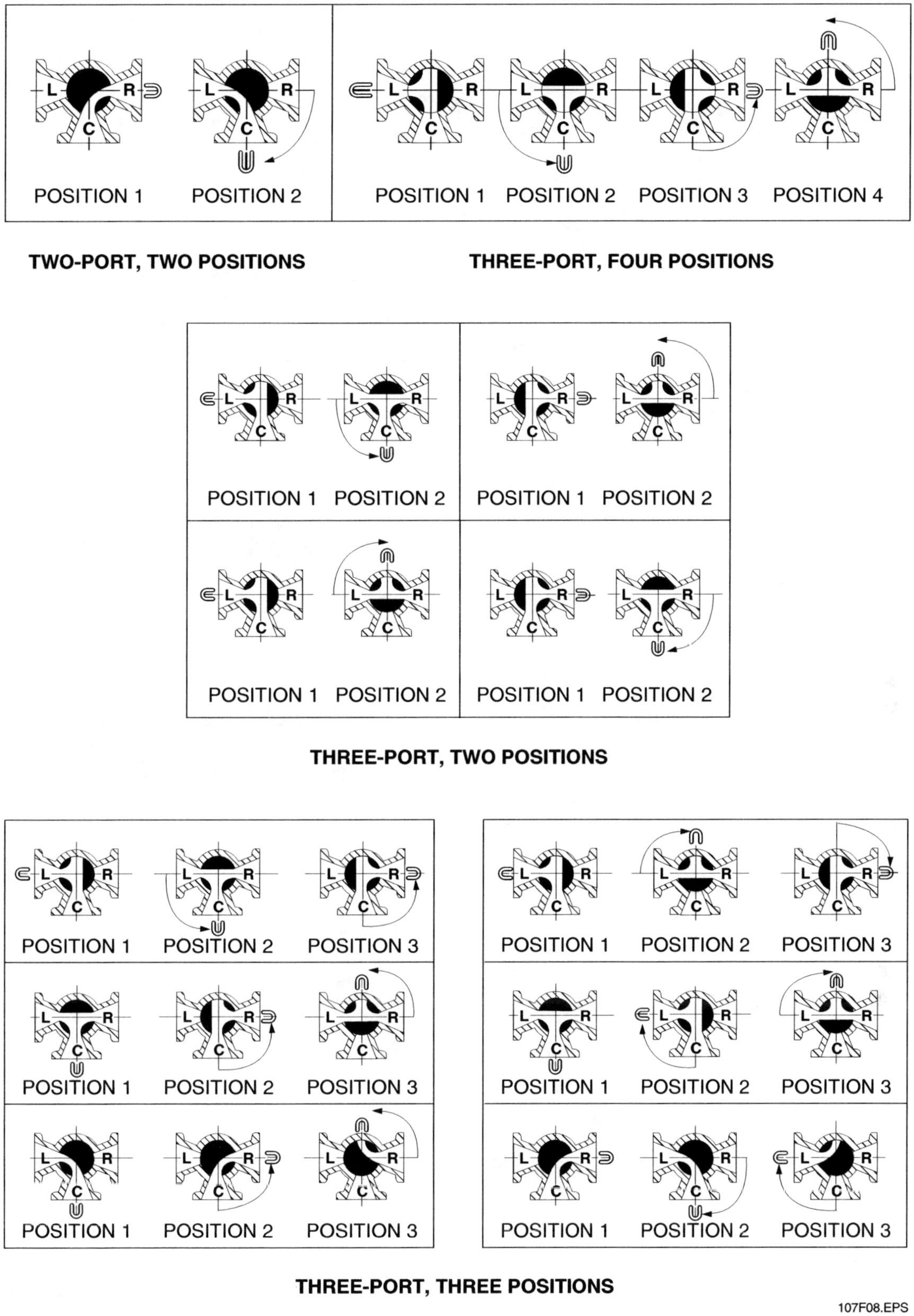

Figure 8 ◆ Applications of three-way valves.

Figure 9 ◆ Four-way valve installed in a pipe prover system.

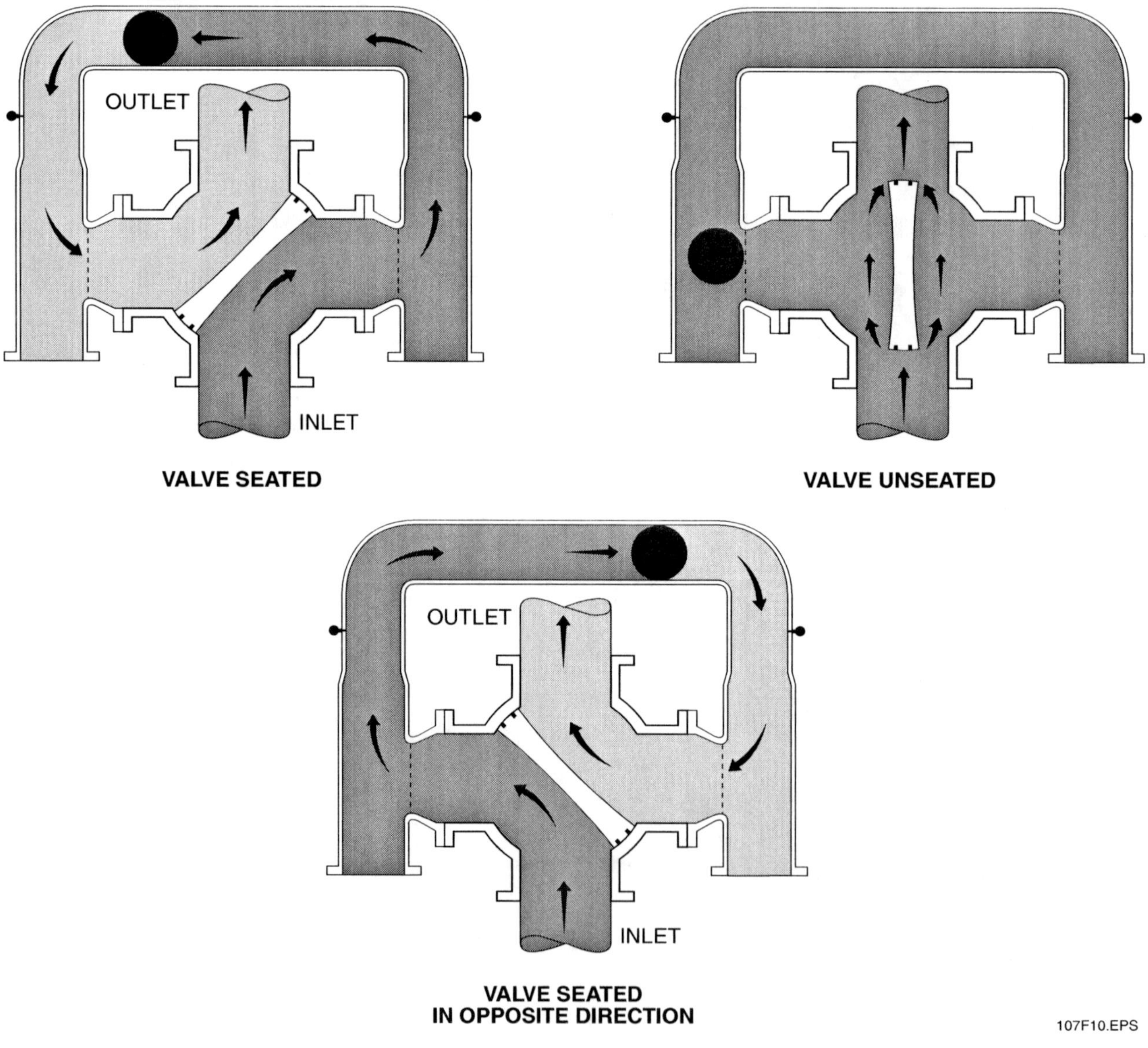

Figure 10 ◆ Four-way valve operating a prover system.

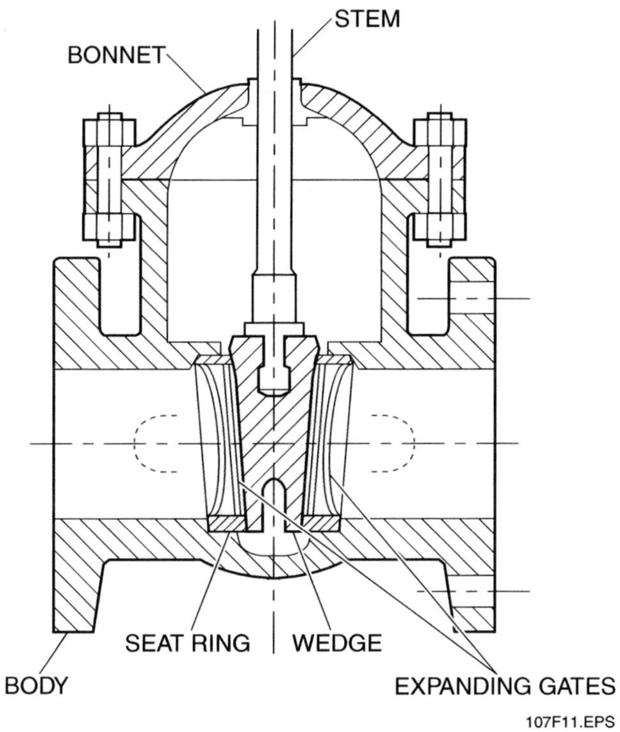

Figure 11 ◆ Expanded gate valve.

3.0.0 ◆ VALVES THAT REGULATE FLOW AND PRESSURE

Many valves are designed to provide accurate flow control through a system. The valves used to regulate flow through a pipeline system include the following:

- Globe
- Angle
- Y-type
- Butterfly
- Diaphragm

3.1.0 Globe Valves

The most common valve used to control flow is the **globe valve** (see *Figure 12*). Globe valves are usually installed in lines that are ½ to 4 inches in diameter but are available in sizes up to 12 inches. Globe valves permit close control of flow because the size of the seat-to-globe opening is directly proportional to the number of turns of the handwheel. The number of turns of the handwheel is directly related to the amount of the opening.

The globe valve has a disc that can be partially opened with very little friction to minimize damage to the valve. The design of the globe valve permits flow all around the disc when the valve is partially open, preventing wear or erosion of just one part of the disc. Globe valves are used in pipelines to regulate flow. They are also installed where frequent operation of the valve is required. The major components of a globe valve are the body, bonnet, disc, and stem.

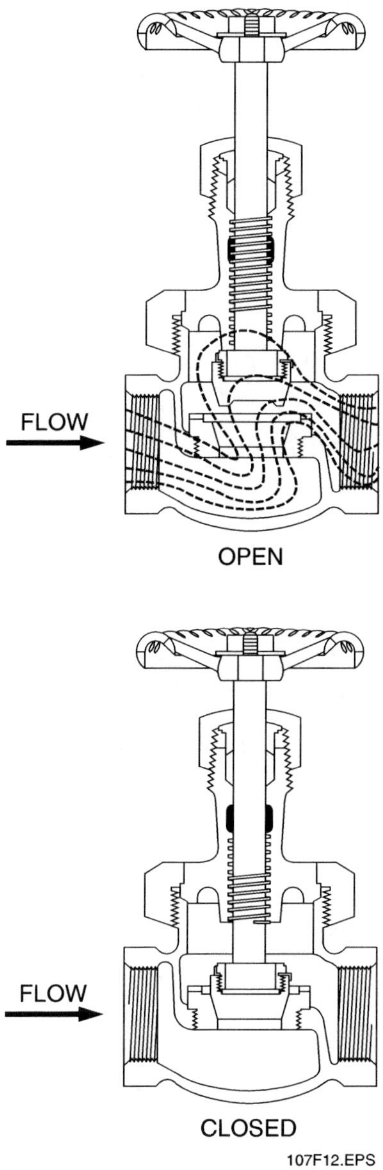

Figure 12 ◆ Globe valve.

3.1.1 Body

Globe valve bodies have the same general appearance as gate valves. They can be manufactured by the same casting process as gate valve bodies. They differ from gate valve bodies mainly in seat design. In a globe valve, the seat opening is in line with the valve stem and the seats are parallel with the line of flow through the valve. Most small valves are built with the seat as a fixed part of the body, and larger, more expensive globe valves are usually made with removable seat rings.

3.1.2 Bonnet

Globe valves are made with several different bonnet arrangements. The more common types of globe valves use a screwed-in bonnet, bolted bonnet, or union bonnet. The screwed-in bonnet should not be used for critical services, because if the stem were to become bound to the bonnet by rust, cross threading, or grit in the threads, the bonnet could be accidentally unscrewed while the valve is being opened. Critical service means service in which valve malfunctions could cause the system to shut down or personnel to be injured.

3.1.3 Disc

Globe valve discs are made in three designs: the conventional disc, the plug-type disc, and the composition disc (see *Figure 13*). There are also variations within each of these types because each manufacturer uses different design features.

The conventional disc has a relatively short taper on its seating face and has a high seating pressure. This enables this type of disc to crush hard deposits that may form on its seating face. The conventional disc is useful in controlling the flow through a pipeline.

The plug-type disc has a much larger seat contact area than the conventional disc but is otherwise quite similar. Its main advantage over other types of discs is that it offers better flow control of the fluid through the valve. The plug-type disc can also withstand more damage from dirt, scale, and solids than the conventional disc before it has to be replaced.

The composition disc consists of three basic parts: a metal disc holder, a disc, and a retaining nut. Composition discs are available in a number of different materials, including rubber, neoprene, and a variety of plastics. They can easily be replaced if they leak during use, but they are not as suitable for throttling products as the other types of discs.

3.1.4 Stem

Globe valves use rising and OS&Y stems. The number of turns of the handwheel determines how far the valve is opened. If 12 complete revolutions of the handwheel are necessary to open the valve all the way, then 3 revolutions will open the valve a quarter of the way, and 6 revolutions will open the valve halfway.

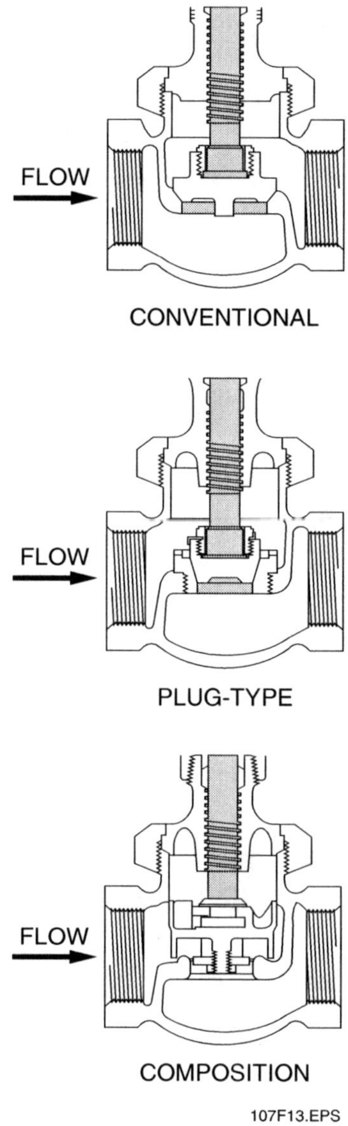

Figure 13 ◆ Types of globe valve discs.

3.2.0 Angle Valves

Angle valves are variations of globe valves (see *Figure 14*). The outlet of an angle valve is at a 90-degree angle from the inlet. There is less resistance to flow in an angle valve than in a standard globe valve, and the angle valve offers a change of direction in a pipeline without using an elbow. These valves are normally used in non-critical services in place of a globe valve and an elbow. You must be careful when using angle valves because bends in piping systems are subject to stresses that should not be placed on valves.

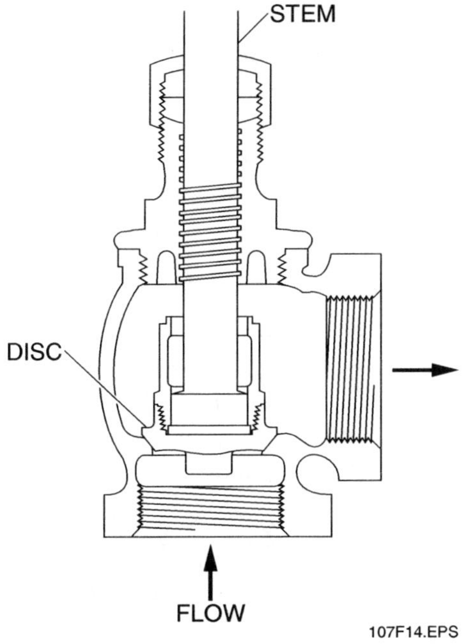

Figure 14 ◆ Angle valve.

3.3.0 Y-Type Valves

A Y-type valve is a cross between a gate valve and a globe valve (see *Figure 15*). A Y-type valve provides straight-through flow with minimum resistance of the gate valve and the throttling ability and flow control of a globe valve. The Y-type valve produces lower pressure drop and turbulence than a standard globe valve and is preferred over a globe valve for use in corrosive services. Y-type valves are also used in many high-pressure applications including boiler systems. All of the disc and seat designs available in globe valves are also available in Y-type valves.

3.4.0 Butterfly Valves

A butterfly valve has a round disc that fits tightly in a mating seat and rotates 90 degrees in one direction to open and allow fluid to pass through the valve. The butterfly valve can be operated quickly by turning the handwheel or hand lever one-quarter of a turn, or 90 degrees, to open or close the valve.

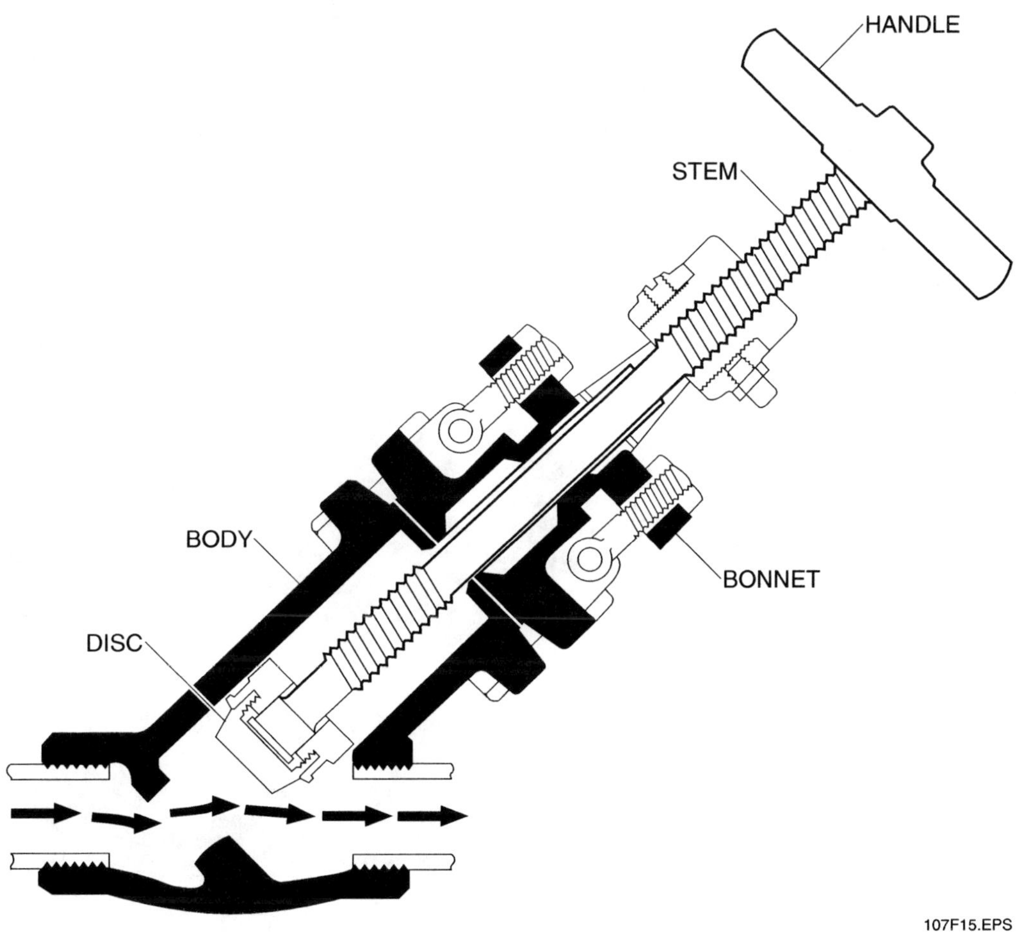

Figure 15 ◆ Y-type valve.

Butterfly valves can be used completely open, completely closed, or partially open for non-critical throttling applications. Butterfly valves are used in low- to medium-pressure and low- to medium-temperature applications. When the butterfly valve is equipped with a hand lever, the position of the hand lever indicates whether the valve is opened, closed, or partially open. When the lever is parallel with the flow line through the valve, the valve is open. When the lever is perpendicular to the flow line through the valve, the valve is closed.

Butterfly valves that are 12 inches in diameter or larger are usually equipped with a handwheel or gear-operated actuator. This is required because of the large volume of product flowing through the valve and the great amount of pressure pushing against the seat that must be overcome when the valve is being closed.

Butterfly valves have an arrow stamped on the side, indicating the direction of flow through the valve. They must be installed in the proper flow direction or the seat will not seal and the valve will leak. Three types of butterfly valves are the wafer, wafer lug, and two-flange.

3.4.1 Wafer Valves

The wafer-type butterfly valve is designed for quick installation between two flanges (see *Figure 16*). The valve seat is lapped over the edges of the body to make contact with the valve faces. Boltholes are provided in large wafer valves to help line up the valve between the flanges. During installation of a wafer valve, the valve must be in the open position before beginning to apply torque to the bolts to prevent damage to the valve seat.

3.4.2 Wafer Lug Valves

Wafer lug valves are the same as the wafer valves except that they are designed with bolt lugs around the valve body (see *Figure 17*). The lugs on some wafer lug valves are drilled and tapped so that when the valve is closed, a pipeline downstream can be dismantled for cleaning or maintenance while product and upstream pipeline are left intact. When a tapped wafer lug valve is used for pipeline end applications, only one flange is necessary.

3.4.3 Two-Flange Valves

The body of a two-flange butterfly valve is made with a flange on each side (see *Figure 18*). The valve seat is not lapped over the flange ends of the valve, so gaskets are required between the flanged body and the mating flanges of the pipeline. The two-flange valve is made with either machined, flat-faced flanges or raised-face flanges. The mating flanges on the pipeline must match the valve flanges.

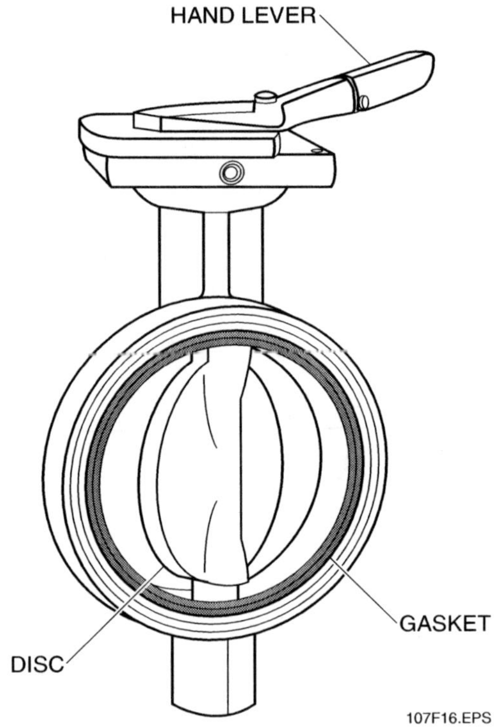

Figure 16 ◆ Wafer butterfly valve.

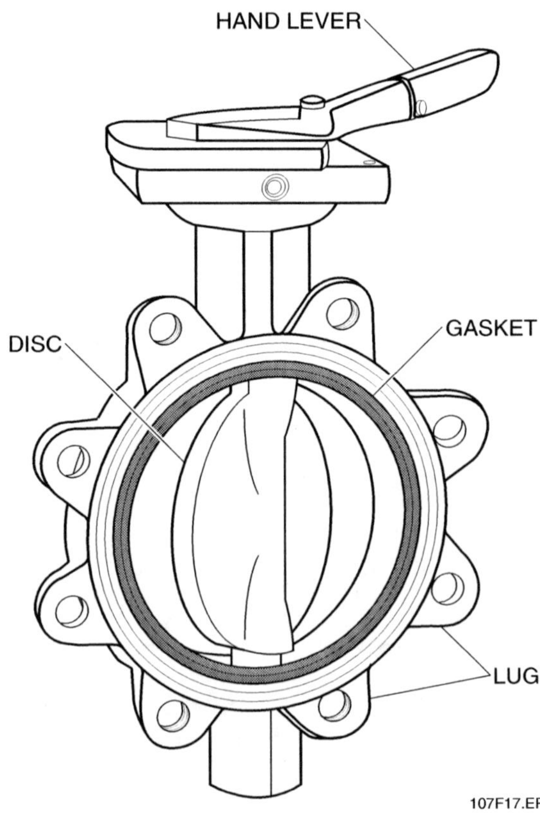

Figure 17 ◆ Wafer lug butterfly valve.

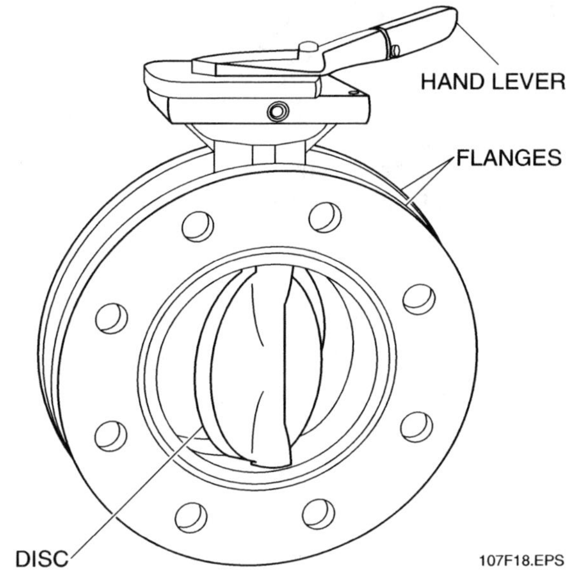

Figure 18 ◆ Two-flange butterfly valve.

3.5.0 Diaphragm Valves

A diaphragm valve controls flow by means of a flexible disc that is connected to a compressor by a stud molded into the disc (see *Figure 19*). The valve stem moves the compressor up and down, regulating the flow. Diaphragm valves have no seats because the body of the valve acts as the seat for the flexible disc. The operating mechanism of the valve is not exposed to the product within the pipeline.

These valves can be used fully opened, fully closed, or for throttling flow. They are good for handling slurries, highly corrosive materials, and materials that must be protected from contamination. Many types of fluids that would clog other types of valves will flow through a diaphragm valve.

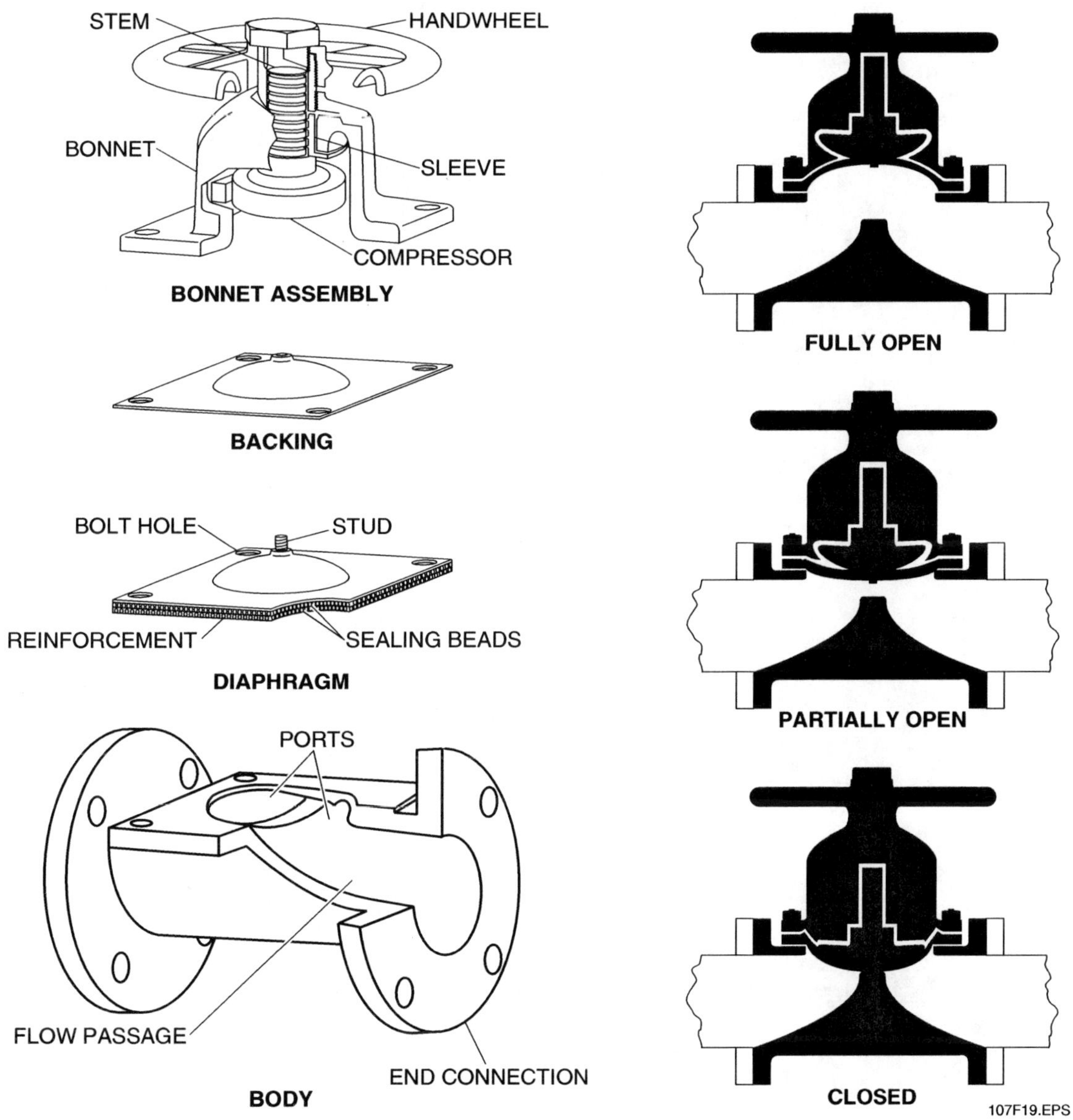

Figure 19 ◆ Diaphragm valve.

4.0.0 ◆ VALVES THAT RELIEVE PRESSURE

Valves used to relieve pressure in a pipeline, tank, or vessel are known as safety valves or pressure-**relief valves**. These are installed in pipelines to prevent excessive pressure from rupturing the pipeline and causing an accident or damage to the environment. Safety valves and pressure relief valves are adjustable and operate automatically.

4.1.0 Control Valves

Control valves are variations of angle and globe valves that are controlled by pneumatic, electronic, or hydraulic actuators. Control valves normally operate in a partially open position. They are commonly used on the pipeline for pressure and flow control. As explained earlier, control valves are used to control flow but they also relieve pressure in some applications. Control valves are precision-built for increased accuracy in flow control. The control valve is usually smaller than the pipeline to avoid throttling and constant wear of the valve seat (see *Figure 20).*

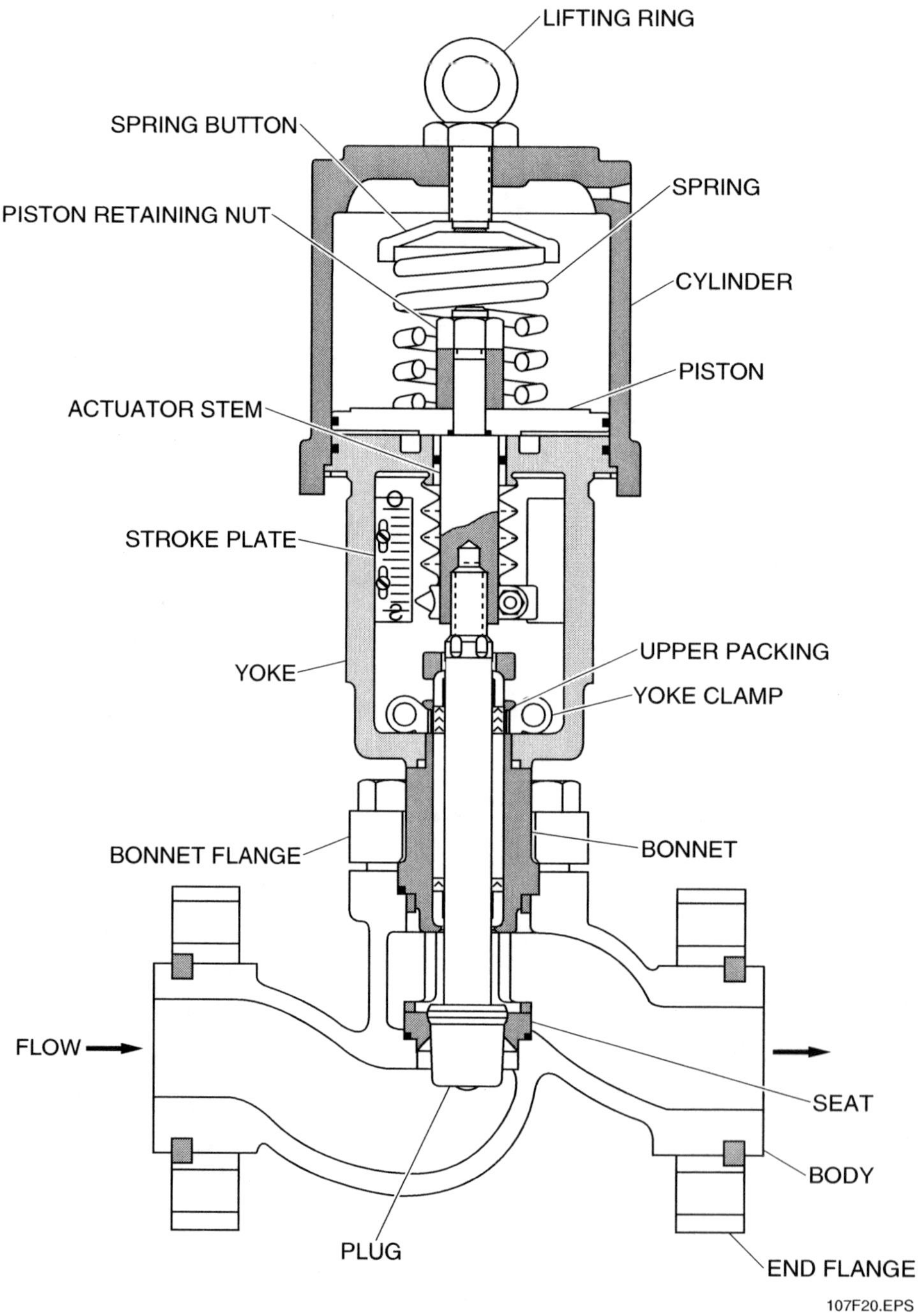

Figure 20 ◆ Control valve.

Control valves are used with pressure-sensing elements that measure the pressure drop at a given point in the pipeline. The flow rate is directly related to the pressure drop. The sensing element sends a pressure signal to a controller that compares the actual pressure drop with the pressure drop for the desired flow rate. If the actual flow is different, the valve controller sends an alarm signal to the control center console. The operator can adjust the valve from the control center through the controller and valve actuator to increase or decrease the flow. *Figure 21* shows a schematic for a control valve in a pipeline.

4.2.0 Globe-Type Safety Valves

Safety valves are used in steam, air, and gas pipelines. Safety valves operate in the closed position until the pressure in the line or unit rises above the preset pressure limit of the valve (see *Figure 22*). At this point, the valve opens fully to relieve the pressure, remains open until the pressure drops, and then snaps shut. Because of this fully open and tight-closing feature, safety valves are commonly referred to as *pop-off valves*.

Safety valves are most commonly installed in the vertical position. Safety valves should be mounted directly to the tank or vessel being protected, using connecting piping and block valves. Many safety valves are fitted with an outside lever so that pressure can be relieved for maintenance.

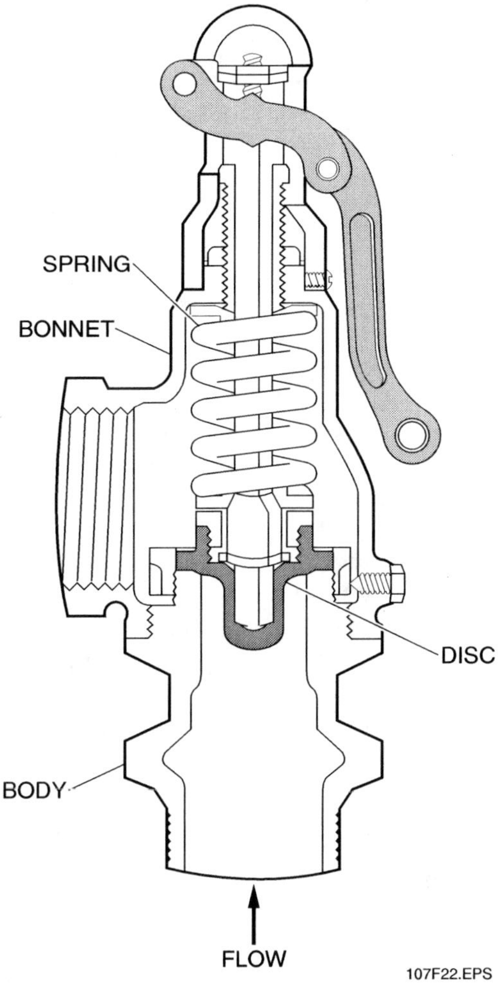

Figure 22 ◆ Safety valve.

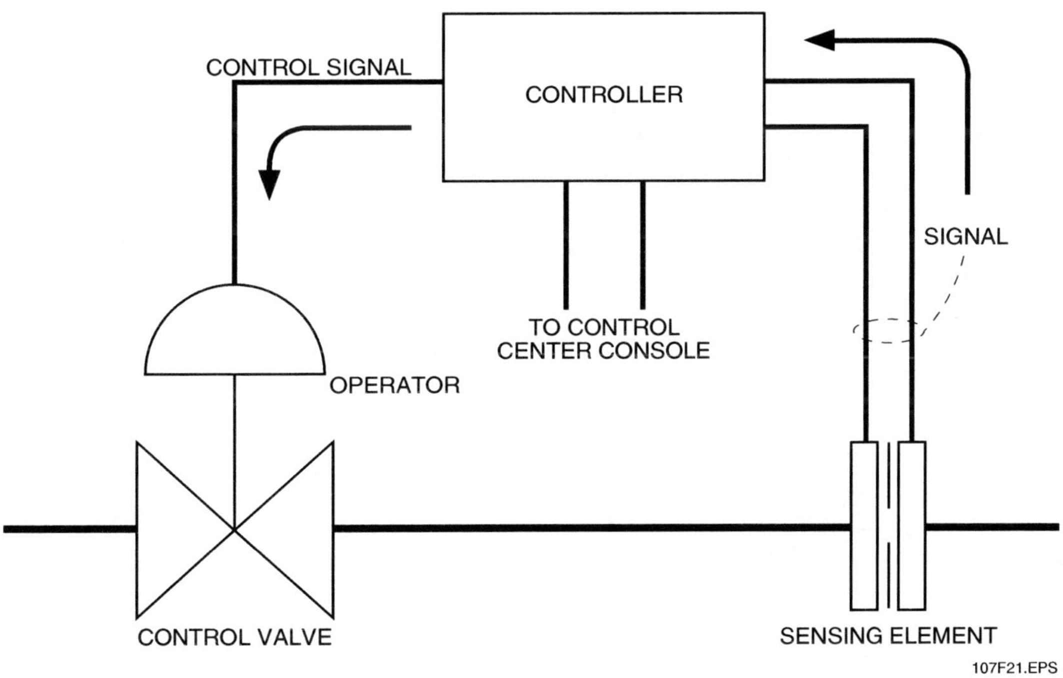

Figure 21 ◆ Schematic for a control valve in a pipeline.

Before a safety valve can close, its discharge pressure must drop several pounds below the opening pressure. Safety valves are normally factory calibrated and adjusted for appropriate operation.

4.3.0 Pressure-Relief Valves

Relief valves are used to relieve liquid or gas pressure on pipelines. Relief valves in small installations are operated by pressure acting directly on the bottom of a disc that is held on its seat by a spring (see *Figure 23*). The amount of spring pressure on the disc is determined at the factory when the valve is manufactured. When pressure becomes great enough to overcome the force of the spring on the disc, the disc rises, allowing the liquid or gas to escape. After enough liquid or gas has escaped to allow the pressure to drop, the system pressure becomes too low to overcome the force of the spring on the disc. The spring then pushes the disc back onto its seat, stopping the flow of liquid or gas through the valve. The force of the spring is called the compression of the valve.

Spring-loaded relief valves are installed with the stem in the vertical position. The piping connected to these valves must be as large or larger than the valve inlet and outlet openings. A reducer may be used to reduce a larger relief pipe down to the size of the outlet of a relief valve. However, the line size to the inlet of the relief valve cannot be increased above the size of the inlet. Relief piping must be well supported so that line strains will not cause the valve to leak at the seat.

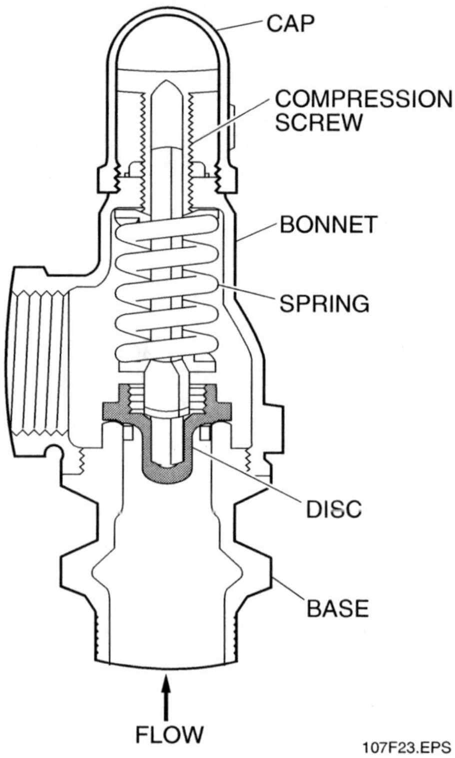

Figure 23 ◆ Pressure-relief valve.

4.4.0 Rupture Disc Devices

Rupture disc devices are pressure-relief devices that are installed to limit the amount of pressure on a line. They are used to protect equipment from overpressure in many manufacturing process installations as well as steam and gas.

The pressure-sensitive part of the device is called a rupture disc. Rupture discs are constructed of a very thin metal membrane, designed and manufactured to withstand predetermined levels of pressure. Discs must be selected for the exact application needed. Installation must be in line with expected flow because they are directional in function. If any backpressure is expected, a support member must be included.

Rupture discs are one-time devices because they operate by rupturing or bursting. When pressure exceeds their design limit and rupture occurs, they must be replaced (see *Figure 24*). Manufacturers now provide many models in quick-change units.

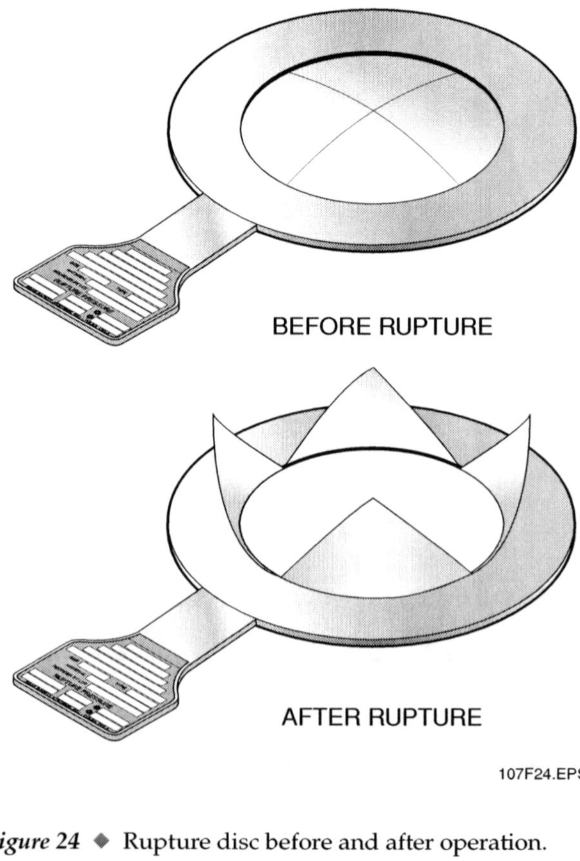

Figure 24 ◆ Rupture disc before and after operation.

4.5.0 Pipeline Relief Valves

Because of the great volumes of product moving at tremendous pressure through a pipeline, pressure relief on the pipeline is a normal occurrence: regular, repetitive, and ongoing. Pipeline relief valves help alleviate excess pressure that builds up in the pipeline.

Pipeline relief valves have a cylindrical shape that allows them to be installed in-line (see *Figure 25*). The inner piston is held in position against the seat with pressure supplied by inert gas. This gas supply comes from a unit that is often buried nearby. When a surge wave approaches with enough force to overcome the gas pressure, the piston is forced open to allow excess product to flow to tankage, thereby relieving the pressure. The valve must open instantaneously to allow the surge to pass into a relief tank or sump. Pipeline relief valves are adjustable through the amount of gas pressure supplied. When the surge pressure has reduced, the piston is designed to close slowly and smoothly so as not to cause a water hammer effect. As with other pipeline equipment, these valves must be strong, durable, and dependable because they are cycling repetitively.

107F25.EPS

Figure 25 ◆ Pipeline relief valve.

Pipeline relief valve design is much more involved than the design of smaller valve units. Piston-type valves are used because they seal completely sealing after a relief event. The piston moves in a cylinder, and flow is cut off before the valve seat is reached. When the valve is being closed, the piston helps to clean any solid material that may have remained in the bore. The seat and bore of piston valves can be replaced very quickly and easily, making it an attractive choice for this application.

Pipeline relief valves are electronically monitored, sending constant signals to the control center console. The operator monitors pipeline relief as it occurs. If relief becomes excessive and a relief valve fails during such a critical time, the operator must respond immediately and decisively. Failure of a pipeline relief valve has the potential for disaster. Such occurrences become abnormal operating conditions.

5.0.0 ◆ VALVES THAT REGULATE DIRECTION OF FLOW

Valves that regulate the direction of flow are called **check valves**. Check valves regulate flow by preventing backflow of the product flowing in the pipeline, permitting flow in only one direction. Check valves are directional valves and must be installed as indicated by an arrow stamped on the side of the valve. A check valve installed in the wrong direction will simply block the pipeline. Force of flow and the action of gravity cause the valve to open and close.

Check valves are required to operate in such a way that they do not create extremely high surge pressure when closing. They are also required to open and close smoothly with no erratic behavior of the closing member. Because of differences in operations, even on the same pipeline, valve selection and placement must be carefully planned. For example, when several main-line pumps are on line and one pump shuts down, the check valve on that pump line must close immediately. On the other hand, if a single main-line unit pumping over a long expanse failed, a slow-acting check valve would be sufficient.

Check valves are manufactured in several different designs, including the following:

- Swing
- Lift
- Ball
- Butterfly
- Foot

5.1.0 Swing Check Valves

The most commonly used type of check valve is the swing check valve (see *Figure 26*). This valve has a disc that swings, or pivots, to open and close the valve. Because check valves provide straight-line flow through the valve body, they offer less resistance to flow than other types of check valves. Swing check valves can be mounted on horizontal installations or vertical installations where product flow is upward.

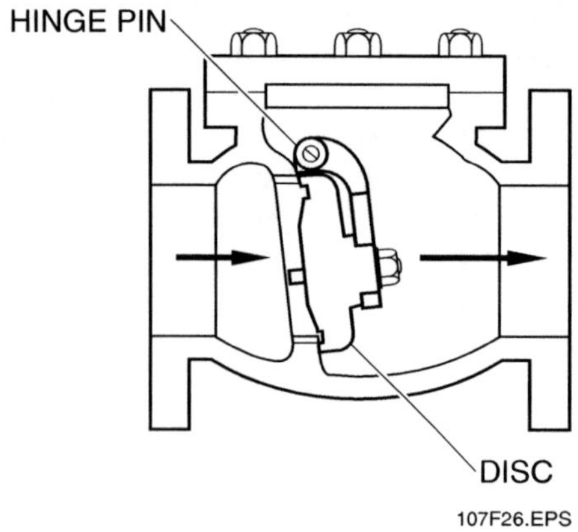

Figure 26 ◆ Swing check valve.

The disc, which is hinged at the top, seats against a machined seat in the tilted wall-bridge opening. The disc swings freely from the fully closed position to the fully open position, which provides unobstructed flow through the valve. The flow through the valve causes the disc to stay open. The amount that the disc opens depends on the volume of product moving through the valve. Gravity or the reversal of flow causes the disc to close, preventing backflow through the valve.

Swing check valves can also be equipped with an outside lever and counterweight to assist the valve in closing. These valves are often referred to as *weighted check valves* (see *Figure 27*).

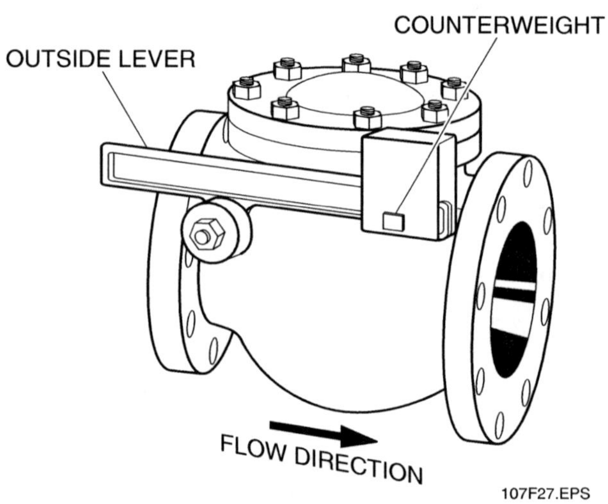

Figure 27 ◆ Weighted check valve.

5.2.0 Lift Check Valves

A lift check valve is designed like a globe valve, with an indirect line of flow through the valve. The lift check valve works automatically by line pressure and prevents the reversal of flow through the valve. A lift check valve consists of a disc that is held against the disc seat by gravity or by the pressure of product flowing in the opposite direction through the valve (see *Figure 28*). When product flows through the valve in the correct direction, the force of product lifts the disc away from the seat, allowing product to pass through the valve. Lift check valves can be designed for horizontal or vertical installations but must only be used for the type of installation for which they are designed.

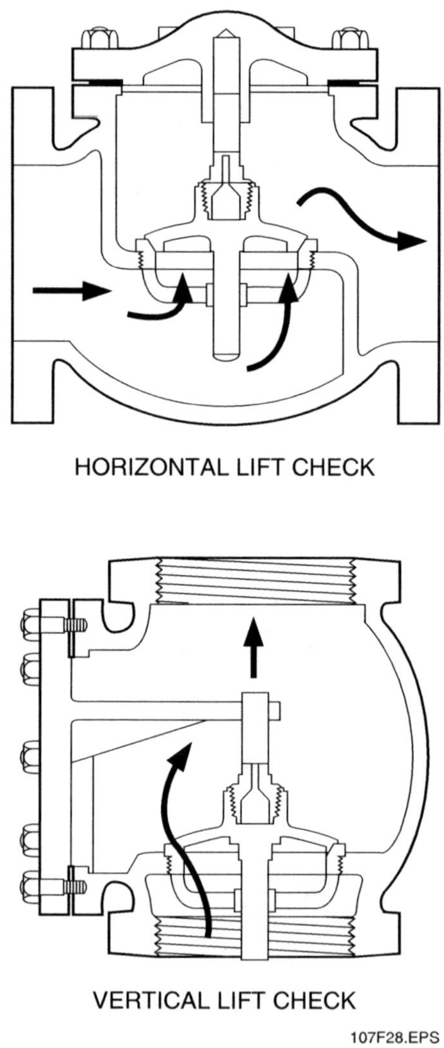

Figure 28 ◆ Lift check valves.

5.3.0 Piston Controlled Check Valves

Piston controlled check valves (see *Figure 29*) are designed for the positive prevention of backflow in gas, air, or liquid systems. Flow is directed under the valve piston and lifts it upward. When flow is stopped, the piston descends, closing the valve to prevent flow reversal.

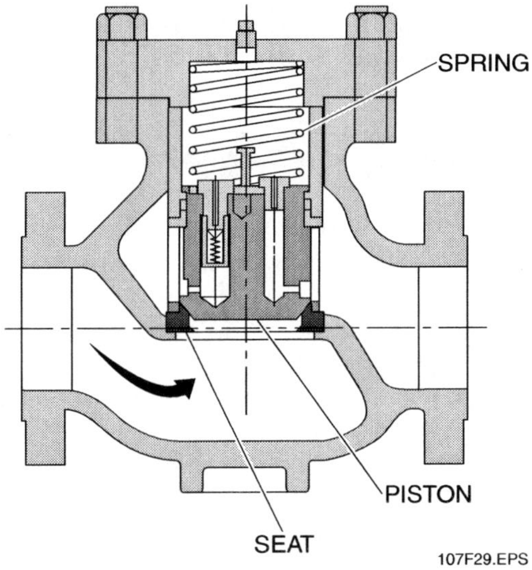

Figure 29 ◆ Piston controlled check valve.

Pulsating flow through the valve, if unrestricted, will cause the piston to follow the rapidly cycling line pressure, resulting in vibration, chattering, and possibly slamming of the piston. The ball check and orifice plug built into the piston prevent this from happening.

As the piston rises, the pressure in the Teflon®-sealed cavity above the piston increases, forcing open the ball check. This reduces the cavity pressure and allows the piston to rise smoothly. As line pressure and flow decrease, the piston moves downward. The ball check closes, and descent of the piston is restricted by fluid flowing through the orifice plug. This equalizes the pressure in the cavity above and below the piston and gives stability to the valve.

5.4.0 Ball Check Valves

Ball check valves operate in the same manner as lift check valves except they use of a ball instead of a disc to allow product to pass through the valve. Some ball check valves are designed to work by gravity while others are spring-loaded (see *Figure 30*). Ball check valves that are not spring-loaded should only be installed in the vertical position. Gravity and the reversal of flow cause the ball to rest against the valve seat, restricting flow through the valve.

Ball check valves are recommended for use in lines in which the pressure changes rapidly, because they stop flow reversal more rapidly than other types of check valves. The action of the ball is practically noiseless. The ball rotates during operation and therefore tends to wear at a uniform rate over its entire surface.

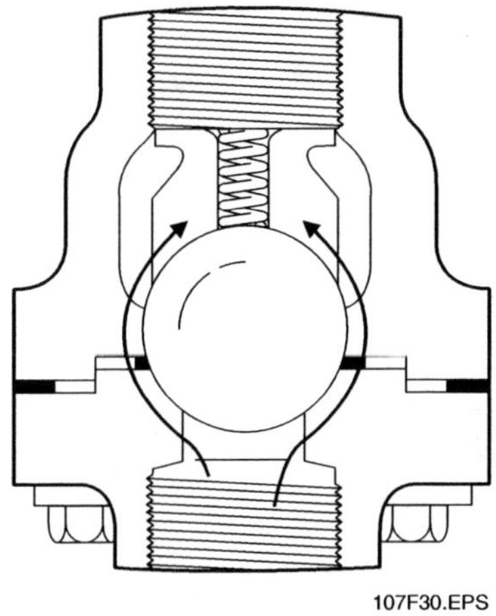

Figure 30 ◆ Spring-loaded ball check valve.

5.5.0 Butterfly Check Valves

A butterfly check valve has two vanes that operate in a fashion resembling the wings of a butterfly (see *Figure 31*). The vanes fold back from a central hinge to open and allow flow but close to prevent backflow. Most butterfly check valves operate completely from the reversal of flow.

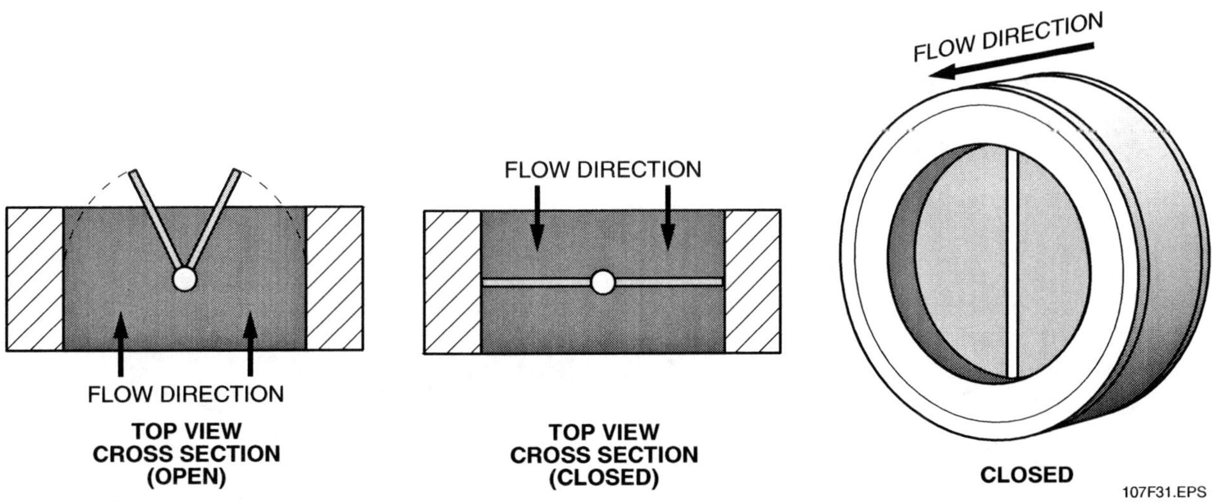

Figure 31 ◆ Butterfly check valve.

5.6.0 Foot Valves

Foot valves are used at the bottom of the suction line of a pump to maintain the prime of the pump. This type of valve is common in domestic shallow well applications and industrial applications that do not require a high head. This type of valve operates similarly to the lift check valve. On the inlet side of the foot valve is a strainer that keeps foreign material out of the line. The weight of the product in the pipeline between the pump and the foot valve, sometimes aided by a spring, keeps the valve closed so that the pipeline remains full when the pump is shut down. When the pump starts, the force of the suction pulls the valve open and causes the product being pulled into the valve to push against the outside of the seat to open the valve fully (see *Figure 32*).

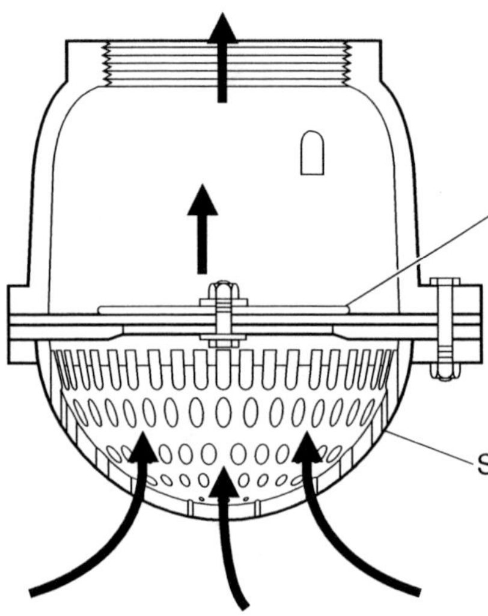

Figure 32 ◆ Foot valve.

6.0.0 ◆ PERFORMING PREVENTIVE MAINTENANCE

Every facet of pipeline operation and safety depends on the reliable operation of valves. Valves on the pipeline must be maintained regularly to ensure long life and dependable service; they must be in **good working order**. When performing any type of maintenance activities, the valve manufacturer's maintenance procedures must be followed. Because very little work on valves is done in the field, removal and reinstallation should be done carefully so the valve will not be damaged. Valves must not be dropped or hit. Valves and materials must never be stored on top of other valves.

Many valves are supplied with warranties. If a valve is not maintained according to the manufacturer's instructions, the warranty may be voided. Before any work is attempted, the particular valve should be checked to see if the warranty is still in effect. If it is, the work will need to be carried out in an authorized manner. The entire valve may need to be returned to the manufacturer to be replaced or repaired. If it is repaired by company personnel, they must be trained and qualified and make certain that the manufacturer's instructions are followed meticulously. It they are not, the valve's warranty may be voided.

When performing maintenance, close attention must be given to any valve that requires removal from the pipeline because many valves are directional in nature and must be replaced exactly as they were removed. This especially applies to safety valves, pressure relief valves, butterfly valves, and others. Such valves should have an arrow on the casting or labels on the ports to designate direction. If the valve is not marked, the technician must determine on which side of the valve the pressure should be located. Gate valves typically require no particular orientation.

Preventative maintenance involves those smaller tasks that, if performed regularly, keep serious problems from occurring as time and service accumulate. Most valves on the pipeline are outside. Many are exposed to harsh conditions from weather, dirt, salt, and corrosion from the elements and products. Most companies carry on a regular schedule of valve maintenance. Most companies use a computerized valve maintenance scheduling system that generates timely work orders for valve maintenance so that work can be scheduled. *Figure 33* shows a valve maintenance log with appropriate abbreviations.

Some of the procedures scheduled to maintain the efficient operation of such valves include:

- Inspection
- Adjusting glands
- Replacing stem O-rings
- Replacing **packing**
- Replacing gaskets
- Lubricating valves
- Repairing/replacing disks
- Painting

NOTE

Before performing any maintenance on a valve in the line, specific company procedures pertaining to work permits, work orders, shutdown procedures, notifications, and documentation must be strictly followed.

Rev. 12/97

Service Performed	Condition of Valve/Site	Deficiency	Action Required	Action Taken
VI - Visual Inspection	AC - Acceptable Condition	DO - Does Not Operate	RP - Replace Part	PR - Part Replaced
MO - Mechanical Operation		LK - Leaking	RR - Repair Required	RP - Repair Performed
LO - Local Mode Operation	NR - Needs Repair / Work	PM - Part Missing		
RO - Remote Mode Operation		DP - Defective Paint	PR - Paint Required	RE - Repainted
RM - Routine Maintenance	RR - Replacement Required	NS - Security	SS - Secure Site	SP - Security Provided
AS - All Services Listed		WO - Wash Out	DW - Dirt Work Required	DP - Dirt Work Performed
	OS - Out of Service	SU - Site Upkeep	SC - Site Clean Up	SM - Site Cleaned-Up
		OT - Other (describe)	OT - Other (describe)	OT - Other (describe)

Date	Print Order	Valve Tag No.	Serial Number	Location / Service	Service Performed	Condition of Valve/Site	Deficiency	Action Required	Action Taken
11/01/01	111	42-37B	BR549	#3 Pump Room	VI	NR	DP	PR	RE
11/13/01	112	54-18C	XJ242	Pump Station 14	MO	RR	LK	RP	PR

SAMPLE

REMARKS: ______

DIVISION: ______ STATION: ______ INSPECTED BY: ______

107F33.EPS

Figure 33 ◆ Valve maintenance log.

6.1.0 Inspection

49 CFR 195 and *49 CFR 192* require that pipeline companies maintain each valve that is necessary for the safe operation of pipeline systems in good working order. Inspections are to be performed at least twice each calendar year on liquid valves, not exceeding intervals of longer than 7½ months, to make certain that such valves are functioning properly. Inspections on gas pipeline valves must be performed once a year, not to exceed 15½ months. Scheduled inspection standards do not usually apply to valves installed only for maintenance purposes, but they are often inspected as a matter of company policy. Inspection is a general procedure that applies to most installations on the pipeline and are carried out much the same anywhere the need is located.

All valves require a complete seal in four places:

- The valve mechanism, itself
- The installation at the pipe ends
- The installation of the bonnet
- The valve stem

A stem leak is the first sign of a valve problem. The sooner it is noticed and reported, the more quickly the problem can be solved. Often with stem-type valves, the only action required is to cycle the valve completely by running the stem in and out several times and wiping it off each time. Afterward, observe the valve to see if the leak has stopped. Such a simple procedure might be performed on a smaller yard valve without requiring a call for maintenance. On a main-line valve that might require as much as eight or ten minutes to cycle in one direction, a maintenance operation may be required to have a work order approved by the appropriate departments.

6.2.0 Isolating and Purging a Valve

Valves requiring maintenance must be isolated and purged. Essentially, the valve must be completely cut off from the system, and the product in the valve and the pipe sections adjoining it must be drained down. All pressure must be relieved.

Maintenance work orders must be requested, initiated, and approved by management before any procedure that affects pipeline safety and flow can be performed. The work order is requested by personnel who discover a specific problem or by the regular maintenance program of the company. When a valve requires maintenance, it must be isolated and purged of product before work begins.

Company procedures must always be followed when isolating and purging a valve. Some companies require that the standard operating procedure (SOP) manual be physically on site with the technicians while such operations are being performed. Such procedures often have checklists that must be checked off as each step in the procedure is accomplished. Always refer to the material safety data sheet (MSDS) for the product in the line. Verify that all precautions and specifications being used match those on the MSDS before beginning procedures.

Isolating and purging a valve typically involves performing the following types of tasks:

Step 1 Notify control center operator or other designated personnel that the procedure is beginning.

Step 2 Verify that controls have been switched to manual.

Step 3 Close the valve immediately upstream of the valve to be isolated.

Step 4 Lock out and tag the upstream valve and any power source associated with it.

Step 5 Close the valve immediately downstream of the valve to be isolated.

Step 6 Lock out and tag the downstream valve and any power source associated with it.

Step 7 Lock out and tag all energy sources associated with all three valves.

Step 8 Apply chain and lock to all handwheels associated with three valves.

Step 9 Collect samples of product in line and test for type of product.

Step 10 Verify that the product is comparable to scheduled shipment.

Step 11 Attach hose connections to drain valves if possible.

Step 12 Extend drain hoses to sump if possible, or place containers of appropriate size under the valve if sump not convenient.

Step 13 Open drain(s) or bleeds in bottom of the valve to be purged.

Step 14 Cut off drain and empty containers as necessary to catch all product possible.

Step 15 When line drains stop, fill valve with solvent compatible with product in line.

Step 16 Allow product to drain back out of valve, noting amount returned.

Isolating and purging a valve, in some cases, is considered a major operation on the pipeline. Such operations require changes in operational proceedings, interruption of flow, and in some events, a shutdown of the pipeline. This must be coordinated through the control center and management. Maintenance records that must be kept for the isolation and purging of a valve may include some or all the following:

- Work order for valve work (including specifications)
- Special work permits
- Control center log
- Excavation work plans

6.3.0 General Procedures

The following procedures are general. Valve manufacturer's guidelines and company procedures vary, as do valve design details. Some companies have specific policies concerning maintenance. Many modern valve designs may allow you to skip many steps in general procedures. Any person attempting to perform any kind of service or maintenance on a valve should first consult company policy and procedure and also refer to the manufacturer's specific instructions for the valve model being considered.

Some modern innovations and valve designs allow many maintenance operations in the field that previously required shop work. With modern double gate valves, four-way valves, and others, some formerly major repairs are now done in a few minutes in the field. It is not possible to cover all these in a module this size, so the guidelines covered here are simply the basics.

6.3.1 Packing Nut Adjustment

If cycling and wiping the stem several times does not eliminate a valve steam leak, the next step is to tighten the packing nut or gland bolts as applicable (see *Figure 34*). With passing time, the packing may have dried out and become smaller. This is common in valves that are seldom used. A simple adjustment of the packing nut or bolts may be all that is required. It is probably better to do so in increments so as not to damage the existing packing by overtightening. After tightening the nuts or bolts, the valve should be observed to make certain the leak is stopped.

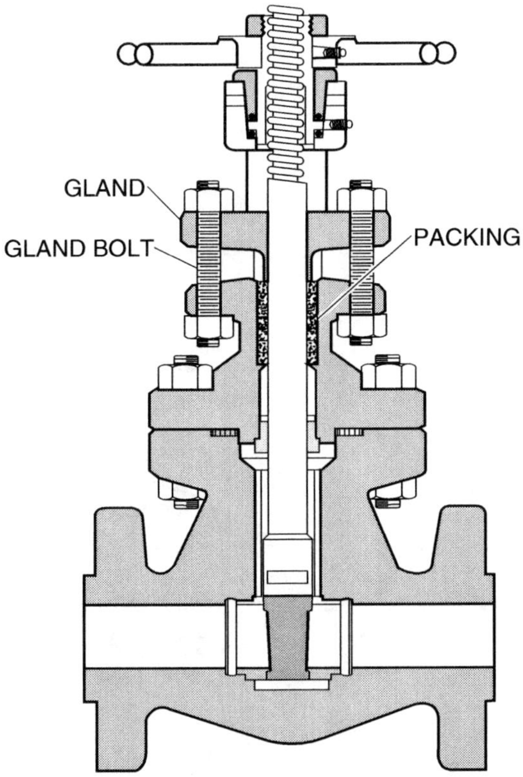

Figure 34 ◆ Gland bolt.

6.3.2 Replacing O-Rings

Some stem valves employ O-rings that create a leak-proof seal and hold lubricant around the valve stem. O-rings are fitted into machined grooves cut around the stem. When such an operation is required to seal the leak, the top valve operator must be disassembled and the bonnet removed. Company procedures and the valve manufacturer's specific maintenance procedures must be followed when replacing O-rings in any valve. The typical procedure may include some or all of the following steps, depending on the valve location, valve duty, and size.

Step 1 Lock out and tag the valve and any energy sources associated with it.

Step 2 Follow company procedure for draining down pressure.

Step 3 Remove the jam nut from the stem using an appropriate wrench.

Step 4 Remove the operator assembly or handwheel.

Step 5 Remove the retaining nut using an appropriate wrench.

Step 6 Remove the yoke nut from the top of the bonnet.

Step 7 Lift the assembly until the O-rings are visible.

Step 8 Inspect the O-rings for wear and damage.

Step 9 Inspect the stem for roughness and pitting.

Step 10 Smooth the valve stem with fine-grit emery cloth.

Step 11 Select the proper O-rings for replacement.

Figure 35 shows an O-ring assembly.

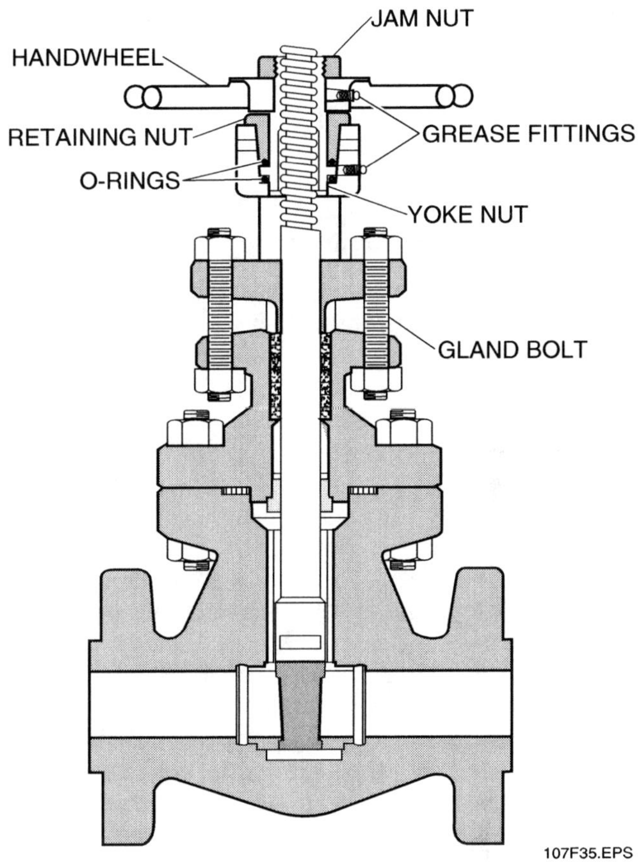

Figure 35 ◆ O-ring assembly.

Once the correct replacements have been found, the following steps should be taken to return the valve and the pipeline to normal operations.

Step 1 Lubricate the new O-rings with light machine oil.

Step 2 Slip the O-rings carefully over the shaft and into the appropriate stem grooves.

Step 3 Slide the stem back into the bonnet and valve.

Step 4 Replace the yoke nut and tighten very carefully.

Step 5 Replace the retaining nut on the yoke nut and tighten.

Step 6 Replace the operator assembly or handwheel (as applicable).

Step 7 Replace the jam nut on top of the shaft and tighten appropriately.

Step 8 Grease all grease fittings with appropriate products.

Step 9 Remove lockouts and tags and re-establish energy sources (if applicable).

If the valve does not work properly, all steps must be followed again. Once it is confirmed that the valve is working properly, operations personnel must be notified that the valve is returned to service.

6.3.3 Replacing Packing

A complete seal around the valve stem is difficult to achieve because the stem is required to move. Various types of packing are employed, depending on valve design and product. The packing around the stem prevents the pressurized product in the pipeline from migrating up the stem. Packing material is compressed into the stuffing box, and a fitting called a gland presses the packing tightly against the stem. The stuffing box requires tightening on a regular basis. Valves that are not used often require tightening before or during periodic use. *Figure 36* shows a valve packing assembly.

Packing is made of solid, braided, or granulated fibers. Cross-sectional shapes include square, round, and wedge-shaped. Packing is composed of rubber, asbestos, natural, synthetic, metal, and plastic in a variety of pure and mixed blends. The packing used is determined by the following factors:

- Product in the pipeline
- Maximum operating pressure
- Normal operational temperature encountered
- Minimum temperature encountered
- Metals used in stem construction

In the unexpected event that tightening the valve does not stop the leak, the packing must be inspected and possibly replaced. The procedure is better accomplished in the maintenance shop or other repair area. For large valves where removal from the pipeline would be prohibitive, the standard company policy and procedures must be followed. Repair includes some or all of the following steps:

Step 1 Lock out and tag the valve and all energy sources associated with it.

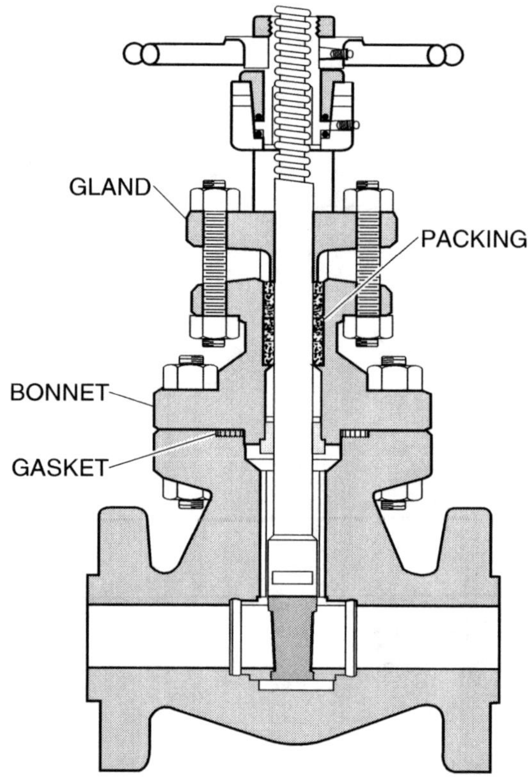

Figure 36 ◆ Valve packing assembly.

Step 2 Follow company procedure for draining down pressure.

Step 3 Remove the valve from the system if permissible and advisable.

> **NOTE**
> Placing a valve in a vise requires that some type of protective material be used to cover the vise jaws to protect valve surfaces.

Step 4 Turn the valve to full open position.

Step 5 Remove the jam nut from the stem using an appropriate wrench.

Step 6 Remove the operator assembly or handwheel.

Step 7 Remove the retaining nut using appropriate wrench.

Step 8 Remove the gland bolts from the bonnet.

Step 9 Slide the gland flange off the end of stem.

Step 10 Slide the packing gland off the end of the stem to reveal the packing.

Step 11 Use a valve packing-removal tool to pull the packing from gland (see *Figure 37*).

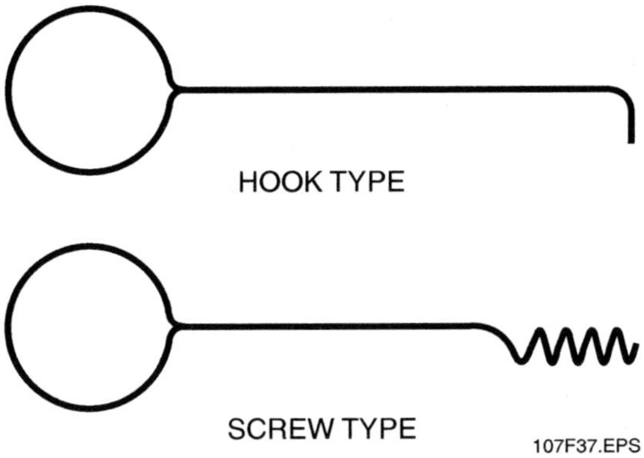

Figure 37 ◆ Valve packing-removal tools.

Step 12 Clean all areas of the stem, gland, and stuffing box with an emery cloth.

Step 13 Remove all residues of old packing from the casting and machined areas.

Step 14 Select manufacturer's recommended packing material.

Step 15 Wrap a strip of the new packing material around the stem.

Step 16 Mark two 45-degree lines where the packing strip overlaps, as shown in *Figure 38*.

Step 17 Remove the packing from stem.

Step 18 Cut at lines with a very sharp knife to obtain a clean cut and fit.

Step 19 Test the cut around the stem to check fit.

Step 20 Use the first strip as a pattern to cut others needed.

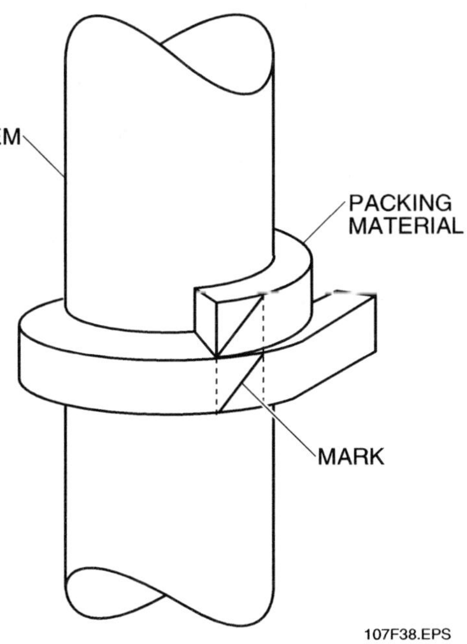

Figure 38 ◆ Measuring valve packing material.

Step 21 Push the first piece into the stuffing box.

Step 22 Continue adding pieces until the stuffing box is full, staggering the joints (see *Figure 39*).

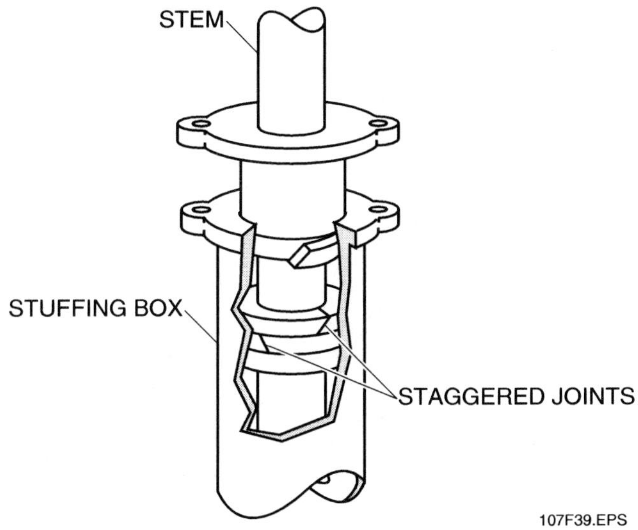

Figure 39 ◆ Staggered packing joints.

Step 23 Slide the packing gland onto the stem.

Step 24 Slide the gland flange onto the stem and valve body.

Step 25 Replace the bolts and tighten them.

Step 26 Replace the handwheel.

Step 27 Replace and tighten the handwheel nut.

Step 28 Clean and lubricate the bolt threads.

Step 29 Reinstall the gland bolts.

> **NOTE**
> Alternate bolt tightening and perform the final tightening of the bolts with a torque wrench according to the manufacturer's specifications.

Step 30 Oil the stem lightly.

Step 31 Test the operator or handwheel for proper operation.

Step 32 Test the valve as required by government standards.

Step 33 Lubricate all grease fittings with grease gun.

Step 34 Return the valve to service.

6.3.4 Injecting Stem Packing

Some valves are furnished with an injector fitting as a secondary sealing feature. This fitting allows plastic, nonflammable stem sealant to be injected into the valve. There are many different seal materials, each tailored to specific valve services and operating temperatures. These types of valves use both the packing contained around the valve stem and the secondary stem sealant to provide a long-life primary seal.

The injectable packing-type gate valve (*Figure 40*) is equipped with two injector fittings in the stem area of the valve. One fitting is used to inject the packing, while the other fitting is used to allow the old injectable packing to bleed from around the stem. The injectable packing seals the valve stem between the primary packing materials in an area known as the lantern ring.

When injecting packing, always inject until all of the old packing comes out and fresh new packing material starts to flow from the second fitting. Follow these steps to inject sealant into a valve stem.

Step 1 Verify the location of the valve to be inspected; ensure the proper valve is located.

Step 2 Verify the accessibility of the valve; ensure there are no obstructions.

Step 3 Verify the valve number and nameplate data. If the nameplate is missing, it is to be replaced per company specifications.

Step 4 Verify the valve type and manufacturer. Verify documentation to ensure proper maintenance procedures for the valve being maintained.

Step 5 Notify the control center and/or affected personnel. For example, you may be required to notify the upstream pump station and downstream pump station when working on main-line valves. You may also need to obtain a work permit or complete a work order.

Step 6 Perform maintenance per manufacturer's or industry recommendations. Following the manufacturer's or industry recommendations will prolong valve life and reduce the possibility of internal valve damage.

WARNING!

If carbon dioxide or an unexpected hazardous liquid is encountered at the valve, try to isolate the leak. If you cannot safely isolate the leak, contact the control center for help or to shut down the line. If necessary, evacuate the area. Follow company policy.

WARNING!

In the case of an excessive stem seal leak, it is possible that enough hydrocarbon vapors could accumulate to present an ignition hazard. In case of fire or explosion, evacuate the area and take only defensive action while using the proper PPE. Contact the control center to isolate or shut down the line. Follow company policy.

Step 7 Identify the appropriate type and amount of injection product per manufacturer's and company specifications. Stem packing varies per manufacturer and application, so ensure that both the proper type and amount of product are used.

Step 8 Operate injection equipment with appropriate products.

Step 9 Inject appropriate products into stem per manufacturer's or industry recommendations.

CAUTION

Be alert for possible component failure. For example, a valve seat could fail after work has begun, which would allow the valve body to pressurize. In such a case, you must try to seal the valve with sealant. If this doesn't work, contact the control center for help or to shut down the line. If necessary, evacuate the area. Follow company policy.

Step 10 Notify the control center and/or affected personnel in order to verify the system's integrity and return system to normal operation.

Step 11 Document results.

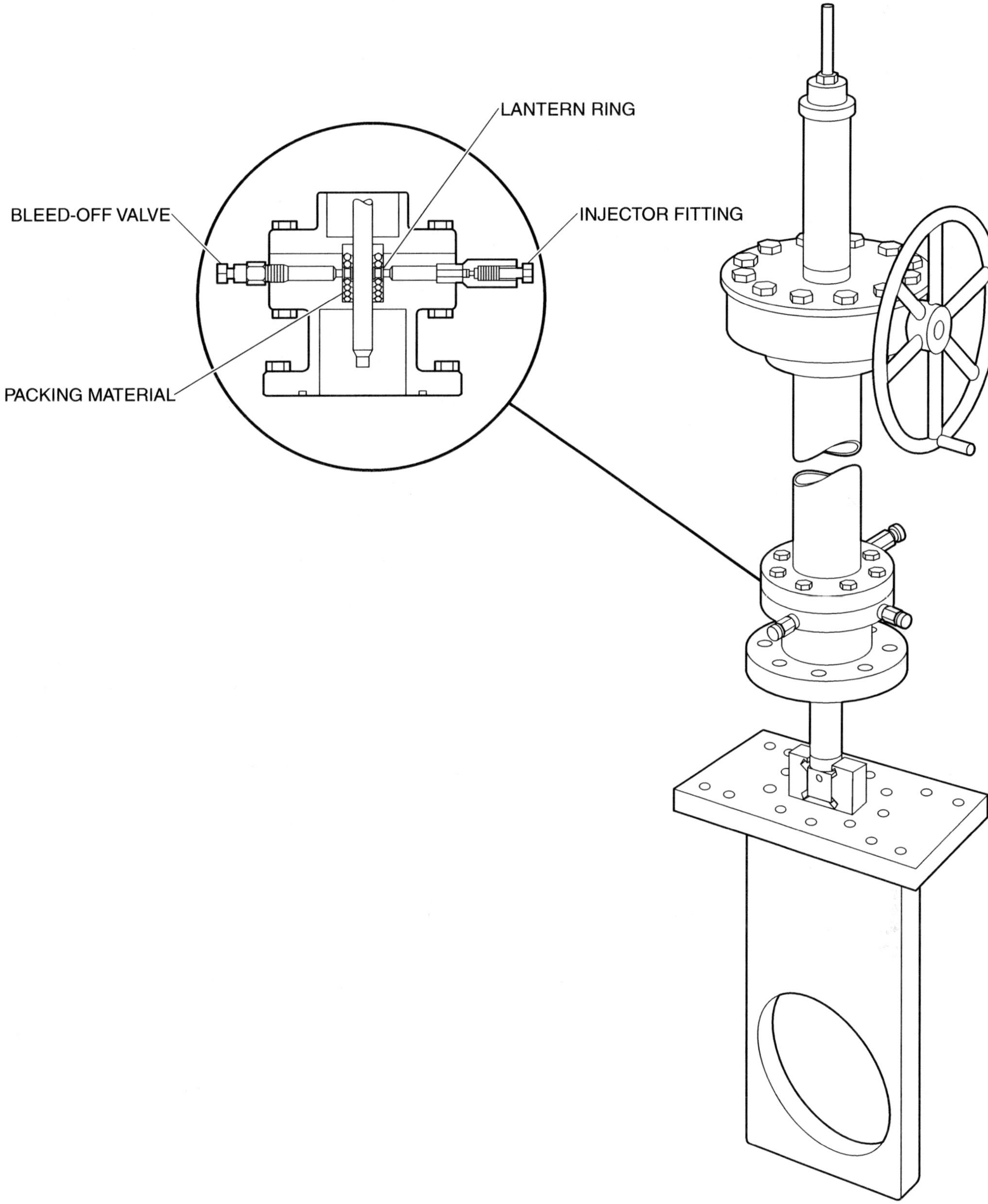

Figure 40 ◆ Injection fitting for packing on gate valve.

6.3.5 Replacing Gaskets

Valve designs with bolted or flanged bonnets use gaskets to achieve a complete seal (see *Figure 41*). Gaskets may be of the flat or ring type. Valves with bolted bonnets are used for pipe sizes over 1¼ inches and are suitable for high pressure and temperature up to 925°F. The flanged bonnet with ring-type gasket is the better choice for high temperature and pressure. When a leak occurs between the bonnet and valve body, the gasket must be replaced.

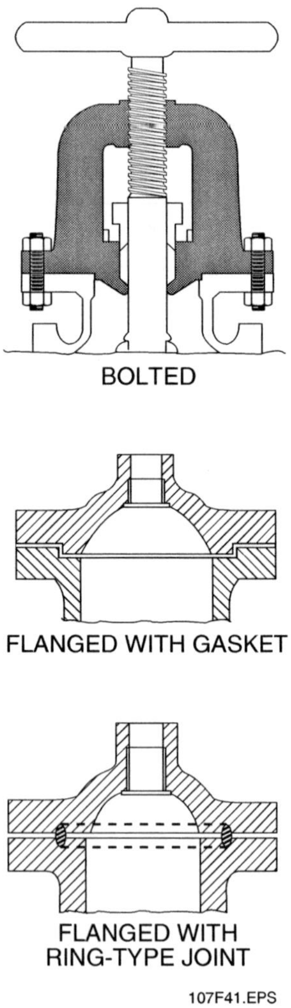

Figure 41 ◆ Typical bolted and flanged valve bonnets.

Follow these steps to replace the gasket on a valve:

Step 1 Lock out and tag the valve and all energy sources associated with it.

Step 2 Follow company procedure for draining down pressure.

Step 3 Remove the valve from the system.

Step 4 Secure the valve ends in a vise using protective covers to protect the valve.

Step 5 Turn the handwheel to open the valve (gate valves), relieving mechanical pressure.

Step 6 Remove the bolts from the bonnet.

Step 7 Remove the bonnet from the valve body.

Step 8 Remove the gasket and all residues from the mating surfaces with a putty knife, gasket scraper, or single-edge razor blade.

> **NOTE**
> The old gasket should be removed as intact as possible to be used as a pattern for cutting the new gasket.

Step 9 Select the proper material to be cut for the new gasket.

> **NOTE**
> Manufacturer's specifications must be followed to achieve proper fit and service. Some gaskets must be ordered from the manufacturer and cannot be cut on the job.

Step 10 Trace the outline of the old gasket in pencil by laying the old gasket on top of the new material.

> **NOTE**
> Try to trace lines smoothly in damaged or torn areas so that the shape of the new gasket will be as near the shape of the original as possible. If the old gasket is destroyed, use the bonnet as a pattern.

Step 11 Cut the new gasket using a gasket cutter and hole punches or a very sharp knife.

> **NOTE**
> A test fit is recommended so that it will be easier to trim or adjust before lubricants or sealers are applied.

Step 12 Lubricate or apply sealer to the gasket according to manufacturer's guidelines.

Step 13 Place the gasket on the valve, taking care to position it properly.

Step 14 Place the bonnet on the valve body.

Step 15 Clean and lubricate the bonnet bolts and threads.

> **NOTE**
> Replacement of corroded, badly rusted, or galled parts should be considered.

Step 16 Install bonnet bolts and hand-tighten.

Step 17 Install the handwheel and tighten it to torque specifications.

Step 18 Tighten bonnet bolts using manufacturer's pattern or cross-over method.

Step 19 Perform final tightening with a torque wrench to manufacturer's specifications.

Step 20 Test the valve as required.

Step 21 Reinstall the valve and return to service.

Step 22 Lubricate all grease fittings with grease gun.

Step 23 Remove all lock-out devices and tags.

6.3.6 Injecting Valve Seat Sealant

Some valves have a sealant injection fitting (*Figure 42*) located on the valve body that allows sealant to be injected into grooves surrounding the valve seat. Injecting sealant into the valve seat assists in forming a leak-proof seal around the valve seat. This can restore damaged seats and drop-tight shutoff on hard-to-hold fluids. Valve seat sealant is commonly injected to seal or reduce a valve internal **leak-by** or **leak-through**.

When injecting sealant into valve seal seats, the valve must be fully open or fully closed. Sealant cannot be injected when the valve is partially opened. After the need to maintain a tight seal has passed, you may need to flush sealant from injection ports and seats with grease because sealants can dry out and plug injection passages. Always follow company policy.

Follow these steps to inject sealant into a valve seal seat.

Step 1 Verify the location of the valve to be injected with seat sealant; ensure the proper valve is located.

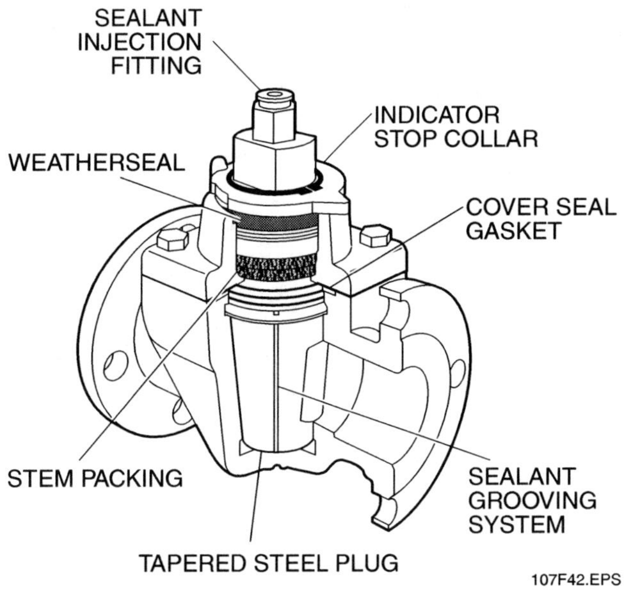

Figure 42 ◆ Injection fitting for seal seat sealant.

Step 2 Verify the accessibility of the valve; ensure there are no obstructions.

Step 3 Verify the valve number and nameplate data. If the nameplate is missing, it is to be replaced per company specifications.

Step 4 Verify the valve type and manufacturer. Verify documentation to ensure proper maintenance procedures for the valve being maintained.

Step 5 Notify the control center and/or affected personnel. For example, you may be required to notify the upstream pump station and downstream pump station when working on main-line valves. You may also need to obtain a work permit or complete a work order.

Step 6 Perform maintenance per manufacturer's or industry recommendations. Following the manufacturer's or industry recommendations will prolong valve life, reduce likelihood of internal valve damage, and provide a safe work environment when sealant are used to seal a valve for work maintenance.

Step 7 Verify proper valve position. The valve cannot be sealed if it isn't fully open or fully closed.

Step 8 Blow down (depressurize) valve body. This will help you determine if valve has internal leak-by.

Step 9 Check for leak-by and leak-through sealing of the valve. Determine the success of different sealing efforts.

> **WARNING!**
> If carbon dioxide or an unexpected hazardous liquid is encountered at the valve, try to isolate the leak. If you cannot safely isolate the leak, contact the control center for help or to shut down the line. If necessary, evacuate the area. Follow company policy.

Step 10 Identify the appropriate type and amount of injection products per manufacturer's/company specifications. Sealants vary per manufacturer and application, so ensure that both the proper valve injection product and the proper amount of product are used.

Step 11 Operate injection equipment with appropriate products.

> **WARNING!**
> Two different kind of components can fail while injecting sealant: the grease fitting and high-pressure hose. If the grease fitting fails and product is released, try to put the hose back on the fitting. If you cannot, evacuate the area and contact the control center as soon as possible. Follow company policy. If the high-pressure hose fails, the button head should prevent product from escaping.

Step 12 Inject appropriate products into seats per manufacturer's or industry recommendations.

Step 13 Notify the control center and/or affected personnel in order to verify the system's integrity and return system to normal operation.

Step 14 Document results.

6.4.0 Painting

Valves must be painted when new and repainted on a regular schedule. Paint is a strong defense against rust and corrosion. Manufacturers coat valves with some type of corrosion-resistant, oxide-type primer, protecting them until delivery. Once valves are delivered, companies paint them according to company policy for resistance to the weather and for product coding (see *Figure 43*). In hot climates, companies paint much of their equipment white. In the colder areas, tanks and equipment are often painted black or some other heat-absorbing color. Some pipelines signify products flowing by color-coding the pipe with paint. During inspections, close attention must be paid to the condition of paint and coatings. Damaged paint allows rust and corrosion to begin. If there is a chip, a ding, or flaked area, it must be touched up before rust and corrosion advance. The operator must arrange to have it done as soon as possible. *Figure 44* shows two color-coded pipelines.

107F43.EPS

Figure 43 ◆ Contrasting paint scheme.

107F44.EPS

Figure 44 ◆ Color-coded pipelines.

7.0.0 ◆ LUBRICATING VALVES

Objects that move produce friction and wear. Equipment exposed to the elements tends to experience the drying out, hardening, and deteriorating of important components as time passes. Dirt and grime collect in unacceptable places and cause problems with movement, increasing wear, corrosion, and damage to machined and polished surfaces. Periodic lubrication of certain parts of a valve prevents or at least slows down the wearing process, greatly extending the life and efficiency of valve components.

Most periodic lubrication is accomplished in several simple procedures. However, there are other serious considerations that must be included in the work including:

- Climate
- Exposure of component
- Application of component
- Product to be used within the valve

For example, some types of grease become very stiff as the temperature drops. If a valve is outside in a cold climate, a completely different type of grease is required than that required for a valve exposed to a warm environment.

The manufacturer's instructions should be consulted first. If the wrong type of grease is used, the valve may be impossible to turn in cold weather. Some valves require lubrication from a simple grease gun. Others are constructed with a factory-designed lubrication system that only need a bolt or screw to be turned to force lubrication from an inner chamber to the appropriate locations. The primary function of the lubricant is for sealing, and the secondary function is for lubricating. Some lubricants do not endure moisture or rain as well as others. If the valve is exposed to rain, snow, ice, or immersion, moisture may destroy the lubricant if the wrong grease is chosen. In areas prone to flooding or tidal activities, harsh environmental influences such as blowing sand or dust, or severe temperature, special lubricants will often be required.

Operators and technicians must become completely familiar with the types of valves being used and the types of maintenance products required for their job assignments. They must also become familiar with the types of lubrication systems each valve type employs.

7.1.0 Applying Lubricants

Applying grease with a brush and/or grease gun is a common procedure performed on stem-type valves on the pipeline. This procedure is performed on a variety of stem-type valves that are accessible to personnel. Keep in mind that a grease gun uses hydraulic pressure to move grease from the pump into the fitting. In cold weather, greater pressure may be required to apply grease to the valve. Pumping may require extra effort. *Figure 45* shows a stem requiring lubrication with a brush.

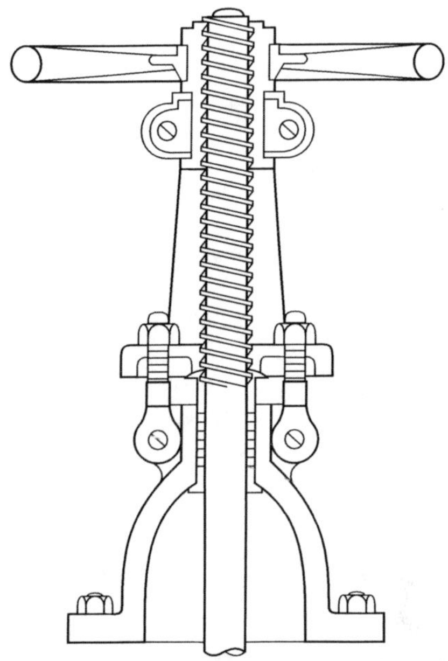

Figure 45 ◆ Stem requiring lubrication with a brush.

Steps to lubricate a stem valve may include:

Step 1 Verify the location of the valve to be lubricated; ensure the proper valve is located.

Step 2 Verify the accessibility of the valve; ensure there are no obstructions.

Step 3 Verify the valve number and nameplate data. If the nameplate is missing, it is to be replaced per company specifications.

Step 4 Verify the valve type and manufacturer. Verify documentation to ensure proper maintenance procedures for the valve being maintained.

Step 5 Notify the control center and/or affected personnel. For example, you may be required to notify the upstream pump station and downstream pump station when working on main-line valves. You may also need to obtain a work permit or complete a work order.

Step 6 Perform maintenance per manufacturer's or industry recommendations. Following the manufacturer's or industry recommendations will prolong valve life, reduce likelihood of internal valve damage, and provide a safe work environment when sealant are used to seal a valve for work maintenance.

Step 7 Identify appropriate injection and lubrication product. Lubricants vary per manufacturer and application, so ensure that the proper type is used.

> **WARNING!**
> If carbon dioxide or an unexpected hazardous liquid is encountered at the valve, try to isolate the leak. If you cannot safely isolate the leak, contact the control center for help or to shut down the line. If necessary, evacuate the area. Follow company policy.

Step 8 Turn the valve to the completely open position so that the stem threads extend.

Step 9 Clean the threads of all rust, dirt, and grime with a wire brush, solvent, and rags.

Step 10 Wipe off all residues from moving parts and seal areas with rags.

Step 11 Locate and remove the safety caps from grease fittings on sides of valve body.

> **NOTE**
> For those valves that do not have grease fittings, apply the recommended grease heavily to the outer threads of the stem with a paintbrush and cycle the valve several times.

Step 12 Pump the grease gun several strokes and operate the valve an inch in each direction to spread grease on the seals.

Step 13 Pump the grease gun lever several more times on each fitting. *Figure 46* shows a valve with grease fittings.

Step 14 Operate the valve completely several times to spread the lubrication.

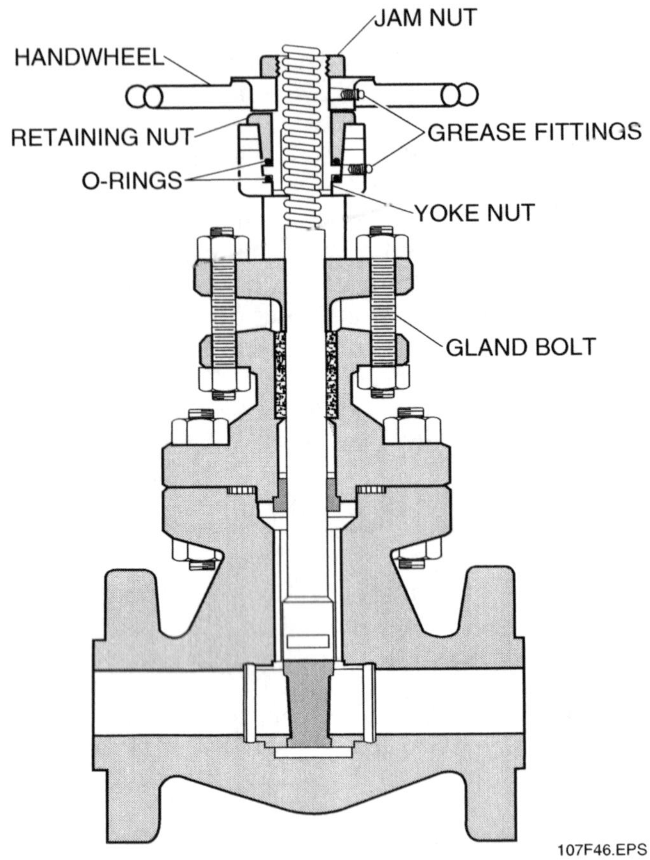

Figure 46 ◆ Valve with grease fittings.

Step 15 Move the valve to full open position and pump the grease gun several more strokes to fill cavity.

Step 16 Remove handwheel nut and handwheel, exposing bearing housing (*see Figure 47*).

Step 17 Remove the setscrew holding bearing assembly in place.

Step 18 Remove the individual bearings and races from the housing.

Step 19 Clean and inspect the handwheel bearings and races.

Step 20 Hand pack the handwheel bearings and races.

Step 21 Reassemble the races and bearings into the housing.

Step 22 Slide the bearing housing back over the stem.

Step 23 Tighten the assembly-housing setscrew.

Step 24 Reinstall the handwheel and nut.

Step 25 Torque the handwheel nut acco[illegible] the manufacturer's specifi[illegible]

Step 26 Remove all lock out devices and tags.

Step 27 Notify the control center and/or affected personnel in order to verify the system's integrity and return system to normal operation.

Step 28 Document results.

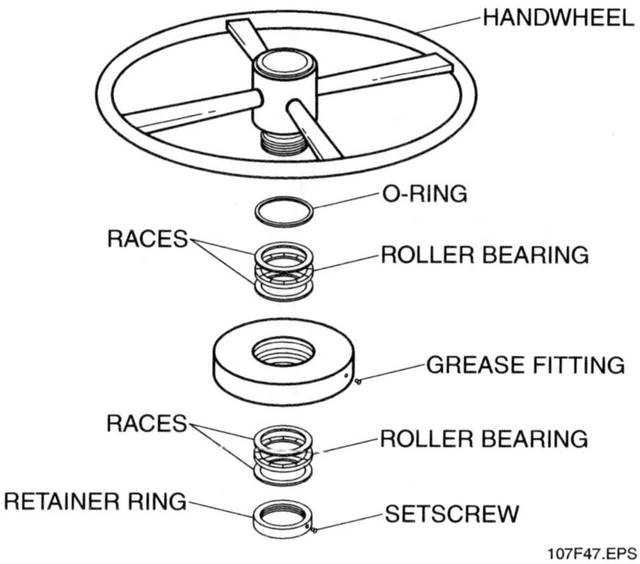

Figure 47 ◆ Typical operator bearing assembly.

7.2.0 Maintaining Lubricated Valves

Lubricated valves have an inner lubricant supply or cartridge within the body of the valve. At specified intervals, maintenance technicians need only to turn the lubricating screw. Lubricant is forced into grooves in the plug, to the bottom lubricant chamber, and to other mechanisms as required. These valves are especially important to critical operations that require the plug to be repacked with lubricant under pressure. The negative aspects of lubricated valves are:

- Product in the pipeline may become contaminated.
- Extra maintenance may be required.
- Lubricant system may determine the useful life of the valve.
- Lubrication intervals must be effectively tracked.

Figure 48 shows a lubricated, tapered plug valve.

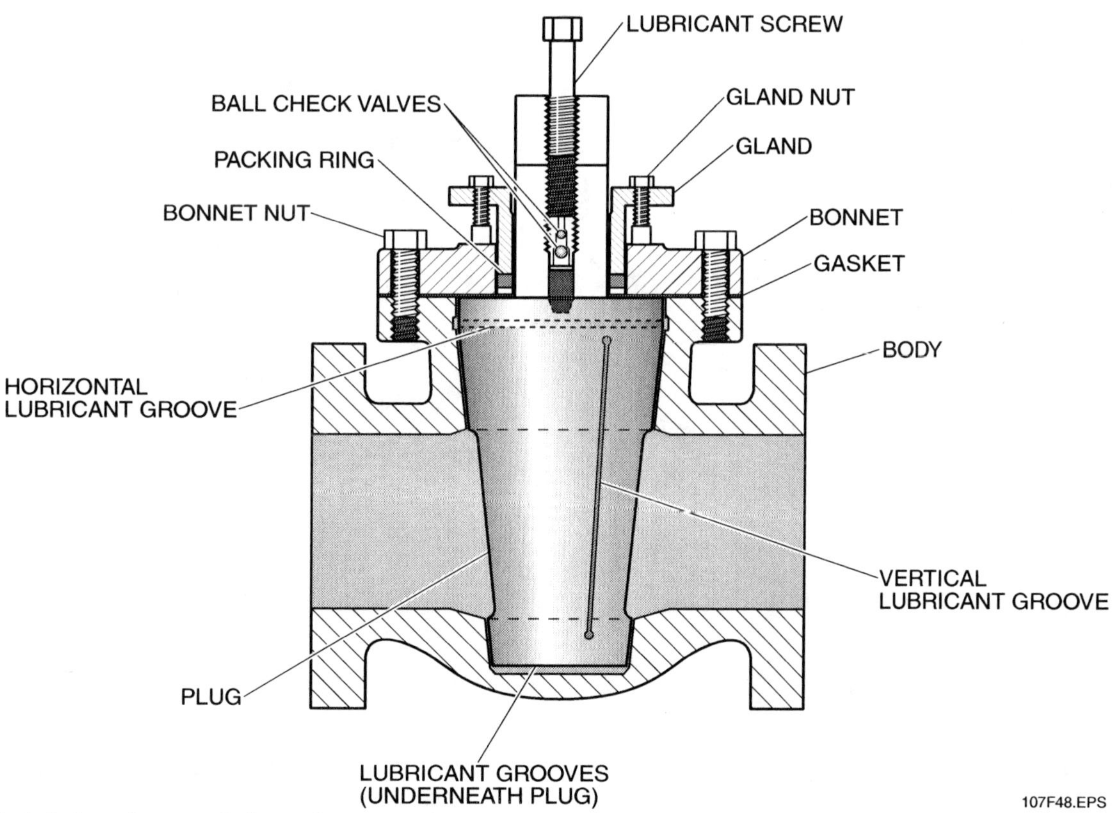

Figure 48 ◆ Lubricated, tapered plug valve.

8.0.0 ◆ WINTERIZING VALVES

Winterizing valves is an extremely important part of maintenance on the pipeline if the temperature is expected to drop below freezing, even for a short time. Winter temperatures present serious hazards on the pipeline. When the temperature drops below 32°F, equipment that is exposed can experience serious and destructive damage from ice forming inside the unit. Valves are very susceptible to damage and are far more expensive to repair or replace than sections of pipe. If a valve is allowed to freeze and rupture, not only is equipment and production lost, but there is danger to personnel and to the environment. The damage may not be obvious at first sight, because the leak will not be revealed until the valve has thawed.

The solution to this problem is to prepare valves for cold weather before the event occurs. After the temperature is below 32°F, any effort might prove useless, because damage may already have occurred. Even if the casting of the valve is not ruptured, serious damage may have been done to the mechanical parts and seals inside. Certain maintenance procedures will eliminate these occurrences, saving much damage, time, and expense.

Larger valves have cavities in which moisture collects. This moisture may come from wet product, residues of water left in the line after an operation, or condensation. Moisture must be eliminated from the inside of the valve before cold temperatures are experienced. Technicians must check every valve. Valves prone to moisture collection have drain plugs located in lower parts of the castings so that they may be drained (see *Figure 49*).

In some locations, draining valve moisture may be a daily task. Company policy must always be followed. Because opening the drain plug will release water contaminated with product, use a receiving vessel to catch the drainage so that it can be disposed of in a proper way. If you are draining a large in-service valve, the pipeline may have to be shut down for a short time to relieve pressure on the valve for service.

Follow these steps to winterize a valve:

Step 1 Verify the location of the valve to be winterized; ensure the proper valve is located.

Step 2 Verify the accessibility of the valve; ensure there are no obstructions.

Step 3 Verify the valve number and nameplate data. If the nameplate is missing, it is to be replaced per company specifications.

Figure 49 ◆ Valve with a bottom drain plug.

Step 4 Verify the valve type and manufacturer. Verify documentation to ensure proper maintenance procedures for the valve being maintained.

Step 5 Notify the control center and/or affected personnel. For example, you may be required to notify the upstream pump station and downstream pump station when working on main-line valves. You may also need to obtain a work permit or complete a work order.

Step 6 Lock out and tag all sources of energy associated with the valve.

Step 7 Place a receptacle under the drain plug or valve to be opened.

Step 8 Blow down (depressurize) valve body, if **body bleed** is available. Isolate the valve from the pipeline flow and drain down the pressure.

Step 9 Drain non-petroleum liquids (such as water) from the valve body, if applicable. Open the bottom drain plugs or valves, as appropriate, to release drainage.

WARNING!
If carbon dioxide or an unexpected hazardous liquid is encountered at the valve, try to isolate the leak. If you cannot safely isolate the leak, contact the control center for help or to shut down the line. If necessary, evacuate the area. Follow company policy.

Step 10 Operate injection equipment and inject appropriate antifreeze, if applicable, to provide freeze and/or corrosion inhibition.

CAUTION
If a component fails, such as an injection hose that bursts or leaks, clean up will be necessary.

Step 11 Close the drain plugs or valves after all moisture has drained from valve.

Step 12 Open the pipeline valves and return to service.

Step 13 Remove all lock-out devices and tags.

Step 14 Remove the receiving vessel and empty it into a sump or other appropriate holding vessel.

Step 15 Notify the control center and/or affected personnel in order to verify the system's integrity and return system to normal operation.

Step 16 Document results.

9.0.0 ◆ INSTALLING THREADED VALVES

Correctly installed valves are essential to the efficient operation of the pipeline system. Valves can be installed in piping systems with threaded or flanged joints. The procedure for installing a small valve is the same as for installing a fitting. However, there are added factors to consider when installing valves on the pipeline other than the immense size of some valves.

The location of a valve in a pipeline system is crucial to operation and safety. Most pipeline drawings indicate exactly where to install each valve. The orientation in which the stem and handwheel are to be located when working with small-bore piping is often the technician's responsibility. When working with large-bore pipe, as on the pipeline, the stem direction and orientation are normally shown on the drawings. Regular maintenance procedures require the valve to be easily accessible. The type of stem action must also be considered because a rising stem must have room to operate. The height of the valve must be accessible without causing hazards. The best installation height for valve handles is between 2 feet and 4 feet 6 inches off the floor or working level. Valves installed below this area expose operators to a tripping hazard. Valves installed above this area up to 6 feet 6 inches expose operators to a face hazard. Handwheels on such valves must have some type of guard around them. *Figure 50* shows the order of preference for valve location.

Valves that are installed with the stem in the upright position tend to work best. The stem can be rotated down to the horizontal position, but should not point down. If the valve is installed with the stem in the downward position, the bonnet acts like a trap for sediment, which may cut and damage the stem. Water trapped in the bonnet in cold weather may freeze and crack the body of the valve.

Another factor that must be considered when installing valves is the direction of flow through the valve. Butterfly valves, safety valves, relief valves, and some other valves are designed to work in only one direction. These valves have arrows stamped on the side, or the ports are labeled indicating direction of flow. When the valve is not marked, the technician must determine which side of the disc should face the pressure. Gate valves can have the pressure on either side. Globe valves should be installed so that the pressure is below the disc unless pressure above the disc is required in the job specifications. *Figure 51* shows pressure above and below the disc.

WARNING!
Threaded valves are heavy enough to cause serious personal injury if dropped on feet, toes, or fingers.

9.1.0 Isolation and Purging

Proper installation of valves can only be performed when the system is isolated. The technician must make the proper preparations before the work can be performed. A location associated with the operation of the pipeline may require a section of the pipeline to be shut down. It is very important for the technician to follow company procedures.

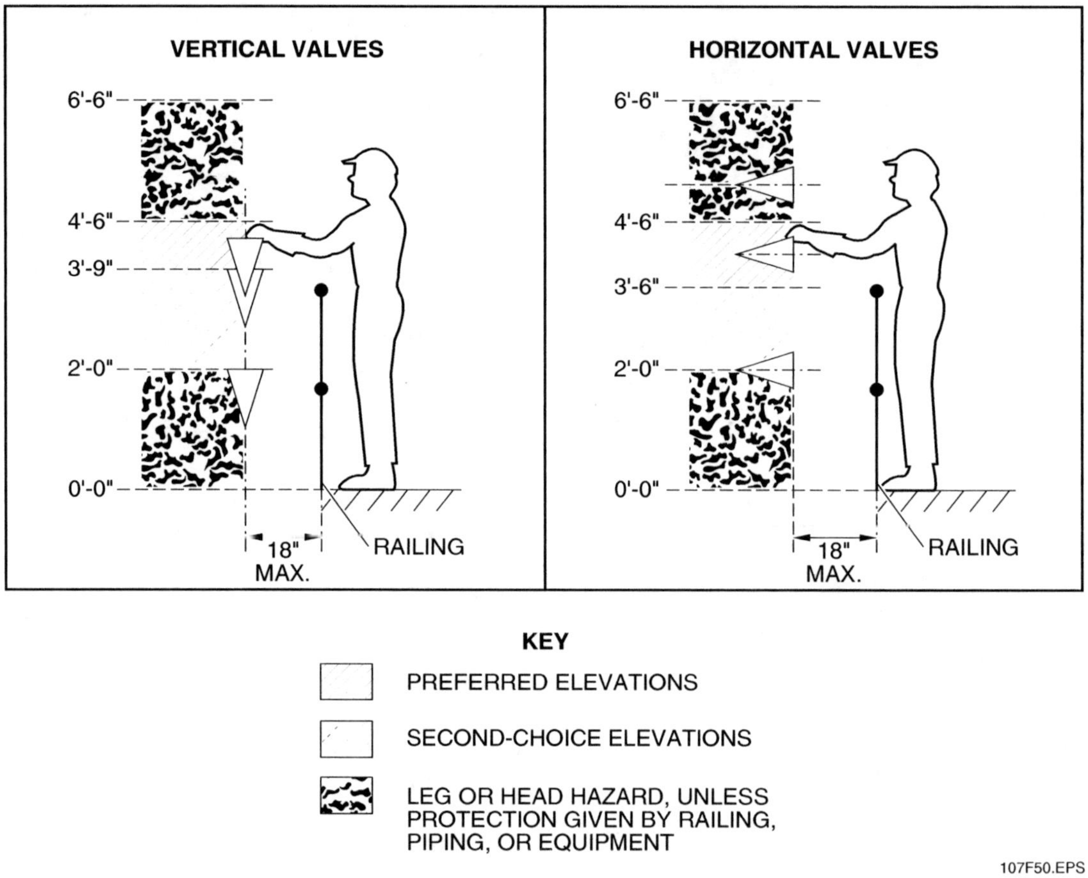

Figure 50 ◆ Order of preference for valve location.

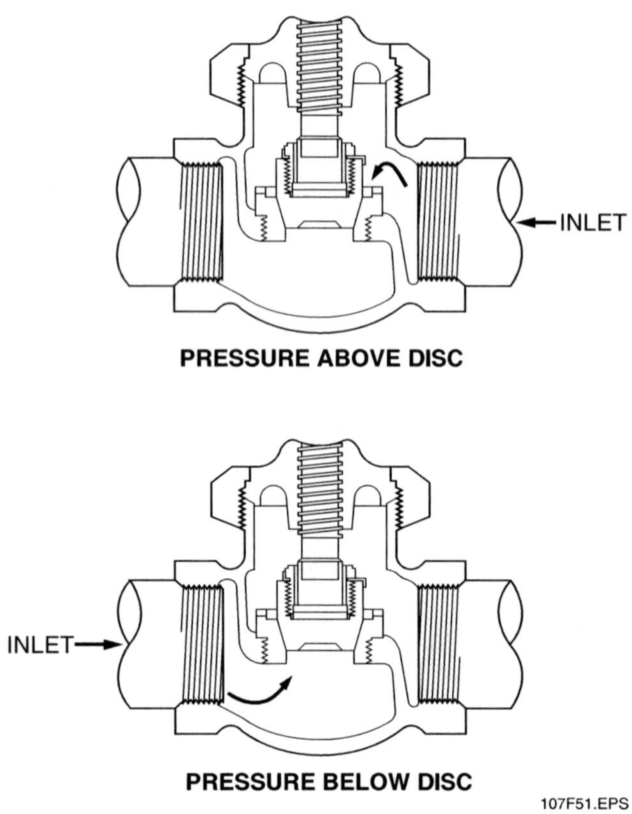

Figure 51 ◆ Pressure above and below the disc.

9.2.0 Making Up the Replacement

The threaded joint is a common method of joining pipe to smaller-sized valves. In order to have a strong, leak-proof threaded joint, the threads must be clean and smoothly cut. They must have the correct pitch, lead, taper, and form. The threads must be correctly sized. If these conditions are not met, the seal will leak.

The joining, or makeup, of a taper-threaded pipe connection is performed in two distinct operations: hand engagement and wrench makeup. Hand engagement refers to how far the fitting can be tightened by hand. Wrench makeup is the additional turning of the fitting with a wrench to completely make up the joint. Hand engagement on properly cut threads is normally between three and four complete revolutions. Hand engagement should be easy to turn at first and is a good way to check the quality of threads after cutting. Wrench makeup is usually about three turns, or approximately seven revolutions for complete makeup. A greater number of turns may be required for larger pipe. *Figure 52* shows hand engagement and wrench makeup.

Special precautions must be taken when installing valves with threaded ends to avoid

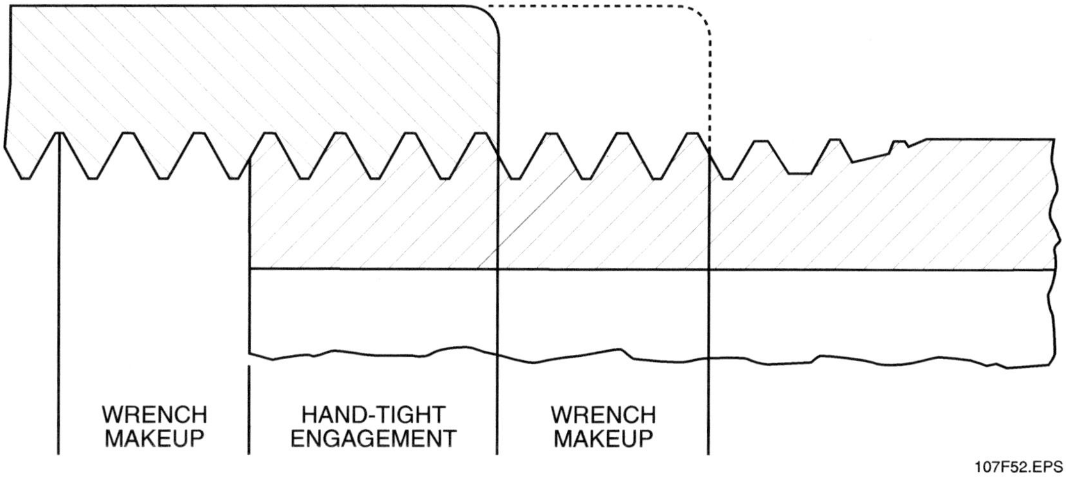

Figure 52 ◆ Hand engagement and wrench makeup.

damaging the valve. These guidelines should be kept firmly in mind when working with valves with threaded ends:

- Never place the valve in a vise.
- Secure the pipe in the vise, and screw the valve onto the pipe.
- Grip the valve with the pipe wrench on the flange of the valve body.
- Use a strap wrench on valves with brass bodies to avoid excessive marring of the valve.
- Apply pipe sealant/lubricant only to the male threads of the pipe.
- Do not lubricate the female threads inside the valve.
- Do not lubricate the thread nearest the end of the pipe.
- Do not turn the male end of the pipe all the way into the valve, because this will damage the seat.
- Leave approximately three threads showing outside the valve.

9.3.0 Installation

Follow these steps to install a threaded valve:

Step 1 Determine the type, size, and schedule of pipe being used in the system.

Step 2 Determine the length of pipe needed, including threaded length.

Step 3 Measure the length of pipe carefully.

(Old rule of thumb: measure twice, cut once. Much error and waste is avoided.)

Step 4 Cut the pipe to the desired length.

Step 5 Identify the type of valve to be used in the system.

Step 6 Ream the pipe to remove all internal burrs.

Step 7 Thread the pipe, being careful not to cut the threads too deep (*Figure 53*).

Step 8 Clean the pipe and valve thoroughly, both inside and out.

> **NOTE**
> All sand, dirt, and oil must be removed from the inside of the pipe and fittings to avoid contaminating the system or clogging the line. The valve and pipe ends can be cleaned with a clean rag soaked in nonflammable solvent.

Step 9 Check the threads on the pipe to make sure they are cut properly and not damaged.

Step 10 Secure the pipe in a pipe vise.

Step 11 Apply joint compound or Teflon® tape to the pipe threads.

Step 12 Start the valve on the pipe by hand.

> **NOTE**
> It should be possible to turn the valve onto the pipe three and one-half revolutions. If not, the threads are not properly cut and therefore must be re-cut.

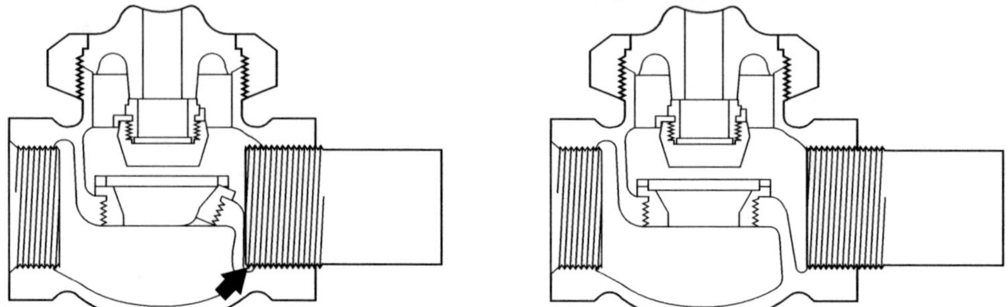

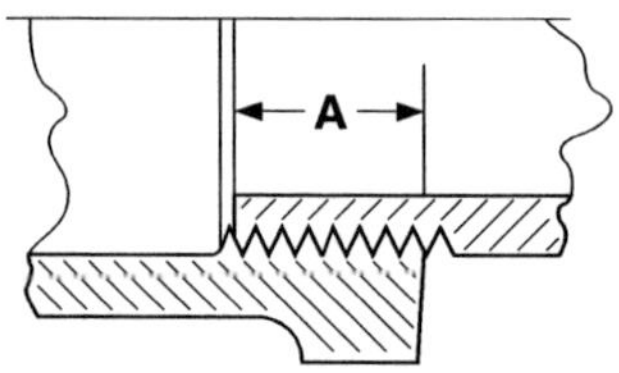

DIMENSION OF PROPER THREAD MAKEUP

SIZE	1/8	1/4	3/8	1/2	3/4	1	1 1/4	1 1/2	2	2 1/2
A	1/4	3/8	3/8	1/2	9/16	11/16	11/16	11/16	3/4	15/16

SIZE	3	3 1/2	4	5	6	8	10	12	14
A	1	1 1/16	1 1/8	1 1/4	1 5/16	1 7/16	1 5/8	1 3/4	

DIMENSIONS IN INCHES

DIMENSIONS GIVEN DO NOT ALLOW FOR VARIATIONS IN TAPPING OR THREADING.

107F53.EPS

Figure 53 ◆ Proper thread makeup.

Step 13 Tighten the valve slowly and smoothly using a pipe wrench or a strap wrench.

> **CAUTION**
>
> Do not allow the threads to bottom out in the valve because this damages the seat and ruins the valve.

Step 14 Turn the valve several times to verify proper movement through cycle.

Step 15 Notify operator to return the pipeline to full pressure to test the valve.

10.0.0 ◆ INSTALLING FLANGED VALVES

Flanged valves are constructed with a flange-type fitting with boltholes on the ends. Valves are available with two types of flange face styles. Each type of flange face style uses a specific gasket. A set of two flanges, one on the valve and one on the pipe end, of the same face style are called *companion flanges* and are used with the proper gasket. *Figure 54* shows a flanged butterfly valve.

Never use two different face styles at one joint unless unavoidable when connecting to manufactured equipment. If a flat-face flange must be connected to a raised-face flange, it must be confirmed that the flat-face flange is not cast iron. The pressure of the bolts can crack or break the cast-iron flange. If the flat-face flange is cast iron, the engineering department must be consulted for approval or recommendations before attempting to connect the raised-face flange to a flat-face flange. Valves and fittings with raised flanges are usually identified in pipeline installations by the noticeable space between bolted flanges. *Figure 55* shows raised face and flat-faced flanges.

Before attempting to replace a flanged valve, appropriate preparations must be made. The technician must follow company procedure. The procedure may include some or all the following steps:

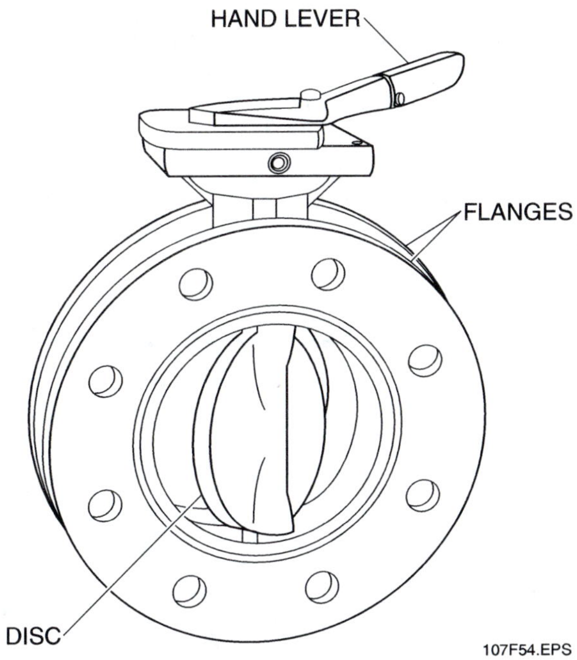

Figure 54 ◆ Flanged butterfly valve.

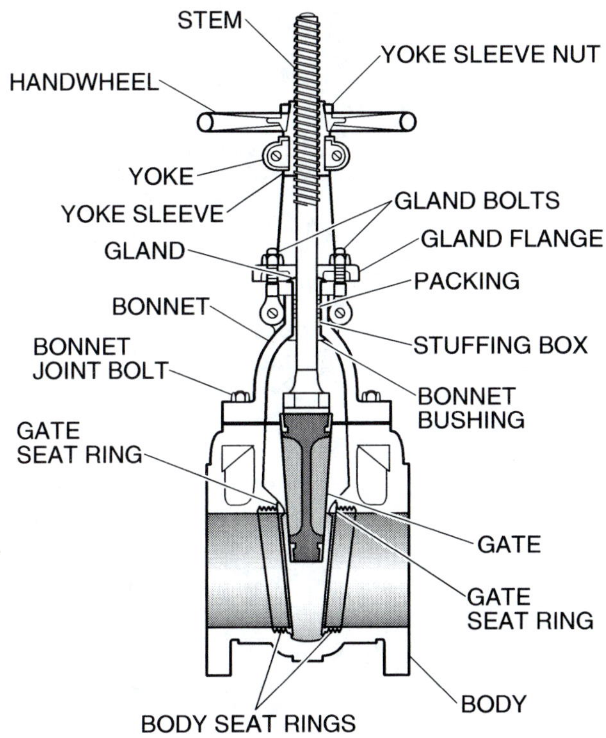

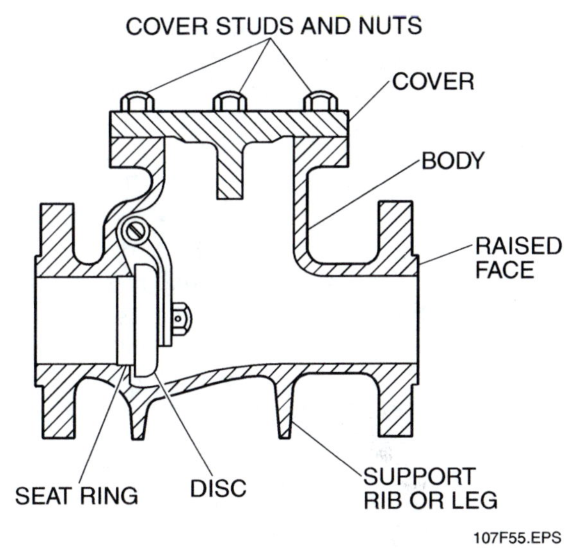

Figure 55 ◆ Raised face and flat-faced flanges.

Step 1 Verify that the pressure has been shut off.

Step 2 Cut off all energy sources associated with the valve.

Step 3 Install appropriate lock out devices and tags.

Step 4 Cut off the valves on the upstream and downstream side of the valve to be replaced.

Step 5 Drain down or purge the segment of pipeline associated with the valve being replaced.

Step 6 Secure proper rigging to the valve to support it once it is unsecured from the pipeline.

> **WARNING!**
> Never begin removal of a pipeline valve until it is solidly supported. Some valves are tremendously heavy and dangerous. Do not allow the valve being removed to simply fall to the surface.

Step 7 Use appropriate methods for removal of valve from pipeline.

10.1.0 Cleaning

All parts in a piping system must be cleaned and inspected before assembly. This must be done before the valve is put in place because access to inner areas will be very limited. Dirt, sand, loose scale, and other materials can clog strainers, contaminate fluids, and wear away parts. Use a clean rag soaked in solvent to clean the inside ends of the pipe, the flanges, and the gaskets. Loose scale may need to be brushed off and then swabbed out.

10.2.0 Unpacking Valves

Large pipeline valves arrive from the manufacturer or re-builder packaged to protect the valve and parts from damage in shipping and handling. Packaging includes mounting on skids and in crates. Protruding parts such as indicator rods, stem protectors, and operator assemblies are usually disassembled and packaged separately. Some arrive in the crate with the valve. Others arrive packaged separately. Operators must exercise care to properly unpack the valve and all parts packages included in the shipment. Some valves are shipped with installation kits that must not be lost. The valve must be closely inspected for any damage or lost parts due to shipping, and, if this has happened, the carrier must be notified immediately. A responsible person must check off the shipping list to make certain all parts are included before installation is attempted.

The shipping frames and pads must be dismantled properly. Manufacturer's unpacking instructions should be followed closely so that the unpacked valve is not damaged after shipment. Some of the packing material or bracing timber may have to be retained in place to aid in storing and moving a large valve. The manufacturer's instructions for moving and the order of assembly should be followed.

10.3.0 Installing New Valve

The replacement valve must be put in place and properly supported and aligned in the pipeline. Any jacks, shackles, spreader beams, stands, timbers, slings, or other rigging must be adjusted until the flanges are properly aligned. *Figure 56* shows properly and improperly aligned flanges.

WARNING!
Never allow personnel under a valve while being installed. Pipeline valves are extremely heavy and dangerous.

Manufacturers install I-bolt-type lifting lugs in the tops of some valves. Other valves have holes in the casting or frame for lifting. Large valves have lifting lugs formed in the casting of the valve. Whatever the design, manufacturer's lifting lugs or holes must be used so that the valve is not distorted or damaged. If no lug is included, flanged valves are lifted with special chain hooks inserted into the top boltholes on the flanges.

When the valve is in place, the boltholes in the flanges must be aligned. A drift punch, metal rod, or tool can be used to align the boltholes. Size required will depend on the size of the flange to be aligned.

WARNING!
Never attempt to align flange boltholes with fingers.

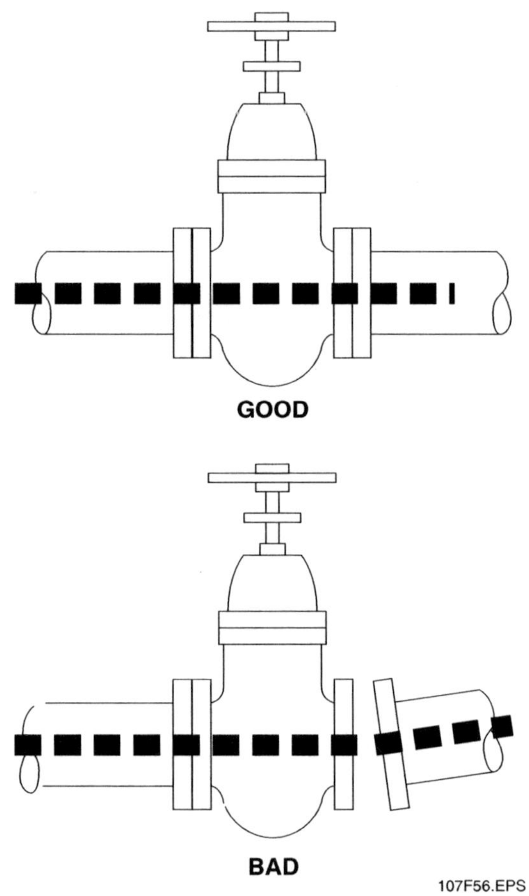

Figure 56 ◆ Properly and improperly aligned flanges.

Flange bolts must be tightened carefully to avoid warping or breaking the flange. This will also prevent damage to the new gasket by incorrect or uneven pressure. The manufacturer's instructions must be followed for proper tightening sequences for the specific valve. *Figure 57* shows typical manufacturer's recommended tightening sequences for a various flange sizes.

Unless the bolts are tightened correctly and in the right sequence, the flange may warp, crack, and leak. Pressure must be applied evenly and in

	SEQUENTIAL ORDER	ROTATIONAL ORDER
8 BOLT HOLES	1–2 3–4 5–6 7–8	1 5 3 7 2 6 4 8
12 BOLT HOLES	1–2 3–4 5–6 7–8 9–10 11–12	1 5 11 4 8 10 2 6 12 3 7 9
16 BOLT HOLES	1–2 3–4 5–6 7–8 9–10 11–12 13–14 15–16	1 16 11 8 3 13 10 5 2 15 12 7 4 14 9 6
20 BOLT HOLES	1–2 3–4 5–6 7–8 9–10 11–12 13–14 15–16 17–18	1 13 5 17 9 3 15 7 19 11 2 14 6 18 10 4 16 8 20 12
24 BOLT HOLES	1–2 3–4 5–6 7–8 9–10 11–12 13–14 15–16 17–18 19–20 21–22 23–24	1 9 17 5 13 21 3 11 19 7 15 23 2 10 18 6 14 22 4 12 20 8 16 24

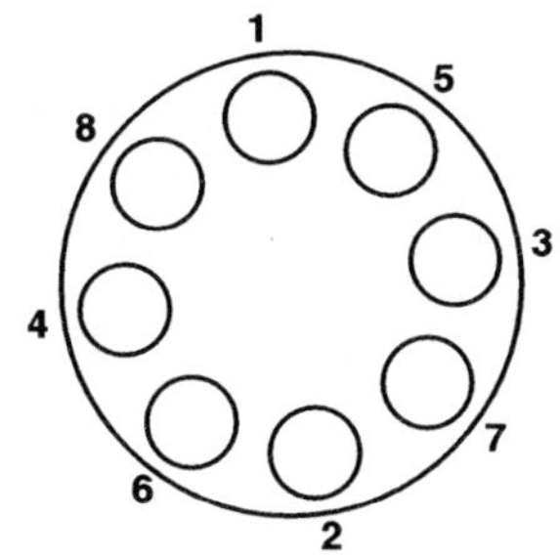

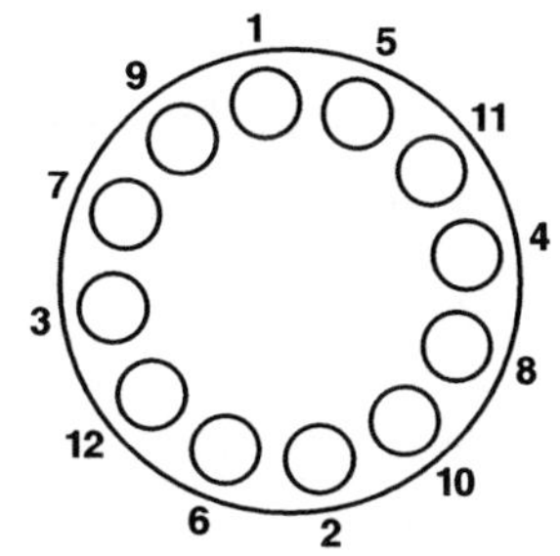

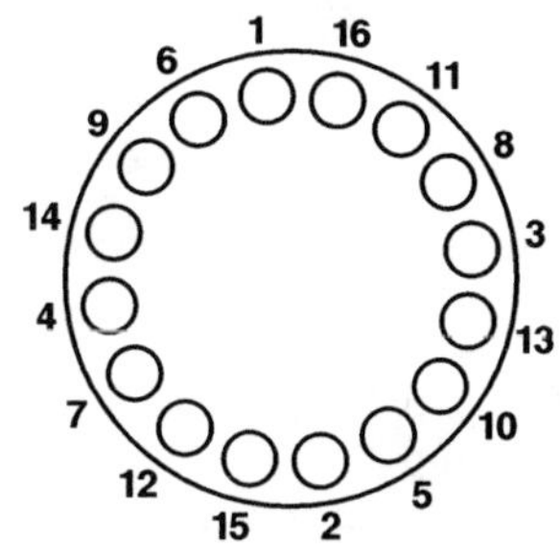

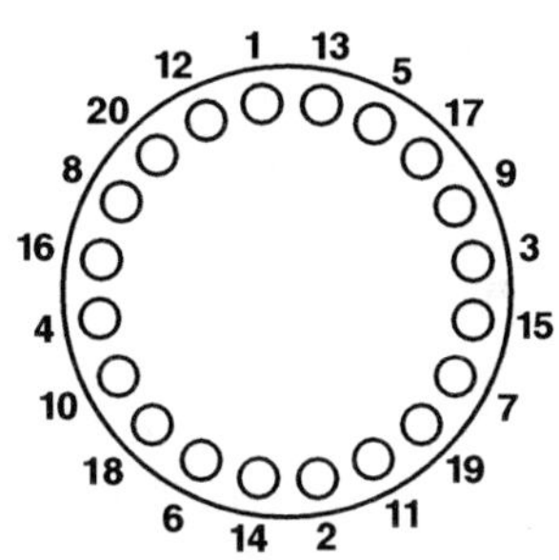

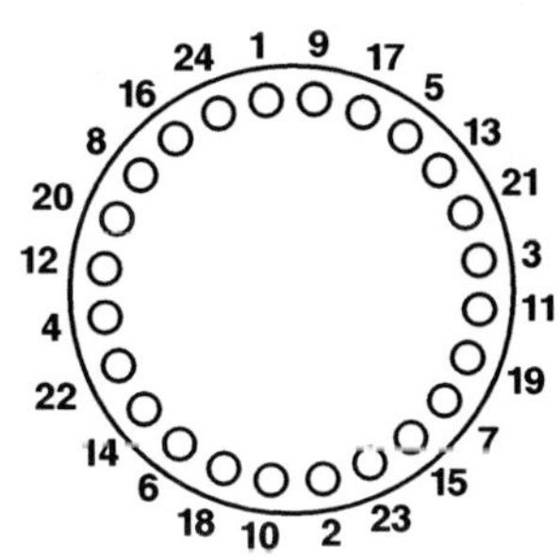

107F57.EPS

Figure 57 ◈ Typical manufacturer's recommended tightening sequences for a various flange sizes.

the correct amount to make a good seal. A torque wrench must be used to apply equal pressure at all points around the flange. Different valve flanges require different torque settings depending on the size of flange and bolts, pressure and temperature in the system, and the type of gasket used. Manufacturer's specifications must be consulted for the proper torque setting. In some areas, valves installed in summer may have to be tightened again during colder weather. The threads of the flange, nut, and bolt must be cleaned and lightly oiled or coated with pipe dope as directed by specifications. The land or face of the nut that will be tightened against the flange should also be lightly lubricated.

CAUTION

Do not apply full torque to bolts on first tightening of the flange. Only apply full torque on subsequent tightening. This will prevent unequal pressure on some areas of the flange.

The following steps are necessary to install a gasket and tighten the bolts correctly:

Step 1 Refer to specifications to determine the appropriate gasket, or use the kit supplied with valve.

Step 2 Find the specified gasket.

Step 3 Place the gasket between the flanges.

Step 4 Hold the gasket in place, and insert the bottom bolts into the flanges.

NOTE

When joining raised-face flanges, the bottom bolts are inserted before the gasket.

Step 5 Insert all the bolts through the flange bolt holes.

Step 6 Place the nuts on the bolts, and tighten them finger-tight.

Step 7 Set the torque wrench at 20% of the required torque.

Step 8 Tighten the top, center bolt to this torque.

Step 9 Tighten the bottom center bolt to this torque.

Step 10 Tighten a side center bolt to this torque.

Step 11 Tighten the bolt directly opposite to this bolt.

Step 12 Continue process of alternating bolts until all bolts are at the 20% torque level.

Step 13 Reset the torque wrench to 40% of the required torque.

Step 14 Tighten all bolts at this torque level following the same crossover method.

Step 15 Reset the torque wrench to 60% of the required torque.

Step 16 Tighten all bolts to this torque level following the crossover method.

Step 17 Reset the torque wrench to 80% of the required torque.

Step 18 Tighten all bolts to this torque following the crossover method.

Step 19 Reset the torque wrench to 100% of the required torque.

Step 20 Tighten all bolts to this torque moving clockwise around the flange.

Step 21 Repeat *Step 20* for a final tightening.

The following guidelines should be kept firmly in mind when tightening a flange with a torque wrench and any other times bolts must be tightened to exact torque.

- Loosen and retighten any flange bolts that have been inadvertently tightened when tightening in sequence. Loosen bolts one full turn, using an open-end or socket wrench.
- Verify that the torque wrench has been recently calibrated.
- Verify that the torque wrench is set to the correct torque setting.
- Place the torque wrench on the nut, not on the bolt. Use an open-end or socket wrench to hold the bolt while torquing.
- Use a smooth, even motion when using the torque wrench. A hurried, uneven motion does not give a correct measurement.

10.4.0 Cycling the Valve

After the valve installation is completed, the technician must operate the valve to verify that the mechanism is installed properly. Verify that the valve is operational before notifying the control center operator. The technician should turn the valve to the completely closed position and then back to the completely open positions three to four times to make certain that it operates satisfactorily. If there is any kind of problem, the valve should be disassembled promptly and the problem corrected immediately.

Summary

Pipelines use valves to start and stop flow, regulate flow, relieve pressure, and prevent reversal of flow. The type of valve used is determined by the type of application. Valves are chosen based on the nature of the product in the pipeline, the temperature and pressure of the system, and the desired operation of the system. Valves are manufactured from various materials that are selected to offer the best service for the product being handled.

Representing a sizeable portion of pipeline investment, valves must be cared for and maintained so that their service life and efficiency is preserved as long as possible. Required by law in many cases, valve maintenance is one of the most important tasks on the pipeline because it ensures the safety, efficiency, and economics of operation. Maintenance begins with receiving, handling, and proper storage until the installation of each valve. Once the valves are installed, proper care ensures that they operate efficiently and safely through all conditions. Special maintenance procedures to prepare for winter ensure their safe operation in cold weather.

Once the valve is put in place, care exercised in handling and installation pays large dividends in the service life and dependability of the unit. Proper support of the valve, seating of gaskets, torquing of bolts, and care in following manufacturer's instructions all work together to provide extended, leak-free operation, ensuring the continued safety and efficiency of the pipeline.

Review Questions

1. The size of the pipe and the size of the _____ must be the same.

a. bonnet
b. valve body
c. gate
d. flange

2. The preferred gate valve to use in severe conditions that requires frequent dismantling is a valve with a _____ bonnet.

a. flanged
b. bolted
c. screwed
d. pressure seal

3. A gate valve that provides the maximum internal opening is called _____.

a. a body valve
b. through port
c. full port
d. large

4. Where space is limited, valves with _____ stems are used.

a. right angle
b. rising
c. nonrising
d. long

5. The handle of a ball valve is attached to the valve stem, which turns _____ of a turn to move the hole in line with the flow through the valve.

a. one-quarter
b. one-third
c. one-half
d. three-quarters

6. A valve that is used for service where lubrication is inconvenient is a _____.

a. nonlubricated plug valve
b. lubricated plug valve
c. right angle plug valve
d. gate valve

7. Valves that are used to more easily accomplish block-and-bleed operations are _____ valves.

a. expanding gate
b. nonlubricated plug
c. three-way
d. lubricated plug

8. Globe valves are available in sizes up to _____ inches.

a. 10
b. 12
c. 14
d. 16

9. Globe valves discs are made in _____ basic designs.

a. two
b. three
c. four
d. five

10. The valve that is similar to a globe valve and is designed to give fine control of flow in small-diameter piping is the _____ valve.

a. Y-type
b. plug
c. wafer lug
d. needle

11. The body of a two-flange valve has a _____ on each side.

a. valve
b. flange
c. disc
d. wafer

12. The amount of spring pressure of a relief valve is set _____.

a. during installation
b. at system startup
c. at the factory
d. by maintenance

13. A pipeline relief valve relieves pressure _____.

a. when opened manually
b. when opened by remote control
c. in the field
d. automatically

14. Ball check valves operate in the same manner as _____ valves.

a. spring
b. relief
c. lift check
d. butterfly

15. All valves require a complete seal in _____ places.

a. two
b. three
c. four
d. five

16. If cycling and wiping the stem several times does not eliminate a valve leak, the next step is to _____.

a. remove the valve
b. install a new valve
c. disassemble the valve
d. tighten the packing nut

17. Valve gaskets can be ring type or _____.

a. square
b. flat
c. fiber
d. packing

18. Lubricated valves have an _____ lubricant supply.

a. external
b. endless
c. inner
d. forced

19. The procedure for installing a small valve is the same as installing a _____.

a. large valve
b. pipe
c. flange
d. fitting

20. After a valve installation is complete an operator must _____ the valve.

a. tighten
b. loosen
c. cycle
d. lockout

GLOSSARY

Trade Terms Introduced in This Module

Ball valve: A valve that uses a ball with an opening through the center that is rotated one-quarter turn to start and stop flow.

Block and bleed: The ability to obtain a seal across the upstream and downstream seat seals of a valve when the body pressure is bled off through blow-down valves or vent plugs. Used to test the integrity of seat seals and when making minor repairs under pressure.

Body: The principle pressure-containing part of a valve containing disc, seat, and valve ports. It is connected directly to the pipe by threaded or flanged connections.

Body bleed: Opening a body drain or vent to bleed off (reduce) internal body pressure or double-seated valves in either the full open or closed position.

Bonnet: The part of a valve that is located above the valve body and contains the trim.

Check valve: A valve that permits flow in only one direction also called a one-way valve.

Control valve: A globe valve that is automatically controlled to regulate flow through the valve.

Disc: The moving part of a valve that directly controls flow through that valve.

Double block and bleed: A valving arrangement or valve type that ensures no flow in a line, even though a valve may leak. It consists of two block valves in the main line with a small bleeder valve draining the line between the block valves.

Downstream: In pipeline operations, downstream refers to any point beyond the origination.

Full port: The maximum internal opening for flow through a valve that matches the inside diameter of the pipe used.

Gate valve: A valve with a disc-shaped closing element that fits over an opening through which fluid passes. It is designed for completely open or closed service.

Globe valve: A device for regulating flow in a pipeline, consisting of a moveable disc-type element and a stationary ring seat in a generally spherical body.

Good working order: All valves are required by *49 CFR Part 195 Paragraph 195.420(a)* to be in *good working order*, which can be interpreted to mean that a valve's performance, based on in-service criteria, meets all of the necessary functions of the valve.

Leak-by: In double-seated valves, a leak-by is an internal valve leak. In gate and ball valves, a leak-by is a condition in which hazardous liquid can leak past either the upstream or downstream seal into the valve by, thereby pressurizing the valve body.

Leak-through: In single-seated valves such as check valves, a leak-through is a condition in which hazardous liquid can leak by the valve seat, causing the valve to internally leak past the valve when the valve is in the closed position. In gate and ball valves, a leak-through is a condition in which hazardous liquid can leak past both valve seats, causing the valve to leak from the high-pressure side of the valve to the low-pressure side when the valve is closed.

Packing: The material between the valve stem and bonnet used to provide a leak-proof seal around the stem, preventing fluid from leaking up around the stem.

Plug valve: A valve that contains a plug with a hole in the center. It rotates to start and stop flow.

Port: The internal opening for flow through a valve.

Relief valve: A valve that automatically opens when a preset amount of pressure is exerted on the valve disc.

Seat: The nonmoving part of a valve upon which the disc rests to form a seal and close the valve.

Stem: The part of a valve that connects the valve operator to the disc. Stems can have rotary, helical, or linear movement.

Throttling: Reducing or controlling the fluid flow in a pipeline.

Trim: A collective term applied to the internal parts of a valve that receive the most wear and can be replaced. The trim includes the stem disc, seat ring, disc holder or guide, wedge, and bushings.

Upstream: In pipeline operations, upstream is a relative terms that refers to any point before your current location.

Figure Credits

Invensys-Nordstrom Audco	107F42
Limitorque	107F49
Serck Audco Pty. Ltd.	107F48

CONTREN™ LEARNING SERIES — USER UPDATES

The NCCER makes every effort to keep these textbooks up-to-date and free of technical errors. We appreciate your help in this process. If you have an idea for improving this textbook, or if you find an error, a typographical mistake, or an inaccuracy in NCCER's Contren™ textbooks, please write us, using this form or a photocopy. Be sure to include the exact module number, page number, a detailed description, and the correction, if applicable. Your input will be brought to the attention of the Technical Review Committee. Thank you for your assistance.

Instructors – If you found that additional materials were necessary in order to teach this module effectively, please let us know so that we may include them in the Equipment/Materials list in the Instructor's Guide.

Write: Curriculum Revision and Development Department
National Center for Construction Education and Research
P.O. Box 141104, Gainesville, FL 32614-1104

Fax: 352-334-0932

E-mail: curriculum@nccer.org

Craft ________________ Module Name ________________

Copyright Date ________ Module Number ________ Page Number(s) ________

Description

____

____

____

____

(Optional) Correction

____

____

____

(Optional) Your Name and Address

____

____

____

Module 63108-02

Identify Types of Valve Actuators/Operators

Introduction to the Pipeline Operator Qualification Rule, API, and NCCER

It could be said that petroleum products keep America running. Petroleum products like oil and natural gas are the principal sources of energy in the United States and are used in many manufacturing and chemical processes. Petroleum products are primarily transported by pipeline. There are more than a million miles of liquid and natural gas pipeline in this country, ranging in diameter from two inches to four feet. These pipelines bring crude oil and natural gas from production fields to processing plants and refineries—and ultimately to the consumer.

Pipeline Operator Qualification

Petroleum products are considered hazardous, so pipeline personnel must take extra precautions and follow specific procedures when working with them. To help ensure the safe handling and transportation of petroleum products, the United States Department of Transportation (DOT) mandates safe practices as federal regulations.

One such regulation, the Operator Qualification (OQ) rule, requires pipeline operators to develop a written qualification plan to evaluate their personnel and their contract personnel (see *49 CFR Part 192 Subpart N—Gas*, and *49 CFR Part 195 Subpart G—Liquid*). The evaluation hinges on the operator's ability to perform certain covered tasks and to be able to recognize and respond to abnormal operating conditions that may be encountered while performing these covered tasks.

Covered Tasks

A pipeline performance task is deemed *covered* if the activity passes the following four-part test:

1. Is it performed on a pipeline facility?
2. Is it an operations or maintenance task?
3. Is it performed as a requirement of *49 CFR Part 195 (Liquid)* or *49 CFR Part 192 (Gas)*?
4. Does it affect the operation or integrity of the pipeline facility?

If the answer to all four questions is *yes*, then the activity is considered *covered* and must figure into the evaluation process of pipeline personnel.

American Petroleum Institute

The OQ rule and its related covered tasks provided the catalyst to develop a unified pipeline training and assessment program—so that pipeline operators and owners have ready access to the assessment and training tools they need. This effort was the direct result of the vision and leadership provided by The American Petroleum Institute (API) and API members.

Efforts to unify the industry are nothing new to API. Indeed, the impetus for forming the Institute in 1919 was the need to standardize engineering specifications for drilling and production equipment. Today, the most pressing issues revolve around public perceptions and government policies toward the industry. Speaking with one voice on these issues has become as essential as having interchangeable parts in the field. API's members determine the petroleum industry's positions on public policy as well as the standards that govern its day-to-day operations worldwide.

Consortium on Operator Qualification

API assembled a focus group whose primary purpose was to identify covered tasks and provide compliance recommendations to pipeline operators.

The Consortium on Operator Qualification (COOQ) became the guiding force for the identification and development of covered tasks for the liquid pipeline industry. Their work product, API publication 1161, *Guidance Document for the Qualification of Liquid Pipeline Personnel*, formed the basis for the resulting Pipeliner Training and Assessment Program.

The covered task list created in API publication 1161 has now evolved to include gas covered tasks and is maintained by NCCER. The most current covered task list can be downloaded at www.nccer.org.

National Center for Construction Education and Research

With the guidance and support of API, its members, and the Consortium on Operator Qualification, the National Center for Construction Education and Research (NCCER) was able to design and implement the Pipeliner Training and Assessment Program.

NCCER is a not-for-profit 501(c)(3) education foundation established in 1995 by the world's largest and most progressive construction and maintenance contractors and national contractor associations. NCCER was founded to address the

severe workforce shortage facing the construction and maintenance industries and to develop a standardized training process and curricula. Today, NCCER is supported by hundreds of leading construction and maintenance companies, manufacturers, and national associations and developed more than 35 construction and maintenance titles.

NCCER also maintains a National Registry, which contains the records of trainees and instructors who have received NCCER training. All completion information for every trainee is sent to the NCCER and filed within the Registry. The Registry can then confirm training and skills for workers as they move from company to company, state to state, or even within their own company.

Standardized Curriculum (NCCER's Contren™ Learning Series)

NCCER curricula are developed using the expertise of industry subject matter experts (SMEs). The use of SMEs as an integral part of every new development or revision ensures that the curriculum is current, industry-driven, and indicative of industry practices. The development and/or revision process is facilitated from start to finish by NCCER project managers, who work closely with the SMEs and a select team of technical writer(s) on the development and presentation of the curriculum.

Pipeline: Five Training Paths

The pipeline curriculum focuses on essential skills and competencies within five skill areas:

- Corrosion Control
- Electrical and Instrumentation
- Maintenance
- Mechanical
- Operations (Field and Control Center)

SMEs from the pipeline industry have determined the skills and competencies needed based on API's covered tasks and the consensus of the group/industry. Specific company information is generally not incorporated into the curriculum unless the SMEs determine that the information is the industry standard.

Training Modules and Covered Tasks

The smallest unit of instruction in the curriculum is a *module*. Each curriculum is composed of modules (the equivalent of stand-alone chapters) that are logically sequenced into complete *levels* of instruction. Every module is task-oriented and requires the successful completion of a written (knowledge-based) examination and a performance (hands-on) test. Modules may be individually selected and/or combined for efficient task training. Modules in the pipeline curricula, to the extent possible, will align with API-developed covered tasks. While not all modules will relate to covered tasks, all covered tasks will relate to at least one module.

Benefits of Training to the Pipeline Marketplace

Applying NCCER's standardized training process allows pipeline owners and contractors in the United States to establish the benchmark for acceptable skill and competency levels within the pipeline industry. This process will assist training organizations in developing and/or improving standardized pipeline training programs that effectively match their resources and operating conditions while responding to regulatory demands and to the needs of the industry. The process will also recognize training organizations that meet the minimum standards of competence, warranting regulatory, public, and professional confidence. This in turn will foster national unity within the pipeline industry for the mutual benefit of individual technicians, owners, contractors, and ultimately pipeline-users and society.

Additional Resources

To review the Pipeliner Training and Assessment Program Implementation and Resource Guide and the latest version of pipeline covered tasks, visit NCCER's Web site at www.nccer.org, or contact:

National Center for Construction Education and Research
PO Box 141104
Gainesville, FL 32614-1104
(888) 622-3720
Web site: www.nccer.org

For more information on API, please contact:

American Petroleum Institute
1220 L Street, Northwest
Washington, D.C. 20005
(202) 682-8000
Web site: www.api.org

To order *API Pub 1161*, contact Global Engineering Documents by calling 1-800-854-7149 or visiting their Web site at www.global.his.com.

Additional sources of information on the pipeline operator qualification regulation include:

DOT – Office of Pipeline Safety
www.ops.dot.gov

Transportation Safety Institute
www.tsi.dot.gov

COURSE MAP

This course map shows all of the modules in the first level of the Pipeline Mechanical curriculum. The suggested training order begins at the bottom and proceeds up. Skill levels increase as you advance on the course map. The local Training Program Sponsor may adjust the training order.

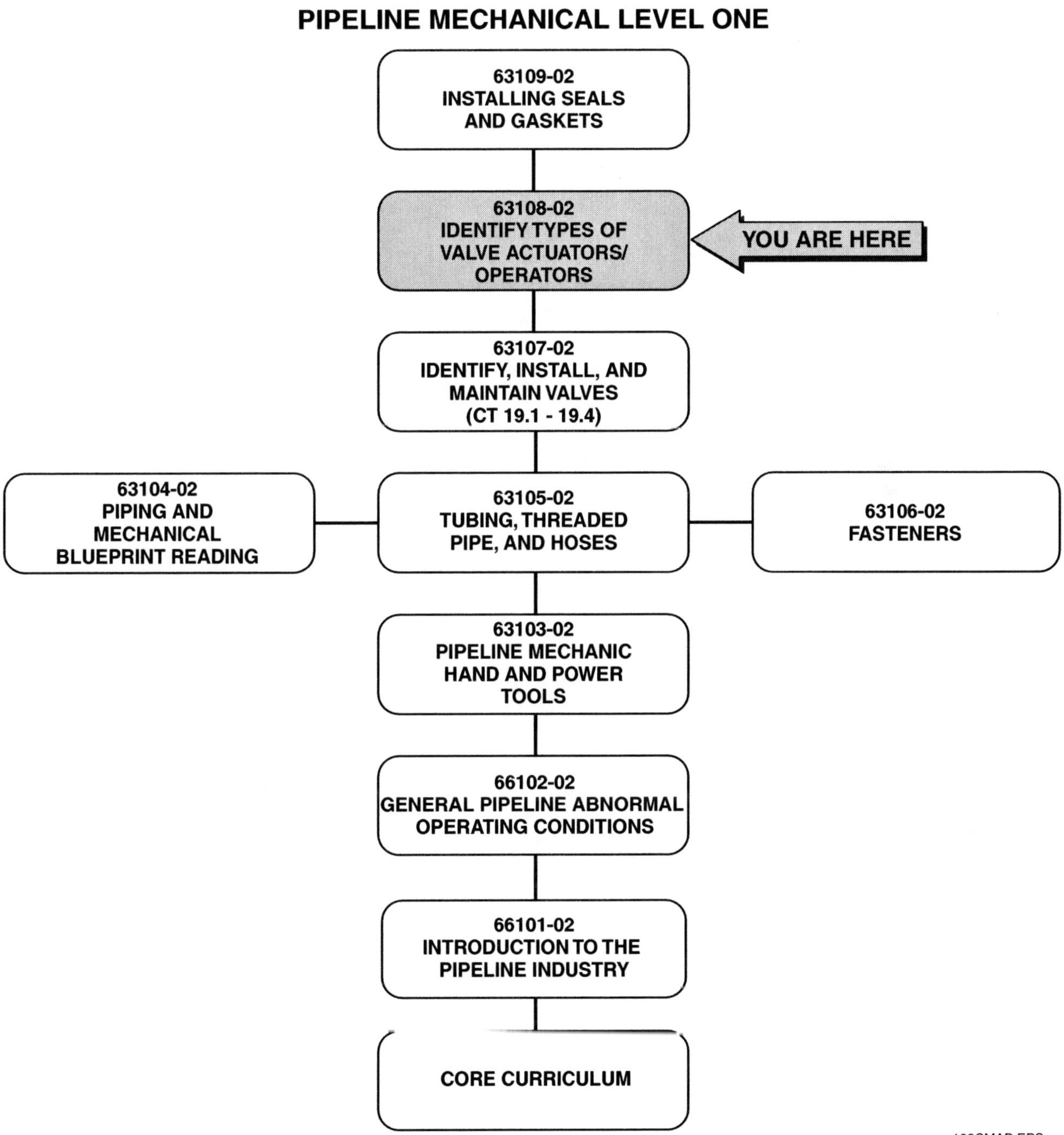

108CMAP.EPS

MODULE 63108-02 CONTENTS

Figures

MODULE 63108-02

Identify Types of Valve Actuators/Operators

Objectives

When you have completed this module, you will be able to do the following:

1. Identify the types of actuators used to open and close valves.
2. Identify the types of controls used with actuators.
3. Explain the principles used for operation of actuators.
4. Perform general maintenance on actuators.

Prerequisites

Before you begin this module, it is recommended that you successfully complete the following: Core Curriculum; Pipeline Core, Modules 66101 and 66102; Pipeline Mechanical Level One, Modules 63103 through 63107.

Required Trainee Materials

1. Pencil and paper
2. Appropriate personal protective equipment

1.0.0 ◆ INTRODUCTION

A **valve actuator**, commonly called an operator, is a device that opens and closes a valve, either manually or automatically. Manual actuators come in the form of handwheels, levers, and wrenches that are attached to the stem. Larger valves that require great effort to turn must be equipped with some type of powered mechanism, such as electric, pneumatic, or hydraulic units mounted on the stem. Some valve actuators use a gearbox with wide reduction ratios. Some manual actuators employ high and low gear ratios to enable personnel to turn the valve.

Valves with automatic actuators are controlled from a console in a control center by remote control, sometimes from many miles away. Such automatic actuators are used when the valve is remote from the main work area, when the frequency of operation would require unreasonable human effort, or if rapid opening and closing of the valve is required. In some installations, the effort required to turn some large valves is beyond the capability of personnel. Some valves, if turned manually, require so much time and effort that the cycling operation would be self-defeating unless it was an emergency. Maintenance crews often carry portable valve actuators to work in outlying installations.

In some installations, automatic and remote control actuators are installed as a matter of convenience or to make possible a reduction of the number of employees required for operation of the facility.

2.0.0 ◆ TYPES OF ACTUATORS/OPERATORS

The types of valve actuators in use on the pipeline are manual actuators and automatic actuators.

2.1.0 Manual Actuators

Manual actuators are used in several common forms. Small valves use the typical round handwheel or wing-shaped handle to turn the valve. It is common for valves up to several inches in diameter to use a lever or portable handle that is carried from one location to another to operate different valves. Hand levers are sometimes combined with a latching mechanism so that the valve can be set and locked into a position. *Figure 1* shows typical manual actuators.

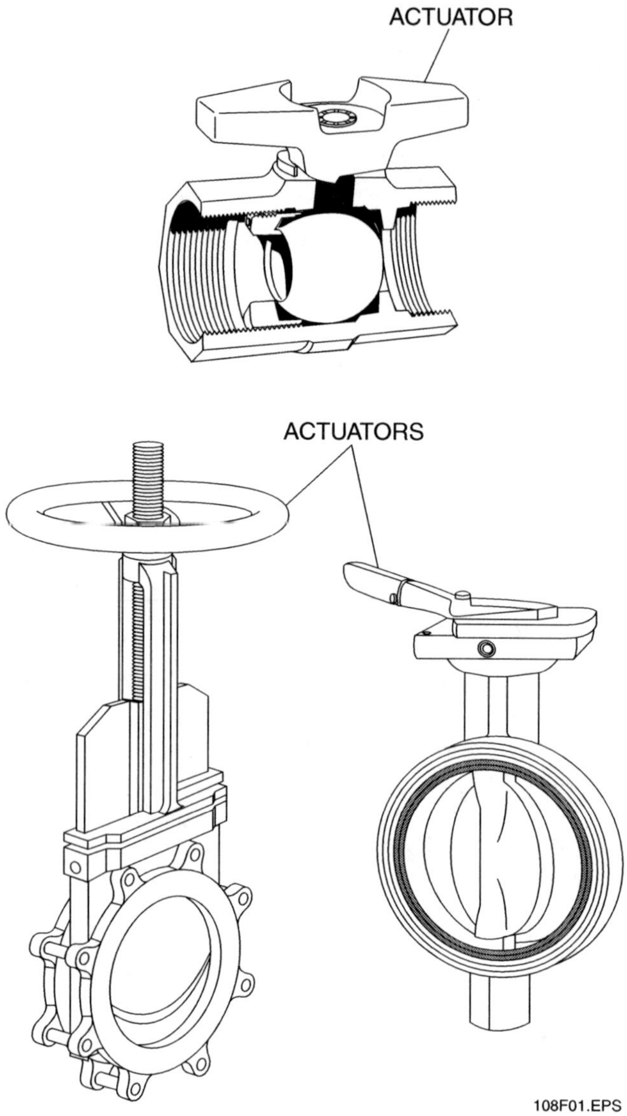

Figure 1 ◆ Typical manual actuators.

2.1.1 Chain Actuators

When necessary, chain actuators are used in some applications where the valve is too high to reach. The stem of the valve is mounted with a chainwheel, and the chain hangs to within 3 feet of the working floor level. *Figure 2* shows a typical chain actuator.

Universal-type chainwheels that attach to regular valve handwheels have been frequently blamed for many industrial accidents. In corrosive atmospheres, where a valve is seldom operated, attachment bolts have failed during attempts to operate the valve. A second safety hazard associated with the chainwheel is that the chain must dangle in the air, causing employees to be accidentally struck or become entangled.

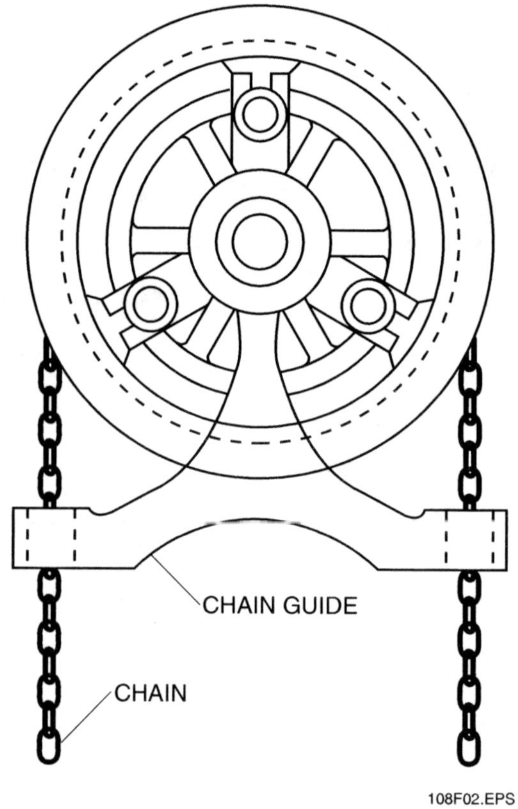

Figure 2 ◆ Chain actuator.

2.1.2 Gear Actuators

Gear actuators minimize the effort required to operate large valves that work at high pressures. Gear actuators are also used to connect to valves located in inaccessible areas. Three basic types of gear actuators are spur-gear actuators, bevel-gear actuators, and worm-gear actuators.

2.1.3 Spur-Gear Actuators

Spur gears are used to connect parallel shafts in order to position the handwheel in a more convenient position and give mechanical advantage to the employee. They have straight teeth cut parallel to the shaft axis and use gears to transfer motion and power from one shaft to another parallel shaft. Spur-gear actuators can be as simple as one gear transmitting power to another gear attached to the valve stem, or they can consist of many gears. The design selected depends on how much mechanical advantage is needed for employees to be able to operate the valve effectively. The mechanical advantage comes from the fact that the handwheel must be turned many times to open or close the valve but is slower to operate than other actuators. *Figure 3* shows a simple spur-gear actuator.

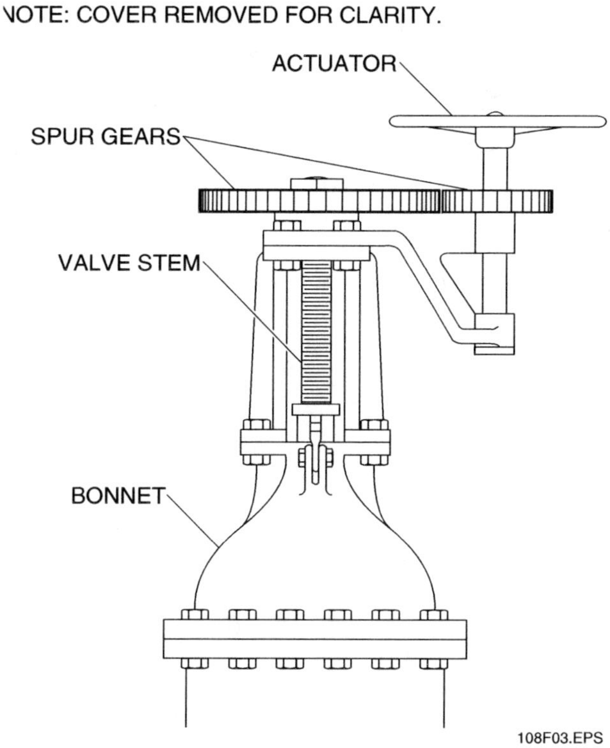

Figure 3 ◆ Simple spur-gear actuator.

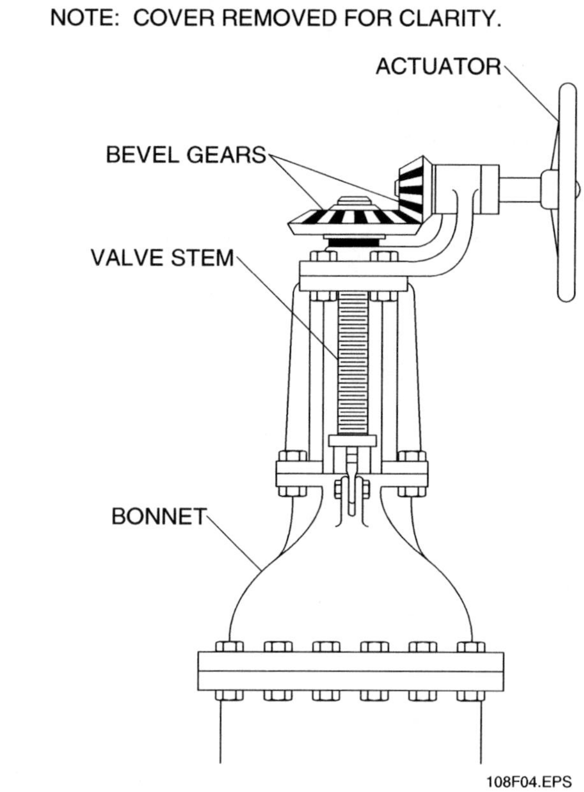

Figure 4 ◆ Bevel-gear actuator.

2.1.4 Bevel-Gear Actuators

Bevel-gear actuators are used to transmit power between two shafts that intersect at a 90-degree angle (see *Figure 4*). Bevel-gear actuators offer the advantage of a more convenient placement of the handwheel combined with the mechanical advantage of a spur gear if desired. The bevel gear resembles a cone in that the teeth on each gear are angled at 45 degrees to mesh with the mating gear. Bevel-gear actuators transmit power from a handwheel and a shaft that is perpendicular to the valve stem. Bevel-gear actuators are slower in operation than some others. Speed depends partially upon how much mechanical advantage is designed into the unit.

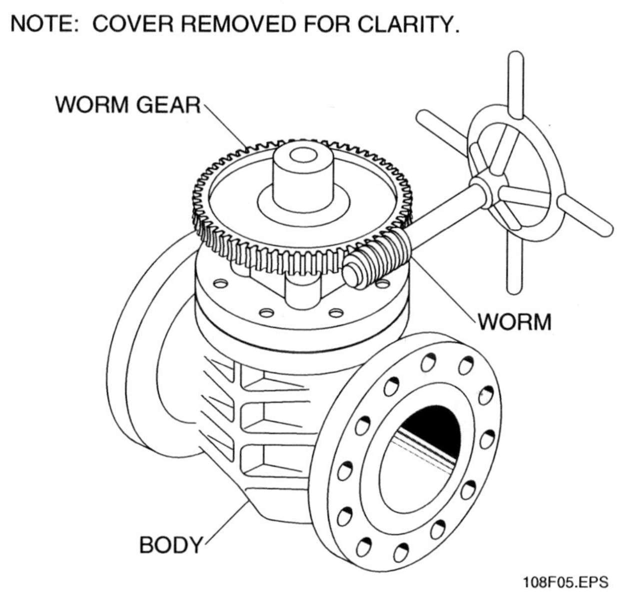

Figure 5 ◆ Worm-gear actuator.

2.1.5 Worm-Gear Actuators

Worm gears are used to transmit power at right angles, allowing a more convenient location of the handwheel (see *Figure 5*). This means that the shafts of the connecting gears are at 90-degree angles to each other. The worm is a cylindrical-shaped gear similar to a screw, and it acts as the driver. The worm gear can be the larger, circular gear in the assembly or only a gear sector. Worm-gear actuators transmit power from the handwheel, which is attached to the worm shaft, to the worm gear, which is attached to the valve stem. Worm gears are set up like bevel gears, with a side-mounted handwheel. A larger speed ratio with much mechanical advantage is obtained with a worm-gear actuator over a bevel-gear actuator because the actuator allows rapid opening and closing of large valves with minimum effort. By design of the worm gear, the unit can be applied to valves turning only 90 degrees or multiple revolutions.

2.2.0 Automatic Actuators

Automatic actuators are motorized or powered units that are operated by remote control. Basic types of automatic actuators used include pneumatic actuators, hydraulic actuators, and electric or air motor-driven actuators. Some actuators are combinations of two systems such as a combination electrohydraulic actuator where an electric motor powers the pump for the hydraulic system.

2.2.1 Pneumatic and Hydraulic Actuators

Pneumatic and hydraulic actuators operate the valve in the same manner that gear-type actuators do. All actuators move valves in the way required to open or close them. Pneumatic actuators move the valve by air pressure applied to the mechanism. Hydraulic actuators move the valve by fluid pressure applied to the mechanism.

A cylinder assembly that contains a piston or diaphragm unit is attached to the valve stem. Air or fluid pressure is applied to one side or the other of the piston to cause the piston to move up or down, opening or closing the valve. In valves that rotate, the piston is attached by an arm or pinion gear to turn the valve. Some pneumatic and hydraulic operators use pressure on only one side of the piston and are spring-loaded on the other side. The spring forces the valve to move to the open or closed position, depending on which position is considered the safe position. This safety feature is required so that the valve will return to its safe position in case of a pressure failure (see *Figure 6*).

Pneumatic actuators are normally powered by pressure from an outside source such as a compressor or storage cylinder. Some pneumatic actuators use a compressed inert gas such as nitrogen, supplied by portable, refillable cylinders or bottles. Pneumatic actuators are especially adaptable to high-pressure gas pipelines because line pressure taps can supply pressure to the actuators. Pneumatic and hydraulic actuators are good choices for installations such as refineries that already have a compressed air supply. Permanent compressed air lines can be installed and control units may be grouped together away from hazardous atmospheres.

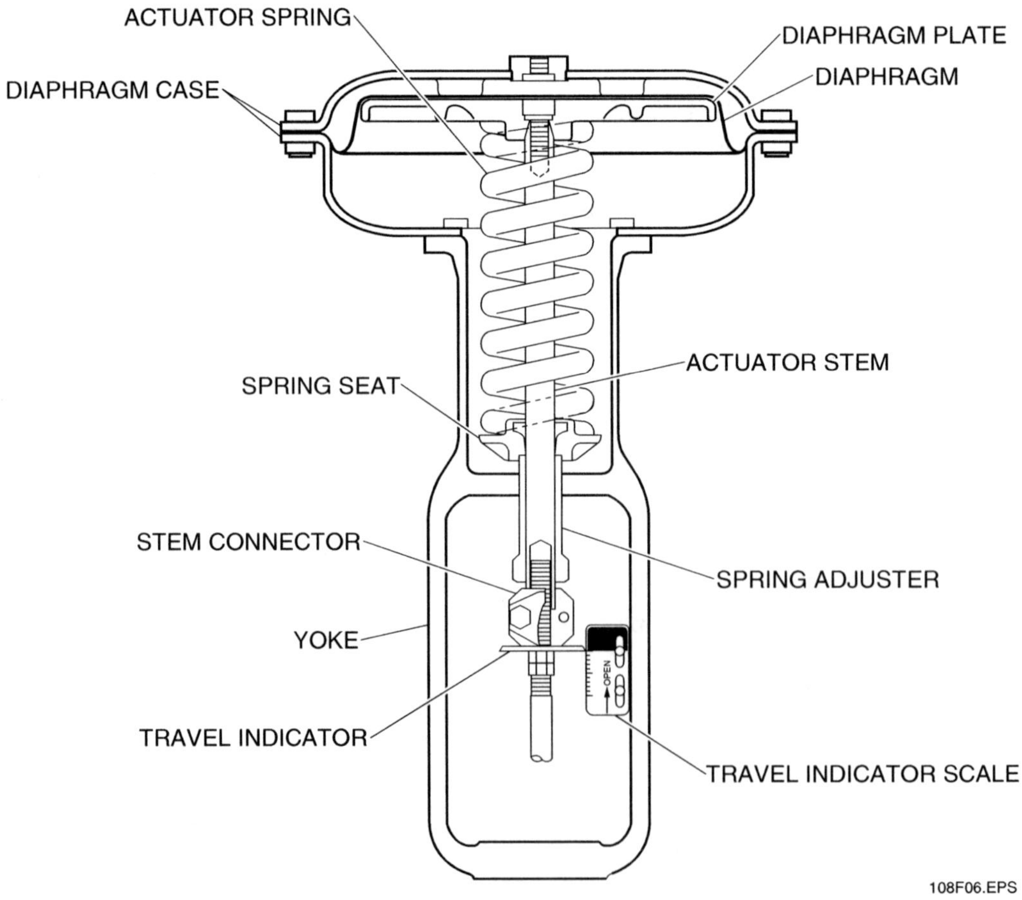

Figure 6 ◆ Spring-loaded pneumatic valve actuator.

2.2.2 *Electric and Air Motor-Driven Actuators*

Electric and air motor-driven valve actuators contain a motor that is linked through reduction gears to the valve stem. The electric motor is equipped with electrical **limit switches** that shut off the motor when the motor has turned the valve stem as far as it should go in either direction. This prevents unnecessary wear and tear on the motor. Electric valve actuators are often controlled from the control center console. The valve can be set at full open or closed, and partially opened for throttling. Electric valve actuators are usually equipped with a handwheel for controlling the valve manually if the power fails. Actuators should only be removed, turned, or adjusted by trained and qualified technicians. *Figure 7* shows an electric motor-driven valve actuator.

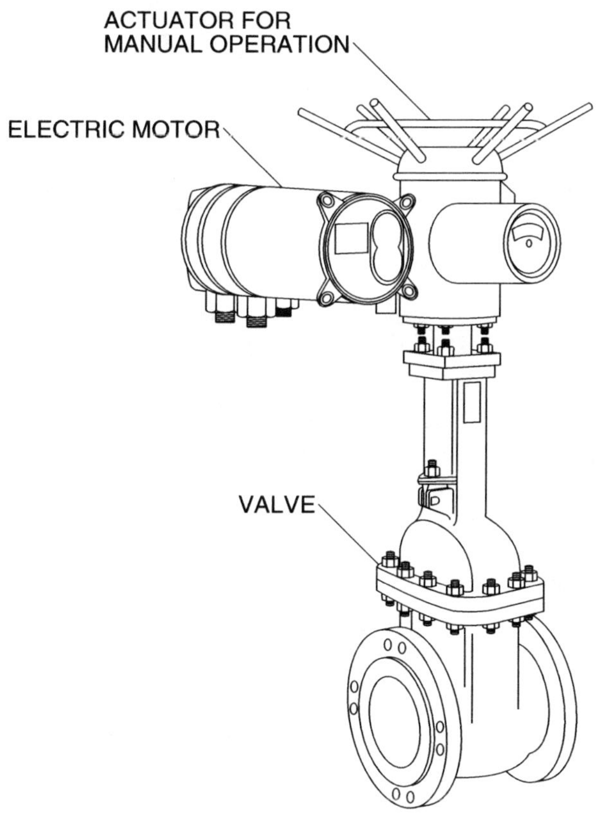

Figure 7 ◆ Electric motor-driven valve actuator.

3.0.0 ◆ ACTUATOR CONTROLS

Valve actuators can be controlled in two ways, remote and local. Actuators can be switched from one control location to the other and then back again. Remote control allows a valve actuator to be operated from the control center hundreds of miles away. Local control allows the valve actuator to be controlled from a panel at the valve site. Remote control is necessary for the control center to operate the pipeline. Local control is necessary for maintenance and times when communications may be lost.

3.1.0 Remote Control

Pipelines are remotely controlled from a control center in the terminal. The control center consoles are connected by a computer system, Supervisory Control And Data Acquisition (SCADA), to every part of the system. The status of the pipeline valves is reported continuously to the operations department on the consoles. *Figure 8* shows analog and digital readouts as they appear on a console.

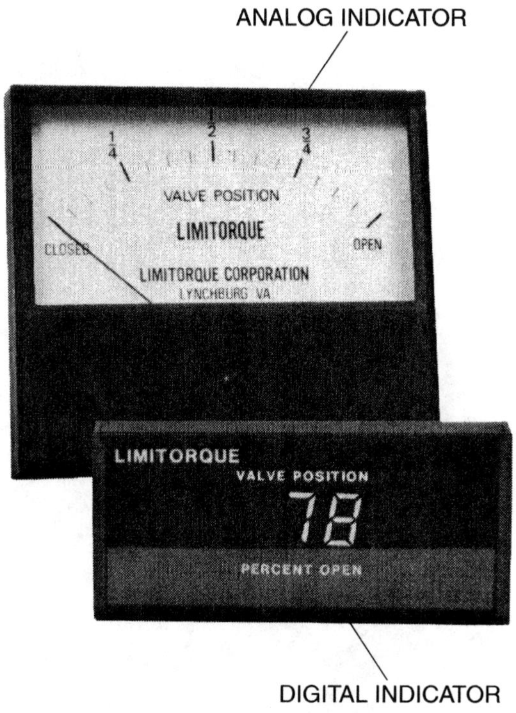

Figure 8 ◆ Analog and digital readouts as they appear on a console.

Adjustments necessary to pressure and flow on the pipeline are made by remotely adjusting valve actuators that control the valves along the pipeline. Changes in the console switches cause valve actuators to open, close, adjust pressure and flow, and shut down or start up the pipeline safely. *Figure 9* shows a control center console with the valve panel.

108F09.EPS

Figure 9 ◆ Control center console with the valve panel.

3.1.1 Redundant Systems

Pipeline control centers use what is called a **redundant system** to back up the regular control system. *Redundant* means "extra or excessive." Because it is so important to be able to control the pipeline safely at all times, the redundant systems are always in hot standby mode so that if the SCADA fails, they will already be on line to control the system. *Figure 10* shows a layout of a pipeline communications system.

3.1.2 Emergency Systems

Companies design actuators with backup systems to operate from digital networks that deliver all pertinent information to the console through field units contained inside the actuator cover. Some modern accessories include a modem to allow the actuator to be controlled by telephone in case the communications system fails.

3.2.0 Local Control

Any time maintenance must be performed, individual units may have to be put under local control. Technicians can operate or **cycle** a valve for inspection or testing. Other valves may need to be operated locally in order to isolate the unit being repaired. At such times, the technicians need to be able to switch the valve actuator from remote control to local control so that they may operate it themselves. When maintenance work is completed, they need to be able to switch the unit back to the remote control of the control center. By the simple turn of a dial, the valve can be switched from one mode to the other. Such a panel would be, of necessity, located near the valve actuator. *Figure 11* shows a remote-local control switch panel with lights to indicate position of the valve.

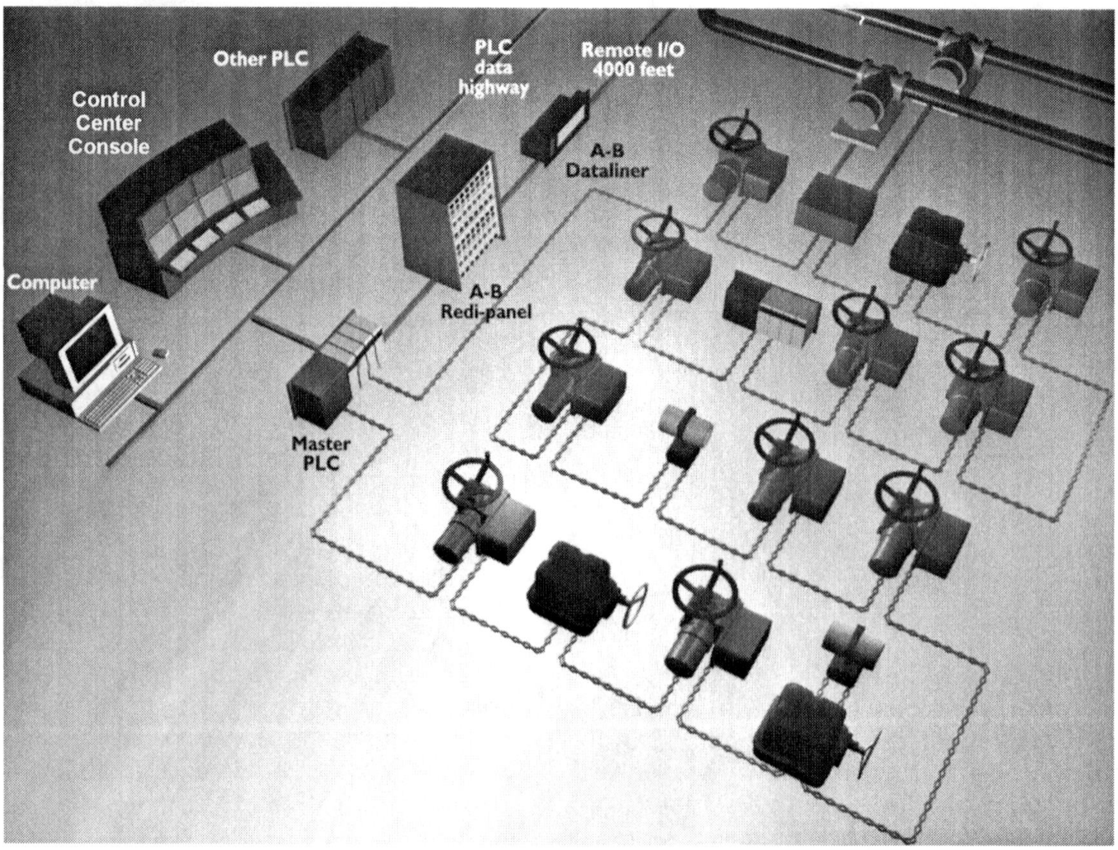

108F10.EPS

Figure 10 ◆ Layout of a pipeline communications system.

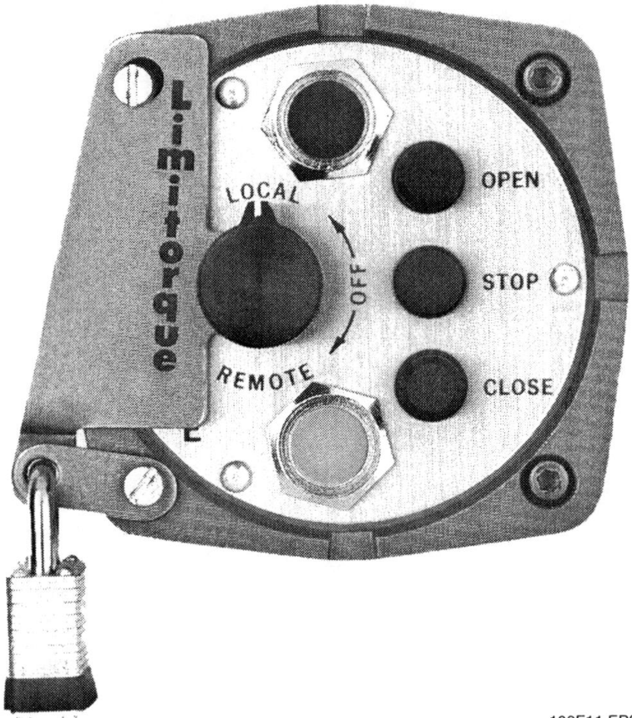

108F11.EPS

Figure 11 ◆ Remote-local control switch panel.

4.0.0 ◆ PRINCIPLES OF OPERATION

In order to control the pipeline with computer systems, automatic valve actuators that can do the job by remote control are essential. Pipeline operations depend on variations on different types of actuators. Manual actuators are still in wide use, and when communications fail, some automatic valves must be operated manually. However, the pipeline relies on automatic actuators for most of its operations. Actuators that operate off one type of power supply in one location may have to be customized to operate from another type of power supply in another location. Some actuators operate off the pressure and flow in the pipeline within which they are installed. Other actuators may require an auxiliary power supply from a generator or batteries. *Figure 12* shows a large, battery-powered, main-line gate valve.

This particular valve installation in *Figure 12* is powered by a 20-horsepower, explosion-proof motor that operates off a 252-volt, DC-battery power supply. The entire motor and actuator setup is enclosed in the top housing. Included in the enclosed unit are the motor, gear drive, valve position indicators, and limit switches. The actuator develops 22,800 foot pounds of torque. The actuator and motor could be modified to operate off other electrical power supplies if available.

The gate valve has a rising stem. The stem is connected to the gear train by means of a worm gear and stem nut/splined drive sleeve on the stem. The worm gear, stem nut, and drive sleeve are designed with lost motion built into the action. This lost motion before the final gear engagement creates a water hammer effect. The gear engagement gives the stem a hard jolt to help start it turning, especially helpful when sealed. Such stem nuts (see *Figure 13*) occasionally require replacement because of this hammer effect.

108F12.EPS

Figure 12 ◆ Large, battery-powered, main-line gate valve.

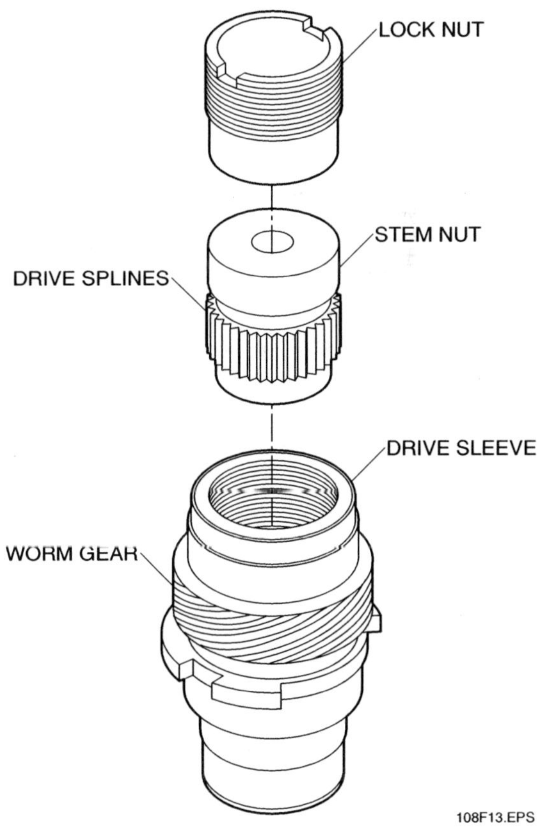

108F13.EPS

Figure 13 ◆ Stem nut assembly.

Once the drive sleeve is engaged by the worm gear, the stem begins turning and opens or closes the valve. The handwheel is set up with a declutching mechanism that disengages the handwheel from the motor so that it can be turned in case the motor will not turn. *Figure 14* shows schematics of several commonly used actuator models.

Actuators are fitted with limit switches, but these are not factory set. They must be set by technicians when the valve is installed. Units are also available with heaters for cold climate operation. If heaters are not used, it is recommended that a **desiccant** be placed in the switch housing to control moisture. Valve actuators are waterproof after installation if the manufacturer's instructions are followed. However, if the particular installation allows the stem to protrude through the outside of the unit, a cover must be secured for the stem. Technicians must replace all plastic covers and plugs with permanent units and make certain all seals and covers are in place and tight.

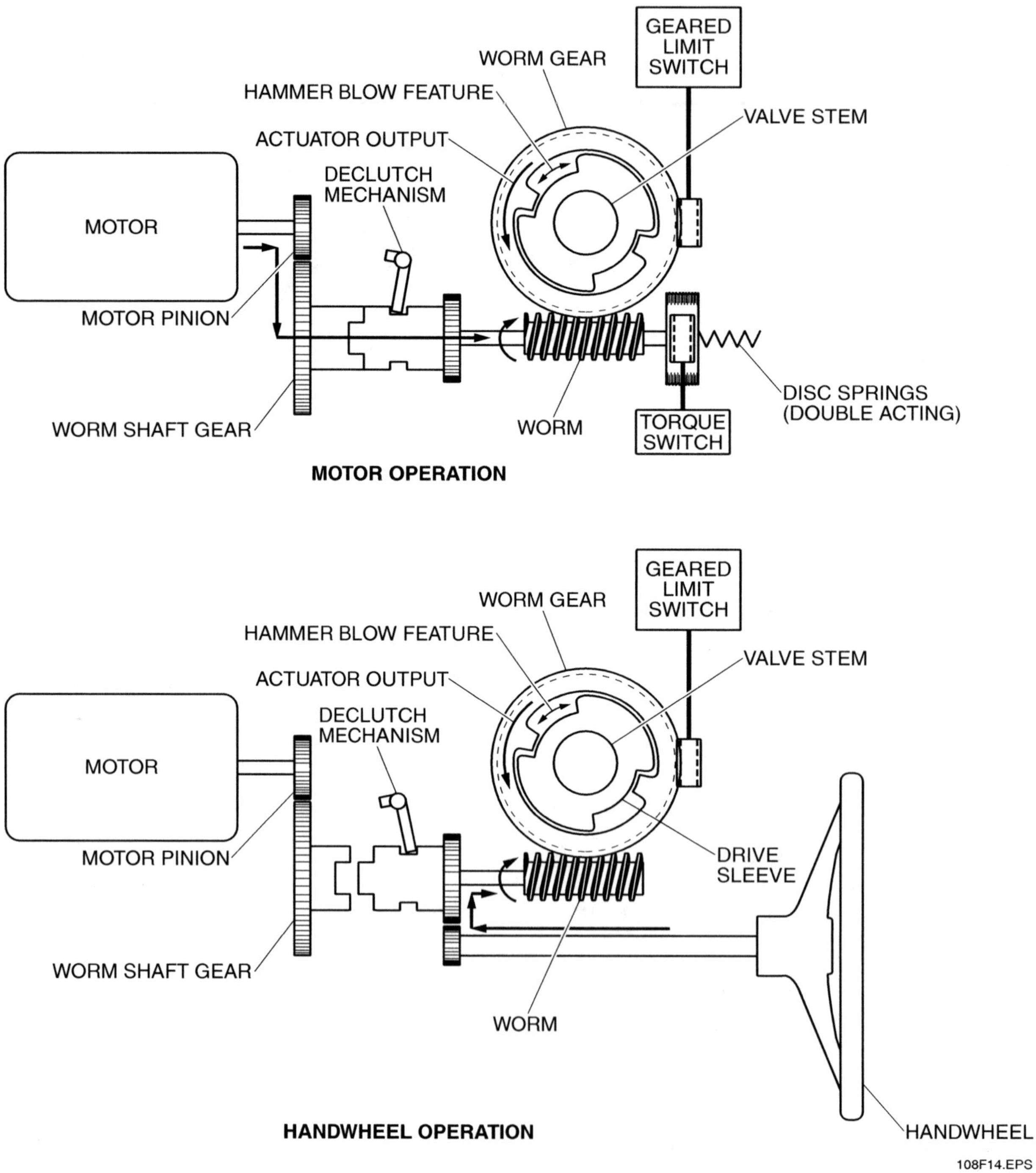

Figure 14 ◆ Actuator schematics.

4.1.0 Worm-Gear Actuators

Worm-gear actuators are commonly used for 90-degree turn, manually actuated valves, including butterfly valves, ball valves, and plug valves. These actuators are designed generally for handwheel use but can be fitted with motor drives for automatic operation. A wide range of gear ratios is available to supply mechanical advantage for high torque loads. These units have little or no backlash and hold the valve firmly in place once it is set, helping to eliminate valve disc vibration and chatter. They are used where exact control is needed and are applicable to a variety of uses. These units are strong, sealed, and weatherproof if installed according to the manufacturer's recommendations. Essentially, they are low-maintenance units. The travel must be adjusted with stop screws for the application. *Figure 15* shows a cut-away view of a worm gear actuator.

108F15.EPS

Figure 15 ◆ Cut-away view of a worm-gear actuator.

4.2.0 Bevel-Gear Actuators

Bevel-gear actuators are used with multiple-turn valves, including gate valves and globe valves. Bevel-gear actuators are used where valve accuracy is needed and are designed to be used where heavy torque loads are required. Bevel gear actuators are designed for manual operations but can be fitted with a variety of motor and remote control applications. Gears are machined to close tolerances to reduce backlash and provide smooth operation. They are self-locking when set. Roller bearings allow for smoother and easier operation.

Bevel gear actuators are sometimes fitted with spur gear accessories that increase both mechanical advantage and available torque for heavy loads. These units are also low maintenance, generally requiring to be greased on a regular schedule. *Figure 16* shows a bevel gear actuator with a spur gear addition.

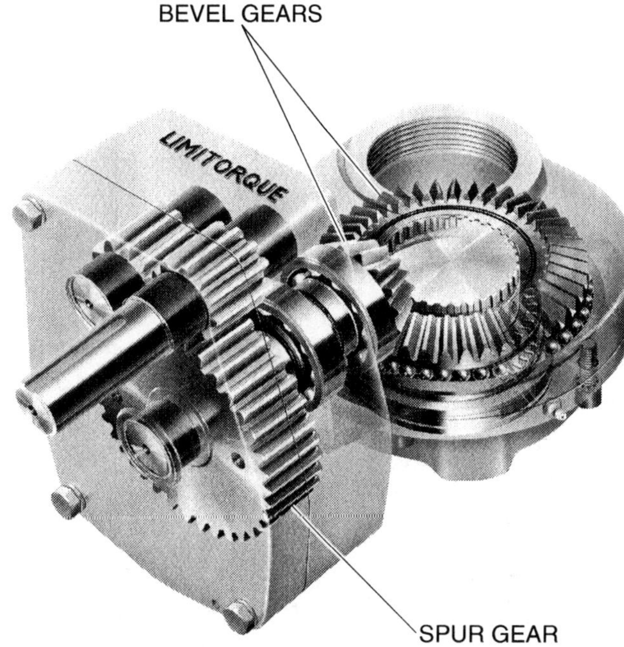

108F16.EPS

Figure 16 ◆ Bevel-gear actuator with a spur-gear addition.

4.3.0 Variable Reduction Actuator

Variable reduction actuators (see *Figure 17*) are specially designed for valves that only turn 90 degrees. The construction design is extremely simple and powerful. The design combines the positive action of a lever and the smooth, powerful action of a worm-gear actuator. They are especially useful where quick opening and closing action is needed to aid in seating and unseating the valve. Units can be fitted with a handwheel or automatic actuator. These actuators are often used with butterfly valves and are available in a wide range of sizes.

Advantages of these units are that the torque of the unit increases at closing. At the same time, the speed of the unit decreases at closing. These two factors help to extend valve life and reduce the effects of water hammer. The threaded shaft of the unit is turned by the handwheel or the power actuator. In pneumatic and hydraulic units, the shaft of the actuator is connected directly to the drive nut, eliminating the threaded shaft.

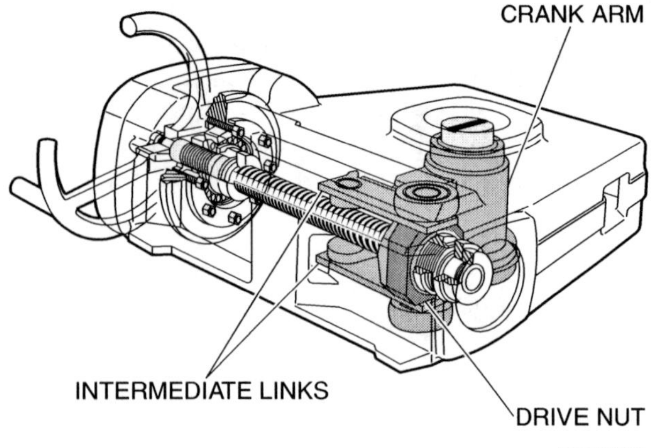

Figure 17 ◆ Variable reduction ratio actuator.

The 90-degree rotation of the valve is achieved by the crank arm that is connected to the threaded drive shaft by intermediate links and the drive nut. As the crank arm turns the valve closer to the full closed position, the design of the intermediate links have the effect of actually increasing the length of the crank arm, increasing the torque.

4.4.0 Gas Actuators

A gas motor actuator is one of the most versatile units available. They are compatible with almost any kind of valve in any application. They are designed to operate off the pressure of the gas in the pipeline in which they are installed. They are designed to operate from pressure up to 1,500 pounds per inch gauge (psig). However, such units can be modified to use corrosive gases or hydraulic fluid from a central supply. Motors can be modified to operate off pressure as low as 15 psig. Users are cautioned to make certain the units are installed with accessories like strainers and inline oilers on the supply line.

Gas actuators are installed with a supply line directly from the pipeline with the strainer, oiler, and line connected to the inlet port of a four-way shear valve. They operate on much the same principle as an old-fashioned waterwheel. *Figure 18* shows a schematic of a gas actuator motor.

When the valve lever is moved automatically or manually, pressure is diverted to the opening or closing side to the motor inlet port. The direction the motor turns is chosen by which port is opened. The opposite port exhausts the gas through another valve to the atmosphere. Valve travel is controlled by a limit switch. The limit switch mechanism causes the gas to be completely shut off. *Figure 19* shows a schematic of a gas-powered actuator.

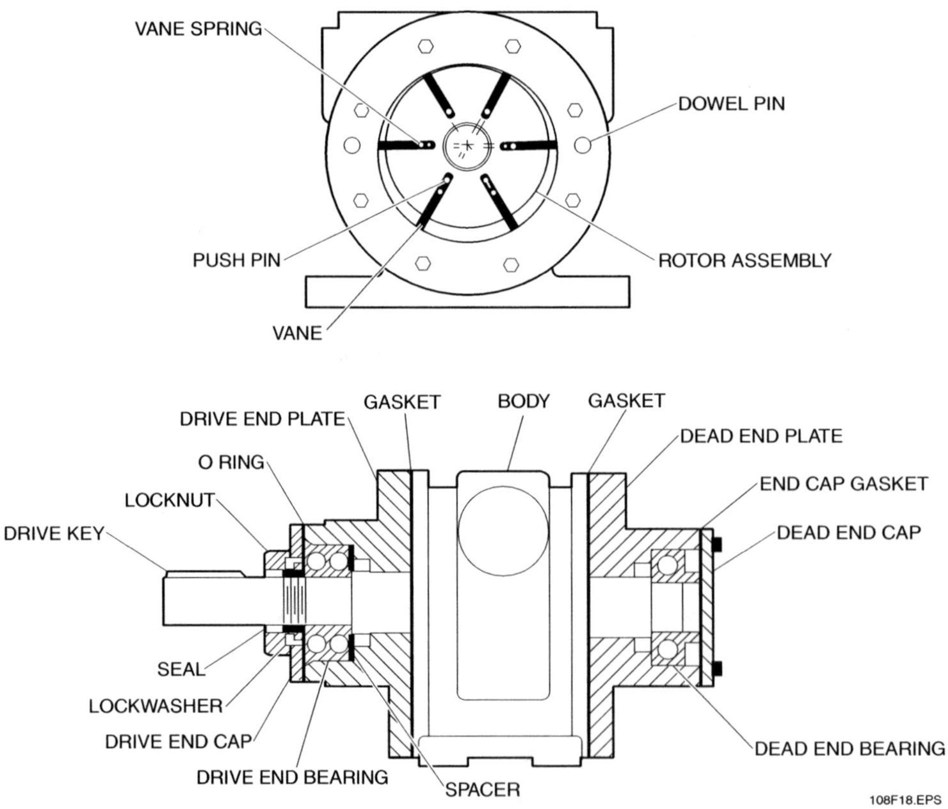

Figure 18 ◆ Schematic of a gas actuator motor.

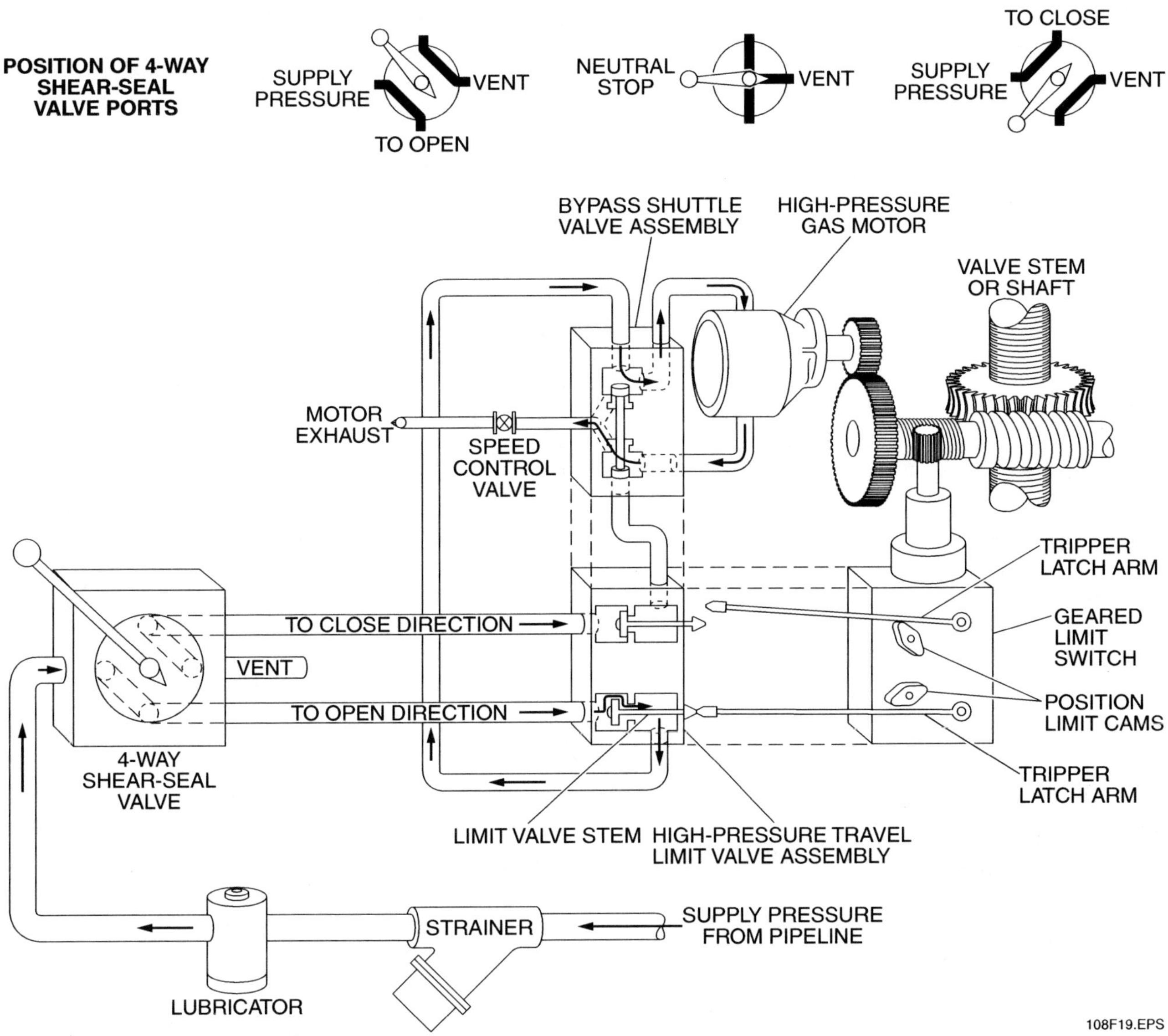

Figure 19 ◆ Schematic of a gas-powered actuator.

4.5.0 Rotary Vane Actuator

The rotary vane actuator is a very compact form of the same principle of operation as the gas actuator, although this unit can be gas-, hydraulic-, or gas/hydraulic-operated. The unique feature about this unit is that it travels through a very short cycle. The motor usually turns no more than 90 degrees. It is used for plug, ball, and butterfly valves.

Although it is small, it is available in many sizes and torque ranges. Its power comes from the design that applies pressure to two opposite vanes at the same time. The pressure is contained against a solid shoe from which the vane on each side is pushed equally. The actuator is very smooth and even in operation because torque is completely balanced throughout the cycle. Such units are very dependable and are low maintenance because there is little movement and there are few components. *Figure 20* shows a rotary vane motor.

4.6.0 Double Cylinder Hydraulic Actuators

Double cylinder hydraulic actuators are commonly used for ball valves and other units requiring a 90-degree turn from open to close. Double cylinder hydraulic actuators are considered to be fast-acting with the initial 20% of rotation being instantaneous and the last 40% of rotation requiring about 6 seconds. This decrease in speed allows extra time for obtaining exact settings in throttling positions. The incremental decrease in speed is controlled by the hydraulic supply system. The quick initial action is designed to place the valve into throttling position as quickly as possible.

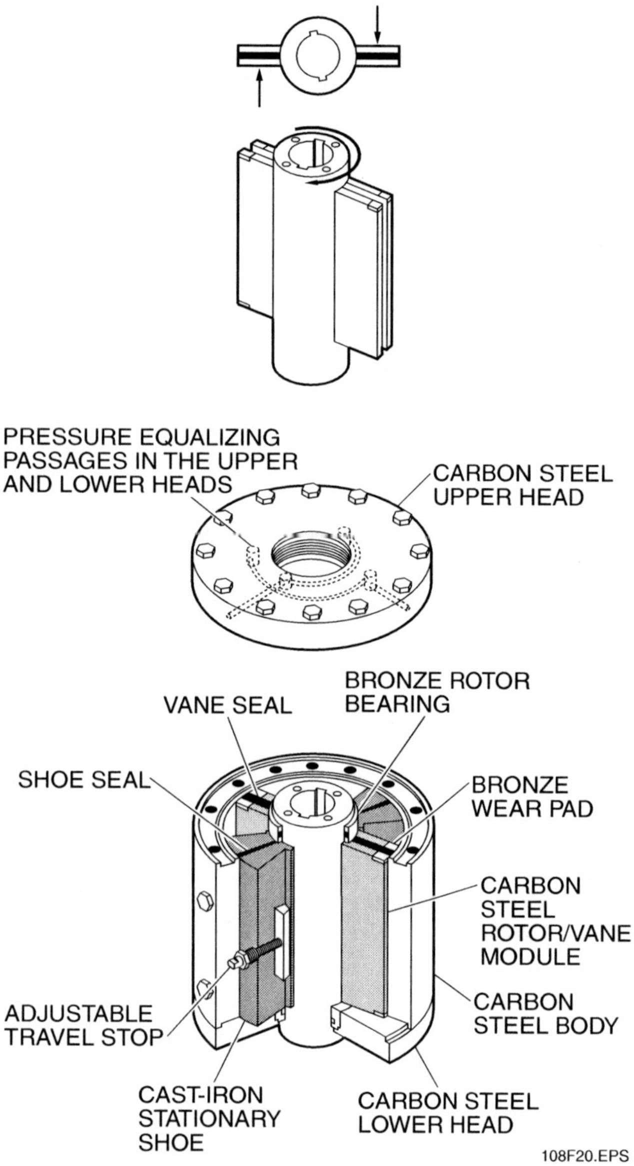

Figure 20 ◆ Rotary vane motor.

However, the speed control setup can be adjusted to an infinite number of settings, as needed. *Figure 21* shows a double cylinder hydraulic actuator.

The action of this actuator is accomplished by two separate hydraulic cylinders, one for each direction of movement. An external hydraulic supply is required. The cylinders are attached to a double-ended crank arm. The crank arm is attached to the valve stem by a splined collar. When installing the actuator, you must carefully bleed all trapped air from all lines, manifolds, and cylinders. Using clean lines and connectors and keeping them clean is of the utmost importance in this type of assembly. The stroking characteristics of the unit can be adjusted by turning the setscrew on the speed control and deceleration valve. Factory recommendations should be followed.

4.7.0 Electrohydraulic Actuators

Electrohydraulic actuators are units that operate from a self-contained hydraulic system for valves with 90-degree movement. The unit employs a hydraulic pump, reservoir, control valves, and other accessories needed to operate the system. When the unit is activated, the pump supplies hydraulic pressure to the designated inlet valve. The fluid is directed into a cylinder to the appropriate side of the piston. The piston is attached to the valve stem by means of a stem connector and crank arm. The actuator is normally remote controlled but can be fitted with a handwheel. When a signal is received, the electric-powered hydraulic pump is started. Pressure supplied to the manifold connection is diverted to the top or bottom side of the piston to open or close the valve as required. *Figure 22* is a schematic of an electrohydraulic actuator.

These actuators are quick-acting and reasonably safe for explosive atmospheres as long as the manufacturer's recommendations are followed. Electrohydraulic units must be installed in a vertical position. As with all hydraulic and electrical units, the connections, switches, and assemblies must be kept free of dirt and corrosion. The hydraulic system must be kept free of dirt, sludge, and air. Excessive noise from the pump indicates air in the system. Technicians must bleed the unit to remove it as soon as possible to prevent pump damage.

4.8.0 Pneumatic Rack and Pinion Actuators

Pneumatic actuators are suited to valves with 90-degree turns and are designed to work off an external air supply and control system. They consist of a cylinder with a piston and opposing spring to return the unit to normal position, or opposed pistons to move the actuator in both directions. The actuator operates the valve stem by means of double rack and pinion gears. The pinion gear is turned as the piston pushes the rack by it. The double rack and pinion construction eliminates backlash. The pinion is connected to the valve stem by means of the machined fitting in the pinion gear.

These units are quick-acting, powerful, compact in size, and dependable. They have very few moving parts, and action is limited to short strokes. They are available in a variety of sizes for wide application. *Figure 23* shows a compact, double rack and pinion actuator.

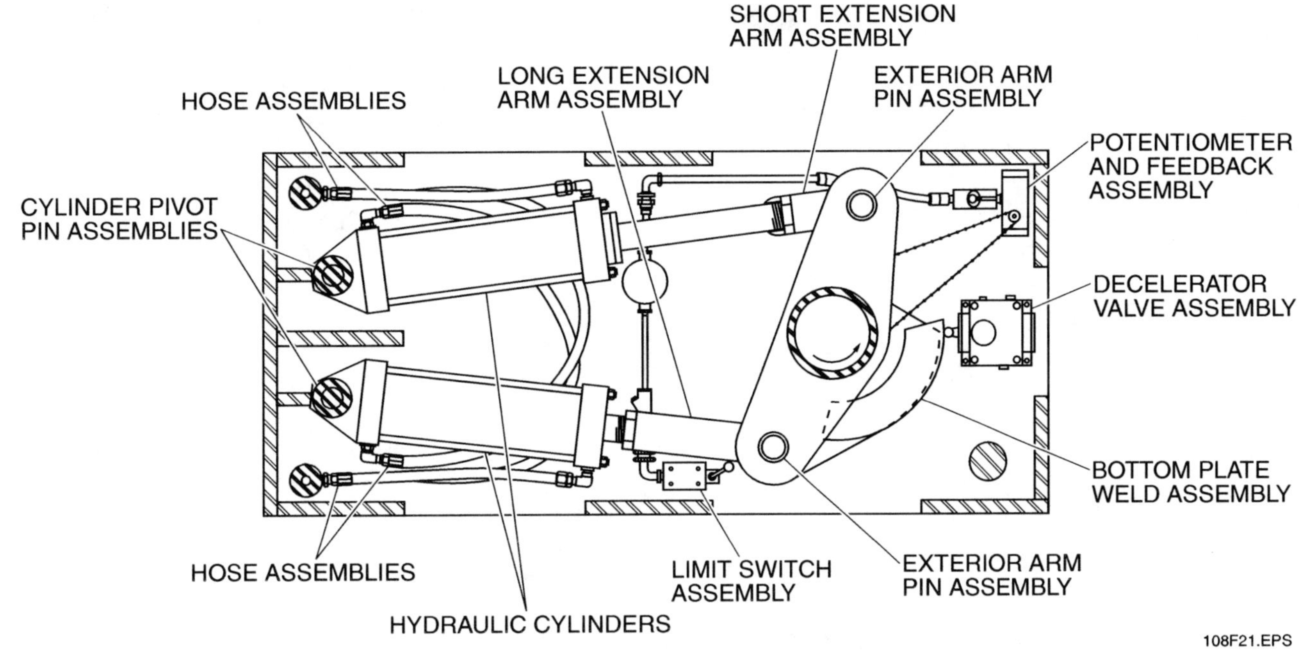

Figure 21 ◆ Double cylinder hydraulic actuator.

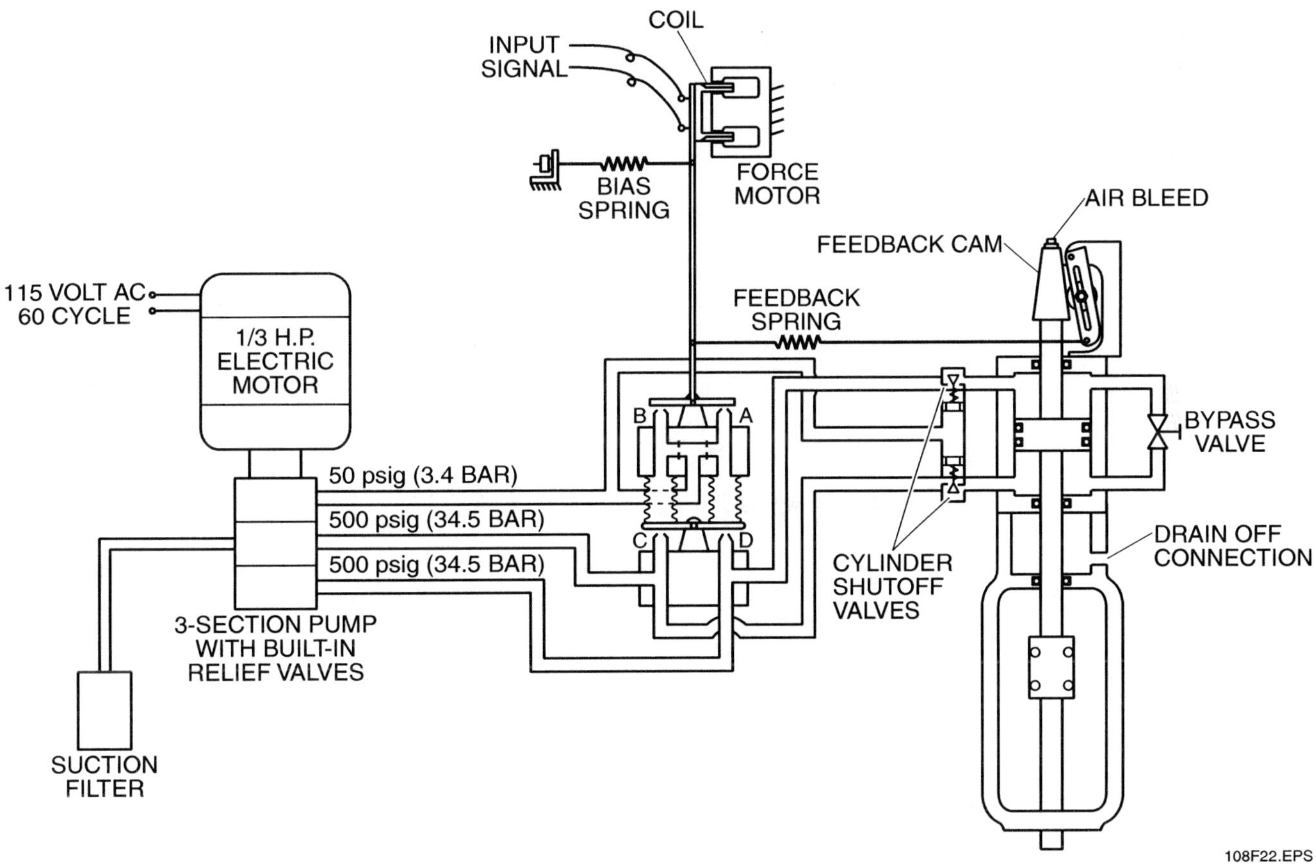

Figure 22 ◆ Schematic of an electrohydraulic actuator.

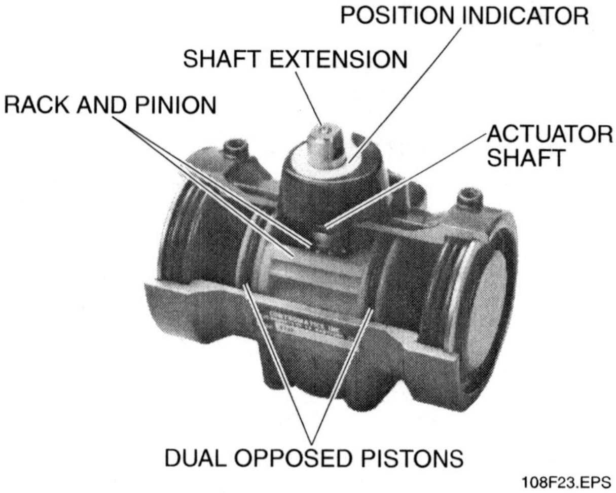

Figure 23 ◆ Compact, double rack and pinion actuator.

4.9.0 Pneumatic Fail-Safe Actuators

Fail-safe valves are important safety features on the pipeline. In emergencies, the actuators are set up to close the valve and shut off the system at various strategic points. In case of pipeline damage or rupture, these valves control the loss of product and protect life and environment. Many fail-safe valves are operated by pneumatic pressure. In case of an abnormal operating condition, the pressure is quickly bled off the piston area and the valve is closed by valve body and spring pressure. These units are designed to operate with a reverse ported gate valve. The valve is open in the down position. In an emergency condition, the actuator forces the gate up and the valve is closed. When the problem is solved and the line is ready for operation, air pressure supplied to the actuator opens the valve again. *Figure 24* shows a pneumatic safety valve.

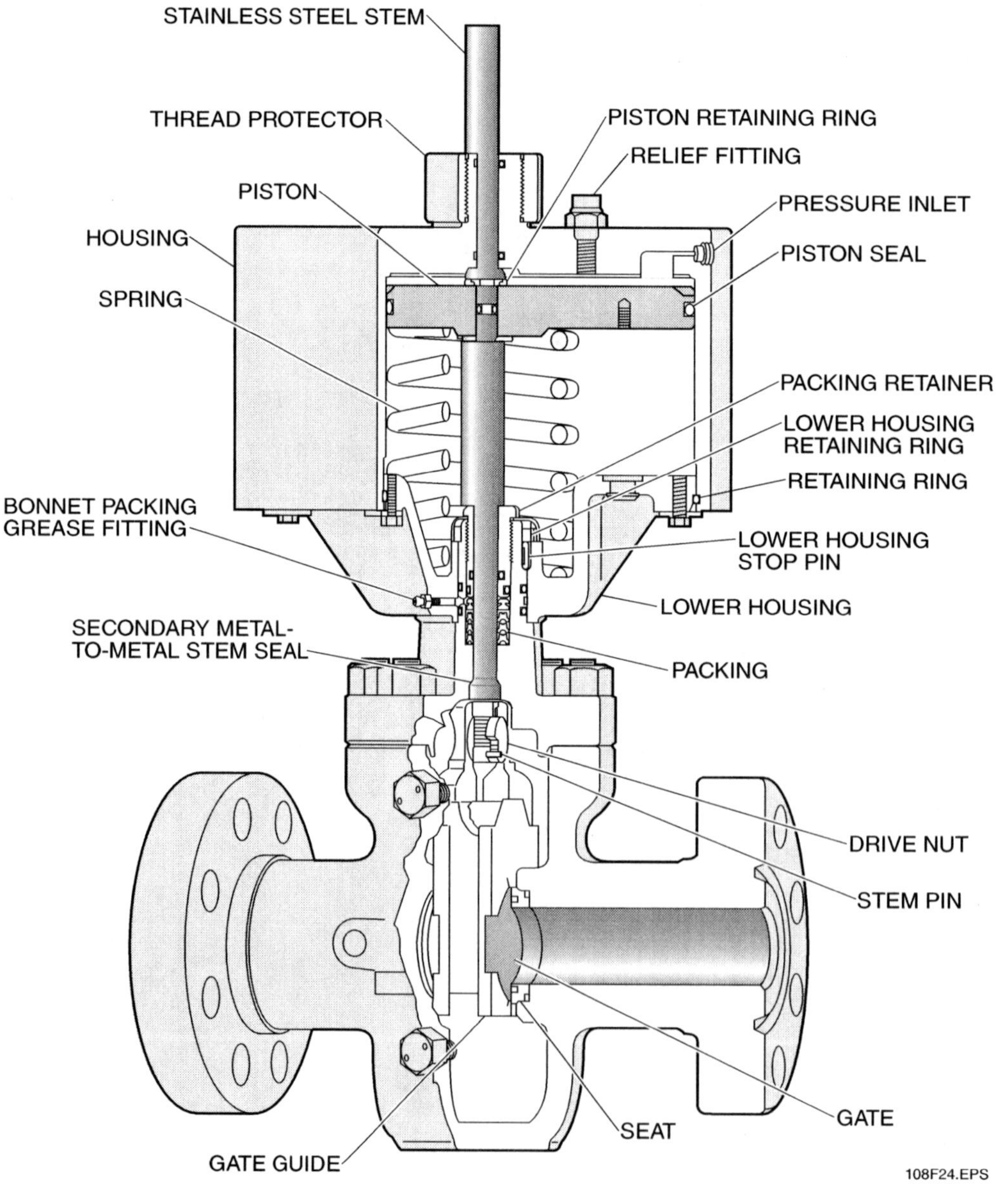

Figure 24 ◆ Pneumatic safety valve.

5.0.0 ◆ GENERAL MAINTENANCE

Maintenance is the ongoing process of inspecting the pipeline equipment and servicing, repairing, or replacing items as needed to maintain and ensure safe operations. Mechanical equipment must be maintained regularly in order to perform efficiently and safely. The principle applies to all types of valve actuators.

Sound management practice requires that all other equipment be maintained in satisfactory condition as well. It is not only a matter of efficient operation but of economic operation as well. Valve actuators represent a sizeable investment that must be protected. Companies use computerized systems to generate work orders for equipment at regular intervals to make certain that legally required schedules are met and that even isolated equipment is included.

Technicians should become completely familiar with the types of actuators being used and the types of maintenance products required for their job assignments. They should also become familiar with the types of lubrication systems each actuator model employs. An ongoing training program with regular updates and reviews is essential. Training should also include proper documentation. *Figure 25* shows a valve maintenance log, including actuators.

5.1.0 Actuator Warranty

Many actuators are supplied with warranties. If an actuator is not maintained according to the manufacturer's instructions, the warranty may be voided. Before any work is attempted, the particular actuator should be checked to see if the warranty is still in effect. If it is still under warranty, the work will need to be carried out as authorized by the manufacturer. Recommended lubricants should be used. If not available, a close replacement should be found. Trained and qualified technicians must make certain that company policy and the manufacturer's instructions are followed. Failure to follow the manufacturer's instructions may void the warranty.

Many manufacturers recommend that units seldom used should be cycled on a monthly schedule. Greasing the fittings with a grease gun and other lubrication must be kept up to date for dependable operation. Manufacturer's recom-

Rev. 12/97

Service Performed	Condition of Valve/Site	Deficiency	Action Required	Action Taken
VI - Visual Inspection MO - Mechanical Operation LO - Local Mode Operation RO - Remote Mode Operation RM - Routine Maintenance AS - All Services Listed	AC - Acceptable Condition NR - Needs Repair / Work RR - Replacement Required OS - Out of Service	DO - Does Not Operate LK - Leaking PM - Part Missing DP - Defective Paint NS - Security WO - Wash Out SU - Site Upkeep OT - Other (describe)	RP - Replace Part RR - Repair Required PR - Paint Required SS - Secure Site DW - Dirt Work Required SC - Site Clean Up OT - Other (describe)	PR - Part Replaced RP - Repair Performed RE - Repainted SP - Security Provided DP - Dirt Work Performed SM - Site Cleaned-Up OT - Other (describe)

Date	Print Order	Valve Tag No.	Serial Number	Location / Service	Service Performed	Condition of Valve/Site	Deficiency	Action Required	Action Taken
11/01/01	111	42-37B	BR549	#3 Pump Room	VI	NR	DP	PR	RE
11/13/01	112	54-18C	XJ242	Pump Station 14	MO	RR	LK	RP	PR

SAMPLE

REMARKS: __________

DIVISION: __________ STATION: __________ INSPECTED BY: __________

108F25.EPS

Figure 25 ◆ Valve maintenance log.

mendations should be consulted for lubricants suited to the climate, products, and type installation involved.

5.2.0 Actuator Lubrication

Proper lubrication protects actuators from rust, corrosion, damage from freezing moisture, and seizure of parts. Properly lubricated actuators operate freely without binding, requiring less energy to turn. Actuator lubrication can be separated into two types, external lubrication and internal lubrication. External lubrication applies mostly to manual valve actuators with threads or working parts exposed to the elements. Internal lubrication applies more to those actuators using gear drive or automatic drive units.

5.2.1 External Lubrication

Objects that move produce friction and wear. Important components in equipment exposed to the elements tends to dry out, harden, and deteriorate as time passes. Dirt and grime collect in unacceptable places and causes problems with movement, increasing wear, corrosion, and damage to machined and polished surfaces. Periodic lubrication of certain parts of the actuator prevents, or at least slows down, the process, greatly extending the life and efficiency of components.

Applying grease with a brush and/or grease gun are common procedures performed on stem-type valves on the pipeline (see *Figure 26*). This procedure applies a variety of stem-type valves that are accessible to personnel. Keep in mind that a grease gun uses hydraulic pressure to move grease from the pump into the fitting. In cold weather, great pressure may be required to apply grease to the actuator. Pumping may require extra effort.

The steps include some or all the following:

Step 1 Turn the valve to the completely open position so that the stem threads extend.

Step 2 Clean the threads of all rust, dirt, and grime with a wire brush, solvent, and rags.

Step 3 Wipe off all residues from moving parts and seal areas with rags.

Step 4 Locate and remove safety caps from grease fittings on the sides of the valve body if applicable.

Step 5 Pump the grease gun several strokes and operate the valve an inch in each direction to spread grease on the seals.

Step 6 Pump the grease gun lever several more times on each fitting. *Figure 27* shows an actuator with grease fittings.

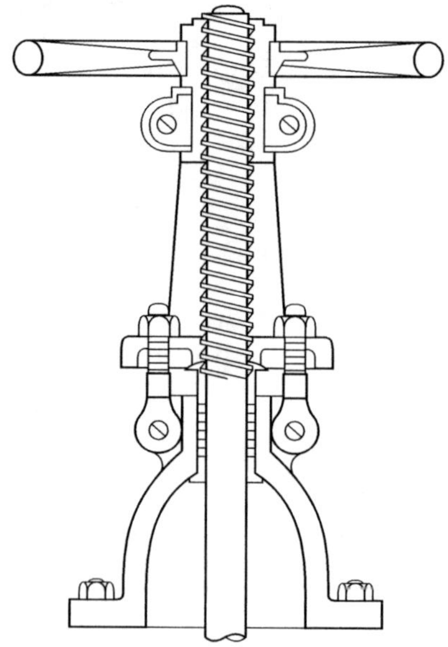

Figure 26 ◆ Stem requiring lubrication with a brush.

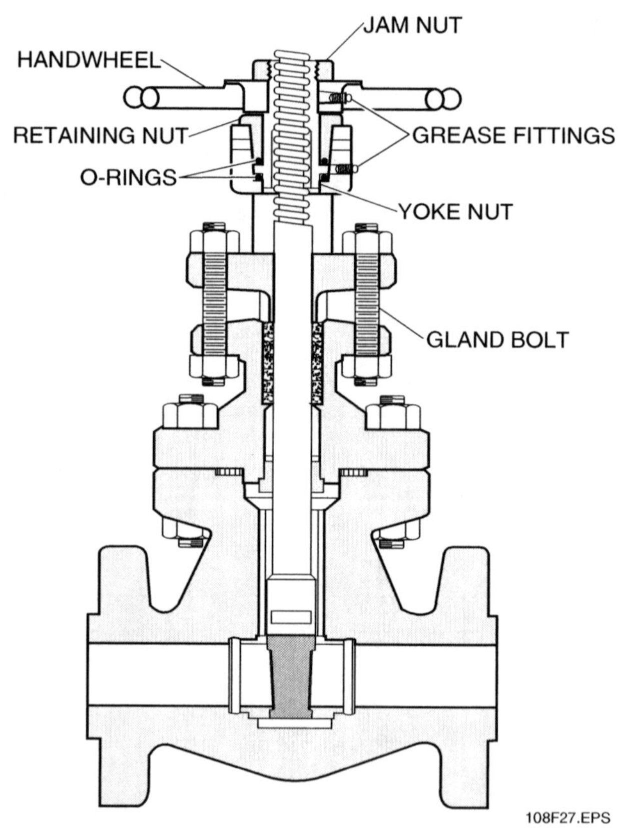

Figure 27 ◆ Actuator with grease fittings.

Step 7 Operate the valve through several complete cycles to spread the lubrication.

Step 8 Move the valve to full open position and pump the grease gun several more strokes to fill the cavity.

> **NOTE**
> Replace the safety caps on the grease fittings.

Step 9 Apply the recommended grease heavily to the outer threads of the stem with a paintbrush for those valves that do not have grease fittings or do have exposed stem sections.

Step 10 Cycle the valve several times.

Step 11 Remove the handwheel nut and handwheel (see *Figure 28*), exposing the bearing housing.

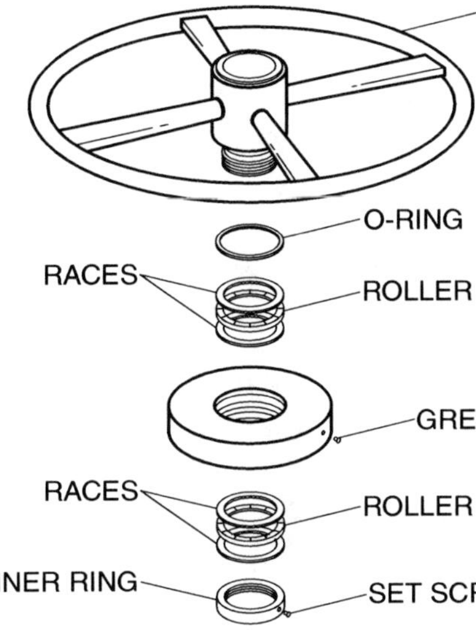

Figure 28 ◆ Typical actuator bearing assembly.

Step 12 Remove the setscrew holding the bearing assembly in place.

Step 13 Remove the individual bearings and races from the housing.

Step 14 Clean and inspect the handwheel bearings and races.

Step 15 Hand-pack the handwheel bearings and race.

Step 16 Reassemble the races and bearings into the housing.

Step 17 Slide the bearing housing back over the stem.

Step 18 Tighten the assembly-housing setscrew.

Step 19 Reinstall the handwheel and nut.

Step 20 Torque the handwheel nut to the manufacturer's specification.

Step 21 Remove all lockout devices and tags.

Once the actuator and valve are returned to service, the control center operator must be notified so that any affected pipeline sections can be started up again. Maintenance personnel must return the work order with all permits, lockout tags, and other documentation to the control center for filing. Appropriate signatures must be secured before filing. The work must be recorded in the control center logs, and any other required documentation must be completed.

5.3.0 Internal Lubrication

Valve actuators employing a closed gearbox design require that lubrication be injected or packed inside the gearbox case in such a way that it reaches all bearings, bushings, sleeves, and gears. It must also be kept at the manufacturer's recommended levels to displace moisture in the case. The lubricant or grease must be hand-packed into the bearings and sleeves as the unit is assembled. Regular lubrication with a gun or re-packing by hand is required to maintain lubrication levels in such designs (see *Figure 29*).

Actuators with sealed cases are designed to operate in a horizontal orientation and to maintain a standard grease level inside the case. Levels are specified in the manufacturer's recommendations and instructions. Levels should be checked regularly by removing a fill plug and verifying that the grease is at operational level. Some actuators can be installed in any direction. In cases where the actuator is installed in a non-horizontal orientation, higher-than-normal lubricant levels may be recommended. Some require the case to be completely filled. Care should be taken to note this condition in the manufacturer's recommendations or instructions. Different grease may also be recommended for this type of installation. A different fill plug will be the designated checkpoint for orientations of units that are not horizontal. *Figure 30* shows valve actuators mounted in positions requiring additional lubricant levels.

Manufacturer's recommendations include checking grease levels on a regular schedule. Before starting a new actuator or one that has been out of service, technicians should remove the plug and check the grease level. You should also note the lubricants' viscosity and contamination. Actuators in service should have the appropriate plug removed and the grease checked every six months.

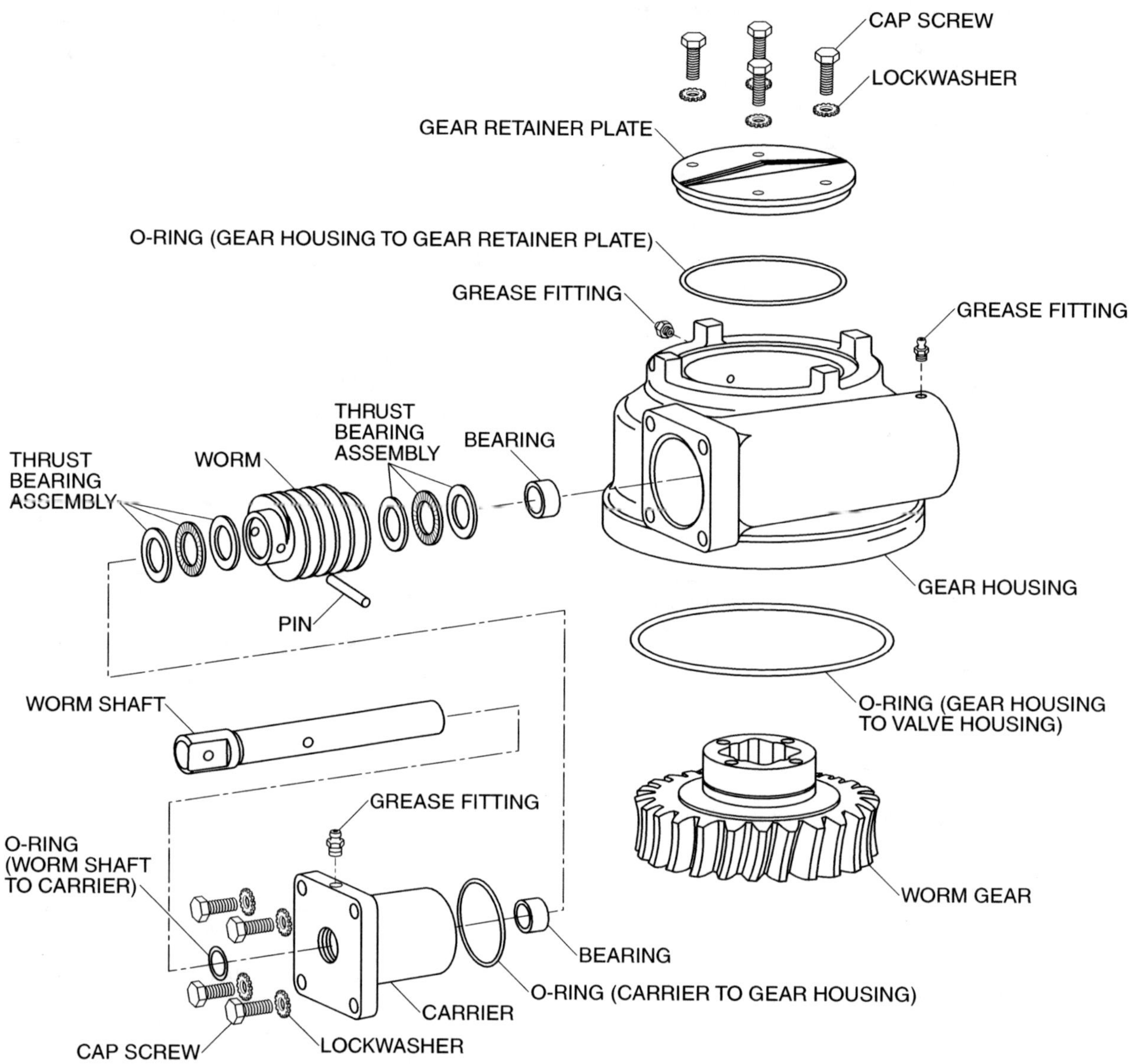

Figure 29 ◆ Schematic of a single reduction worm gear with grease fittings.

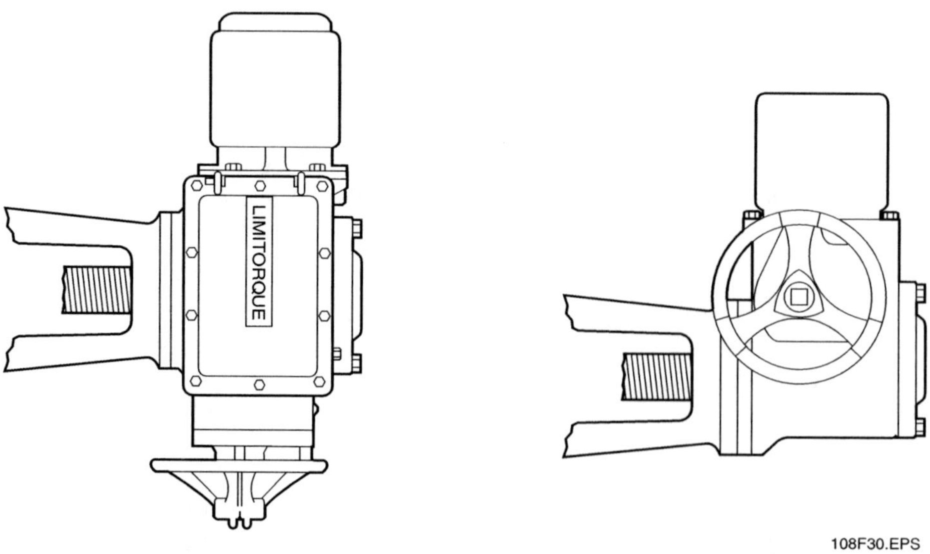

Figure 30 ◆ Valve actuators mounted in positions requiring additional lubricant levels.

Valve actuators are made of the best materials, and most are heavy. You should not attempt to manually lift automatic actuators with drive units. Manufacturers install **lifting lugs** in the tops or specially drilled holes in the casing of some valve actuators for lifting. Manufacturer's lifting lugs or holes must be used so that the valve actuator is not distorted or damaged when removal is necessary. When lifting an actuator, it must be separated from the valve body. Lugs on actuators are not designed to lift the entire assembly. If attempted, serious damage to the actuator and/or valve may result. *Figure 31* shows actuators with typical manufacturer's lifting lugs.

Figure 31 ◆ Actuators with lifting lugs.

5.3.1 Multiple Reduction Actuators

Many gear-type actuators require multiple reduction sets in order to obtain sufficient mechanical advantage to turn the particular valve on which they are installed. Technicians must properly identify the actuator they are servicing. Basic gear designs are used by many companies. Most have their own variations of placement and number of lubrication points and procedures. Manufacturer's recommendations should be consulted for the exact application and the type of grease to be used. Grease guns are used for general application procedures. However, technicians must be sure to locate every grease fitting so fittings obscured by the location of other equipment will not be missed and eventually lead to serious damage. Where a gear case needs larger volumes of lubricant, the gun may need to be pumped a number of times, the valve cycled, and then a few more pumps added. *Figure 32* is a schematic of a double reduction worm gear actuator with several fittings.

5.3.2 Spur-Gear Reduction Units

Spur-gear reduction actuators are used directly on valve stems. Some involve multiple gear trains with many gears, bushings, and bearings in the system. Some of these units must be disassembled and hand-packed. Others have grease fittings for gun use. If a unit has to be disassembled, the manufacturer's instructions must be followed. Each part, seal, and gasket must be accounted for and put back in its exact location for the unit to work well and be leakproof and weathertight again. Each bearing must be hand-packed, and each shaft must be hand-coated with the appropriate grease. The case must be filled to the manufacturer's recommendations in order for the unit to resist moisture buildup. Some manufacturers suggest that the case needs to be at least 80% full. *Figure 33* is a schematic of a double reduction spur gear operator with no grease fittings.

Spur-gear reduction units are sometimes added on to other types of operators in order to multiply mechanical advantage to turn a heavy valve. This is common in many manual applications. Because it is a separate unit, no lubrication passes between the actuator and the reduction unit. Each unit depends on its own internal lubricant supply. Technicians must be careful to recognize these add-on units because they must be serviced as a separate unit and a different set of manufacturer's recommendations and instructions apply.

5.3.3 Bevel-Gear Actuators

Bevel-gear actuators sometimes require hand lubrication. Many of these actuators require disassembly, bearings packed by hand, and the shafts coated with grease by hand. Again, it is important that technicians make a correct identification of the model being serviced. Manufacturer's recommendations and instructions must be followed. Care must be taken with seals because these units usually operate directly on the valve stem. *Figure 34* shows a schematic of a bevel gear operator with no grease fittings.

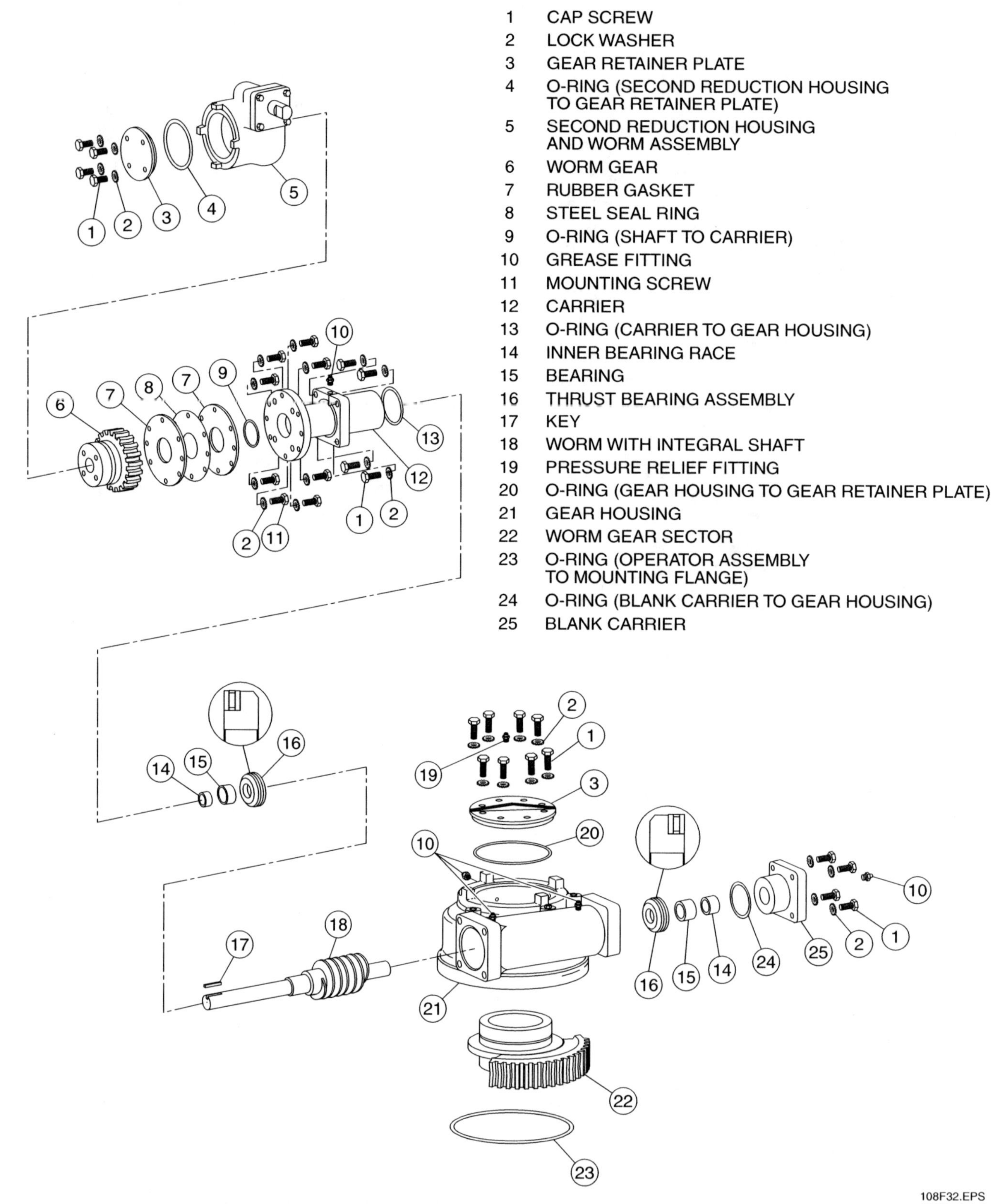

Figure 32 ◆ Schematic of a double reduction worm gear actuator with several fittings.

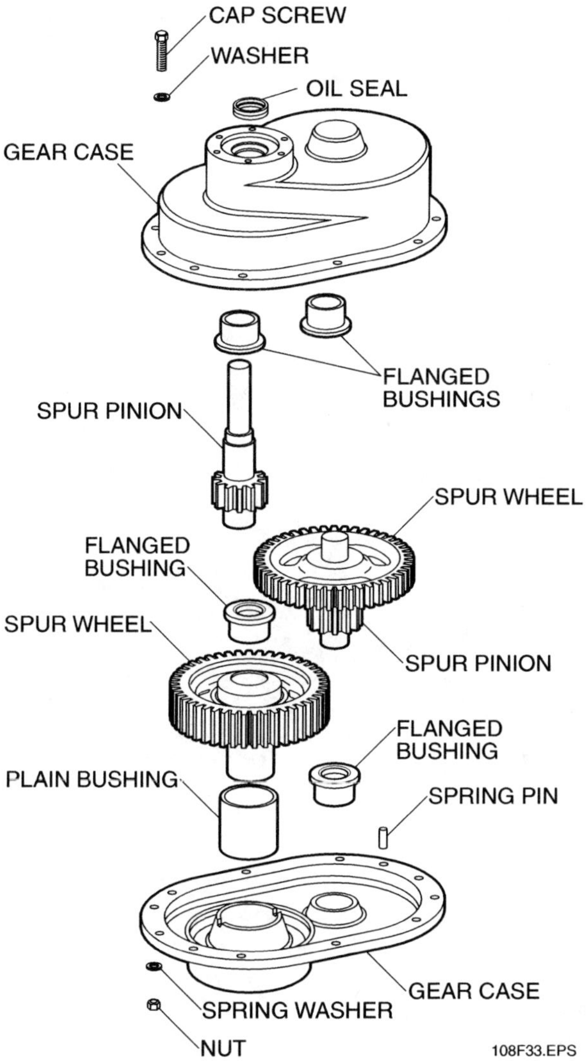

Figure 33 ◆ Schematic of a double reduction spur gear operator with no grease fittings.

5.4.0 Automatic Actuators

Automatic actuators add another dimension to the maintenance of valve actuators because they operate valves automatically and use a source of control and power from another location. Proper lubrication is essential to the longevity of any power transmission equipment, and valve actuators are no different. Some automatic actuators are factory-packed and sealed and, under normal usage, need only be checked once in one and one-half years. They do need to be inspected regularly for seal or bearing failure. Actuators subjected to long storage, a year or more, should be observed closely until it is verified that they are not leaking.

Limit switches are an integral part of an automatic valve actuator because they signal the unit to start and stop at exactly the right time. If a limit switch were not present, the equipment would soon destroy itself. Limit switches are often gear devices that measure the travel or the torque of the valve when in operation. The principle is that the valve must turn just far enough to ensure a tight seal but not far enough to damage the seals, gears, and other components. It must stop before the motor stalls and motor damage results.

The limit switch must be lubricated according to the manufacturer's recommendations because it has moving parts. It is also an electronic device, and the type of lubricant used must be compatible with electronic equipment. Technicians must follow the manufacturer's recommendations. The internal gear lube is not compatible with the electronic components. Some of these units are

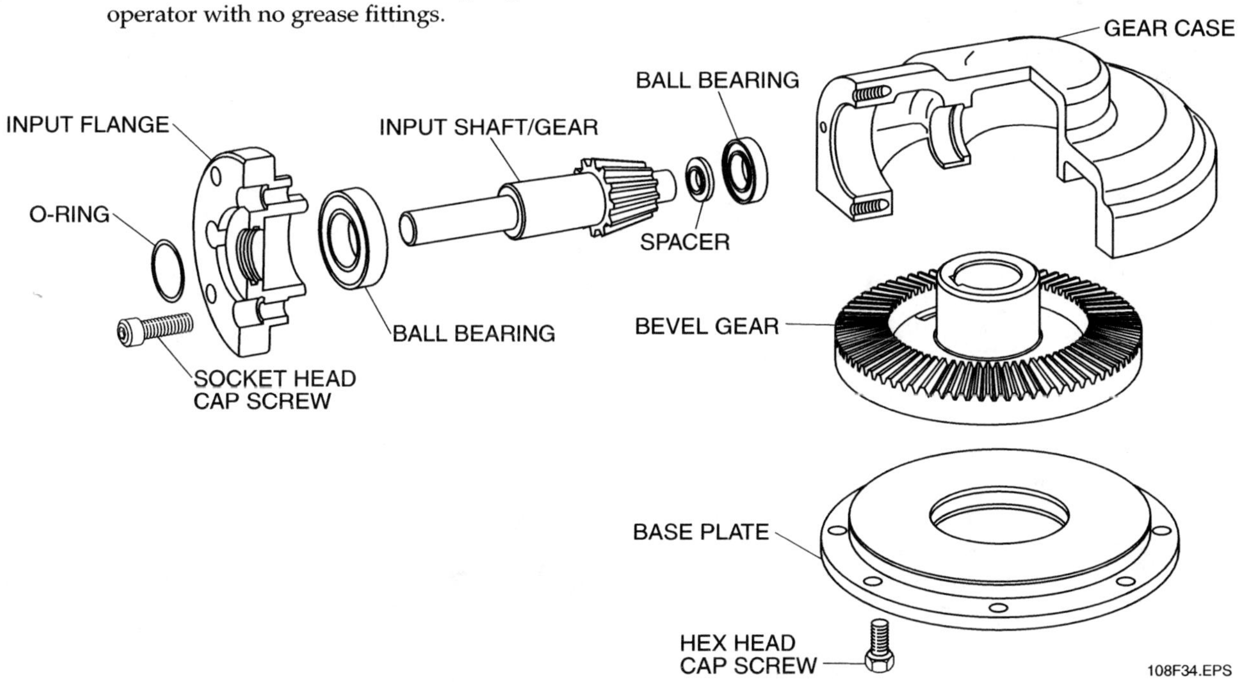

Figure 34 ◆ Schematic of a bevel gear operator with no grease fittings.

installed in an orientation that allows a leak to enter and saturate the motor and switching compartments. The units and compartment of these mechanisms must be flushed and cleaned with an appropriate electrical solvent and dried before use. *Figure 35* is a double torque limit switch.

While performing maintenance on automatic actuators, limit switches should be tested to verify proper settings. The technician should cycle the valve electrically to verify the settings. If the valve movement is wrong, the switch should be adjusted. Manufacturer's instructions must be followed to correct it. Only trained and qualified technicians should adjust the unit on a valve actuator. If the unit itself is of questionable operation, it should be replaced. It cannot be repaired. Technicians should follow company policy because work orders and permits are required. Once the unit operation is confirmed, a small amount of the recommended, electronically compatible grease should be applied to the appropriate parts of the limit switch.

108F35.EPS

Figure 35 ◆ Double torque limit switch.

5.5.0 Painting

Companies use regular painting as a means of protecting expensive pipeline equipment. Paint inspection is usually a point on regular inspection checklists. *Figure 36* shows a contrasting pipeline paint scheme.

Valve actuators should be painted when they are new and repainted on a regular schedule. Paint is a strong defense against rust and corrosion. Manufacturers normally coat new actuators with some kind of corrosion-resistant, oxide-type primer, protecting them until delivery. Some actuators may have the manufacturer's final coat applied. Once actuators are delivered, companies paint them according to company policy for resistance to the weather and for product coding. During inspections, close attention should be paid to the condition of paint and coatings. Damaged paint allows rust and corrosion to begin. A chip, ding, or flaked area should be cleaned and touched up before rust and corrosion advance. Technicians should arrange to have painting scheduled as soon as possible.

108F36.EPS

Figure 36 ◆ Contrasting pipeline paint scheme.

Some companies apply a company paint scheme to every part of their pipelines, including valve actuators. In hot climates, some companies paint much of their equipment white to resist heat, while in the colder areas, equipment is often painted black or some other heat-absorbing color. Some pipelines signify products by color-coding the pipe. The color of paint identifies the type of pipeline, and product can be determined from a distance if necessary. This is especially helpful in time of emergency. *Figure 37* shows two color-coded pipelines.

108F37.EPS

Figure 37 ◆ Color-coded pipelines.

Summary

Pipelines use valves to start and stop flow, regulate flow, relieve pressure, and prevent reversal of flow. Valves are essential for the control of pipeline operations. Without the actuator, the valve would be no more than a clogged piece of pipe.

The type of valve actuator used is determined by the type of application and power source available. Actuators can be manually, electrically, hydraulically, pneumatically, or gas-operated, or a combination. Some actuators offer quarter-turn service. They are used for ball valves, plug valves, and butterfly valves that require 90 degrees of rotation. Other operators incorporate gear drive trains that move the valve stem for many revolutions. Still others only act with a short, sliding movement. Each type is properly suited for very specific valve action and performs that action well.

Because they represent a sizeable portion of pipeline investment, valve actuators must be cared for and maintained so that their service life, safe operation, and efficiency are extended for as long as possible. Actuator maintenance is one of the most important tasks on the pipeline because it ensures the safety, efficiency, and economics of operation. Maintenance begins with receiving, handling, and proper storage until each actuator is installed. Once the actuators are installed, proper care ensures that they operate efficiently and safely through all conditions. Special maintenance procedures to prepare for winter ensure their safe operation in cold weather because moisture can condense and freeze inside gear cases and damage the unit.

Once the actuator is put in place, care exercised in proper maintenance pays large dividends in several areas. Maintaining seals and grease levels, applying grease to outside parts, keeping paint in good condition, and care in following manufacturer's recommendations and instructions all work together to provide extended service life, ensuring the continued safety and efficiency of the pipeline operation.

Review Questions

1. Chain actuators are used in applications where the valve is _____.
   a. too high to reach
   b. too far away
   c. automatic
   d. slow opening

2. Spur gears are used to connect _____.
   a. gearboxes
   b. parallel shafts
   c. valves
   d. actuators

3. A worm gear is a cylindrical-shaped gear similar to a _____.
   a. screw
   b. bevel-gear
   c. spur-gear
   d. chain actuator

4. Pneumatic and hydraulic actuators operate the valve in the same manner as _____-type actuators.
   a. gear
   b. wheel
   c. chain
   d. pressure

5. The status of pipeline valves is reported continuously to the _____.
   a. maintenance department
   b. supervisors
   c. owners
   d. operations department

6. When exact control is needed for various uses, a _____ actuator is used.
   a. chain
   b. bevel-gear
   c. worm-gear
   d. spur-gear

7. The rotary vane actuator is very _____.
   a. compact
   b. large
   c. fast
   d. slow

8. Valve actuators that are seldom used should be cycled on a _____ schedule.
   a. daily
   b. weekly
   c. monthly
   d. yearly

9. A grease gun used _____ to move grease.
   a. electricity
   b. pneumatic pressure
   c. gravity
   d. hydraulic pressure

10. Grease must be hand-packed into a closed gearbox as it is _____.
    a. installed
    b. removed
    c. disassembled
    d. assembled

11. Actuators in service should have the grease checked every _____ months.
    a. two
    b. four
    c. six
    d. eight

12. Multiple reduction sets can be used on _____ valves.
    a. chain operated
    b. handwheel
    c. gear-type
    d. hydraulic

13. Where a gear case heeds larger volumes at lubricant, the valve must be _____ while lubricating.
    a. cycled
    b. removed
    c. disassembled
    d. submerged

14. Factory-packed and sealed automatic actuators need to be checked every _____ months.
    a. six
    b. twelve
    c. eighteen
    d. twenty-four

15. An integral part of an automatic valve actuator that signals the unit to start and stop at exactly the right time is the _____.
    a. spur pinion
    b. limit switch
    c. input flange
    d. gear retainer

GLOSSARY

Trade Terms Introduced in This Module

Cycle: To cause a valve to move completely from one configuration to the opposite and then back to the original configuration, whether open or closed.

Desiccant: Soluble or insoluble chemical substance with such a great affinity for water that it will abstract water from many fluid materials. Soluble chemicals are calcium chloride and glycerol, and insoluble chemicals are bauxite and silica gel.

Lifting lug: A projection, head, or loop on a heavy metal part or unit, serving as a strong or convenient attachment point for lifting and handling the unit.

Limit switch: A switch designed to cut off power automatically at or near the limit of travel of a moving object controlled by electrical means.

Redundant system: A backup remote control system for pipeline control that is kept in hot standby so that if the normal remote control system loses communications, this system can automatically take over.

Valve actuator: A device such as a handwheel, lever, or powered unit attached to the stem of a valve to enable the valve to be turned.

Figure Credits

AMRI Inc.	108F17
Cameron Division Cooper Cameron Corp.	108F24, 108F29, 108F32, 108F33, 108F34
Contromatics	108F23
Custom Controls Company	108F21
Fisher Controls	108F22
Limitorque	108F08, 108F09, 108F10, 108F11, 108F13, 108F14, 108F15, 108F16, 108F18, 108F19, 108F30, 108F35
Mokveld Valves, Inc.	108F31
Shafer Valve Company	108F20

CONTREN™ LEARNING SERIES — USER UPDATES

The NCCER makes every effort to keep these textbooks up-to-date and free of technical errors. We appreciate your help in this process. If you have an idea for improving this textbook, or if you find an error, a typographical mistake, or an inaccuracy in NCCER's Contren™ textbooks, please write us, using this form or a photocopy. Be sure to include the exact module number, page number, a detailed description, and the correction, if applicable. Your input will be brought to the attention of the Technical Review Committee. Thank you for your assistance.

Instructors – If you found that additional materials were necessary in order to teach this module effectively, please let us know so that we may include them in the Equipment/Materials list in the Instructor's Guide.

Write: Curriculum Revision and Development Department
National Center for Construction Education and Research
P.O. Box 141104, Gainesville, FL 32614-1104

Fax: 352-334-0932

E-mail: curriculum@nccer.org

Craft _______________ Module Name _______________

Copyright Date _______________ Module Number _______________ Page Number(s) _______________

Description

____

____

____

(Optional) Correction

____

____

(Optional) Your Name and Address

____

____

Module 63109-02

Installing Seals and Gaskets

Introduction to the Pipeline Operator Qualification Rule, API, and NCCER

It could be said that petroleum products keep America running. Petroleum products like oil and natural gas are the principal sources of energy in the United States and are used in many manufacturing and chemical processes. Petroleum products are primarily transported by pipeline. There are more than a million miles of liquid and natural gas pipeline in this country, ranging in diameter from two inches to four feet. These pipelines bring crude oil and natural gas from production fields to processing plants and refineries—and ultimately to the consumer.

Pipeline Operator Qualification

Petroleum products are considered hazardous, so pipeline personnel must take extra precautions and follow specific procedures when working with them. To help ensure the safe handling and transportation of petroleum products, the United States Department of Transportation (DOT) mandates safe practices as federal regulations.

One such regulation, the Operator Qualification (OQ) rule, requires pipeline operators to develop a written qualification plan to evaluate their personnel and their contract personnel (see *49 CFR Part 192 Subpart N—Gas*, and *49 CFR Part 195 Subpart G—Liquid*). The evaluation hinges on the operator's ability to perform certain covered tasks and to be able to recognize and respond to abnormal operating conditions that may be encountered while performing these covered tasks.

Covered Tasks

A pipeline performance task is deemed *covered* if the activity passes the following four-part test:

1. Is it performed on a pipeline facility?
2. Is it an operations or maintenance task?
3. Is it performed as a requirement of *49 CFR Part 195 (Liquid)* or *49 CFR Part 192 (Gas)*?
4. Does it affect the operation or integrity of the pipeline facility?

If the answer to all four questions is *yes*, then the activity is considered *covered* and must figure into the evaluation process of pipeline personnel.

American Petroleum Institute

The OQ rule and its related covered tasks provided the catalyst to develop a unified pipeline training and assessment program—so that pipeline operators and owners have ready access to the assessment and training tools they need. This effort was the direct result of the vision and leadership provided by The American Petroleum Institute (API) and API members.

Efforts to unify the industry are nothing new to API. Indeed, the impetus for forming the Institute in 1919 was the need to standardize engineering specifications for drilling and production equipment. Today, the most pressing issues revolve around public perceptions and government policies toward the industry. Speaking with one voice on these issues has become as essential as having interchangeable parts in the field. API's members determine the petroleum industry's positions on public policy as well as the standards that govern its day-to-day operations worldwide.

Consortium on Operator Qualification

API assembled a focus group whose primary purpose was to identify covered tasks and provide compliance recommendations to pipeline operators.

The Consortium on Operator Qualification (COOQ) became the guiding force for the identification and development of covered tasks for the liquid pipeline industry. Their work product, API publication 1161, *Guidance Document for the Qualification of Liquid Pipeline Personnel*, formed the basis for the resulting Pipeliner Training and Assessment Program.

The covered task list created in API publication 1161 has now evolved to include gas covered tasks and is maintained by NCCER. The most current covered task list can be downloaded at www.nccer.org.

National Center for Construction Education and Research

With the guidance and support of API, its members, and the Consortium on Operator Qualification, the National Center for Construction Education and Research (NCCER) was able to design and implement the Pipeliner Training and Assessment Program.

NCCER is a not-for-profit 501(c)(3) education foundation established in 1995 by the world's largest and most progressive construction and maintenance contractors and national contractor associations. NCCER was founded to address the

severe workforce shortage facing the construction and maintenance industries and to develop a standardized training process and curricula. Today, NCCER is supported by hundreds of leading construction and maintenance companies, manufacturers, and national associations and developed more than 35 construction and maintenance titles.

NCCER also maintains a National Registry, which contains the records of trainees and instructors who have received NCCER training. All completion information for every trainee is sent to the NCCER and filed within the Registry. The Registry can then confirm training and skills for workers as they move from company to company, state to state, or even within their own company.

Standardized Curriculum (NCCER's Contren™ Learning Series)

NCCER curricula are developed using the expertise of industry subject matter experts (SMEs). The use of SMEs as an integral part of every new development or revision ensures that the curriculum is current, industry-driven, and indicative of industry practices. The development and/or revision process is facilitated from start to finish by NCCER project managers, who work closely with the SMEs and a select team of technical writer(s) on the development and presentation of the curriculum.

Pipeline: Five Training Paths

The pipeline curriculum focuses on essential skills and competencies within five skill areas:

- Corrosion Control
- Electrical and Instrumentation
- Maintenance
- Mechanical
- Operations (Field and Control Center)

SMEs from the pipeline industry have determined the skills and competencies needed based on API's covered tasks and the consensus of the group/industry. Specific company information is generally not incorporated into the curriculum unless the SMEs determine that the information is the industry standard.

Training Modules and Covered Tasks

The smallest unit of instruction in the curriculum is a *module*. Each curriculum is composed of modules (the equivalent of stand-alone chapters) that are logically sequenced into complete *levels* of instruction. Every module is task-oriented and requires the successful completion of a written (knowledge-based) examination and a performance (hands-on) test. Modules may be individually selected and/or combined for efficient task training. Modules in the pipeline curricula, to the extent possible, will align with API-developed covered tasks. While not all modules will relate to covered tasks, all covered tasks will relate to at least one module.

Benefits of Training to the Pipeline Marketplace

Applying NCCER's standardized training process allows pipeline owners and contractors in the United States to establish the benchmark for acceptable skill and competency levels within the pipeline industry. This process will assist training organizations in developing and/or improving standardized pipeline training programs that effectively match their resources and operating conditions while responding to regulatory demands and to the needs of the industry. The process will also recognize training organizations that meet the minimum standards of competence, warranting regulatory, public, and professional confidence. This in turn will foster national unity within the pipeline industry for the mutual benefit of individual technicians, owners, contractors, and ultimately pipeline-users and society.

Additional Resources

To review the Pipeliner Training and Assessment Program Implementation and Resource Guide and the latest version of pipeline covered tasks, visit NCCER's Web site at www.nccer.org, or contact:

National Center for Construction Education and Research
PO Box 141104
Gainesville, FL 32614-1104
(888) 622-3720
Web site: www.nccer.org

For more information on API, please contact:

American Petroleum Institute
1220 L Street, Northwest
Washington, D.C. 20005
(202) 682-8000
Web site: www.api.org

To order *API Pub 1161*, contact Global Engineering Documents by calling 1-800-854-7149 or visiting their Web site at www.global.his.com.

Additional sources of information on the pipeline operator qualification regulation include:

DOT – Office of Pipeline Safety
www.ops.dot.gov

Transportation Safety Institute
www.tsi.dot.gov

COURSE MAP

This course map shows all of the modules in the first level of the Pipeline Mechanical curriculum. The suggested training order begins at the bottom and proceeds up. Skill levels increase as you advance on the course map. The local Training Program Sponsor may adjust the training order.

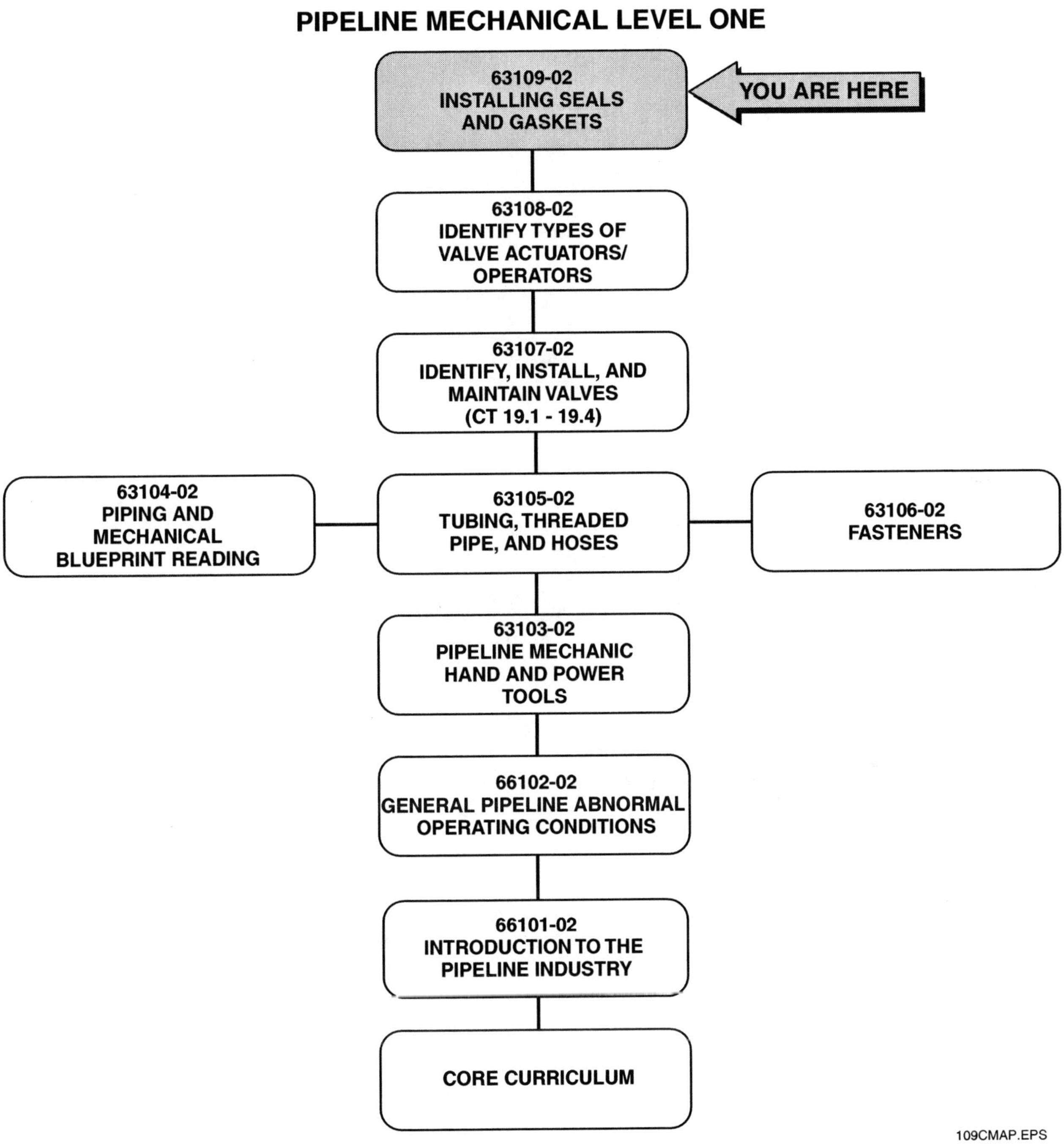

109CMAP.EPS

MODULE 63109-02 CONTENTS

Figures

Tables

MODULE 63109-02

Installing Seals and Gaskets

Objectives

When you have completed this module, you will be able to do the following:

1. Identify and explain types of seals.
2. Remove and install seals.
3. Identify and explain gasket types and materials.
4. Layout, cut, and install gaskets.

Prerequisites

Before you begin this module, it is recommended that you successfully complete the following: Core Curriculum; Pipeline Core, Modules 66101 and 66102; Pipeline Mechanical Level One, Modules 63103 through 63108.

Required Trainee Materials

1. Pencil and paper
2. Appropriate personal protective equipment

1.0.0 ◆ INTRODUCTION

Seals are devices or materials used to prevent or control leakage between two surfaces. They are used between moving and stationary parts or between two stationary parts. Seals come in various types and sizes and are made of various materials for different applications. Some seals require regular replacement to ensure proper performance. Selecting the proper type and size of seal and properly installing the seal are critical to the performance of the seal and the equipment. Incorrect selection or installation of seals can result in premature failure of the seals and can damage the equipment. Pipeline mechanical technicians must be able to identify the types of seals and seal materials and remove and replace seals.

Gaskets also seal joints in pipe systems, gear boxes, pumps, motors, and other types of equipment. There are many types of gasket and gasket materials for different applications. The type of gasket used in an application depends on the operating temperature of the system or equipment, the type of fluid the system or equipment handles, the type of connection being made, and the system or equipment pressure range. Many specialty gaskets, such as pipe flange gaskets and equipment assembly gaskets, are manufactured for specific uses. These gaskets are already cut to the required pattern and the holes are already punched. When specialty gaskets are not available, you must identify and select the proper gasket material, cut the gasket, and punch the holes to match the application. This module explains identifying and selecting gasket types and gasket materials and laying out, cutting, and installing gaskets.

WARNING!
Gasket materials must never be arbitrarily substituted unless authorized by an engineer. Substituting gasket material may cause equipment failure and even personal injury. Always refer to engineering specifications before installing a gasket into a system.

2.0.0 ◆ TYPES OF SEALS

When selecting a seal, you must determine whether the application is static or dynamic. Static applications are those in which there is no movement between the two joining parts or between the seal and the mating part. O-rings are used for static applications. Dynamic seals are used where there is movement between two mating parts or between one of the parts and the seal. The two types of dynamic applications are positive-contact and noncontact seals. Seals used for dynamic applications include O-rings, lip seals, oil seals, and labyrinth seals.

Positive-contact, or rubbing, seals are used where the seal area is continuously flooded. If properly selected and installed, contact seals can prevent all leakage of most fluids. However, because they are sensitive to temperature, pressure, and speed, improper use can cause early failure. Positive-contact seals can be used on both rotating and reciprocating shafts.

Noncontact, or controlled-clearance, seals are those that prevent rubbing between rotating and stationary parts. They operate by controlling fluid movement through narrow passages in the seal. Noncontact seals are frictionless and are not sensitive to temperature and speed. They are most effective for limiting leakage rather than stopping it. Noncontact seals have limited uses when leakage rates are critical. These seals are frequently custom-designed for a particular application and can be expensive.

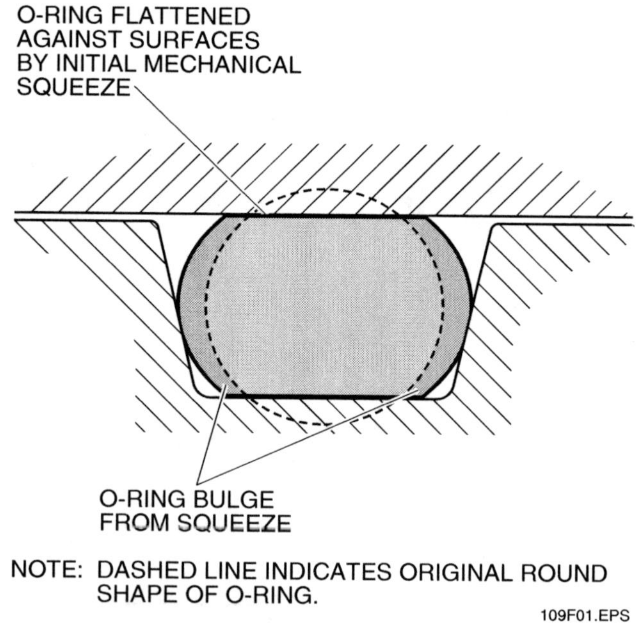

Figure 1 ◆ Mechanical squeeze.

2.1.0 O-Rings

O-rings can be used as either static or dynamic seals for liquid pipeline systems. O-rings cannot be used for seals in natural gas systems. In a dynamic application, an O-ring is usually placed in a groove or joint that is 135 to 150% wider than the cross section of the O-ring. In a static application, an O-ring is installed in a groove approximately 25% wider than the cross section of the O-ring. When the O-ring comes in contact with the areas to be sealed, it is slightly distorted in a motion called *mechanical squeeze* (see *Figure 1*).

A dynamic application requires a running fit, while a static application needs a complete seal and the O-ring should almost fill the O-ring groove. *Figure 2* shows static and dynamic applications.

The pressure caused by mechanical squeeze holds the O-ring in contact with the surfaces to be sealed. This pressure also causes the O-ring to roll and slide to the side of the groove away from the pressure, lubricating the O-ring surface and extending its life (see *Figure 3*).

Mechanical squeeze is necessary to maintain a seal. Too much squeeze wears out the O-ring quickly; insufficient squeeze allows leaks to start. The correct squeeze is usually equal to 10% of the O-ring diameter. General-purpose O-rings are made with a cross section 10% larger than the nominal size. This allows for the initial mechanical squeeze. *Table 1* lists the nominal and actual diameters of standard O-rings. *Figure 4* shows a housing vent with O-rings.

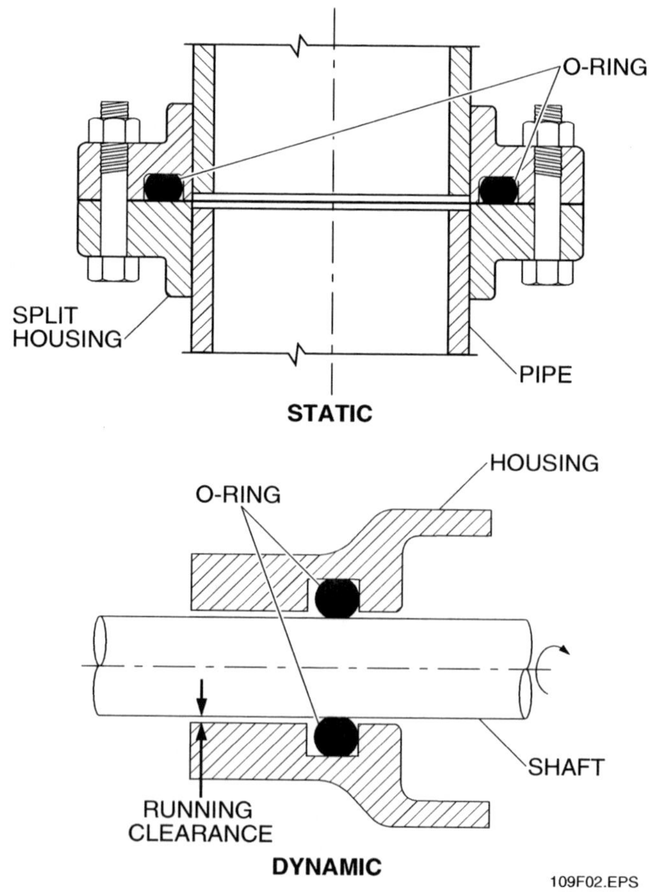

Figure 2 ◆ Static and dynamic applications.

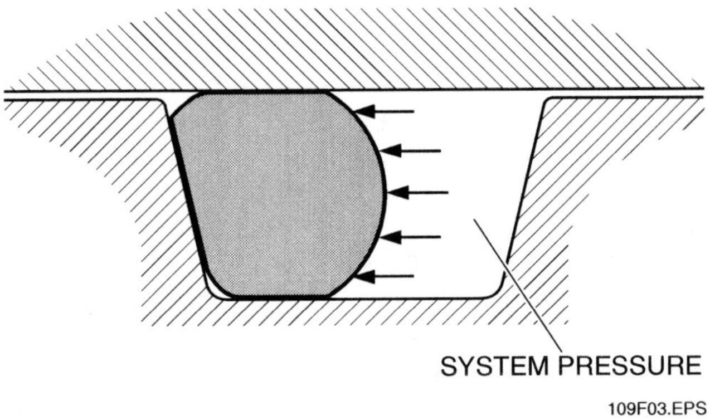

Figure 3 ◆ O-ring sealing a joint.

Table 1 Nominal and Actual Diameters of Standard O-Rings

Nominal Cross Section Diameter (inches)	Actual Cross Section Diameter (inches)
1/32 (0.03125)	0.040
3/64 (0.046875)	0.050
1/16 (0.0625)	0.070
3/32 (0.09375)	0.103
1/8 (0.125)	0.139
3/16 (0.1875)	0.210
1/4 (0.250)	0.275

109T01.EPS

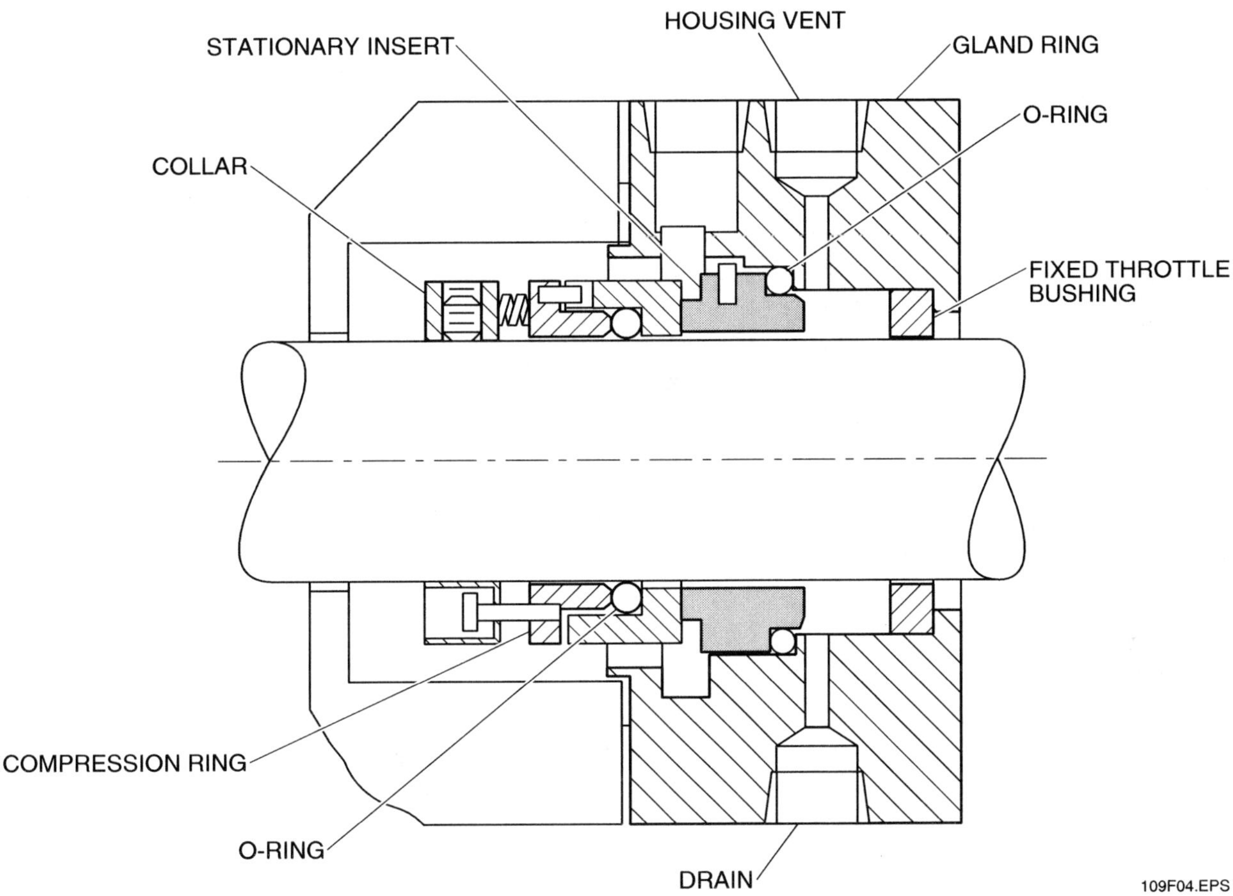

Figure 4 ◆ Housing vent with O-rings.

2.2.0 Lip Seals

Lip seals are low-pressure, positive-contact seals used with rotating shafts. Pressure on the lip or a vacuum behind the lip pushes it against the shaft for a tighter seal. Lip seals usually consist of a housing that is used for aligning the seal when it is installed and for supporting the lip, a lip that fits around the shaft, and a spring that holds the lip in contact with the shaft.

Lip seals are usually installed to contain lubricant in a housing when the outside conditions are relatively clean. In dirty conditions, the seal is installed with the lip facing out to prevent foreign matter from getting into the housing. With the lip facing out, the housing should be vented to prevent the lubricant from leaking. A vent and drain can be applied to any inside seal by drilling and tapping two openings into the gland ring outboard of the stationary insert.

In some applications, the area being sealed changes from pressure to vacuum conditions. Double-lip seals are available for these applications to prevent air or dirt from getting in and lubricant from getting out.

2.3.0 Oil Seals

The oil seal, or radial lip seal, is a positive-contact seal used on rotating or reciprocating shafts (see *Figure 5*). This seal is used to keep fluids in or to keep dirt and other foreign matter out. Oil seals are usually used to keep fluids in and are installed with the lip facing in. When an oil seal is used to keep contaminants out, it is installed with the lip facing away from the housing. Double-lip seals that perform both functions are also available.

Oil seals are manufactured in a wide variety of types for use in various applications. These types can be divided into the following three broad categories:

- *Cased seals* – These contain a sealing element in a manufactured metal case.
- *Bonded seals* – These contain a sealing element bonded to a washer or a formed metal case.
- *Dual-element seals* – These contain a double-sealing element. The double element is used for applications where liquids are present on both sides of the seal or for severe service.

2.4.0 Labyrinth Seals

Labyrinth seals are noncontact dynamic seals that are used on rotating shafts (see *Figure 6*). Labyrinth seals do not actually touch the surface to be sealed. Instead, they contain a narrow path through which fluid must flow before it can leak. Because small amounts of liquid can leak from a labyrinth seal, they work best with thick or semisolid materials, such as grease. However, they can also be used with liquids and gases if some leakage is not a problem.

The noncontact design of the labyrinth seal also provides the following benefits:

- Prevents damage to the shaft
- Helps prevent heat buildup
- Does not require lubrication

2.5.0 Cup Seals

A cup seal is a positive-contact dynamic seal generally used on cylinder pistons (see *Figure 7*). Sealing occurs when pressure forces the cup lip against the cylinder barrel. Cup seals are pressure-driven in both directions. They are supported by a backing plate and can handle operating pressures of 1,000 to 1,500 psi. To work correctly, cup seals must be clamped tightly in place. In a cylinder piston application, the piston actually holds the cup seals.

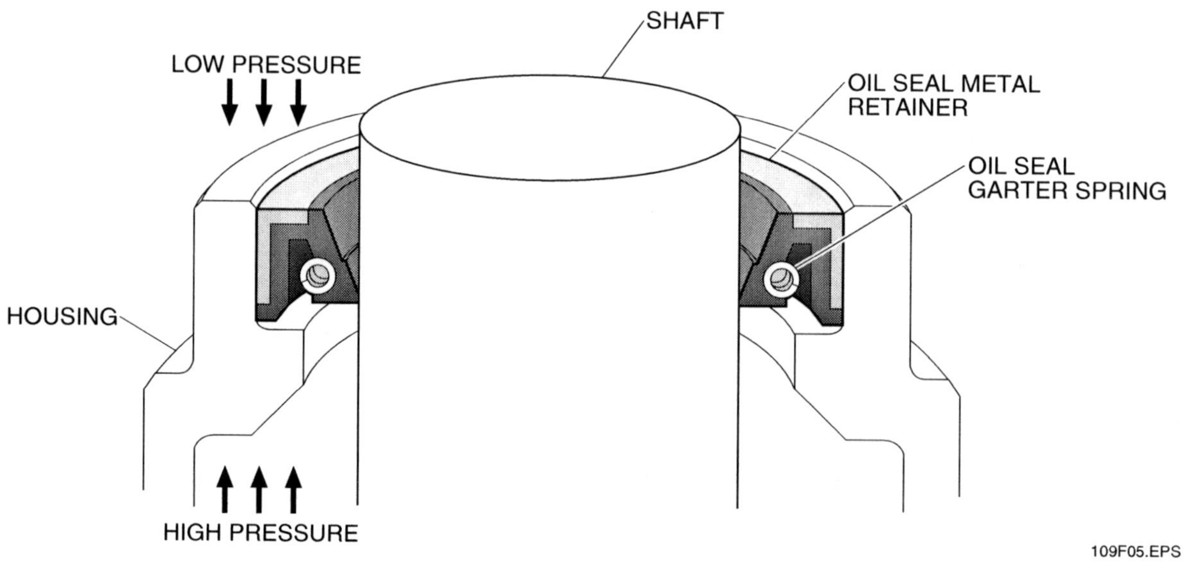

Figure 5 ◆ Typical oil seal.

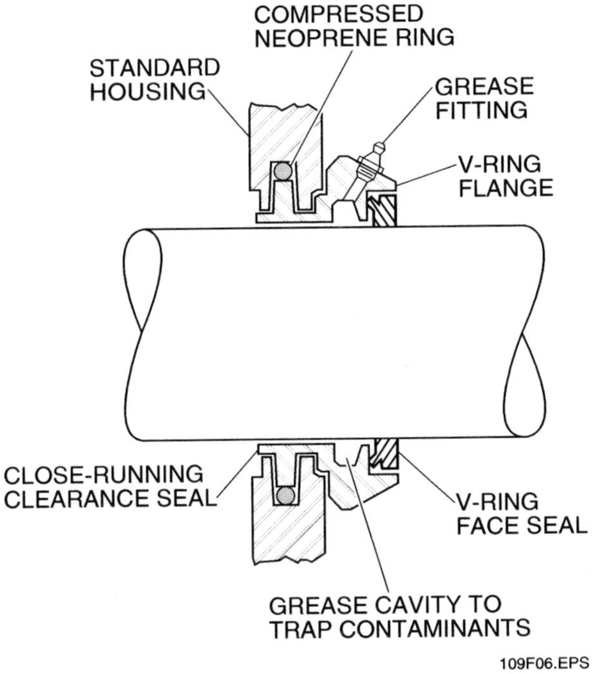

Figure 6 ◆ Labyrinth seal.

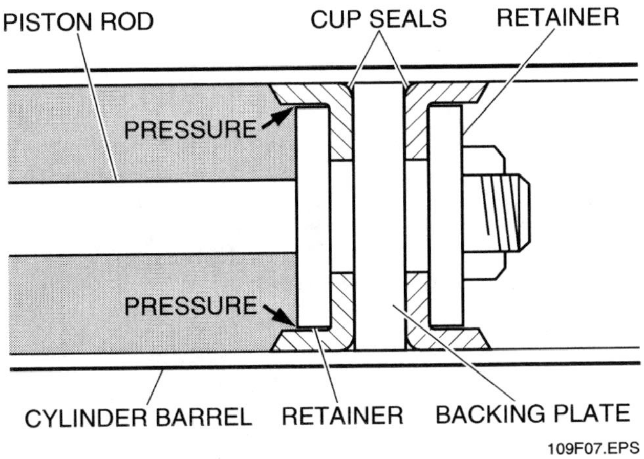

Figure 7 ◆ Cup seals.

3.0.0 ◆ SEAL MATERIALS

The design and composition of seals must be compatible with the fluid used in a system, including its use and the operating conditions. The following points should be considered when choosing a seal material:

- *Fluid compatibility* – Few seals are compatible with all fluids. Some fluids and fluid additives can cause certain types of seals to disintegrate. Contact the fluid supplier for a list of compatible seal materials.
- *Temperature* – Seals operate within specific temperature ranges. At extremely low temperatures, a seal can become brittle. At high temperatures, a seal can harden, soften, or swell.
- *Pressure* – Fluid pressure puts a strain on seals and can cause leaks. Always use a seal rated to withstand the pressure in a given application.
- *Lubrication* – No seal should be installed or operated without lubrication. Some seals must be soaked in fluid before they are installed; others only need to be coated with fluid.

Each seal manufacturer has applications charts to ensure compatibility with each application; therefore, consult the manufacturer's specifications when choosing a seal for any application. Do not arbitrarily substitute seal materials.

3.1.0 Buna-N

Buna-N, or nitrile, is the most widely used sealing material. It is an **elastomer** that wears well and is inexpensive. Buna-N is compatible with petroleum oil and works at temperatures ranging from –40 to 230°F. At high temperatures, it retains its shape in most petroleum oils where other materials tend to swell. It does swell in some synthetic fluids.

3.2.0 Leather

Leather is an inexpensive and tough seal material used for some cup and lip seals. Some leather seals are saturated with an elastomer to improve the sealing ability. The disadvantages of leather are a tendency to squeal when dry and a limited temperature range. Most leather seals cannot function above 165°F; however, they function well to temperatures of –65°F.

3.3.0 Silicone

Silicone is an elastomer used for rotating shaft seals and static seals in systems with wide temperature ranges. Silicone retains its shape and sealing ability at ranges from –60 to 500°F. It is compatible with most fluids. At high temperatures, silicone absorbs oil and swells, but this is not a disadvantage for static applications. It is not used for reciprocating applications because it tears too easily.

3.4.0 Neoprene

Neoprene is an elastomer used only for low-temperature systems using petroleum fluids. It is only suitable at temperatures below 150°F because it tends to **vulcanize** above that temperature.

3.5.0 Plastic and Elastomer Compounds

Several sealing materials are formed by combining fluorine gas with an elastomer or plastic, including Kel-F®, Viton A®, and Teflon®. Another synthetic material with similar properties is nylon. Nylon is often used in combination with the elastomers to reinforce them. All of these materials can be used at temperatures up to 500°F and are compatible with most fluids.

3.6.0 Room Temperature Vulcanizing (RTV) Seals

Room temperature vulcanizing (RTV) seals are liquid-based seals that can be used in a variety of applications. RTV sealing compounds deliver a tight mechanical joint seal, filling voids, gaps, scratches, and flatness imperfections on metal surfaces. By achieving a metal-to-metal seal, RTV seals distribute stresses uniformly over a large area, dampen vibration, resist thermal shock, and provide excellent clamp load. These sealants adhere to metals, glass, ceramics, silicone rubber, and most plastics. Always refer to the manufacturer's application guidelines and engineering design standards when selecting RTV seals.

4.0.0 ◆ REMOVING AND INSTALLING SEALS

The working life of a seal depends on the type of seal used and the application for which it is used. The most common indication of seal failure is excessive leakage. When a seal fails, it must be removed and replaced. Removing seals is a relatively simple procedure, but it is important that the procedure be performed properly. The procedure for removing seals varies slightly from one piece of equipment to another, but the steps are basically the same. Refer to the equipment manufacturer's manual for particular procedures.

Following proper installation procedures for a seal is critical to the life expectancy and efficiency of the seal. Improperly installing a seal can result in a loss of up to 75% of its working life and may cause excessive wear to the equipment. The following sections explain how to remove a seal and how to install O-rings and lip and oil seals. Always follow manufacturer's installation instructions when removing and installing seals.

4.1.0 Removing Seals

Before a new seal can be installed, the old seal must be removed. Follow these steps to remove a seal:

Step 1 Clean the shaft and exterior of the seal, using a clean, soft cloth.

Step 2 Remove any burrs or sharp edges from the shaft.

Step 3 Remove the old seal.

> **NOTE**
> To remove the old seal, tap it gently using a flat hammer, or cut into the seal using a rounded, blunt chisel.

> **CAUTION**
> Do not cut all the way through the seal. Cutting through the seal can damage the seal **bore**.

4.2.0 Installing O-Rings

Follow these steps to install an O-ring:

Step 1 Select the proper size O-ring for the application.

Step 2 Ensure that the O-ring is not damaged.

Step 3 Inspect, polish, deburr, and clean the shaft and housing in the seal area.

> **NOTE**
> Repair any damaged areas of the housing shaft.

Step 4 Cover any threads with a cone made of brass shim stock, aluminum, or stiff plastic sheeting.

Step 5 Lubricate the O-ring with the lubricant specified by the manufacturer.

Step 6 Place the O-ring in the groove.

Step 7 Reassemble inspect for leaks, and return to service.

4.3.0 Installing Lip and Oil Seals

Follow these steps to install lip and oil seals:

Step 1 Select the proper size seal and an installation tool, such as a press ram or driving tool.

NOTE
The outside diameter of the installation tool should not be more than 0.010 inch smaller than the bore diameter. If an installation tool is not available, a tube or short length of pipe that is slightly smaller in diameter than the bore can be used. If a tube is not available, a block of wood placed squarely on the back of the seal can also be used.

Step 2 Ensure that the seal has not been damaged and that its spring is in place.

Step 3 Inspect, polish, deburr, and clean the shaft and housing in the seal area.

NOTE
Repair any damaged areas of the housing shaft.

Step 4 Place a mounting thimble on the shaft to protect the seal (see *Figure 8*).

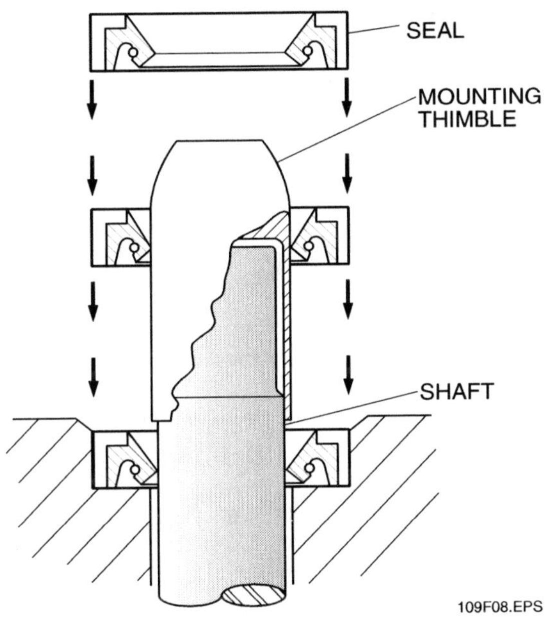

Figure 8 ◆ Preparing to mount the seal.

NOTE
If a mounting thimble is not available, cover the edge or shoulder with a protective covering, such as brass shim stock, aluminum, or stiff plastic sheeting. Wrap the protective covering around the shaft and over the end that could damage the seal.

Step 5 Lubricate the shaft and seal lip, using the lubricant specified in the seal catalog.

Step 6 Slide the seal over the shaft and into place.

Step 7 Place an installation tool against the seal (see *Figure 9*).

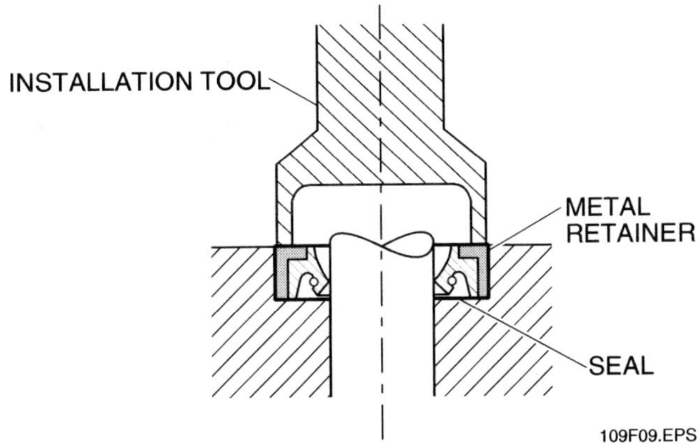

Figure 9 ◆ Using an installation tool.

Step 8 Tap the installation tool, using a mallet to seat the seal.

CAUTION
To prevent damaging the seal, never hit it directly with a mallet.

Step 9 Ensure that the seal is installed squarely with the shaft centerline.

NOTE
Press the seal flush with the front of the bore or bottom it against a shoulder.

Step 10 Reassemble, inspect for leaks, and return to service.

5.0.0 ◆ GASKET TYPES

The two basic types of gaskets are flat and O-ring, of which there are many varieties. A gasket has one basic function: to create a positive seal between two relatively stationary parts. The thinnest gasket that provides this seal is the most efficient and lasts the longest. It is also the most economical. Gaskets are chosen for their pressure rating, their temperature rating, and their resistance to the process on which they are used. Special care should be taken when handling or storing any gasket to prevent bending, creasing, or otherwise damaging them. *Figure 10* shows some common gasket types.

The various types of gaskets explained in this section are:

- Ring
- Spiral wound
- Full face
- Jacketed
- Envelope
- Split ring
- Strip

5.1.0 Ring Gaskets

Ring gaskets, which are commonly used, are flat-type flange gaskets made of various materials for different applications. They are made to fit inside the bolt circle of a flange, and they have no bolt holes. They usually have a pressure range of 150 to 200 psi.

5.2.0 Spiral Wound Gaskets

Spiral wound gaskets are flat-type flange gaskets, commonly made of stainless steel with a graphite insert. They are used on very high-temperature, high-pressure systems. A spiral wound gasket is a crushable gasket with high **resilience** that allows it to adjust automatically to changes in line pressure, thermal shocks, vibration, and minor flange separation.

5.3.0 Full Face Gaskets

The full face gasket, another commonly used gasket, is a flat-type flange gasket. These gaskets are very similar to ring gaskets, but are made to fit the flange and have holes for the flange bolts. It is made from various materials for different applications.

5.4.0 Jacketed Gaskets

Jacketed gaskets have a metal exterior cover with an internal filler. The filler can be of various materials. The outer jacket prevents the fluid in a system from contacting the inner filler material and prevents the filler material from contaminating the system fluid. These gaskets are commonly made with a stainless steel jacket and an asbestos filler.

5.5.0 Envelope Gaskets

An envelope gasket is made using the same principle as a jacketed gasket. The outer cover is usually Teflon® with various types of filler. Envelope gaskets have a limited temperature range and are commonly used where flange imperfections exist because of their superior sealing ability.

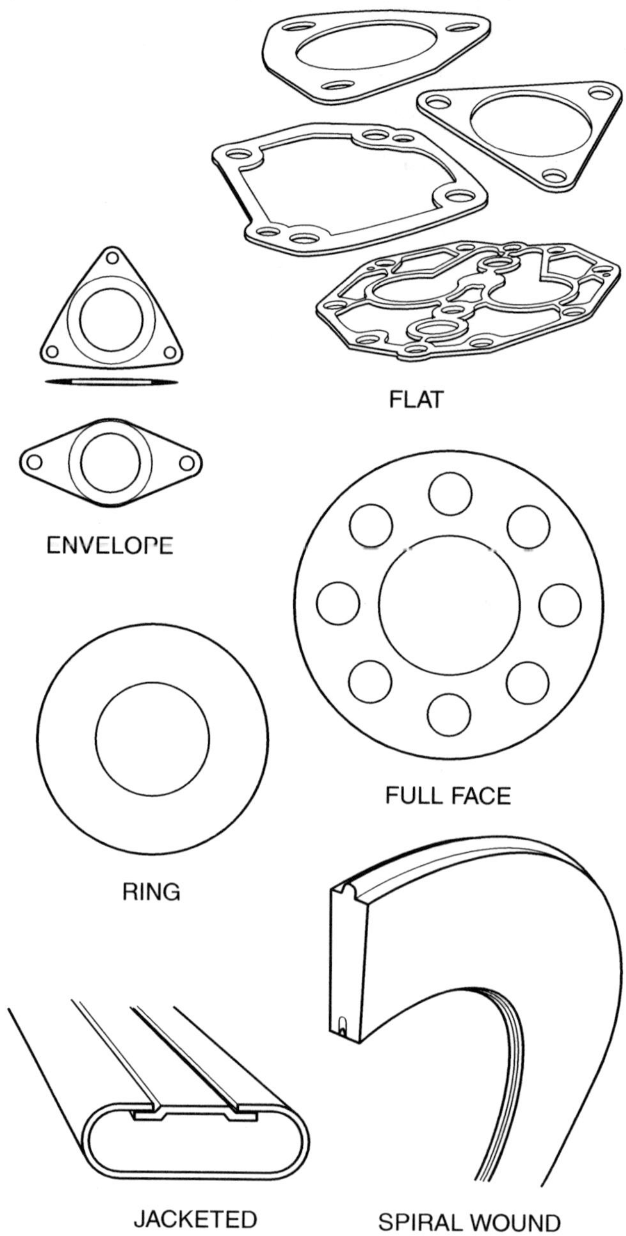

Figure 10 ◆ Common gasket types.

5.6.0 Split Ring Gaskets

Split ring gaskets, which are not commonly used, are made in two pieces and are used where space is limited. They can be slipped into a flange without removing all of the bolts.

5.7.0 Strip Gasket

Strip gasket material, usually made of Teflon® material, comes in a roll. The gasket is pulled from the roll and placed on the flange in one or more strips. This gasket is only good for medium-pressure, low-temperature systems. Strip gaskets are considered a "bandaid" fix and are not commonly used for permanent service.

6.0.0 ◆ GASKET MATERIALS

Gaskets, which are made from a wide range of materials, are manufactured from natural substances as well as man-made substances. Some materials, called *composite materials*, have both. The material chosen to make a gasket is very important. If the wrong material is chosen, the gasket will fail. There are many qualities to consider when choosing gasket material. The following are some of the criteria that should be met when selecting gasket material:

- The material must be impervious to the fluid that the system handles.
- The gasket must have adequate chemical **resistance** at the fluid contact areas to prevent adverse effects on the physical properties of the gasket.
- The material must not contaminate the fluid in the system.
- The material must have adequate elasticity to maintain an adequate seal in case there is joint movement.
- The material must develop sufficient friction in contact with the flange surface to resist excessive **creep** or **extrusion**.
- The material must not promote corrosion of the flange with which it comes in contact.
- The material must be able to withstand the required bolt torque without crushing or undergoing excessive **plastic flow**.
- In order to compensate for normal irregularities, the material must be sufficiently deformable to distribute the applied pressure evenly over the flange surfaces.
- The material must be rated to handle the maximum system pressure or a **blowout** may occur.
- The material must be rated to handle the maximum system temperature.

In addition to the above criteria, the pressure rating of a gasket must be matched to the system. The pressure rating of a gasket varies with the grade of material used and the type of gasket.

Many gasket suppliers provide a physical properties chart to be used to match gasket material to the application. You should use a chart of this type or contact the system engineer before installing a gasket in a system. You must be sure that the gasket is compatible with the system. *Figure 11* shows a sample physical properties chart.

Different gasket materials have very different properties. This section explains some of the properties of the following common gasket materials:

- Natural rubber
- **EPDM**
- Neoprene
- Nitrile
- Silicone
- Viton
- Gylon

> **WARNING!**
> Asbestos gaskets used to be very popular; however, gaskets made of asbestos should be avoided because asbestos is carcinogenic and causes lung inflammation.

6.1.0 Natural Rubber

Natural rubber is seldom used in installations today. It has low solvent and oil resistance. It is excellent when used with water. Natural rubber has excellent resilience and may be used at temperatures up to a maximum of 175°F.

6.2.0 EPDM

Ethylene propylen dieneterpolymer (EPDM) is very good in an oxygen environment. It has good ozone resistance. EPDM has low resistance to solvent and oil. It is a good material to be used with water and chemicals and has a maximum temperature rating of 350°F. It also has excellent flame resistance.

6.3.0 Neoprene

Neoprene, which has good resilience, is only used in noncritical conditions. Neoprene has fair oil and solvent resistance but has poor chemical resistance and is not recommended for water service. It has a maximum temperature rating of 250°F.

COMMON NAME			NATURAL RUBBER	SBR	BUTYL	EPDM	NEOPRENE	NITRILE	POLYSULFIDE	POLYURETHANE	SILICONE	HYPALON	FLUOROELASTOMERS
WEIGHT OF BASE ELASTOMER		LB/CU IN.	0.033	0.034	0.033	0.031	0.044	0.036	0.048	0.039	0.036	0.040	0.05-0.070.93
		SPEC. GR.	0.93	0.94	0.92	0.85	1.23	1.00	1.34	1.05	0.95	1.10	1.4 TO 1.95
PHYSICAL PROPERTIES FOR ELASTOMER COMPOUNDS	DUROMETER, RANGE		30-100	40-100	30-100	30-90	45-95	20-90	20-80	25-90	25-80	50-95	65-90
	RESILIENCE		EXCELLENT	GOOD	FAIR	GOOD	EXCELLENT	GOOD	FAIR	GOOD TO EXCELLENT	POOR TO EXCELLENT	GOOD	FAIR
	TENSILE STRENGTH, PSI (REINFORCED)		4000+	2000+	2000+	2000-3000	3000+	1000-3500+	500-1500	4000-8000	600-1500	1500-2500	1500-3000
	ELONGATION, % (REINFORCED)		500	450	300-800	500	650-850	400-600	200-550	250-800	90-800	250-500	100-450
	DRIFT, ROOM TEMP		EXCELLENT	EXCELLENT	FAIR	FAIR	FAIR TO GOOD	GOOD	POOR	GOOD TO EXCELLENT	FAIR TO EXCELLENT	FAIR	GOOD
	COMPRESSION SET		GOOD	GOOD	FAIR	FAIR	FAIR TO GOOD	GOOD	POOR TO FAIR	EXCELLENT	GOOD TO EXCELLENT	FAIR TO GOOD	GOOD TO EXCELLENT
	ELECTRICAL RESISTIVITY		EXCELLENT	EXCELLENT	EXCELLENT	EXCELLENT	FAIR	POOR	FAIR	GOOD	EXCELLENT	GOOD	GOOD
	IMPERMEABILITY, GAS		GOOD	FAIR	EXCELLENT	GOOD	GOOD	GOOD	EXCELLENT	GOOD	FAIR	EXCELLENT	EXCELLENT
RESISTANCE PROPERTIES — MECHANICAL	RESISTANCE TO	IMPACT	EXCELLENT	EXCELLENT	GOOD	GOOD	GOOD	FAIR	POOR	EXCELLENT	POOR TO FAIR	FAIR TO GOOD	POOR TO GOOD
		ABRASION	EXCELLENT	EXCELLENT	GOOD	GOOD	GOOD TO EXCELLENT	EXCELLENT	POOR TO FAIR	EXCELLENT	POOR TO GOOD	GOOD	GOOD
		TEAR	EXCELLENT	FAIR	GOOD	POOR	GOOD	GOOD	POOR TO FAIR	EXCELLENT	POOR TO GOOD	FAIR TO GOOD	POOR TO GOOD
		CUT GROWTH	EXCELLENT	GOOD	EXCELLENT	GOOD	GOOD	GOOD	POOR	FAIR TO EXCELLENT	POOR TO GOOD	FAIR TO GOOD	POOR TO GOOD
TEMPERATURE	TENSILE STRENGTH; PSI, AT	250F	1800	1200	1000	2000	1500	700	700	1800	850	500	300-800
		400F	125	170	350	400	180	130	UNDER 25	200	400	200	150-300
	ELONGATION %, AT	250F	500	250	250	300-500	350	120	140	300	350	60	100-350
		400F	80	60	80	0-120	0-100	20	UNDER 25	140	200	20	50-160
	DRIFT AT 212F		GOOD	GOOD	FAIR	FAIR	FAIR TO GOOD	EXCELLENT	POOR	EXCELLENT	EXCELLENT	FAIR	GOOD TO EXCELLENT
	HEAT AGING AT 212F		GOOD	GOOD	EXCELLENT	EXCELLENT	GOOD	GOOD	GOOD	FAIR TO GOOD	EXCELLENT	EXCELLENT	EXCELLENT
	FLAME RESISTANCE		POOR	POOR	POOR	POOR	GOOD	POOR TO FAIR	POOR	POOR TO FAIR	FAIR TO GOOD	GOOD	EXCELLENT
	TEMPERATURE: MAXIMUM (¡F)		200	275	325	350	250	250	200	180	550	325	400
	LOW TEMPERATURE	STIFFENING, F	-20 TO -50	0 TO 50	10 TO 40	-20 TO -50	+10 TO -20	+30 TO -20	-10 TO -45	-10 TO -30	-60 TO -160	-30 TO -50	+20 TO -30
		BRITTLE POINT, F	-80	-80	-80	-90	-45	-65	-60	-60	-90 TO -180	-60	+10 TO -60
ENVIRONMENTAL	WEATHER		FAIR	FAIR	EXCELLENT	EXCELLENT	EXCELLENT	GOOD	EXCELLENT	EXCELLENT	EXCELLENT	EXCELLENT	EXCELLENT
	OXIDATION		GOOD	GOOD	EXCELLENT	GOOD	GOOD	FAIR TO GOOD	EXCELLENT	EXCELLENT	EXCELLENT	EXCELLENT	EXCELLENT
	OZONE		POOR	POOR	EXCELLENT	EXCELLENT	EXCELLENT	POOR	EXCELLENT	EXCELLENT	EXCELLENT	EXCELLENT	EXCELLENT
	RADIATION		FAIR TO GOOD	GOOD	POOR	POOR	FAIR TO GOOD	FAIR TO GOOD	FAIR TO GOOD	GOOD	FAIR TO EXCELLENT	FAIR TO GOOD	FAIR TO GOOD
	WATER		EXCELLENT	EXCELLENT	EXCELLENT	GOOD TO EXCELLENT	GOOD	EXCELLENT	GOOD	GOOD	GOOD	GOOD	GOOD
	ACID		FAIR TO GOOD	FAIR TO GOOD	EXCELLENT	GOOD TO EXCELLENT	GOOD	GOOD	FAIR	POOR TO FAIR	POOR TO GOOD	EXCELLENT	GOOD TO EXCELLENT
	ALKALI		FAIR TO GOOD	FAIR TO GOOD	EXCELLENT	GOOD TO EXCELLENT	GOOD	FAIR TO GOOD	GOOD	POOR TO FAIR	POOR TO FAIR	EXCELLENT	POOR TO GOOD
	GASOLINE, KEROSENE, ETC. (ALIPHATIC HYDROCARBONS)		POOR	POOR	POOR	POOR	GOOD	EXCELLENT	EXCELLENT	EXCELLENT	POOR TO FAIR	FAIR	EXCELLENT
	BENZOL, TOLUOL, ETC. (AROMATIC HYDROCARBONS)		POOR	POOR	FAIR TO GOOD	FAIR	POOR	GOOD	EXCELLENT	POOR TO FAIR	POOR	POOR TO FAIR	EXCELLENT
	DEGREASER SOLVENTS (HALOGENATED HYDROCARBONS)		POOR	POOR	POOR	POOR	POOR	POOR	FAIR TO GOOD	FAIR TO GOOD	POOR TO GOOD	POOR TO FAIR	GOOD
	ALCOHOL		GOOD	FAIR	EXCELLENT	POOR	FAIR	EXCELLENT	GOOD	GOOD	GOOD	GOOD	EXCELLENT
	SYNTHETIC LUBRICANTS (DIESTER)		POOR TO FAIR	POOR	FAIR	POOR TO FAIR	POOR	FAIR TO GOOD	GOOD	POOR	POOR TO FAIR	POOR	FAIR TO GOOD
	HYDRAULIC FLUIDS	SILICATES	POOR	POOR TO FAIR	FAIR	FAIR TO GOOD	GOOD	FAIR	POOR TO FAIR	-	POOR	GOOD	GOOD
		PHOSPHATES	POOR TO FAIR	POOR	GOOD	GOOD TO EXCELLENT	POOR	POOR	POOR TO FAIR	POOR	GOOD	POOR TO FAIR	POOR
SUBJECTIVE PROPERTIES	TASTE		FAIR TO GOOD	FAIR TO GOOD	FAIR TO GOOD	GOOD	FAIR TO GOOD	FAIR TO GOOD	POOR TO FAIR	GOOD	GOOD	FAIR TO GOOD	FAIR TO GOOD
	ODOR		FAIR TO GOOD	GOOD	GOOD	GOOD	FAIR TO GOOD	GOOD	POOR	GOOD	GOOD	GOOD	GOOD
	NONSTAINING		POOR TO GOOD	POOR TO GOOD	GOOD	GOOD	GOOD TO EXCELLENT	POOR TO GOOD	POOR TO FAIR	GOOD	EXCELLENT	EXCELLENT	POOR TO GOOD
BONDED TO RIGID MATERIALS			EXCELLENT	EXCELLENT	FAIR TO EXCELLENT	POOR	GOOD TO EXCELLENT	GOOD TO EXCELLENT	FAIR TO GOOD	EXCELLENT	FAIR TO EXCELLENT	FAIR TO GOOD	POOR TO GOOD

109F11.EPS

Figure 11 ◆ Physical properties chart.

6.4.0 Nitrile

Nitrile is used with medium-pressure oil and solvent services. Nitrile cannot be used with acetone or methyl ethyl ketone (MEK). Nitrile has a maximum temperature rating of 250°F. Nitrile has poor ozone resistance, but is an excellent material to use with water or alcohol. Nitrile also has good chemical resistance but poor flame resistance.

6.5.0 Silicone

Silicone is a rubber-type material that is widely used. There are more grades of silicone material than any other rubber-type material. It has a maximum temperature rating of up to 550°F. Silicone is not recommended for use with oil or solvents. It has fair to good chemical resistance, excellent ozone resistance, and fair to good flame resistance. It has poor to excellent resilience depending on the grade used. It is not recommended for use on water systems.

6.6.0 Viton

Viton is a fluoroelastomer material. It has excellent chemical resistance and is the most commonly used chemical-resistant material. It has a maximum temperature rating of 400°F. Viton, a hard material, has excellent oil and solvent resistance and also has excellent ozone resistance. Viton is a relatively inexpensive gasket material.

6.7.0 Gylon

Gylon is a Teflon® product. It is chemically **inert**, has a high resilience (45-percent recovery), and has a maximum temperature rating of 500°F. Gylon 3510 has one of the best sealability of any gasket material in the industry. It is one of the most commonly used gasket materials. It is used heavily in the chemical industry and the paper industry, which also uses a lot of chemicals.

7.0.0 ◆ LAYING OUT AND CUTTING GASKETS

Preformed, factory-made gaskets need not be cut before installation. When preformed gaskets are not available, gaskets must be laid out and cut as needed. The two types of gaskets that a pipeline mechanical technician may need to lay out and cut are flange gaskets and machine gaskets.

7.1.0 Flange Gaskets

Many times a flange gasket may be directly marked by laying the flange on the gasket and marking the inside diameter, the outside diameter, and the hole pattern on the gasket, using a sharp pencil. When this is not possible, a takeoff must be done to determine the size and shape of the gasket, and the gasket must be laid out to fit the flange. Follow these steps to lay out a flange gasket:

Step 1 Measure the inside diameter and the outside diameter of the flange, using a scale or measuring tape.

> **CAUTION**
> It is very important that the correct inside and outside flange diameters be used when laying out a gasket, because a gasket of the wrong size may restrict fluid flow or the gasket may leak.

Step 2 Lay out the center lines of the flange on the gasket (see *Figure 12*).

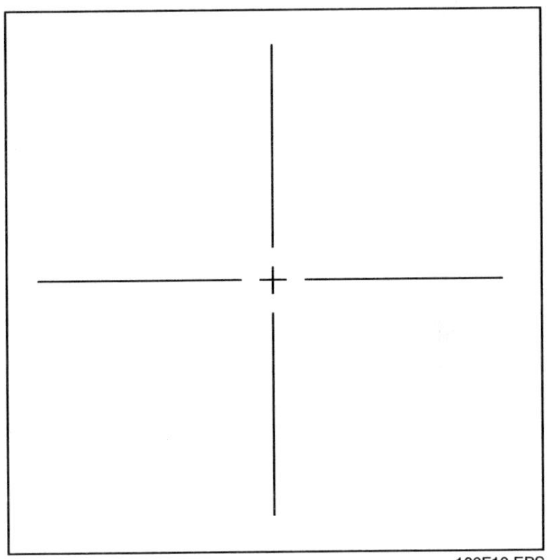

Figure 12 ◆ Laying out center lines.

> **NOTE**
> When cutting gaskets, you should cut the material in a manner that will conserve material.

Step 3 Lay out the inside diameter and the outside diameter of the flange on the gasket (see *Figure 13*).

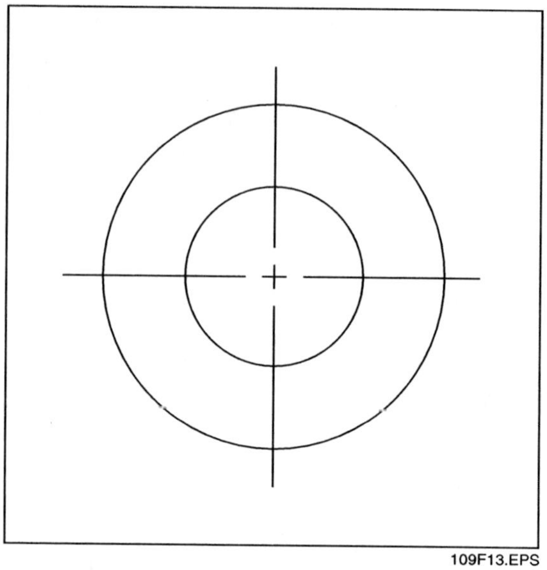

Figure 13 ◆ Laying out flange diameters.

Step 4 Measure the diameter of the flange bolt circle.

Step 5 Lay out the bolt circle on the gasket (see *Figure 14*).

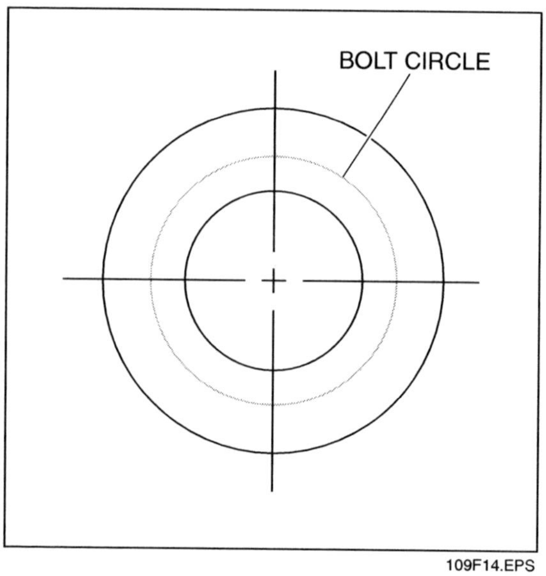

Figure 14 ◆ Laying out bolt circle.

Step 6 Divide the bolt circle into the required number of equal spaces to match the number of bolt holes in the flange (see *Figure 15*).

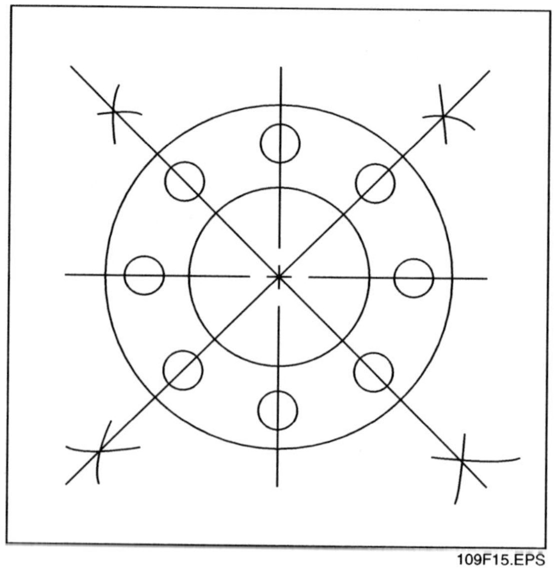

Figure 15 ◆ Laying out bolt holes.

Step 7 Adjust a gasket cutter to the outside diameter and material thickness, and cut out the gasket.

CAUTION

When using a gasket cutter, be sure to follow all manufacturer's recommendations for proper use.

Step 8 Adjust the cutter to the inside diameter, and cut out the center (see *Figure 16*).

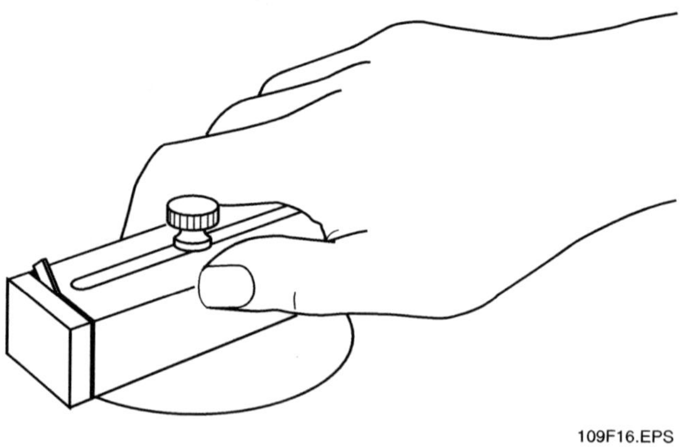

Figure 16 ◆ Cutting out gasket.

Step 9 Select a hole punch (see *Figure 17*) of the correct size to cut the bolt holes.

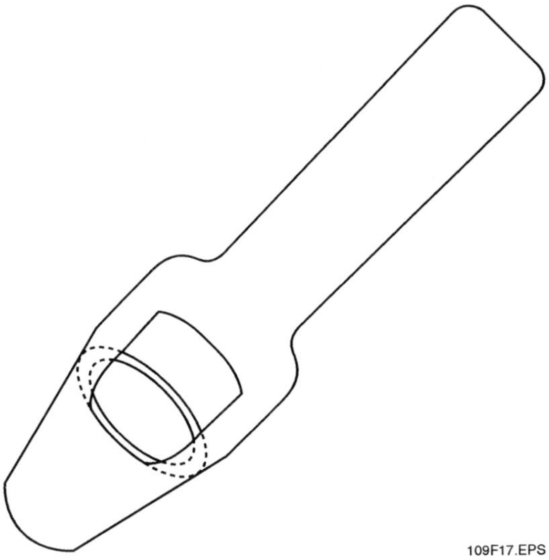

Figure 17 ◆ Hole punch.

> **NOTE**
> The hole in the gasket should be $\frac{1}{32}$ of an inch larger than the hole in the flange to allow the flange bolts to turn easily upon installation. For instance, a hole for a $\frac{3}{8}$-inch diameter bolt would be $\frac{13}{32}$ inch. You would use a $\frac{13}{32}$-inch punch to cut the holes.

Step 10 Place the gasket on a piece of wood.

> **CAUTION**
> Punching holes on a metal table or a piece of steel damages or dulls the punch.

Step 11 Punch the bolt holes in the gasket, using a hammer and the selected punch.

Step 12 Place the gasket against the flange, and check it for proper fit. Once the gasket fits well, proceed to install the gasket.

> **NOTE**
> If the gasket does not fit properly, discard it and cut another one.

7.2.0 Machine Gaskets

Machine gaskets come in many configurations. It is not practical to try to lay out a machine gasket in the same manner as you lay out a flange gasket. You must transfer a pattern from the machine to the gasket. This is done using bluing ink. Follow these steps to lay out and cut a machine gasket:

Step 1 Clean the machine surface where the gasket fits (flat) thoroughly to remove the old gasket and any foreign matter, using solvent if needed and prescribed.

Step 2 Ensure that the machine surface is completely dry.

Step 3 Cut a piece of gasket material a little larger than the area to be covered by the gasket.

Step 4 Apply an even coat of bluing ink to the machine flat.

Step 5 Lay the gasket on the machine flat.

> **CAUTION**
> Use extra care when placing the gasket on the machine flat to prevent smearing the gasket with ink. The ink should make a distinct impression on the gasket.

Step 6 Rub the gasket with your finger along the machine flat to ensure that the ink contacts the gasket at all points.

Step 7 Carefully remove the gasket from the flat.

Step 8 Allow the ink to dry.

Step 9 Cut out the gasket, using scissors or a utility knife.

Step 10 Punch the bolt holes, using the proper size hole punch.

Step 11 Place the gasket on the machine flat, and check the fit. Once the gasket fits well, proceed to install the gasket.

8.0.0 ◆ INSTALLING GASKETS

Once the gasket has been laid out and cut, it is ready to be installed on the flange or machine. The steps for installing a gasket may vary depending on the type of gasket material and the type of joint. Normally, the basic steps for installing gaskets are much the same. Follow these steps to install a gasket:

> **NOTE**
> When replacing gaskets, the old gasket must be removed and the flanges cleaned. The flanges and old gaskets should be inspected for irregularities. Any irregularities should be reported to your supervisor.

Step 1 Clean the surface where the gasket is to be installed, using solvent if needed.

Step 2 Apply any adhesive, sealer, or lubricant to the gasket as specified.

> **CAUTION**
> Normally, adhesives, sealers, or lubricants should be used only when specified. In some cases, adhesives can be used to hold the gasket in place during installation. If any of these are used, ensure that they are compatible with the gasket material and will not cause gasket breakdown or contaminate the fluid in the system.

Step 3 Place the gasket in position on the flange or machine flat.

Step 4 Install the mating flange or machine part.

> **CAUTION**
> Make sure to leave approximately ¼ inch of the gasket protruding past any intersecting mating surfaces and trim the gasket after the mating flange has been bolted down.

Step 5 Install the hold-down bolts.

Step 6 Tighten the hold-down bolts, following the proper tightening sequence.

> **CAUTION**
> Each bolt should be tightened a little at a time, following the proper sequence. This prevents warping or damaging the flange or machine part and prevents leaks. Tighten the bolts only until they are snug. Do not overtighten the bolts.

Step 7 Tighten the bolts to the specified torque, using a calibrated torque wrench, and following the proper tightening sequence (see *Figure 18*).

> **CAUTION**
> Do not overtighten the bolts because this could damage the gasket, the flange, or the bolts. The required torque for an application depends on the type of gasket used, the size of the flange, and the grade bolts used. Contact the design engineer if you are not sure of the torque requirements.

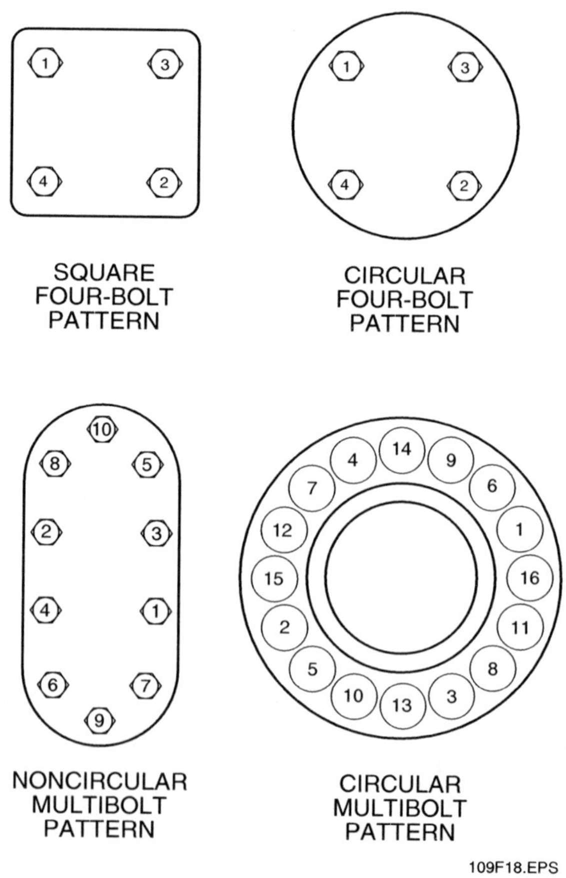

Figure 18 ◆ Proper tightening sequence.

SUMMARY

Seals are devices or materials used to prevent or control leakage between two surfaces. Seals are used between moving and stationary parts or between two stationary parts. Pipeline mechanical technicians must be able to identify the types of seals and seal materials and remove and replace worn-out seals.

There are many types of gaskets for different applications. The purpose of a gasket is to create a positive seal between two relatively stationary parts. The thinnest gasket that accomplishes this purpose is the best to use. The gasket used must be matched to the application. Gaskets are chosen for their pressure rating, their temperature rating, and for their resistance to the process on which they are used. Preformed, factory-made gaskets need not be cut before installation. When preformed gaskets are not available, gaskets must be laid out and cut as needed. Gaskets must be cut to close tolerances. Torquing specifications must be followed.

Review Questions

1. Applications in which there is no movement are called _____ applications.
   a. positive-contact
   b. controlled-clearance
   c. dynamic
   d. static

2. In a dynamic application, an O-ring is usually placed in a groove or joint that is _____% wider than the cross section of the O-ring.
   a. 135 to 150
   b. 125 to 135
   c. 115 to 125
   d. 105 to 115

3. Low-pressure, positive-contact seals used with rotating shafts are _____.
   a. cup seals
   b. O-rings
   c. labyrinth seals
   d. lip seals

4. The most widely used sealing material is _____.
   a. Buna-N
   b. silicone
   c. neoprene
   d. Teflon®

5. Silicone is not used for _____ applications.
   a. rotating
   b. reciprocating
   c. static
   d. dynamic

6. Plastic and elastomer compound seals can be used at temperatures up to _____°F.
   a. 500
   b. 400
   c. 350
   d. 250

7. The flat-type flange gasket commonly made from stainless steel with a graphite insert is the _____ gasket.
   a. split ring
   b. strip
   c. spiral wound
   d. envelope

8. The gasket used with medium-pressure oil and solvent services is _____.
   a. nitrile
   b. natural rubber
   c. EPDM
   d. Gylon

9. The hole in a flange gasket should be _____ of an inch larger than the hole in the flange.
   a. $\frac{1}{32}$
   b. $\frac{1}{16}$
   c. $\frac{1}{8}$
   d. $\frac{1}{4}$

10. When installing gaskets, tighten each bolt _____.
    a. completely tight in the proper sequence
    b. hand tight, working clockwise around the gasket
    c. after applying adhesive or sealer
    d. a little at a time in the proper sequence

GLOSSARY

Trade Terms Introduced in This Module

Blowout: A pressure failure of a gasket. The gasket is blown out of the flange by pressure.

Bore: The interior diameter of a shaft.

Creep: The loss of thickness of a gasket that results in bolt torque loss and leakage.

Elastomer: A synthetic material with the elastic qualities of natural rubber.

Ethylene propylene dieneterpolymer (EPDM): A gasket material and general purpose polymer that is heat, ozone, and weather resistant. EPDM is not oil resistant.

Extrusion: A gasket protruding out of a flange.

Inert: Unreactive. Exhibiting no chemical activity.

Plastic flow: The flowing of gasket material under stress.

Resilience: The ability of a gasket to return to its original shape after being compressed.

Resistance: In this context, the ability of a gasket to withstand the effects of chemicals and other substances without damage or change.

Vulcanize: To improve the strength and resiliency of a material by combining it with additives and subjecting the mixture to heat and pressure.

CONTREN™ LEARNING SERIES — USER UPDATES

The NCCER makes every effort to keep these textbooks up-to-date and free of technical errors. We appreciate your help in this process. If you have an idea for improving this textbook, or if you find an error, a typographical mistake, or an inaccuracy in NCCER's Contren™ textbooks, please write us, using this form or a photocopy. Be sure to include the exact module number, page number, a detailed description, and the correction, if applicable. Your input will be brought to the attention of the Technical Review Committee. Thank you for your assistance.

Instructors – If you found that additional materials were necessary in order to teach this module effectively, please let us know so that we may include them in the Equipment/Materials list in the Instructor's Guide.

Write: Curriculum Revision and Development Department
National Center for Construction Education and Research
P.O. Box 141104, Gainesville, FL 32614-1104

Fax: 352-334-0932

E-mail: curriculum@nccer.org

Craft ____________ Module Name ____________

Copyright Date ____________ Module Number ____________ Page Number(s) ____________

Description ____________

____

____

____

(Optional) Correction ____________

____

____

(Optional) Your Name and Address ____________

____

____

Index

Index

Index

Index